Perception of the Visual Environment

Springer
New York
Berlin
Heidelberg
Barcelona
Hong Kong
London
Milan
Paris
Singapore
Tokyo

Ronald G. Boothe

Perception of the Visual Environment

With 472 Illustrations

 Springer

Ronald G. Boothe
Yerkes Regional Primate Center
Emory University
Atlanta, GA 30322 USA
boothe@rmy.emory.edu

Library of Congress Cataloging-in-Publication Data
Boothe, Ronald G.
    Perception of the visual environment / Ronald G. Boothe.
        p.   cm.
    Includes bibliographical references and index.
    ISBN 0-387-98790-8 (alk. paper)
    1. Visual perception.   I. Title.
    BF241 .B66 2001
    152.14 – dc21   00-053768

Printed on acid-free paper.

Production coordinated by Chernow Editorial Services, Inc., and managed by
MaryAnn Brickner; manufacturing supervised by Jacqui Ashri.
Typeset by Best-set Typesetter Ltd., Hong Kong.
Printed and bound by Edwards Brothers, Inc., Ann Arbor, MI.
Printed in the United States of America.

9 8 7 6 5 4 3 2 1

ISBN 0-387-98790-8   SPIN 10715364

Springer-Verlag   New York Berlin Heidelberg
A member of BertelsmannSpringer Science+Business Media GmbH

To my family

Preface: Why Did I Write This Book?

Visual perception is by its very nature an interdisciplinary topic that requires a basic understanding of concepts derived from diverse fields, including physiological optics, neuroscience, cognitive science and psychology, clinical optometry and ophthalmology, philosophy, information theory, neural networks, artificial intelligence, and robotics. After teaching an undergraduate course on the topic of perception for several years, I remain frustrated in my attempts to find a textbook that covers an appropriate range of these topics. Several years ago, I began preparing a number of lecture handouts for my own students. Over the years, these handouts expanded into the current manuscript, which will serve as a textbook for my course on visual perception.

The primary audience consists of juniors and seniors majoring in psychology, biology, or neuroscience. Many are planning to attend graduate school in psychology or neuroscience or to enter medicine or a related health profession. Graduate students and postgraduates in related fields who have an interest in learning more about perception form a secondary audience.

The book introduces a number of diverse topics and concepts that pertain to perception. No background knowledge of any of these topics is assumed except for basic concepts learned in introductory courses in biology and psychology.

Such a text requires many hard choices, sometimes compromises, to keep the scope of the material within the limits of what can reasonably be covered in a one semester undergraduate course. One choice is how much of the primary scientific literature based on studies of animals to include. A huge scientific literature in the field of visual neuroscience is derived from studies of many species of animals. This text largely limits the discussion of animal studies to results obtained from monkeys, even in cases where the monkey studies are simply a confirmation of a much larger prior literature based on work with other animals. Although this approach fails to give appropriate credit for many discoveries that were first made in other species and only later confirmed in monkeys, the primary purpose of this book is pedagogical. It is not to provide a detailed historical description of how the facts presented here were originally discovered. The text strives to explain in a manner that can be grasped by undergraduates the major ideas and scientific discoveries that have allowed us to understand perception as it occurs in humans.

A similarly hard choice was deciding how many citations to include to the primary scientific literature. Almost every paragraph in this book could have numerous citations to published studies or references to the names of individuals who developed the ideas being discussed. However, specific citations are not included in the text, which only refers to a few individual scientists by name, thereby neglecting a vast army of other researchers who made the discoveries summarized here. In some cases, particularly when describing a very specific finding or set of studies by contemporary workers, the name of the author in whose lab the study was conducted is mentioned. However, the inclusion of numerous citations or names of scientists interfered with the primary goal of introducing the major concepts and facts in an uncluttered manner. Also, at times, complex results had to be simplified in a manner that remains true to the spirit of the original research findings, but may differ enough in some specific details that a specific citation might give offense rather than credit. To compensate, a limited list of suggested readings appears at the end of each chapter. These readings, along with key words or phrases, allow the interested student to locate relevant original literature through bibliographic searches.

I would like to thank all of my students, graduate and undergraduate, over the past several years who provided considerable feedback about previous drafts of the material. I also owe a particular debt of gratitude to a few colleagues who were kind enough to read drafts of one or more chapters and provide comments: Velma Dobson, Larry Barsalou, Dolores Bradley, Jim Wilson, Michael Mustari, and Valab Das. Many improvements were made to the manuscript by taking their feedback into account. Of course I did not always heed the good advice that was given to me, and any deficiencies and errors that remain in the final version are my own.

I thank Emory University for granting me a sabbatical leave during which the basic framework of the text was outlined and drafted. I spent the sabbatical leave working at the University of Washington, away from the normal distractions of my professorial duties. I thank Davida Teller for sponsoring me for a visiting scholar appointment in the Department of Psychology at the University of Washington during my leave. I owe a debt of gratitude to Jean Torbit for providing me with innumerable assistance in obtaining permissions to republish figures from other sources. Finally, I am grateful to my family for their ongoing support and inspiration, and in particular my wife Marilyn for providing me with drawings of penguins and for her patience during the writing of this manuscript.

Ronald G. Boothe
Atlanta, Georgia

Contents

1
Conceptual and Philosophical Issues: What Does It Mean to Assert That an Observer Perceives?

Questions

After reading Chapter 1, you should be able to answer the following questions:

1. Provide a working definition of the term perception.
2. Perception can be conceptualized as having 3 components. Characterize each of these components from two or more perspectives or levels of analysis.
3. Percepts convey to observers "knowledge about what," "knowledge about how," and "subjective perceptual experience." Use the example of the phenomenon of binocular rivalry to illustrate what is meant by these three distinct aspects of percepts.
4. Compare and contrast the philosophical positions of realism, idealism, and phenomenalism regarding the causal relationships that exist between the environment and percepts.
5. What is the mind-body problem with regard to perception, and what types of solutions have been proposed to try to account for this problem?

6. Summarize the historical ideas about how objects in the environment impress themselves onto our sense organs and how activities in the sense organs are transmitted to the brain and processed.
7. How does the theoretical approach of information processing differ from that of causal theories?
8. Describe the major processes and mechanisms involved in constructing percepts.
9. Compare and contrast direct and indirect psychological theories of perception.
10. Describe some of the major approaches and levels of description that have been applied to the topic of perception and the kinds of explanations that can be derived from each.
11. Differentiate between simplifying and realistic models of perception and give some examples of each.
12. What are some of the special conceptual issues that arise when applying theories of perception to non-human observers?

This introductory chapter summarizes some of the major conceptual and philosophical issues that pertain to the topic of perception. These issues have been discussed for millennia in natural philosophy and played a central role in the theories of psychologists and brain scientists. We will be able to highlight only a few major themes here.

1 Major Components of Perception

Perception is defined in this text as *the act of using only one's own sense organs to gain knowledge about, interact with, and experience the environment.* The act of perceiving can be conceptualized in terms of three major components: 1) inputs from the environment, 2) processing mechanisms, and 3) output products (Figure 1.1). Each of these components can be described from various perspectives or levels of analysis.

The Environment Provides the Input to Perception

The **environment** provides the input to perceptual processing. As I write this text, I am working in an environment. There is a desk in front of me with a computer sitting on top of it. This environment can be described in various ways, three of which are particularly convenient for understanding perception. First are **levels of description derived from physics.** For example, the light waves emitted from the computer screen are a form of electromagnetic radiation that can be measured and characterized by instruments. Second are **formal descriptions of the information in the environment** that is potentially available to be picked up by a perceiving observer. For example, a computer display can be analyzed in terms of the quality and quantity of the information it transmits to the eye of an observer viewing it. Third are levels of description derived from an **ecological perspective** that emphasizes the **functional significance of the environment** for a particular observer. For example, the screen functions to allow me the opportunity to look at the text produced as I hit the keys on the keyboard.

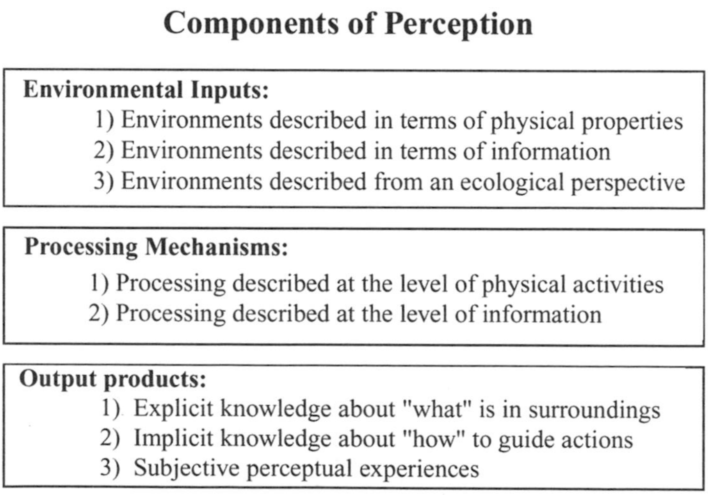

Components of Perception

Environmental Inputs:
  1) Environments described in terms of physical properties
  2) Environments described in terms of information
  3) Environments described from an ecological perspective

Processing Mechanisms:
  1) Processing described at the level of physical activities
  2) Processing described at the level of information

Output products:
  1) Explicit knowledge about "what" is in surroundings
  2) Implicit knowledge about "how" to guide actions
  3) Subjective perceptual experiences

FIGURE 1.1. Perception can be conceptualized has having three major components, each of which can be described from different levels of analysis or perspectives.

Perceptual Processing Mechanisms

Perceptual processing mechanisms can also be described in different ways. One level of description is in terms of the **physical activity in the sense organs and brain** of the perceiving observer. For example, sound waves from the humming computer are vibrating my eardrum. Mechanoreceptors in the skin of my fingers are being deformed each time I strike a letter on the keyboard. Light waves emitted from the screen are entering my eye and causing electrical activity to be sent up the optic nerve to my brain. Sensory physiologists and neuroscientists often have study of these physical activities as their primary interest in the topic of perception.

Perceptual processing can also be described in terms of **abstract operations performed on information.** For example, in addition to characterizing the physical activities taking place in sensory organs, we can also try to specify the sensory information that is being coded and analyzed by these activities. An active area of perception research involves trying to formalize the inherent **information-processing operations** that are carried out during perceptual processing. Another active area involves trying to relate the physical and information-processing levels of description of perceptual processing to one another. The term **sensory coding** is used to refer to the way information about the environment is coded in the physical activities of the sense organs and the brain.

Percepts Are the Output Products of Perceptual Processing

In the text, we will use the term **percept** when referring to the outputs of perceptual processing that are utilized by an observer at a given point in time or in a given context. It is convenient to differentiate three aspects of percepts.

One aspect of percepts is that they provide an observer with **knowledge about *what* is present in the immediately surrounding environment.** In other words, perception serves to inform observers about "what is out there." This can be illustrated with a simple exercise. Go stand outside the doorway of a room in a public building that you have never entered without looking in. Ask yourself a series of questions about the room, such as:

- How are the chairs arranged?
- Is there a table present in the room?
- What kind of lighting fixtures are present?

Before you enter you will not be able to reliably answer any of these questions. Now enter the room and look around. Suddenly, and seemingly without effort, you are able to answer all of them. As an output product, your perceptual system has provided you with lots of knowledge about the room that was not in your possession prior to this perceptual act.

It is relatively easy to use **behavioral reports** as a probe to study this type of perceptual knowledge. Since the knowledge involves **explicit facts,** observers can simply be asked to us **words** or **symbols** to describe what they perceive to be present in their environment. A simple example would be to turn on a light bulb and ask an observer to report whether she sees the light (Figure 1.2).

Similarly, we could probe this aspect of percepts with symbols by having the observer use a keyboard or a pencil. Our hypothetical observer might produce the output:

'l' 's' 'e' 'e' 't' 'h' 'e' 'l' 'i' 'g' 'h' 't'

A second aspect of percepts that we will be concerned with is that they provide observers with **knowledge about *how* to guide immediate actions.** Consider the actions performed by an outfielder in the game of baseball while trying to catch a fly ball. It is perception that provides the information about how to guide these actions. This form of information is often **implicit** rather than explicit. For example, an outfielder may be very proficient at catching fly balls, yet when using words or symbols be unable to provide more than a rudimentary description of how perception allows this to be accomplished. Thus, behavioral reports in the form of **actions performed during perception** often serves as a better probe than words or symbols for studying this form of perceptual knowledge.

"I see the light."

FIGURE 1.2. One way to study perception is to simply ask an observer to report what is seen. In this example, an observer reports that she sees a light.

A third, and final, aspect of percepts that we will be concerned with is subjective **perceptual experience,** one of the major subcategories of the types of consciousness described by philosophers. Our perceptual systems output perceptual experiences continuously while we are conscious. As I sit at my desk now, I *hear* the sound of the computer humming on my desk. I *feel* the touch of the computer keys as I press them with my fingers. I *see* the text emerging on the screen of the computer as I type. Subjective perceptual experience evolves continuously over time. By **subjective** we mean perceptual experience reflects the private and personal point of view of a given observer.

When dealing with a normal human adult observer, we can use behavioral reports of the observer in the form of words and symbols to probe characteristics of subjective perceptual experience. However, conclusions we can draw from scientific studies based on these behavioral reports differ in some fundamental ways from conclusions that can be drawn from behavioral reports used to probe an observer's perceptual knowledge of what and how. These differences are elaborated in Chapter 2.

There are a number of formal disciplines in addition to scientific studies of perception that attempt to deal with various aspects of subjective perceptual experience, including the arts (e.g., music, painting, film), Zen, mysticism, and phenomenology. Levels of description derived from these disciplines are rich and deserving of extended study. However, in this text, the main emphasis is not going to be on trying to understand subjective perceptual experience per se but in trying to relate these experiences to other aspects of percepts (knowledge of *what* and *how*) and to the other components of perception (inputs and processing mechanisms).

Distinctions among the three aspects of percepts (explicit perceptual knowledge about *what*, implicit perceptual knowledge about *how*, and conscious perceptual experience) can be made more concrete with an example of a perceptual phenomenon called **binocular rivalry.** You can demonstrate binocular rivalry to yourself by employing the simple methods described in Box 1.1.

Box 1.1
Binocular Rivalry

Stand this book on edge on a table opened to this page and place a piece of cardboard or similar blocking material oriented vertically in front of the page and perpendicular to it. View the stimuli shown at the bottom of this box in Figure 2 while placing your nose in front of the cardboard as shown in Figure 1. Your left eye should be able to only view the left stimulus and the right eye only the right stimulus.

Look directly at the circles in Figure 2. As you stare at these stimuli, you should notice changes in appearance over time. During some periods of time you might experience a percept of rightward-tilted stripes, at other times leftward-tilted stripes. At other times your percept is likely to include various combinations of leftward- and rightward-titled stripes. You might be able to accentuate these unstable changes in appearance by closing one or both eyes momentarily. The instability you are experiencing in the percept is called binocular rivalry, a topic we will describe in more detail in Chapter 9.

Viewing Situation

FIGURE 1. The stimulus in Figure 2 is to be viewed as illustrated here, with the head positioned such that the left eye can see only the left stimulus and the right eye only the right stimulus.

FIGURE 2. When this stimulus is viewed under the conditions illustrated in Figure 1, binocular rivalry will be experienced as described in the text.

FIGURE 1.3. The philosophical position of
realism asserts that our percepts are caused
by the physical environment.

Realism

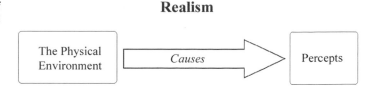

As you view the stimulus in the text box, think about the various properties of the percept that is being generated. Note that your percept is providing you with some **explicit facts about** *what* is present in your surroundings. For example, it is providing you with knowledge that there are some oriented black and white lines on the page in front of you. Your percept is also providing you with some **implicit knowledge about** *how* to interact with this environment, for example, while looking at the stimuli, you automatically know how to guide your arm to reach out and touch the stimuli should you desire to do so. Finally, your percept is providing you with a subjective **perceptual experience** that is dynamically changing over time in a manner that is strictly personal to you. At each instant in time, your percept is of leftward- or rightward-oriented stripes or some combination of the two. No one else, even when looking at the exact same stimuli, will experience the identical pattern of changes over time that you are experiencing.

2 Causal Theories

Causal theories of perception are formal statements regarding **causal relationships** that occur within and among the input, processing, and output components. Issues regarding causal relationships that might exist between the major components of perception have been considered by philosophers for hundreds of years. Three topics that have received extended discussion include presumed causal relationships between 1) the environment and percepts, 2) physiological processing in our brains and percepts, and 3) the environment and physiological processing in our brains.

The Environment and Percepts

Philosophical positions regarding causal relationships between the environment and percepts take many forms, but the dominant position assumed (implicitly, if not explicitly) by scientific theories of perception is referred to as **realism.**

Realism

The realist position assumes that there is an external world that exists independently of ourselves, and that this external world causes our percepts (Figure 1.3).

The realist position seems so obvious to most of us with intellectual backgrounds shaped by Western civilization that we seldom even consider that alternative causal relationships are logical possibilities.

Idealism

One example of an alternative position, advocated by the eighteenth-century philosopher **George Berkeley,** is **idealism,** which asserts that **percepts cause the existence of the physical world** (Figure 1.4).

An implication of the idealism position is exemplified by the question:

Can you prove that you are not somebody else's dream, and that when that person wakes up everything you are currently calling reality will not disappear?

Most of us raised in Western culture do not take this question seriously and treat it as simply a silly riddle. Nevertheless, idealism cannot be rejected on logical grounds, and there is no empirical scientific evidence that would allow us to rule it out. This is

Idealism

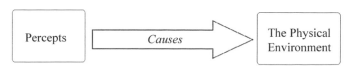

FIGURE 1.4. The philosophical position of
idealism asserts that the physical environment is caused by our percepts.

also the case for a number of other historical alternatives to realism. However, these alternative positions will not be dealt with further here, as they have had little influence on contemporary scientific theories of perception.

Phenomenalism

There is one modern alternative to realism, called **phenomenalism,** that has implications for perception important enough that it deserves mention, even though it has not yet had much practical impact on scientific theories of perception.

It comes as a shock to those of us accepting an implicit form of realism in our everyday thinking to learn that the realist position is demonstrably false. Quantum physics, developed early in the 20th century, informs us that any act of observation has the potential to change that which is being observed (Figure 1.5). Thus, the **phenomena** we perceive following an act of observation do not necessarily have properties identical to those of the physical reality that existed prior to the observation. Thus, modern-day theoretical physicists, have been forced to reject Realism, since the debates between Albert Einstein, and Werner Heisenberg as elaborated in Box 1.2.

An act of observation during which photons of light are absorbed in the eye is an essential step in visual perception. Thus, any theory of visual perception that wants to remain compatible with modern theoretical physics must concede that our percepts can inform us only about properties of phenomena, not about properties of the environment as it existed prior to being observed. Perception cannot escape the epistemological constraint that some facts about the environment are unknowable, as elaborated in Box 1.3.

The terminology involved in continually making a distinction between properties of phenomena and

> ### Box 1.2
> ### Einstein and Heisenberg Debate Realism Versus Phenomenalism
>
> Einstein was one of the last great theoretical physicists to hold out for a theory of physics that was based on realism and not limited to phenomena. He debated this issue in letters to Heisenberg, the founder of quantum mechanics. In his letters, Einstein argued the position that a theory of physics should be able to account for objective reality, i.e., everything that exists in the physical universe, whether or not it is being observed. He argued that such a theory would be "more beautiful" than modern theories, such as quantum physics, that settle for trying to account only for observed phenomena. In a letter responding to Einstein, Heisenberg stated eloquently:
>
> > If I have understood correctly your [Einstein's] point of view, then you would gladly sacrifice the simplicity [of modern quantum physics] to the principle of [objective reality]. Perhaps we could comfort ourselves [with the idea that] the dear Lord could go beyond [quantum physics] and maintain [objective reality]. I do not really find it beautiful, however, to demand more than a physical description of [phenomena].

the unobserved environment seems unduly cumbersome for general usage, and this text will continue using the framework and terms associated with realism. However, this is done with the understanding that, technically, the term *environment* does

Phenomenalism

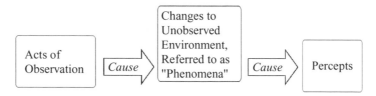

FIGURE 1.5. The modern philosophical position of phenomenalism asserts that the act of observing the environment can cause its properties to change. The properties of the observed environment are called phenomena. Our percepts always reflect properties of phenomena, and these are not necessarily the same as properties of the physical environment as it existed before it was observed.

Box 1.3
Perception, Epistemology, and Trees Falling in Forests

Epistemology is a general theory of human knowledge about the world and the conditions under which that knowledge is acquired. Any theory of perception, when considered broadly, must be compatible with an epistemology regarding how we come to have knowledge of the world we live in. Phenomenalism asserts that some facts about the physical universe can never be known in principle. For example, phenomenalism provides an answer of sorts to the question: "Does a tree in the forest make a sound if it falls when there is no one present to hear it?" The answer provided is that the answer is unknowable. The most we can hope to attain, in principle, is what the "sound" would have been like if it had been observed.

---

not generalize beyond "properties of what is out there while it is being observed."

Direct and Indirect Forms of Realism

The simplest form of the realist position, called **direct realism,** assumes that the external environment causes our perceptual experiences directly, without any intervening processes. Since there are no intervening processes, the direct realist position leads to the conclusion that we should perceive the world **veridically,** i.e., *directly as it is*. Consequently, the direct realist position is hard to reconcile with observations suggesting that our perceptions are sometimes **illusory.**

Examples of apparent illusory perception have been noted since the times of the ancient Greeks. For example, a straight stick submerged partway in water appears to bend. In other words, the stick has two appearances. Under some conditions (when completely in air or water) it appears straight, but under others (when partly in air and partly in water) it appears bent. One explanation would be that the stick actually changes its shape when it is submerged partway into the water. An alternative explanation, and one that is more in line with everyday common sense, would be that the stick is always straight but that our perception of the stick is not always veridical.

Another example of apparent nonveridical perception is **sensory adaptation,** a phenomenon you can demonstrate to yourself as described in Box 1.4.

Box 1.4
A Demonstration of Sensory Adaptation

Take three small bowls and fill each with water: one with cold water, the second with warm water, and the third with water at room temperature. Now put your hands into the warm water for a minute or so, and then immediately dip your fingers into the bowl of water at room temperature. You will experience that the water feels cool. Now place your hands into the cold water for about a minute and then dip your fingers into the same bowl of room temperature water. The water will now feel warm. One explanation of these findings would be that the water in the bowl mysteriously changed its temperature based on what you had done with your hands before you dipped your fingers into the bowl. However, that position is hard to reconcile with the following demonstration. Hold your left hand in the cold water and, simultaneously, your right hand in the warm water. Now put both of your hands into the same bowl of room temperature water simultaneously and report its temperature.

Examples such as these led philosophers such as **John Locke,** writing in the 17th century, to advocate an **indirect realist** position. The external world is assumed to cause some intervening state that in turn leads to our experience (Figure 1.6).

The indirect realist position is able to account more easily for cases of nonveridical perceptions. These are attributed to imperfections or peculiar properties of the intervening state. For example, a potential explanation for the sensory adaptation demonstration described in Box 1.4 is that perception of temperature is mediated by our sensory system (our nerves and our brain). Thus, conditions that change the adaptation state of the sensory system (such as holding the fingers in warm or cold water for an extended period) can alter our subsequent perception of temperature irrespective of the true state of affairs in the environment.

Indirect Realism

FIGURE 1.6. The philosophical position realism takes two forms. This diagram illustrates one form, called indirect realism, in which the environment does not directly cause our percepts, but does so indirectly via some intervening steps.

Biology and Percepts

Modern biologically based theories of perception are based on an indirect realist position. It is assumed the external world leads to neural activity in the brain that in turn causes percepts. Thus, these theories need to be concerned about two classes of causal relationships, one between the external environment and the brain, the second between the brain and percepts. We will deal with issues of the causal relationships between the physical environment and neural processing in the brain in the next section. Here we deal with some special problems associated with the causal relationship between neural processing and percepts.

Recall that percepts convey both (explicit and implicit) knowledge and subjective experiences to observers. It is the subjective experiential aspects of percepts that are most problematic for causal theories. The causal chains that lead to either explicit or implicit perceptual knowledge can be traced, in principle, to physical alterations in the brain or to physical activities in muscles. However, the causal chains that lead to perceptual experiences are less well understood.

The causal powers of the brain that allow it to produce subjective perceptual experiences and other forms of consciousness have been considered troublesome by many philosophers and scientists since the influential writings of **René Descartes**

in the 17th century. Descartes argued for a **dualism** in which mental activities belong to the realm of the mind and bodily activities to the realm of the physical. Scientists and philosophers who accept some form of this dualism have difficulty with the idea that the brain, part of the **physical body**, causes conscious perceptual experiences that are part of the **mind.** At the least, many find it hard to accept that the term "cause" has the same meaning for the arrow shown in Figure 1.7 connecting the brain with subjective perceptual experience as for the other arrow that connects strictly physical entities. The difficulty in interpreting the meaning of this causal arrow is called the **mind-body problem.**

Most modern biologically based theories of perception that attempt to deal with the mind-body problem do so by adopting, explicitly or implicitly, some variation on one of four approaches: **homunculus, bridge locus, emergent property,** or **limited scope.**

Homunculus Solutions

Homunculus solutions to the problem of how activities in our brain get conveyed to our mind propose that this is accomplished by a little person inside our head. Our eyes and optic nerves function to form a picture in our brains of "what is out there." Then the homunculus looks at the picture,

The Mind-Body Problem

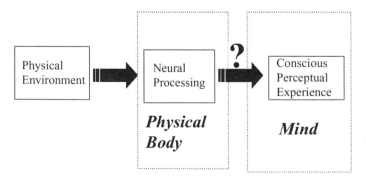

FIGURE 1.7. The causal chain for many biologically oriented theories of perception is based on a form of indirect realism in which the physical environment causes activity in our brains, and the brain activity leads to our perceptual experiences. Some question whether it is meaningful to talk about physical activities in our brain causing conscious mental events such as perceptual experiences. This concern is referred to as the mind-body problem.

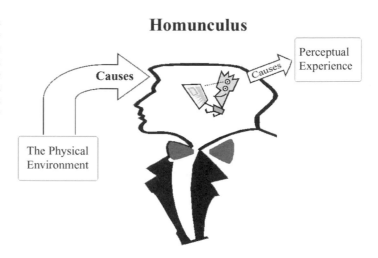

FIGURE 1.8. Homunculus theories are a form of indirect realism in which the intervening process that operates between the environment and our percepts is a homunculus. The eye causes a picture to be formed in the brain. The homunculus is a little person sitting our brain who looks at the picture. Our percepts are caused by what the homunculus sees.

and what it sees provides our perceptual experiences (Figure 1.8).

The idea of a "little person in the head" seems silly when we state if formally and explicitly. Furthermore, the basic notion of a homunculus was rejected by philosophers a long time ago on logical grounds because it leads to an infinite regress. There would have to be another little person inside the head of the homunculus to interpret the picture in its brain, etc., etc., ad infinitum.

However, our everyday thinking about perception often includes an implicit notion that there is a little person in our heads that accounts for our percepts. Similarly, many biologically based theories of perception, when they are made explicit, also end up either being some form of a homun-culus theory or having similar philosophical problems. For example, neuroscientists have been able to demonstrate that various brain regions contain neural representations of the visual environment (for example, see Figure 12.1 in Chapter 12). An examination of the properties of these neural representations is valuable in terms of trying to address questions about how sensory information is coded as it is processed in the brain. However, when evaluating the significance of these neural representations to perceptual processing, we need to be careful not to be lulled into thinking that these provide an explanation for subjective perceptual experience. Such thinking only makes sense if we have adopted an explicit or implicit homunculus theory of perception. It makes no difference whether a picture is present in the brain unless there is someone present to look at it.

Bridge Locus Solutions

Bridge locus solutions to the mind-body problem take seriously the idea that it is the brain that causes our conscious perceptual experiences and designate specific structure(s) in the brain as being responsible. These neural correlates of consciousness are assumed to have some privileged, and as yet mysterious, causal powers that allow conscious perceptual experiences to the produced.

Bridge locus theories vary in terms of how localized the structure is that is purported to give rise to a perceptual experience. In its **weak forms,** a bridge locus is considered to be a distributed property involving extensive regions of the brain, and in its **strong forms,** it is believed to be restricted to more localized brain structures. In its weakest form, the bridge locus could be the entire nervous system, and in its strongest form, it could be localized within a single neuron. Intermediate forms involve neural circuits consisting of many highly interconnected neurons.

Bridge locus theories have a long history. In 1637 Descartes argued that conscious experiences were the result of activities in the pineal gland, which served as a bridge locus to the mind (Figure 1.9).

Modern scientific theories do not take the idea that the pineal gland provides a bridge locus to the mind seriously. However, current theories of perception still make claims that are conceptually similar in many ways to the ideas of Descartes. A common modern variation on this idea proposes that different brain regions, each responsible for a particular kind of perceptual experience, operate

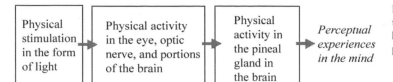

FIGURE 1.9. The view of Descartes that the pineal gland acts as a bridge locus between activity in our brain and percepts in our minds is illustrated.

in parallel, as illustrated in cartoon form in Figure 1.10.

We will discuss some specific examples of this approach in subsequent chapters of this text. For example, particular regions of the brain proposed as bridge loci for the experience of color will be discussed in Chapter 7, those for form in Chapter 8, those for depth in Chapter 9, and those for motion in Chapter 10.

When a bridge locus theory is stated in its strongest form, the concept that a single neuron provides the bridge locus, the neuron performing this function is sometimes called a **grandmother cell** and theories incorporating grandmother cells are after refered to generally as **grandmother cell theories.** The rationale for this colloquial name is that those theories assume that for every percept we can experience there has to be a bridge locus cell that gives rise to it. So, it is argued, if we can experience our grandmother, then there has to be a "grandmother cell" somewhere in our brain. We experience our grandmother if and only if that cell

is active. Ordinarily that grandmother cell becomes active when our grandmother is present and we are looking at her. But note that the grandmother cell concept also predicts that whenever the cell is active we should experience our grandmother, even if she is not present.

The mapping between grandmother cells and a given percept could be **one-to-one** or **many-to-one.** If it is one-to-one, then any condition that destroyed the grandmother cell or caused it to stop functioning would cause a loss of ability to perceive grandmother. However, the capability to perceive grandmother could be more robust if the relationship is many-to-one, thus providing redundancy.

One potential criticism of grandmother cell theories is that our brains would need enough grandmother cell neurons to account for all of our percepts. If the mapping is many-to-one, the situation becomes potentially even more problematic. Not everyone agrees with that criticism, as discussed in Box 1.5.

A variant of bridge locus approaches is to consider consciousness as arising from a **dynamic neural process** rather than from a particular anatomical structure. Neural activity in widely distributed regions of the brain appears to be highly synchronized under some conditions, and there has been speculation that this distributed, synchronized activity provides the neural correlate of consciousness. We will discuss these ideas further in Chapter 12.

Bridge Loci Operating in Parallel

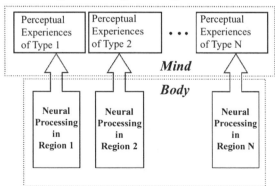

FIGURE 1.10. Modern neuroscience theories of perception often follow a scheme along the lines shown. Neural processing of perceptual information is considered to take place in parallel in different parts of the brain, and the processing that takes place within each brain area is responsible for certain qualities of percepts. For example, processing in one brain region might give rise to color, another to shape, and so on.

Emergent Properties Solutions

Emergent properties solutions take a different approach in trying to deal with the mind-body problem. Rather than considering mind to be a causal product of matter, mind is considered to be simply a property of matter that emerges under certain conditions in which there is sufficient complexity. As a complex perceptual system is operating (i.e., the environment is causing neural activity to reverberate throughout the nervous system of a perceiving organism, producing behavior), it takes on a set of properties that can properly be described as its mind. This mind is a property of the entire

Box 1.5
How Many Percepts Do
You Have?

In the undergraduate course I teach on the topic of perception, I have encountered strong resistance to the concept of grandmother cells from many students. One of the most prevalent arguments is, "There could not possibly be enough brain cells to account for all of my perceptual experiences." In response to that criticism, I have asked my students to make a list of all of their "distinguishable percepts." For example, if a student asserts that she has 3,000 distinguishable percepts, then she would have to be able, in principle, to reliably identify each of the 3,000 with no confusion among them. To date, no student has given me a list of as many as 100,000,000,000, which is the approximate number of neurons in our brains.

The number of neurons in our brain is so huge that intuition fails to provide us with reliable information about what is feasible. Suppose a human were to start forming percepts on the day of birth at the rate of 5 per second, 24 hours per day, 7 days per week, and used one brain cell to store each percept. The person would reach about 700 years of age before running out of neurons that could be utilized to serve as grandmother cells.

Box 1.6
Melodies and Gas Pressures as
Emergent Phenomena

An analogy is sometimes made to a melody, which is an emergent property of a group of notes when they are played together. A melody cannot be found in any of its constituent notes. Similarly, it is argued that a perceptual experience cannot be found in any of the constituent parts of a perceiving organism.

Analogies are also made to examples from other branches of science, including physics, in which emergent properties have been discovered. A widely cited example involves the laws of physics that govern pressure of a gas in a container. These laws have no strictly causal relationship based on the activities of the individual molecules that form the gas, but emerge when the molecules interact in a complex manner within a restricted volume.

system and is not to be found isolated in any of the component parts.

The concept of emergent properties as an explanatory device for conscious perception has an intuitive appeal, based partly on analogies with emergent properties that have been discovered in other domains, as described in Box 1.6.

Although emergent property solutions have strong intuitive appeal for many scientists, there are limitations to analogies. Even if one accepts the argument that one could never find a melody in a single note, it is not clear that the relationship between perceptual experience and the neural elements that make up the brain is analogous. Similarly, the fact that some laws in physics are emergent properties does not prove that similar laws will be found within the scientific study of perception. Furthermore, bridge locus solutions

have had a greater impact, at least to date, in terms of design and interpretation of perceptual experiments. As a prototypical example, consider the perceptual phenomenon of **phantom limbs** that is described in Box 1.7.

Limited-Scope Solutions

One final potential class of solutions to the mind-body problem needs to be at least mentioned, because these kinds of solutions are adopted implicitly by many neuroscientists. I will refer to these approaches as **limited-scope solutions,** because they try to simply ignore the mind-body problem by limiting the scope of questions that are asked about perception to ones that do not involve conscious perceptual experience. Limited-scope solutions are based on an implicit argument that, as a practical matter, many interesting questions about perception can be addressed without getting into problematic issues such as consciousness. The problem with this class of solutions is that an experiential component is strongly embedded in what most of us mean when we refer to "perception." Thus, any theoretical approach to perception that completely ignores

---

Box 1.7
Phantom Limbs

People who have had limbs amputated often have a perceptual experience that the missing limb is still present. This neurological condition is sometimes characterized by stating that the patient has a perceptual experience of a **phantom limb.** Consider the following explanation for phantom limbs that has been formulated within the context of a bridge locus theoretical framework:

Presumably, the neurons in the brain that previously served as the bridge locus for the perceptual experience of the limb are still present even though they have lost their sensory input from the limb. Some amount of neural reorganization takes place following the injury. As a result, the bridge locus neurons have made new connections that cause them to sometimes become active even though they no longer receive their original input from the limb. Whenever those neurons become active, for whatever reason, they still give rise to the "experience" of the limb.

This hypothesis has been put to a direct test in patients with a phantom limb who had to undergo electrical stimulation applied to parts of their brain as part of therapy for chronic pain. When the neurons in the portions of the brain that previously received input from the limb were stimulated, patients reported a sensation in the phantom limb.

Furthermore, sensations in the missing limb can sometimes be evoked by touching "trigger zones" on other parts of the body. For example, touching a particular location on the face of an arm amputee may produce sensations of both the face and the missing hand. The interpretation of these trigger zones from the perspective of a bridge locus theoretical framework is that during neural reorganization following the injury, a bridge locus neuron for experiences of the hand has formed new connections to nerve endings in the face so that it can now be activated by touching the face.

Note how easily these kinds of detailed facts can be explained within a general theoretical framework involving bridge locus neurons. Although not impossible, those facts are much harder to explain within a framework based on emergent properties.

---

conscious perceptual experience is prima facie inadequate.

The Environment and the Brain

Causal theories regarding relationships between the environment and the brain try to address issues about how the environment manages to impinge on the brain. There are two issues with regard to visual perception. First, how does the environment impress itself onto the eye? Second, how does information from the eye make it to the brain for further processing?

Environment to Sense Organ

The ancient Greeks prior to Aristotle introduced two ideas about how the environment interacted with the visual sense organ. The first was that the mind extended out from the body until it met objects. The second was that objects emitted copies of themselves that were transmitted to the senses.

Aristotle disliked both ideas – the first because he did not like the idea that the mind could leave the body and bump into things, the second because objects do not dwindle away over time, as would be expected if they were constantly giving off copies. He developed a general idea that light activates a medium that allows the eye to receive the sensible forms of things, or **sense impressions,** without making direct contact with either objects or their replicas.

The development of geometrical optics was needed to account for how these sense impressions preserved spatial relations. The origin of a geometrical optic theory in the form of rays of energy that travel through space in straight lines and operate on the eye to cause vision can be traced to **Ibn al-Haitham** in the 1st century AD. He thought vision depended only on rays traveling from the direction where the eye was pointing, and this allowed the directions of objects with respect to the eye to be specified. **Euclid** had a similar theory, but in his case, the rays were supposed to pass out of the eye and through space in straight lines.

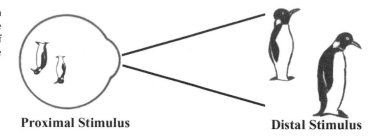

FIGURE 1.11. The distinction between distal and proximal stimuli is shown. The distal stimulus in this example is a pair of penguins. The proximal stimulus is the retinal image of the penguins.

Proximal Stimulus   **Distal Stimulus**

Until **Johannes Kepler's** discovery of the retinal image in the 17th century, the sense impressions were generally assumed to be received by the lens near the front of the eye. Since Kepler, the part of the body that receives visual sense impressions has been known to be the retina, at the back of the eye. The sense impression is in the form of an image formed on the retina, referred to as the **proximal stimulus.** This proximal stimulus is distinct from the object in the environment that has been imaged, called the **distal stimulus.** This distinction is illustrated in Figure 1.11.

The distal stimulus in this example is a pair of penguins. The corresponding proximal stimulus is the image of the penguins formed on the retina at the back of the eye. This distinction between the proximal and distal stimuli has had a major impact on theories of perception. It is generally assumed the brain has no direct access to information about properties of the external environment; it has access only to whatever information is available in the retinal image. Perceptual processing must try to glean information about properties of the external world based on information in the retinal image.

Sense Organ to Brain

The information available in the retinal image has to be transduced, encoded, and transmitted to the brain before it can lead to percepts. Models of the way this takes place also have a long history. Aristotle proposed a generalized hydraulic model in which sensible impressions were transmitted from the eye into the body by movements of a fluid. With various modifications, this general view held for centuries. At the time Descartes was writing, in the 1600s, the generalized model had changed from fluids to mechanical movements such as occur in a machine.

By the 1800s, this had been replaced by a model based on electrophysiological activity in neural tissue. Neural models have continued to dominate perceptual theories up until the present day. A pro-

totypical example of a modern generalized neural model is the **Neuron Doctrine,** published by **Horace Barlow** in 1972. Barlow models the neural machinery as single neurons that act as electrical devices. Neurons convert stimulation of the retina by the retinal image into patterns of electrical firing that encode complex perceptual experiences.

3 Information Processing Theories

Information-processing theories of perception differ in emphasis from the causal theories described in the previous section, because they are not primarily concerned with trying to detail all of the steps along the causal chain between the environment and percepts. Instead, they emphasize the kinds of information potentially available at various stages of processing.

The causal and information-theoretical approaches to the topic of perception are not necessarily incompatible, and there have been some historical attempts to incorporate both. The most notable was by **Hermann von Helmholtz** in the 19th century. However, since Helmholtz, there has been a general divergence such that most biology-based theories of perception have emphasized details of the causal chain, while theories derived from psychology have grappled with information-processing issues. In recent years, scientific studies of perception carried out from within the relatively new discipline of **cognitive neuroscience** have once again begun to incorporate both approaches.

Constructive Theories

Psychological information-processing theories are often called **constructive theories,** because their primary emphasis is on trying to explain how perceptual processing is able to "construct" our percepts from the information that is input from the environment. The distinction between distal and proximal stimuli has had a large influence on these theories. The task at hand for a perceptual system is considered to be construction of percepts that reflect properties of the environment, even though the only information provided at the input stage comes from the retinal image. These theories put a major emphasis on trying to account for the fact that our percepts are usually veridical, even when information in the retinal image seems to be inadequate or inaccurate in terms of specifying the distal stimulus.

This emphasis can be illustrated by an example. The phenomenon of **perceptual constancy** refers to

the fact that we perceive properties of objects to remain constant even when the retinal image is undergoing change. Box 1.8 illustrates the operation of one particular form of perceptual constancy, called **size constancy.**

A scheme illustrating how a constructive theory might conceptualize the construction of percepts is illustrated in cartoon form in Figure 1.12.

The arrow along the bottom illustrates events as they might be conceptualized in a causal theory. The environment causes stimulation of the sensory organ and initial sensory processing in peripheral neural pathways. This leads to additional percep-

Box 1.8
Size Constancy

Consider the relationship between the sizes of the distal and proximal stimuli as distance from the observer is varied as illustrated in Figure 1. In the top panel, the observer is looking at a penguin from a close distance and the retinal image of the penguin is relatively large. In the bottom panel, the observer is looking at the same penguin from a longer distance and the retinal image has shrunk, Despite this fact, observers usually exhibit size constancy, i.e., they do not report that objects shrink and grow large as they move farther away and closer.

The general form of the explanations for size constancy offered by constructive theories is that observers can make use of other information to

"figure out" that the distal object has not really changed in size. For example, perhaps the observer has some way to estimate how far away the distal object is and considers this distance information when constructing a percept of object size. Or perhaps the observer knows, from learning or innate knowledge, that objects such as penguins do not actually change in size when they move near or far away and so discounts the change in size of the proximal image. Regardless of the specific mechanisms that are responsible, size constancy is assumed to reflect a property of percepts that comes about due to perceptual processing rather than from properties of the proximal stimulation.

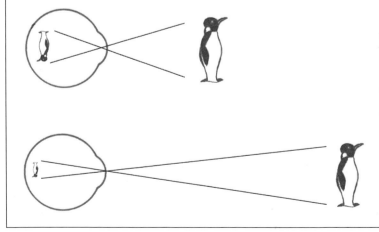

FIGURE 1. Illustration of size constancy. The penguin (distal stimulus) remains the same size regardless of the distance from which it is being viewed. However, the size of the retinal image formed by the penguin (proximal stimulus) varies with viewing distance. The brain receives its information from the proximal stimulus. However, our percepts of size ordinarily do not change with viewing distance. Thus, our percepts are more similar to properties of the distal than the proximal stimulus.

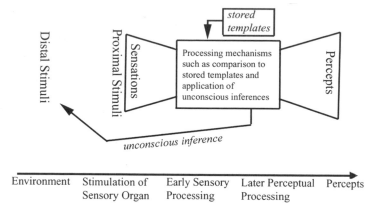

FIGURE 1.12. Constructive theories of perception emphasize the extent to which our percepts have to be constructed during perceptual processing. At early stages of processing, sensations are formed from the stimulation of the sensory organ. These sensations are primitive and reflect properties of the proximal stimulus more than those of the distal stimulus. During processing, additional sources of information, such as memories, are used to enhance the information carried in the sensations. In addition, processes such as use of logic and hypothesis testing are applied to the information to form percepts whose qualities reflect inferred properties of the distal stimulus.

tual processing in higher-order brain areas and finally to percepts. The top of the figure illustrates these same events as they might be conceptualized by a constructive theory. **Sensations** are formed from the information present in the proximal stimulus. The sensations are then subjected to information-processing mechanisms, such as comparison to **stored memory templates** (see Box 1.9).

In the course of this processing, **sensations fade** and **percepts emerge.** Additional information-processing mechanisms are applied to the per-cepts as they are being formed to give them properties related to the distal stimulus. An example is the **application of unconscious inferences** (see Box 1.10).

Sensations Versus Percepts

Constructive theories are part of what is sometimes referred to as the **sensationalist tradition** in the scientific study of perception. This tradition expresses, often implicitly, the general view that perceptual processing can be separated into two stages. Initial processing operates on relatively simple, or primitive, inward-directed **sensations** that reflect properties of the stimulation falling on the sense organ (proximal stimuli). Later stages of processing involve more complicated or higher-order **percepts** that reflect properties of objects in the environment (distal stimuli).

Similar distinctions along these lines have been made by philosophers for a long time. For example, the Scottish philosopher **Thomas Reid** made a distinction in the eighteenth century between **sensations,** which reflect sensory processes, and **perceptions** (percepts in our terminology), which he asserted are more closely related to distal stimuli. Scientific theories of perception have been influenced by these same kinds of distinctions, although there has been some ambiguity and confusion in the ways these distinctions have been characterized.

Box 1.9
Stored Memory Templates

Suppose you look at an apple and store a memory of what it looked like. Similarly, you store a memory of an elephant. Later, you are confronted with a small red object that cannot be seen clearly. The information provided to the sense organ by the small red object may be insufficient for you to tell that it is an apple. However, by comparing the information to the stored memory templates, you can arrive at the conclusion that the object is more likely to be an apple than an elephant.

Box 1.10
Helmholtz' Unconscious Inferences

The concept of unconscious inferences was championed by Helmholtz. He argued that when we have the impression our percepts reflect "what is out there," this impression is actually a claim based on an elaborate system of unconscious inferences. The operation of unconscious inferences can be illustrated with a simple perceptual phenomenon that involves perception of motion. Place your finger along the side of your eyeball and gently press at a rate of a few times per second. Note that the environment appears to oscillate with each push. When the eye is pushed to the left the environment appears to move to the right and vice versa. Now remove your finger from your eyeball and voluntarily look from side to side. Note that the environment does not appear to move under these conditions.

Why did the brain produce a percept of a moving environment in the first condition but not in the second? After all, a similar pattern of motion of the retinal image was produced by the voluntary eye movement and when the eye was moved passively by the finger. An explanation based on unconscious inferences is as follows: When retinal motion is perceived in the absence of voluntary eye movements, our perceptual systems make an inference that the world must have moved. However, the same retinal motion accompanied by a voluntary eye movement leads to an inference that the motion was caused by the eye movement rather than by movement of the world. Motion perception is discussed in more detail in Chapter 10.

In some formulations, the distinction between sensations and percepts is between strictly **psychological phenomena.** For example, in the school of experimental psychology founded by Wilhelm **Wundt** in Germany in the late 1800s, sensations were considered **primitive elements of the mind,** while percepts were more **complex mental entities** built up out of these primitive sensations. A percept of an apple could be considered to be a combination of sensations such as "redness" and "apple-shape." The psychologist Edward Bradford **Titchener** carried this tradition to American schools of psychology early in the 20th century. Titchener insisted that sensations should be observable in consciousness, but despite his efforts, empirical studies eventually forced the conclusion that primitive sensations, if they exist, are often not available for conscious inspection.

In other formulations, the distinction between sensations and percepts is related to stages of **physiological processing.** Sensations are related to **physiological activities generated at early stages of processing in peripheral neural pathways.** Percepts are related to **later stages of physiological processing in pathways that are more central.** The 19th-century writings of Helmholtz and Ewald **Hering** provide many good examples of this characterization. For example, Helmholtz characterized sensations as

what we call the impressions on our senses, in so far as they come to our consciousness as states of our own body, especially of our nervous apparatus.

He contrasted sensations with percepts,

in which we form out of these impressions the representation of outer objects.

The basic distinction between sensations and percepts has continued to the present day and plays a prominent role in modern cognitive neuroscience approaches to perception, although the terminology has changed some. The modern terminology for sensations characterizes them in terms of **information coded in the domains of space, time, and spectral composition that can be extracted from retinal images.** These include **low-level primitives,** such as oriented contours, velocity, and spectral composition. Perceptual processing is supposed to turn these into percepts of shape, motion, and color (Figure 1.13).

4 Direct Perception Theories

Direct perception theories derive their name from the fact that they deny the importance of perceptual processing in forming properties of percepts and emphasize the similarity of percepts to properties of the environment. Proponents of direct perception argue that the input from the environment contains enough information to specify properties of the environment without help. Perceptual

Modern Formulation That Sensations Lead to Percepts

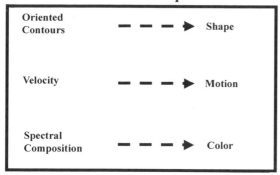

FIGURE 1.13. Modern neuroscience theories of perception reflect the idea that perceptual processing converts low-level information derived from the retinal image (e.g., oriented contours, velocity, and spectral composition) into higher-order qualities of percepts (e.g., shape, motion, and color).

systems can simply pick up the available information and do not need elaborate mechanisms to enhance or elaborate this information to construct percepts.

The most prominent theory of **direct perception** was formulated by **James J. Gibson.** Gibson claimed that we perceive the environment directly with no need for intervening psychological or biological processes to mediate what we perceive. Proponents of direct and indirect theories of perception generally adopt an adversarial position and consider the basic tenets of the two types of theories to be mutually incompatible. However, Gibson, even in the most radical formulations of his theory of direct perception did not deny that there is a causal chain passing from the environment through brain activity to our percepts. What he denied was that this causal chain played a significant role in "constructing" properties of our percepts. He preferred

to think of the causal chain that is involved in perceptual processing as functioning by simply "picking up" the rich perceptual information that was already present in the environment.

Gibson's arguments were directed primarily against psychological constructive theories in which sensory stimuli on the retina were considered to be too impoverished to produce percepts without cognitive processes such as using memory templates and forming inferences. He reacted in particular against many of his contemporaries, who modeled perception as hypothesis testing. These theories proposed that observers form hypotheses about the environment, then sample and evaluate stimuli from the environment, and finally accept or reject the hypotheses based on the evaluation.

Despite the contentious nature of much of the discussion from proponents of the two extreme camps associated with constructive theories and direct perception theories, there is potentially much common ground between the extremes, as illustrated schematically in Figure 1.14.

It is possible to adopt a pragmatic point of view between the extremes, in which specific properties of percepts are examined simultaneously from two points of view: 1) To what extent are properties of percepts reflections of properties of environmental stimulation (direct perception)? 2) To what extent are properties of percepts constructions based on intervening processes?

Theories of direct perception share the problems of any theory grounded in a philosophical position of direct realism. Since they deny the very presence of intervening mechanisms that might distort the information that is picked up, they must try to account for why apparent nonveridical, or illusory, perception sometimes occurs. Gibson argued that cases in which perception appears to be illusory occur mostly when using artificial stimuli in laboratory settings and that perception is, in fact, usually veridical in natural environments. He proposed that the appro-

FIGURE 1.14. Direct theories and constructive theories of perception can be conceptualized as falling along a continuum. The extreme direct perception point of view is that all properties of percepts are determined exclusively by the environment, with no contribution from processing. The extreme constructive point of view is that all properties of percepts are constructed during processing.

Direct Perception Theories

Constructive Theories

Properties of percepts should be determined exclusively by the environment.

Properties of percepts should be determined by combination of environment and intervening processes.

Properties of percepts should be determined exclusively by intervening processes.

Causal Relationships Among Major Components

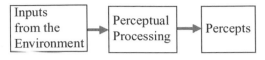

FIGURE 1.15. This figure illustrates causal relationships described at a relatively molar level of analysis in which only major components are delineated.

priate way to study perception is from an **ecological approach,** in which observers are studied in their natural environments.

Proponents of direct perception nevertheless find themselves in a tenuous intellectual position when trying to explain examples of apparent nonveridical perception that occur in natural environments. For this reason, theories of direct perception have been generally out of fashion in recent decades, and the primary conceptual approach driving most current perception research is based on constructive theories. The main impact of direct perception has come from its emphasis on use of the ecological approach in analyzing the kinds of rich perceptual information that are present in natural environments.

5 Where Should We Search for Explanations of Perception?

The topic of perception is characterized by statements that come from widely different levels of analysis. Take, for example, the following pair of statements:

- The observer says that the sky looks blue.
- While the observer is looking at the sky, activated rhodopsin within the photoreceptors of the eye triggers the cGMP cascade by interacting with a G-protein.

Each statement describes something that relates to this observer's perception of the sky, but the statements provide descriptions from strikingly different **levels of analysis.**

Levels of Analysis

The causal chain of perception from a relatively molar level of analysis involving only its major components is illustrated in cartoon form in Figure

1.15. Each of these major components can be broken down into **subcomponents.** Figure 1.16 shows an example in which the single component labeled "Inputs from the Environment" in Figure 1.15 has been broken down by a Finer analysis into four subcomponents, connected by causal arrows.

Obviously, these subcomponents could be further broken down if we had an interest in describing causal relationships at an even finer level. As a general principle, the causal chain can be described at levels of analysis varying from molar to molecular, depending on the question of interest.

Are Some Perspectives or Levels of Description Special?

Perceptual theories differ not only in their levels of analysis, but also **qualitatively in the choice of descriptive terms that are used to characterize the components of perception.** We have already encountered several examples in this chapter, starting with Figure 1.1. It is interesting to ask what relationships, if any, hold between theories that use different qualitative types of descriptive terms and/or provide different levels of description. Theories that are built from one or more of these perspectives or levels often make the claim that their particular approach is special and provides the most fundamental, or ultimate, explanation of what is described at the other levels. Probably the most well known example is **reductionism,** which proposes that the various social, biological, and physical sciences can be ranked from high to low in

More Detailed Causal Relationships Within Environmental Input

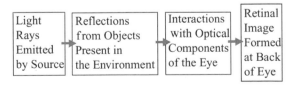

FIGURE 1.16. This figure illustrates causal relationships described at a finer level of analysis than is shown in Figure 1.15. The individual boxes in this figure correspond to subcomponents that would all be part of the box in Figure 1.15 labeled "Inputs from the Environment."

Reductionism

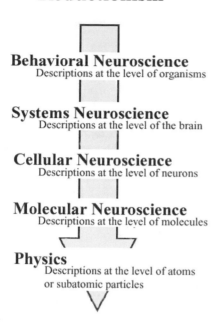

FIGURE 1.17. The philosophy of science position called reductionism asserts that explanations for phenomena described at any given level of description are to be found in levels of description at a finer spatial scale. The ultimate explanation is supposed to be found at the spatial scale used to describe subatomic particles.

terms of the level of description they provide (Figure 1.17).

According to reductionism, the explanation for each level of description is to be found at the next lower level, and an ultimate explanation is to be found at the lowest level. Levels of description for perception at a high level can be found within **behavioral neuroscience** or **psychology.** Our previous example, "The observer says that the sky looks blue," is an example of a description of perception that might be made at this level. The explanations for descriptions at this level are, according to reductionism, to be found at the next lower level, **systems neuroscience.** An example might be, "The sky appears blue because the part of the brain that processes color information is activated when the observer looks at the sky." Explanations for systems neuroscience levels are to be found within **cellular neuroscience.** For example, "The amount of electrical activity is very high in a population of cells in the observer's brain that respond to short-wavelength light." The explanation for cellular neuroscience descriptions is to be found in **molecular neuroscience.** Our earlier example, would fall

within this category. Ultimate explanations, according to reductionism, will be found at the level of **physics,** in the form of descriptive statements about what happens to atoms or perhaps subatomic particles.

The reductionistic approach has a long history in the scientific study of perception. Prominent examples include Helmholtz and Hering in the 19th century, who argued that percepts could be reduced to complex physiologic processing in the sense organs, and the psychologist **William James** near the turn of the 20th century, who asserted that the best way of formulating many facts about perception is physiological. Modern neuroscience research has continued to make effective use of reductionistic approaches to many aspects of perception. Philosopher of science **P.S. Churchland** has argued recently that much of psychology may eventually be substantially revised by a reduction to neuroscience, or even replaced.

In this text, we will find it useful to explain many perceptual phenomena described at a high level by searching for a description at a lower level. However, as outlined below, our search for explanations between levels of description will not be strictly one-way. Sometimes phenomena described at lower levels will find their explanations at a higher level of description.

Furthermore, we will not pursue a goal of reduction to an *ultimate* level. This is simply because that ultimate level is not very convenient for helping us understand most of the topics covered in this text. Box 1.11 illustrates this limitation of reductionism's ultimate level by using an analogy to computer programming.

Another example of a general approach to understanding perception based on an "ultimate' level is **evolutionary theory,** predominant in biology since Charles **Darwin.** Perceptual systems that are present in biological organisms today are the result of about a billion years of evolution of multicelled organisms. Ultimate levels of explanation for perception from an evolutionary perspective must include an analysis of function. It is assumed that biological perceptual systems evolved for the purpose of facilitating propagation of the genome by promoting survival of the individual and its kin in the environment. Note that the ultimate explanations advocated by evolutionary theory and reductionism are near the opposite ends of spatial and temporal scales. Ultimate explanations from an evolutionary perspective are to be found only by considering groups of organisms interacting with their environments over long periods that span many generations.

Box 1.11
Limitations of Ultimate Levels of Description

The computer that I use in my laboratory has a database that includes the visual acuities for many subjects. I can write a program in a high-level database programming language that contains the statement "select subjects whose acuity is normal." After I write this program, a compiler on my computer translates the high-level language into code in a computer language called "C." This results in a few hundred lines of C code. Internally, the computer then executes another compiler that translates the C language code into a long sequence (perhaps hundreds or thousands) of statements in assembly language code. Finally, the assembly code is translated into a binary code that consists of a very long sequence (perhaps millions) of ones and zeros. If you were to examine the memory of the computer while it is executing the code to search for subjects whose acuity is normal, you would find that only this binary code is present. Thus, it could be argued that this binary code provides an ultimate explanation about what the computer program does. The statements at the levels of assembly code, C, and database languages are simply higher-level descriptions of this binary code (Figure 1).

However, suppose you tried to use the database to find all of the subjects with normal acuity but discovered that every time you tried to run the program the computer crashed and had to be restarted. What would your reaction be if someone then told you: "The ultimate explanation about why your computer is crashing is to be found in the list of millions of ones and zeros that make up your binary code?" Would you immediately set your printer to work printing out a tall stack of sheets of computer paper filled with the ones and zeros of your

binary code? Probably not! Even though, in principle, the ultimate explanation about what is causing your computer to crash is to be found in that huge list of ones and zeros, to fix the problem that concerns you this "ultimate" explanation is largely irrelevant. You would instead have your printer make a listing of your computer program at whatever higher level of description you find most convenient to try to figure out what is causing the computer to crash.

Computer Programs

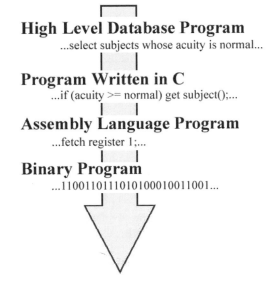

High Level Database Program
...select subjects whose acuity is normal...

Program Written in C
...if (acuity >= normal) get subject();...

Assembly Language Program
...fetch register 1;...

Binary Program
...11001101110101000010011001...

FIGURE 1. Explanations for how computer programs operate can be sought at various levels of description in a manner that is roughly analogous to reductionism.

Reductionistic and evolutionary approaches, despite the claim of each about providing the ultimate explanation, do not have to be treated as being mutually exclusive. Explanations for perception can arise out of interplay between these approaches. Evolutionary theory defines the problems that a perceiving system has to have solved in order to survive in an environment. Reductionistic

approaches can discover the mechanisms that evolved to solve these problems.

Arguments that certain levels or perspectives are special are not limited to grand philosophy of science positions like reductionism and evolution. They can also permeate discussion of perceptual theories at smaller scales. A prominent example is much of the discussion in the current scientific literature between proponents and opponents of direct perception. At an even smaller scale are differences of opinion among individuals regarding what aspects of perception are most important to take into account. Some examples are illustrated in Box 1.12.

Box 1.12
What Aspects of Perception Are Most Important?

- Psychologist: "In diagnosing a psychosis it is important to make a judgment about the extent to which the patient's perception is reality based."
- Interior decorator: "The color of a room is very important because the color can make its perception vary from warm to cold."
- Television ad executive: "The most important aspect of a commercial is how quickly it grabs the viewer's perceptual attention."
- Religious mystic: "Perceptions must be holistic with the universe."
- Eye doctor: "Visual acuity is best quantified along the Snellen acuity dimension of coarse (20/200) to fine (20/20)."
- Neuroscientist: "It is very important to understand how perception varies with regard to the extent it is mediated primarily by cortical or subcortical regions of the brain."
- Student taking a course on the topic of perception: "The most important aspect of perception is the likelihood that specific facts about it will appear on the final exam."

An alternative to searching for the ultimate level that can explain all of perception is to search for **explanations** in terms of **relationships between levels of description.** The task at hand for this approach is to try to build bridges that allow phenomena described at one level to be related to descriptions at other levels. This position is similar to that adopted by the philosopher Churchland and **Terrance Sejnowski,** a neuroscientist, about the general discipline of computational neuroscience. As they put it, research at one level provides correction, constraints, and inspiration for research at higher and lower levels, and phenomena that appear mysterious at one level of description are often found to have simple explanations when related to another level. This can be in the direction of a reductionistic explanation, such as when a complicated conscious perceptual experience has a simple explanation in terms of brain function. However, the explanations can also go in the opposite direction, as, for example, when an enigmatic brain characteristic becomes understood once its function is known.

Marr's Integrative Approach

The prototypical example of an attempt to integrate across approaches and levels to understand perception comes from **David Marr,** culminating in his book titled *Vision*, published in 1982. He argued that a complete explanation of perception requires the simultaneous use of three approaches, or levels, which he called **computational, algorithmic,** and **implementation.** The computational level is a description of the problem that the system is trying to solve along with an analysis of why it is important for the organism to solve this problem. For example, a problem that might be important for some observers is determining how far away a particular object is located. The algorithmic level describes formal solutions to this problem. For example, one potential way an observer with two eyes might be able to determine distance would be to carry out a specific set of numeric calculations involving disparities in the images from the left and right eyes. The implementation level describes the physical instantiation of how the problem has been solved by an organism. For example, neurons have been found in the visual cortex of the brains of monkeys that respond to disparities between images in the two eyes.

Each of the approaches delineated by Marr can be tackled somewhat independently by different teams of investigators. However, Marr made the argument that in order to truly understand perception one needs to consider all of these approaches simultaneously.

Simplifying and Realistic Models

One approach taken by scientists who want to understand human perception is to utilize **models.**

These models do not incorporate all of the details of the system that is being modeled. Rather, a model tries to incorporate certain aspects of the system that are deemed important from a **particular level of abstraction.** These levels of abstraction can vary from simplifying to more realistic.

Simplifying models of human perception do not attempt to incorporate details of biological and psychological mechanisms. An example is the kind of models derived from the field of **artificial intelligence** that are based solely on a level of abstraction involving **symbol processing.** These models are typically implemented in the form of computer programs in which perceptual processing is simulated as an algorithm operating on symbols that represent perceptual information.

Models of human perception that are somewhat more biologically realistic try to take into account more details of the way brains operate and the constraints neurobiology imposes on perception. The pioneering models of this type were the **formal neuron models** developed by **Warren S. McCulloch and Walter H. Pitts** in the 1940s. Realistic modeling approaches continue to the present day within the relatively new discipline of **computational neuroscience.** The aim of these **neural models** within the field of perception is to specify what computations are possible, necessary, and sufficient to account for the activities accomplished during perceptual processing by brain tissues.

Neural models of perception can be simulated on digital computers. The architecture of the computer does not matter in principle. However, serial computers, even with a very fast processor, are usually too slow to model perceptual functions in real time. A recent trend is to run simulations on hardware devices that mimic more closely some aspects of the architecture of the brain, such as parallel processing and recurrent connectivity. These physical devices can execute the operations of the model fast enough to approximate simulations of some perceptual phenomena in real time.

Parallel distributed processing models that became widely available in the 1980s provide one architecture that maps more realistically onto neural circuits. Small groups of **highly interconnected elements** called **nodes** are used to model perceptual processing. **Neural network models** utilize nodes designed to be analogous to neural circuits in the brain.

Neural network models allow psychological and biological levels of description to be combined by using nodes to model biological neurons, whose activities in turn represent simple psychological properties such as sensations or percepts. This type of modeling has the potential of allowing discoveries at the neuronal and psychological levels to be bridged.

An even more realistic class of neural models takes the form of concrete **physical devices built from electronic components.** Operational amplifiers can take on the function of neurons, while wires, resistors, and capacitors can be used to form connections. An example is the **artificial retina** built out of silicon by **Carmen Mead** and colleagues. This is not a mere simulation of a retina, but a complex information-processing physical system that responds to light in real time. Some expect that these types of human-engineered devices will be able to simulate many aspects of visual processing as it occurs in biological brains, in real time, within the coming decades.

Inspiration and guidance about what biological details are important to include in realistic models of perception can be gleaned from close examination of perception in biological organisms. The most highly realistic model of biological aspects of human perception is provided by the **monkey model.** The biology of the eye and those portions of the brain involved in early processing of visual information are virtually identical in humans and monkeys. Furthermore, when behavioral methods (described in Chapter 2) are used to determine what monkeys perceive, many aspects of their percepts are identical to those of humans.

6 Perception in Machines

If a model completely replicated every aspect of a perceiving human, it would be a perceiving human rather than a model of a perceiving human. It seems indubitable, perhaps a truism, that such a physical system would be able to perceive in the same manner that humans perceive. But **would** *every* **level of detail be** *necessary* in order for such a physical system to perceive in the same manner as humans perceive? If not, then *which* **details will prove to be** *sufficient?* If we knew the answers to these questions, we would know if it makes sense to attempt to construct physical systems that not only serve as models of human perception but can also perceive on their own.

The bias of this text is to assume that perception of the kind performed by humans can ultimately be carried out by physical devices other than human brains. The rationale for this assumption is simply a conviction that the basic principles discovered by the study of perception as it occurs in biological organisms produced by the evolutionary process is likely to be generalizable to other physical systems. Nevertheless, it is important to keep in mind the caveat that human brains differ profoundly from any manufactured physical device in the scale of their intrinsic complexity. Many scientists expect that as the complexity of robots evolves, their underlying mechanisms will gradually converge to the complexity underlying human perception. However, that belief is currently only a promissory note. These questions will be answered in the coming years as roboticists push the limits on what kinds of perceptual activities can be performed by manufactured devices.

However, even if the enterprise of attempting to build perceiving machines does not accomplish the goal of fully replicating human perception, it will serve two functions. First, the attempt is valuable from a scientific perspective in terms of providing better models for testing theories about human perception. Second, machines built with this goal in mind are valuable from an engineering perspective in terms of being able to perform at least some perceptual tasks that could heretofore be performed only by biological organisms.

In the remaining chapters of this text, our primary goal is going to be to try to understand human perception. We will utilize abstract models that range from simplifying to realistic and biological models based on monkeys to help elucidate some of the basic principles on which human visual perception is based. Permeating much of our discussion will also be a consideration of what it would take for models of perception, implemented in the form of robotic machines, to operate according to these same principles.

Summary

Humans are physical systems of a special biological kind that can perceive. The act of human perception can be conceptualized in terms of three components: inputs from the environment, processing mechanisms that operate on those inputs, and output products in the form of percepts. Causal relationships among and within these components of perception are in some cases mysterious, and this has led to extensive discussions of these issues in both the philosophical and the scientific literature. Biological theories have focused on the causal chain that progresses from the environment through neural activity in the brain to our percepts. Psychological theories have attempted to characterize the processes by which percepts are formed from a perspective of information theory. More broad-based integrative approaches try to take into account computational, biological, and psychological levels of description with levels of analysis ranging from molar to molecular. Models that range from simplifying to realistic serve as important tools used by scientists in trying to understand perception. Ultimately, the most realistic models, in the form of perceiving robots, may cease being mere models and take on characteristics of perceiving agents.

Selected Reading List

Barlow, H.B. 1972. Single units and sensation: A neuron doctrine for perceptual psychology? *Perception* 1:371–394.

Boring, E.G. 1942. *Sensation and Perception in the History of Experimental Psychology.* New York: Appleton Crofts.

Churchland, P.S. 1982. Mind-brain reduction: New light from the philosophy of science. *Neuroscience* 7:1041–1047.

Churchland, P.S., and Sejnowski, T. 1992. *Computational Neuroscience.* Cambridge: MIT Press.

Cosmides, L., and Tooby, J. 1995. From function to structure: The role of evolutionary biology and computational theories in cognitive neuroscience. In *The Cognitive Neurosciences,* ed. M. Gazzaniga, pp. 1199–1210. Cambridge: MIT Press.

Descartes, R. [1637] 1965. The optics. Reprinted in *Discourse on Method, Optics, Geometry, and Meteorology,* trans. P. Olscamp. Indianapolis: Bobbs-Merrill.

Glanz, J. 1995. Measurements are the only reality, say quantum tests. *Science* 270:1439–1440.

Helmholtz, H. 1924. *Helmholtz's Treatise on Physiological Optics,* trans. ed. J.P.C. Southall. Rochester, NY: The Optical Society of America.

James, W. 1890. *The Principles of Psychology.* New York: Henry Holt.

Koch, C., and Laurent, G. 1999. Complexity and the nervous system. *Science* 284:96–98.

Marr, D. 1982. *Vision: A Computational Investigation into the Human Representation and Processing of Visual Information.* San Francisco: Freeman.

McClelland, J., Rummelhart, D.E., and the PDP Research Group. 1986. *Parallel Distributed Processing: Exploration in the Microstructure of Cognition,* Vols. 1 and 2. Cambridge: MIT Press.

Mountcastle, V.B. 1986. The neural mechanisms of cognitive functions can now be studied directly. *Trends Nuerosci.* 9:505–508.

Pais, A. 1982. *Subtle is the Lord: The Science and Life of Albert Einstein.* New York: Oxford University Press.

Reed, E.S. 1988. *James J. Gibson and the Psychology of Perception.* New Haven, CT: Yale University Press.

Teller, D.Y. 1980. Locus questions in visual science. In *Visual Coding and Adaptability,* ed. C. Harris. pp. 151–176. Hillsdale, NJ: Erlbaum Associates.

Westheimer, G. 1983. Hermann Helmholtz and origins of sensory physiology. *Trends Neurosci.* 6:5–9.

2
Psychophysical Methods: What Scientific Procedures Can Be Used to Ask an Observer What Is Being Perceived?

Questions

After reading Chapter 2, you should be able to answer the following questions.

1. Describe some special characteristics of the data collected and studied by perceptual scientists that make them different from the data collected and studied in most other scientific disciplines.
2. Distinguish between empirical knowledge and perceptual knowledge.
3. Distinguish between Class A and Class B observations.
4. What are the basic psychophysical methods for answering questions about whether or not an observer has perceptual knowledge about *what*, i.e., "what is out there"?
5. What are the basic psychophysical methods for answering questions about whether or not an observer has perceptual knowledge about *how*, i.e., "how to interact with the environment"?
6. What basic methods are used to study subjective perceptual experience?
7. What are absolute and difference thresholds, and how are they measured?
8. What are diagnostic systems and how do they relate to issues of perception?
9. Describe a model in terms of "what goes on inside the head of an observer" that can account for the fact that observers sometimes give false alarms during detection tasks.
10. Describe receiver operating characteristics (ROC) curves, and indicate with some examples what performance will look like for observers with different sensitivities and biases when plotted on these curves.
11. What are psychophysical linking hypotheses?

The neurophysiologist G. S. **Brindley,** in a book published in 1970, noted that the data collected, analyzed, and interpreted by scientists studying perception have some special characteristics not present in most other scientific disciplines.

"The physiology of sensory pathways differs from that of other parts of the body in that it is based not only on objective observations and measurements which can be described in physico-chemical and anatomical terms, but also on the results of sensory experiments, that is, experiments in which an essential part of the result is a subject's report of his own [percepts]."

The special nature of the raw data have forced scientists to develop and refine procedures that can be used to infer properties of percepts based on measurements of the observer's behavior. If measured and evaluated properly, an observer's behavioral reports (in the form of words, symbols, or actions) can be used by scientists to study properties of percepts.

1 Behavior as a Probe for Learning About Percepts

Psychophysical methods for studying perception can be traced to Gustav **Fechner** in the 19th century. The general rationale underlying psychophysical methods is that behavior can serve as a probe to learn about properties of percepts, roughly analogously to the way a stethoscope can be used as a probe to learn about properties of the heart (Figure 2.1).

Specialized psychophysical methods have been developed for evaluating each of the three aspects of percepts outlined in Chapter 1: **perceptual knowledge about** *what* is present in the environment, **perceptual knowledge about** *how* to interact with the environment, and **subjective perceptual experiences.**

2 Measuring What-Perception

Perceptual acts of looking around convey to observers percepts in the form of a knowledge base of facts regarding their immediately surrounding environment. We will use the term **What-perception** to characterize these perceptual acts.

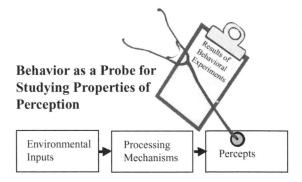

Behavior as a Probe for Studying Properties of Perception

| Environmental Inputs | Processing Mechanisms | Percepts |

FIGURE 2.1. The rationale for using behavior as a probe to study properties of percepts is that it is analogous to the use of a stethoscope to study properties of the heart. Observers use percepts to guide behavior, (in the form of words, symbols, or actions) and thus, carefully designed behavioral experiments can be used as a probe to study properties of percepts.

Balance -- Information about weight

Microscope -- Information about fine structure

Chemistry Lab -- Information about chemical composition

Atom Smasher -- Information about atomic and subatomic structure

Psychological Rating Scale -- Information about how humans relate to an object

Etc., etc., etc.

FIGURE 2.2. Empirical knowledge refers to everything we know about the environment that surrounds us based on making direct observations with our sense organs and by making observations with instruments.

One important class of psychophysical methods is geared towards establishing and characterizing What-perception. Before embarking on a discussion of how these methods are employed, we first need to understand the rationale behind their use.

Rationale Behind Measures of What-Perception

Perceptual Knowledge Versus Empirical Knowledge

Empirical knowledge refers to the database of information about the external world gained by using **instruments** along with and **in addition to one's own sense organs.** Suppose you wanted to make an exhaustive list of everything you know about some object in the world, such as an apple: Its weight could be determined with a balance, its molecular structure by slicing thin sections and examining them under a microscope, its chemical composition by analyzing samples in a chemistry lab, and so on (Figure 2.2).

Furthermore, the term **instrument** is used here in a much broader sense than is typical in everyday language. Books, magazines, movies, and television are all examples of instruments. For example, you can gain knowledge about apples by watching a documentary on television. Other human observ-

ers can also serve as instruments. For example, a group of human subjects might be asked to fill out a psychological rating scale in which they make judgments about "how they feel about" the apple. This rating scale instrument then provides additional empirical information about the apple.

If all of the information that makes up the empirical knowledge base we have about any single object, such as an apple, were to be made explicit and enumerated, it would form a very long, perhaps huge, list. The empirical knowledge we have about the entire environment can be conceptualized as the concatenation of the individual lists about each of the objects in the environment of which we have knowledge.

Ultimately, empirical knowledge is what allows us to differentiate among objects in the environment. Consider two objects present in the environment that are in actuality different from one another but whose differences cannot be picked up by any instrument we have available to use. The existence of those differences – that fact about the world – will necessarily remain hidden from us.

Perceptual knowledge refers to information an observer obtains about properties of objects in the world from her **sense organs alone.** Perceptual knowledge, like empirical knowledge, could be enumerated, in principle, in a list, as illustrated in Figure 2.3.

Some items will be present in both the empirical knowledge and the perceptual knowledge lists. For example, an observer might be able to achieve veridical perceptual knowledge about

Perceived Weight -- Information obtained by lifting

Perceived Shape -- Information obtained by manipulating object with fingers

Perceived Chemical Composition -- Information obtained by smelling and tasting

Perceived Distance -- Information obtained by looking

Etc., etc., etc.

FIGURE 2.3. Perceptual knowledge refers to what an observer knows about the surrounding environment based on using only her own sense organs.

weight of an apple, based solely on proprioceptive sensory input from the muscles and joints of the arm and hand while picking up the apple. In this case, the perceptual knowledge about weight of the apple corresponds to the empirical knowledge that might be gained by placing the apple on a balance.

There are no items on the perceptual list that are not also included on the empirical list, since the definition of empirical knowledge includes information gained from the sense organs as well as from instruments. However, there is information on the empirical list that is not present on the perceptual list. For example, knowledge about molecular structure, determined with the aid of a microscope, is unavailable to a human observer limited to using the unaided eye. Thus, perceptual information is a subset of empirical information.

Class A Versus Class B Observations

Brindley, in his book published in 1970, defined and delineated two special types of behavioral reports one can elicit from an observer following her observation of the environment. He defined as **Class A observations:**

"Observations that can be described as an identity or non-identity of two [percepts]."

Class A observations can be illustrated with a concrete example. Imagine an observer participating in a perceptual experiment. She is shown some pencils and paper clips and asked to look at them and report simply whether they are distinguishable, i.e., nonidentical (Figure 2.4).

Class A Observations

"These two kinds of objects are distinguishable."

FIGURE 2.4. When reporting a Class A observation, an observer states only whether percepts produced by two stimuli can be distinguished or are identical.

Class B Observations

"I see pencils on my right and paper clips on my left."

FIGURE 2.5. When an observer reports anything about an observation that does not qualify as Class A, the observation is relegated to Class B.

Anything else she reports, such as the fact that one group of objects looks like pencils and the other like paper clips, is relegated to **Class B** (Figure 2.5).

We will discuss Class B observations below when we consider measures of subjective perceptual experience.

This distinction between Class A and Class B observations is of fundamental importance in terms of how the raw data collected from an observer during a perceptual experiment can be treated. Class A observations can be treated scientifically as **objective results,** because they can be compared to data obtained with an **external referent** in the form of an **instrument.**

Class A observation can be evaluated in terms of being **correct or wrong.** This can be illustrated with an example of an experiment in which we want to decide whether an observer can perceive the state, on or off, of a light bulb. The following procedure can be defined. The experimenter takes the observer into an otherwise dark room, turns on the light bulb, and gives the following instructions:

"I have an instrument called a photometer that registers the state of this light bulb. The photometer now indicates that the light is 'on.' Please look at the light bulb while it is in this state."

Then the experimenter turns off the light bulb and states:

"The photometer now registers that this light bulb is in the 'off' state. Please look at it."

After this initial training period, during which the observer samples both conditions, the actual experiment begins. The experimenter uses a random procedure, such as flipping a coin, to decide

FIGURE 2.6. When evaluating the performance of an observer that takes the form of
a Class A observation, a scientist can label
the observation as being correct or wrong
based on a comparison to the performance
of an instrument.

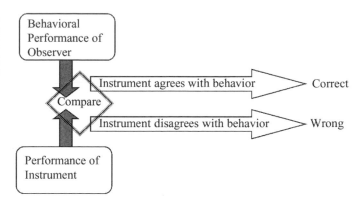

whether the light bulb will be set to the "on" or the
"off" state on the next trial. Then the observer is
given the following instructions:

"Look at the light bulb and tell me whether it is in the 'on'
or the 'off' state."

The observer responds "on" or "off," and the experimenter tallies the response as being "correct"
or "wrong."

We need to analyze carefully what it means
when we state that the experimenter tallies the
observer's response as "correct" or "wrong" in
this example. We have distinguished between two
conditions in the environment ("light bulb is 'on'"
or "light bulb is 'off'") based on **empirical
information** supplied by an instrument in the
form of a photometer. The observer is being asked
to distinguish between the same two conditions. To
state that an observation is **correct** in this situation
is to state that **the observer's response agrees with
the instrument.** Similarly, to state that the observation is **wrong** is to state that the response
of **the observer disagrees with the instrument**
(Figure 2.6).

Equivalent-Perceptual-Knowledge Criterion

We can combine the concepts of Class A observations, perceptual knowledge, and empirical
knowledge to devise an operational definition of
What-perception. Assertions that an observer can
perceive specific facts about her environment can be
decided by translating the assertions into questions
about whether the **observer has perceptual knowledge that is equivalent to empirical knowledge
available from designated instrument(s).** We will
refer to this as the **equivalent-perceptual-knowledge criterion** for establishing that an observer has
What-perception.

Two basic types of psychophysical procedures,
called **forced-choice** and **yes-no,** can be used for
establishing What-perception based on this equivalent-perceptual-knowledge criterion. We will illustrate the use of these procedures by continuing our
example of trying to establish that an observer can
perceive the "on" or "off" state of a light bulb.

Psychophysical Procedures

Forced-Choice Psychophysical Procedures

Forced-choice psychophysical procedures require
an observer to **choose a correct stimulus from
among two or more alternatives.** If the observer
must choose between two alternatives, the procedure is referred to as two-alternative-forced-choice.
Similarly, three alternatives would be three-alterna-
tive-forced-choice, etc.

Regardless of the number of alternatives,
there are two basic variations of forced-choice
procedures, **temporal** and **spatial.** In the temporal
versions, the alternatives are presented one at a
time, sequentially. In our light bulb example, this
might be accomplished by having the experimenter
show first the "light bulb 'on'" condition and
then the "light bulb 'off'" condition, with the
instructions:

"Tell me whether the light bulb was 'on' in the first or the
second interval."

In the case of spatial forced-choice, the alternative
stimuli are presented simultaneously but in different spatial positions. For example, two identical
light bulbs might be placed side by side. Only one
is turned on during each trial, with the instructions:

"Tell me whether the light bulb that is 'on' is located in
the left or the right position."

If an observer is unable to distinguish between the alternative conditions, then performance is expected to be at chance levels, i.e., 50% correct for a two-alternative procedure, 33% correct for a three-alternative procedure, etc. Performance at levels significantly better than expected based on chance satisfies the equivalent-perceptual-knowledge criterion; we would be forced to conclude that this observer can perceive the on or off status of the light.

Yes-No Procedures

Another psychophysical method is to use some variation of a **yes-no procedure,** in which an observer is asked to respond "yes" or "no" to a question that pertains to some aspect of whether two objects are distinguishable.

One variation of the yes-no procedure is to have the observer respond to a question of the form "Are these objects distinguishable?" A response of "yes" on more than half of the presentations of the stimuli is typically taken as satisfying the equivalent-perceptual-knowledge criterion.

A second variation is to supply the observer with a label for one of the alternative stimuli, as was done in our earlier example, in which the observer was initially informed, "This is the 'light bulb on' condition." Then the observer is asked to respond to the question, "Is the light bulb on?" A response of "yes" on more than half of the trials serves as evidence of equivalent-perceptual-knowledge.

Methodological Control Procedures

Two potential methodological concerns arise whenever one tries to establish What-perception based on the equivalent-perceptual-knowledge criterion. Specialized control procedures are needed to eliminate these concerns.

The first concern arises whenever one obtains results from an observer that meet the equivalent-perceptual-knowledge criterion of achieving better than chance levels. It is important to keep in mind that observers participating in a psychophysical study are not necessarily motivated to help the researcher answer the perceptual question of interest. Thus, it is important for the researcher to attempt to rule out any extraneous cues that might have been exploited by the observer to achieve the good performance.

Two kinds of procedures are typically employed to minimize this concern. First, all obvious extraneous cues that the experimenter can anticipate are eliminated. An example would be the introduction

of a white noise sound generator to prevent an observer from learning to exploit incidental sounds made by the experimental apparatus that could provide cues about the correct answer.

A second procedure is to include control conditions in which there is a priori reason to assume the observer should be *unable* to distinguish between the alternative stimuli. For example, in a forced choice task the light might be made so dim that it is known that it cannot be seen (perhaps even turned off). Obviously, if the observer continues to perform at better than chance levels under these conditions, the rationale for attributing good performance during the actual experiment to the perceptual question of interest is also suspect.

A second type of concern arises whenever the observer fails to meet the equivalent-perceptual-knowledge criterion and produces only chance performance. It is then necessary to differentiate chance performance that is due to the stimuli's being perceptually indistinguishable from chance performance due to the fact that the observer does not understand or is unable to perform the task. One control condition that should always be included is a condition in which there is a priori reason to believe the observer can distinguish between the alternatives. If the observer achieves only chance performance on this control task, then the conclusion that poor performance during the original experiment was due solely to lack of perceptual knowledge is suspect.

Another methodological concern is specific to yes-no tasks. Observers sometimes respond "yes" on trials in which the labeled stimulus was not present. These are called **false alarm** responses. A variety of methods have been developed to try to deal with false alarm responses. We will discuss this issue in detail below in reference to the theory of signal detection.

What Kinds of Observers Can Be Tested with These Procedures?

In all of the examples so far, verbal reports from the observer have been used as the response measure. This restricts use of the methods to human observers with language. However, these procedures can all be easily adapted to a nonverbal response measure, such as having the observer press a button, as illustrated in Figure 2.7.

In this hypothetical example, a light bulb is connected to a computer that records both state of the light bulb (on or off) and state of the mouse button (pressed or released). The observer can be given

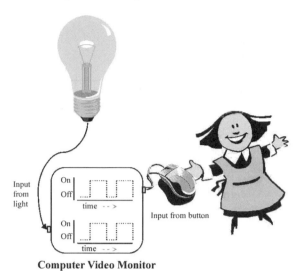

Computer Video Monitor

FIGURE 2.7. An observer can report a Class A observation with a nonverbal response, such as pressing a mouse button attached to a computer. In the example shown, inputs specifying the state of a light bulb are attached to a computer. The observer demonstrates that the "on" and "off" states of the light bulb are distinguishable by pressing the mouse button whenever the light is perceived to be on and releasing it when the light is perceived to be off. Evaluation of the responses of the observer over time compared to the record of the state of the light demonstrates that the observer can distinguish between these two stimuli.

instructions such as, "Press the mouse button whenever the light is on." These procedures can be easily adapted for use in **animal psychophysics** studies (Figure 2.8).

In this example, a monkey has been taught to press a button when it sees a light. Training is accomplished by giving the monkey a **reward,** such as a piece of banana, when it responds in the proper manner in the presence of the stimulus.

Assuming the monkey can in fact see the light, then following a sufficient number of training trials, it will "learn" to press the button only when the light is on. Exactly what it means to state that the monkey learns to perform on this task is an interesting question in itself, but beyond the scope of our interests here. We will just accept that monkeys, as well as other animals, can be taught to perform tasks such as this following training, just as we accept that humans will perform following verbal instructions.

In principle, these same procedures could be applied to address questions regarding perceptual capabilities of nonbiological agents, such as a robot equipped with sensors (Figure 2.9).

It may or may not be meaningful to use these procedures to establish that a robot equipped with a photoelectric sensor can "perceive" whether a light bulb in the environment is on or off. That question raises issues regarding the distinction between methodologically correct versus interesting results.

Methodologically Correct Results Versus Interesting Results

It is important to keep in mind that in this chapter we are concerned only with issues about whether specific psychophysical methods are being carried

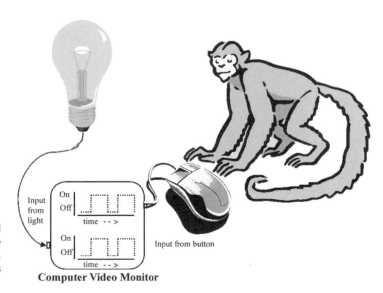

FIGURE 2.8. There is nothing special about the procedures illustrated in Figure 2.7 that restricts their use to humans. Perception can be studied in animals with similar methods.

Computer Video Monitor

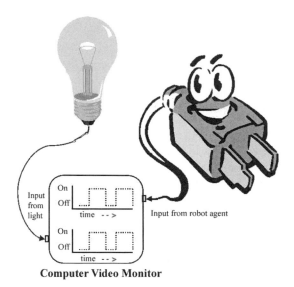

Input from light

On
Off
time -->

Input from robot agent

On
Off
time -->

Computer Video Monitor

FIGURE 2.9. There is nothing special about the procedures illustrated in Figure 2.7 that precludes adopting them for evaluating perceptual abilities of machines.

out properly. The fact that a scientific study establishes What-perception by using one of these procedures is meaningful only in the context of the particular perceiving agent being evaluated, the particular explicit facts being perceived, and the instrument(s) being used to establish the equivalent perceptual-knowledge criterion. Although there are a great number of potential questions one might address with these methods about a given agent's abilities to perceive what is out there, not all questions that can be phrased in this manner are interesting. Most will in fact be quite trivial and boring! The challenge for perceptual scientists is to figure out how to use psychophysical methods for testing interesting and important questions regarding What-perception.

3 Measuring How-Perception

In addition to looking around and picking up explicit facts about what is present in the surrounding environment, biological observers use perception to interact with their environments. Ordinary daily activities such as walking or driving an automobile depend on perceptual knowledge. However, the perceptual knowledge used to guide actions differs in some important ways from the database of explicit facts described in the previous section. Most of us would probably agree that we do not access the same perceptual knowledge base

---

Box 2.1
Malingering

Human patients sometimes claim to have visual disorders that qualify them to obtain benefits such as worker's compensation insurance. Eye doctors are sometimes asked to evaluate these individuals in order to make a determination as to whether the patient is visually impaired or is simply malingering in an attempt to gain unwarranted benefits. Methodologies based on Class A observations are not very convenient for making these kinds of assessment, because a malingering observer may lie and deliberately state that stimuli appear indistinguishable when they are really distinguishable. A prominent eye doctor once described a methodology he used when making these kinds of evaluations to detect malingerers. The eye doctor conducted the evaluation in a standard vision testing room equipped with a patient chair and appropriate charts and optical devices. The doctor would stand near the chair and invite the patient to enter the room. The patient would enter the room and walk to the chair, ostensibly to carry out a test of visual function. However, the real test took place as the patient entered the room. The eye doctor had placed various obstacles between the door to the room and the chair. If the patient avoided the obstacles while crossing the room, these actions were taken as de facto evidence that the patient had functional vision. If the patient then claimed, during formal testing in the chair, an inability to see these same objects, this discrepancy was taken as evidence of malingering. It may be arguable whether (implicit) How-perception necessarily implies (explicit) What-perception. In either case, this example demonstrates a clear application of a methodology based on actions to make inferences about whether an observer can perceive.

---

when looking at the Mona Lisa as when steering our car down the freeway in rush hour traffic. The perceptual activities involved with guiding actions are sometimes characterized by the term **perception-and-action.** In this text, we will use the term

How-perception to characterize perceptual acts in the service of **guiding actions.**

Although it is not exclusively so, knowledge conveyed by How-perception tends to be **implicit** rather than explicit. In other words, observers are often able to provide only vague descriptions of the information being used to guide actions when asked to describe what they are doing in words or symbols. (Recall our example in Chapter 1 of an outfielder catching a fly ball.) The psychophysical methods described for measuring What-perception rely on explicit perceptual knowledge and are not very convenient for measuring How-perception.

Necessary-for-Action Criterion

In cases where specific perceptual information can be shown necessary to perform a particular action, a demonstration of that action proves that the observer has access to the perceptual information. This provides the rationale for deriving operational definitions for How-perception based on a **necessary-for-action criterion.**

The general approach of using perceptually guided actions as a criterion to establish that an observer can perceive is widely used in settings other than scientific studies of perception. An example is presented in Box 2.1.

The psychophysical methodologies for establishing How-perception have not been formalized and systematized to the same extent as the psychophysical methods used for What-perception. Perceptual scientists who use a necessary-for-action criterion in their research have tended to develop individualized methods, highly specialized for

the actions they are studying. These methods are best characterized by describing a few prototypical examples rather than by trying to systematize any general principles on which they are based.

Stereotypical Movements Demonstrate How-Perception in Infants

When a human observer perceives that an object is on a collision course to strike his head, this percept elicits stereotypical movements of the head and limbs and closing of the eyes. This perceptually guided action serves the function of avoiding injuries to the eyes and head. When adults are asked to describe their perceptual experiences associated with this event, they report perceiving a **looming object.**

This stereotypical action has been used by scientists to study specific hypotheses about the nature of the information leading to perception of looming. One hypothesis tested is that the perceptual information guiding this action is based on the characteristic increases in size of the proximal stimulus as an object approaches the head. This has been tested in studies in which various kinds of human and animal subjects, including human infants, were held in front of a video monitor, as illustrated in Figure 2.10.

The size of a displayed object was increased in a manner that mimics the apparent changes in size that would take place if the observer were looking at an actual object on a collision course with his head. Infants looking at this display exhibited the same stereotypical reaction as they would to the impending collision of an actual object. This result

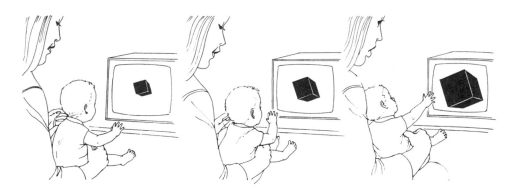

FIGURE 2.10. An illustration of the behavioral response that is elicited in a human infant when a video monitor is used to simulate size changes that appear to take place when an object is on collision course with his head. Reproduced from R. Sekuler and R. Blake, *Perception*, Third Edition, © 1994, by permission of McGraw-Hill, New York.

is taken as evidence that environmental information in the form of characteristic changes in the size of the retinal image formed by an object lead to a perception of looming in human infants, just as it does in adults.

Videotaped Observations Demonstrate How-Perception in Birds

The methodological approaches used to study properties of percepts based on actions can be much more powerful than is indicated by simply demonstrating the presence or absence of a particular type of perceptual information. Consider the actions performed by large sea birds diving into the water to catch fish (Figure 2.11).

As the gannet dives towards the water, it has two incompatible goals that must be balanced. The gannet needs to keep its wings open as long as possible in order to guide its trajectory and optimize its chance of catching its prey. If this goal is not met, the gannet will starve. On the other hand, the gannet needs to make sure that its wings are closed when it hits the water, or it will risk permanent injury. The percepts of a gannet must provide it with the specific information it needs to accomplish a wing-folding action at the appropriate time to accomplish both goals.

David Lee and **P.E. Reddish** published a study in 1981 in which they analyzed How-perception in diving gannets. They first formulated a hypothesis about how gannets guide their actions. Their hypothesis was based on an optical parameter,

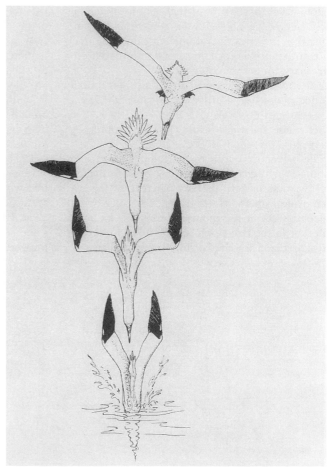

FIGURE 2.11. Photographs (left panel) and drawings (right panel) of gannets during a dive. Adapted from B. Nelson, *The Gannet*, © 1978, by permission of T. & A.D. Poyser Limited, Berkhamsted, UK.

FIGURE 2.12. Measurements of the time-to-contact remaining when wing-folding initiation was observed (data points) compared to predictions from the tau hypothesis (smooth curve). Adapted from D.N. Lee and P.E. Reddish, Plummeting gannets: A paradigm of ecological optics. *Nature* 293:293–294, © 1981, by permission of Macmillan Magazines Ltd.

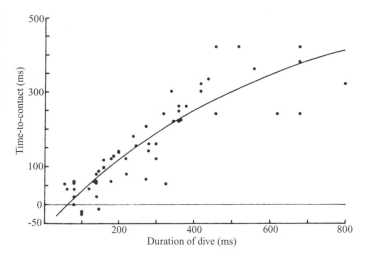

called tau, that can be derived, in principle, from changes in the retinal image that occur during a dive. Lee and Reddish proposed that gannets fold their wings when the value of tau reaches a set value. This hypothesis was used to derive specific predictions about the time-to-contact at which gannets should be observed to fold their wings during dives initiated from various heights. Lee and Reddish made measurements of wing folding from videotapes of dives. The predictions derived from the tau hypothesis provided a reasonably good description of the actual behavior of the gannets, as illustrated in Figure 2.12.

The data points are measurements of the time-to-contact remaining when wing folding actually occurred during videotaped dives. The smooth curve is the prediction about when wing folding should occur based on the tau hypothesis. The good correspondence between the measured and predicted values is considered evidence in support of the hypothesis that gannet wing folding is guided by perceptual information related to the value of tau.

How-Perception Can Be Measured in Machines

Any machine that utilizes sensors to guide its movements in an environment has to be employing at least a rudimentary level of How-perception. In fact, the complexity of How-perception in robotic insects is rapidly approaching that present in animals with simple nervous systems. An example of a robotic insect is illustrated in Figure 2.13.

The same strategies and methodologies used for characterizing and analyzing How-perception in biological organisms can be used to evaluate How-perception in robotic machines.

4 Measuring Subjective Perceptual Experiences

In addition to explicit and implicit perceptual knowledge, percepts convey **subjective perceptual experiences** to observers. In the absence of deception, neurological dysfunction, or other interference, it is generally accepted as an operational definition that human observers have whatever perceptual experiences they report to us in words.

Class B Observations

The term **Class B observations,** as proposed by Brindley, is used to refer to phenomenological reports in which observers describe subjective properties of their own percepts. Defined operationally, Class B observations are something of a "leftover" category, defined in terms of what they are not. They describe properties of percepts that are not compared to an external referent (which would qualify as explicit perceptual knowledge about "what") and are not tied to specific actions (which would allow an inference regarding implicit perceptual knowledge about "how"). Conceptually, Class B observations can be thought of as reports focused on properties of percepts that reflect an

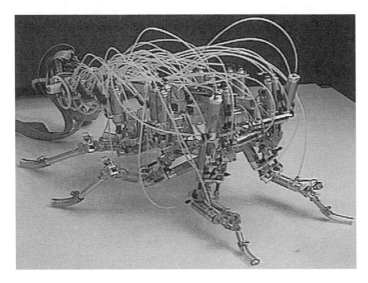

FIGURE 2.13. Photograph of a robot based on a cockroach developed in the Bio-Robotics Laboratory at Case Western Reserve University through a collaboration between Roger Quinn in Department of Mechanical Engineering and Roy Ritzmann in Department of Biology. The robot was built by Richard Bachmann and the controller by Gabriel Nelson. Photograph provided courtesy of Dr. Roger Quinn.

observer's private point of view rather than an objective external environment.

The subjective nature of Class B observations can be illustrated with an example. To make this example concrete, I will play the role of the imagined observer and you, the reader, the role of the experimenter. I want you to imagine that you are standing in front of me holding up a piece of chalk in your left hand and a pencil in your right hand. You ask me to report to you, with a Class B observation, what I see. Here is my reply:

"I see on the left a large elephant that appears day-glow orange with purple stripes and on the right a multicolored rainbow dominated by green."

Consider whether you are in a position to assert that my description of "what I see" is wrong. In order to continue with this hypothetical example, we will assume you do in fact try to assert that the Class B observation I just reported is incorrect.

We need to analyze this situation carefully. I described to you what I perceived when I looked at the two objects. If you assert that my response is "wrong," you are stating that I did not really experience what I described. On what authority do you assert that? I ask! I might thank you for your misguided input, but I would beg to differ with you if you think you know more about my perceptual experience than I do.

This hypothetical example illustrates that each individual observer is the ultimate authority, the supreme court of appeals, if you will, when it comes

to deciding what constitute his or her own personal experiences.

There is no way for a scientist to establish authority for asserting that a Class B observation reported by an observer is either correct or wrong (Figure 2.14). For this reason, scientific data from Class B observations must be treated as subjective rather than objective results. In order to use Class B observations to make scientific inferences about properties of percepts, one has to be willing to simply trust that the response is truly related to the observer's experience.

Psychophysical Procedures for Studying Perceptual Experiences

Methods for studying subjective perceptual experiences are relatively straightforward when studying adult humans, who possess language. Typically, the experimenter simply asks the observer to provide a report in the form of a Class B observation about some aspect of what is being perceived. For example,

"What color do you see?"

However, this example illustrates that subtleties can come into play in differentiating Class A and Class B observations. If the observer interprets these instructions along the lines,

"Tell me the color name that would be applied to this object based on information provided by a spectrophotometer,"

FIGURE 2.14. When evaluating the performance of an observer that takes the form of a Class B observation, a label of correct or wrong depends on a comparison of the report of the observer to the experience of the observer. No one other than the observer is in a position to make that comparison.

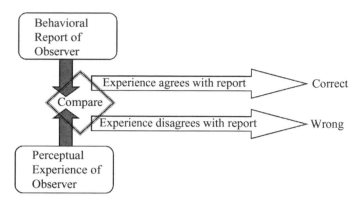

then the response elicited will be conceptually of Class A rather than Class B. Only if the observer interprets the instructions as meaning

"Tell me the color you experience as you look at this object,"

will the elicited response be Class B.

If the instructions given to observers in poorly designed scientific studies are ambiguous, it may be uncertain whether the responses elicited should be treated as Class A or Class B observations, and consequently whether the scientific results of the study are objective or subjective.

Although descriptions of subjective perceptual experiences are most frequently elicited in the form of spoken or written reports, this is not essential. Other behavioral actions can serve the same purpose. For example, the following instructions could be given to a human observer:

"I am going to present you with a series of colors, and I want you to press the left button whenever the color appears 'blue' to you."

In this example, the observer is using a nonverbal behavioral response to report a Class B observation. The situation becomes much more challenging when evaluating humans with insufficient language skills to understand verbal instructions or when studying animals or robots. We will defer discussion of these issues until Chapter 12.

Formal Methods of Estimating Perceptual Magnitude

Class B observations do not have to be strictly qualitative. For example, the psychologist **S.S. Stevens** developed a psychophysical method called **magnitude estimation** that can be used to quantify

certain aspects of perceptual experiences even though the method is based on collecting Class B observations. Stevens' method can be illustrated by considering an example of how one might ask an observer to quantify the intensity of her perceptual experiences while looking at lights that vary in physical intensity. Using magnitude estimation, one would simply ask the observer to look at a light and, without being informed of its physical intensity, report a number that represents its brightness. The observer is free to use any number she chooses. This procedure is then repeated while the observer looks at lights with a number of different physical intensities. When magnitude estimations collected in this manner are plotted on a graph, it is discovered that there are lawful quantitative relationships between physical intensity and perceived magnitude. We will illustrate use of these methods in Chapter 3 when we deal with issues of perceptual scaling.

5 Measuring Perceptual Thresholds

Scientific questions regarding perception are frequently concerned with establishing the limits of perceptual knowledge. Examples would be: What is the smallest object that can be seen? What is the smallest difference in intensity between two lights that can be detected? These kinds of questions involve **thresholds.** Observers are asked to distinguish between two (classes of) objects that are identical except that they vary along a single dimension. The question of interest is the smallest difference along this dimension that can be distinguished, referred to as a **just noticeable difference (JND).** The threshold is the location along this dimension where perception changes from *not* *different* to *different*. There are two basic classes of thresholds, absolute and difference. Difference thresholds are the more general case and will be discussed first.

Difference Thresholds

How much does the intensity of a light bulb have to be dimmed before an observer will notice a change in brightness? This is an example of a question that can be answered by measuring a difference threshold. An observer can be asked to look at one light bulb, called the **standard,** and at others, called **comparisons,** in which the intensity is attenuated by various amounts. On each trial the observer is asked whether the comparison stimulus is distinguishable from the standard. Hypothetical results that might be obtained from such a study are illustrated in Figure 2.15.

The results take the form of a **psychometric function** in which observer performance is plotted as a function of light intensity of the comparison stimulus. Arbitrary units of light intensity are used for this discussion, ranging from zero when the comparison stimulus light bulb is turned off to 1,000 when it is turned fully on to match the standard. (Actual units for measuring light intensity are covered in Chapter 3). Percentages of correct trials on a two-alternative forced-choice task are specified on the vertical axis at the left side of Figure 2.15. For case of comparison, corresponding values one would expect to obtain if the data were collected with a yes-no psychophysical procedure in which the observer is asked whether the two lights are distinguishable are indicated on the right side of this figure.

Data points near the left side of Figure 2.15 demonstrate that chance performance (50%) was obtained when the standard and comparison intensities were identical (1,000 units of intensity). The rightmost data points reveal that performance was near perfect when the comparison stimulus was dimmed to intensities less than 100 units.

These two ends of the psychometric function establish the two necessary control conditions that were described in the section on measuring What-perception when we discussed methodological control procedures. The right end of the function, where performance is near perfect, establishes that the observer knows how to perform the task on conditions that are easy to distinguish. Performance near the left end of the function demonstrates that the observer is not able to use inappropriate cues to achieve better than chance performance when there is little to no physical difference between the comparison and standard stimuli.

Having established these two essential control conditions, performance over the entire span of conditions can be used to interpolate a smooth function over the span between chance and perfect performance. A crude interpolation can be obtained by simply connecting the individual data points, as in Figure 2.15, or a statistical procedure can be used for a more sophisticated interpolation. Detailed treatments of appropriate procedures can be found by consulting the reading list at the end of this chapter. The value of the comparison stimulus where the interpolated smooth curve falls halfway between chance and perfect performance, i.e., 75% correct in the case of two-alternative forced-choice results or 50% yes in a yes-no procedure, is typically defined as the threshold. For the hypothetical data shown in Figure 2.15, threshold occurs at a comparison stimulus intensity of 500 units. The magnitude of the JND in this example is also 500 (1,000 − 500 = 500).

Absolute Thresholds (RL)

An **absolute threshold** is just a special case of a difference threshold measured under conditions in which the magnitude of the standard is zero. In other words, the absolute threshold is simply the **difference threshold from a standard of nothing.**

This can be illustrated with the earlier example of a light bulb. In that example, the difference threshold was measured from a standard of having the

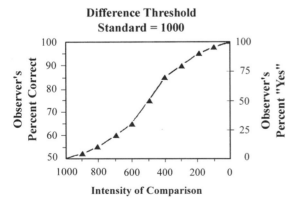

FIGURE 2.15. Illustrates the hypothetical results for an observer performing a task in which a standard stimulus with an intensity of 1,000 must be discriminated from comparison stimuli of various intensities.

Absolute Threshold
Standard = 0

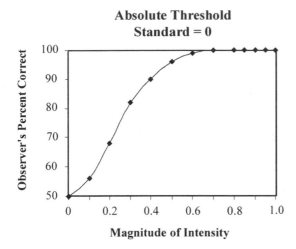

FIGURE 2.16. Illustrates the hypothetical results for an observer performing a task in which comparison stimuli of various intensities must be discriminated from a standard having zero intensity.

light bulb fully on. What if we were interested in the difference threshold at the other end of the range, with the light bulb off? This absolute threshold could be evaluated by measuring another difference threshold, this time using a standard of zero intensity. Hypothetical absolute threshold results are shown in Figure 2.16.

Note that the values on the horizontal axis have been reversed compared to those in Figure 2.15. The direction of the axes on this kind of graphs is arbitrary, but a common convention is to arrange the axes such that best performance is plotted on the right side of the graph. The value of the comparison stimulus on the far left of this graph corresponds to the standard (zero intensity) condition. Also, the horizontal axis has been stretched and truncated to show the intensity range only from 0 to 1.0. There is no need to show the rest of the range because the hypothetical observer performs perfectly for all intensities above 1.0.

The observer's absolute threshold can be defined in the conventional manner as the value of a comparison stimulus that would be distinguished from (a standard of) nothing on 75% of the trials. Interpolation of the results in the figure reveals that the absolute threshold occurs when the comparison intensity has a value of 0.25. The magnitude of the JND is always the same as the value of the threshold in the case of absolute thresholds, since the value of the standard is zero (0.25 − 0 = 0.25).

In What Order Should Comparison Stimuli Be Presented?

When measuring thresholds, the experimenter must decide in what order to present the various comparison stimuli. For example, the conditions that are easy to distinguish (shown on the right sides of Figures 2.15 and 2.16) could be presented first, followed by the harder conditions (shown on the left side). On the other hand, the stimuli could be presented in the opposite order, in random order, or in some other order. These kinds of "nuts-and-bolts" questions about psychophysical methods can make a practical difference in terms of the exact threshold value that is obtained, so they cannot be ignored. A brief synopsis of some of the classic psychophysical methods for measuring thresholds is presented here. More detailed treatments can be found by consulting the reference list at the end of this chapter.

Method of Limits

One traditional procedure for measuring thresholds is the **psychophysical method of limits.** This method has two forms, **ascending** and **descending.** In an ascending series, one presents the conditions that are hard to distinguish first. Over several trials, the conditions are made gradually easier until the observer can distinguish the stimuli on every trial. The responses from the observer on a yes-no task are expected to take the form

"... no, no, yes, yes ..."

The stimulus where the transition from "no" to "yes" takes place demarcates the location of the threshold. In a descending series, the easy stimuli are presented first, and these become gradually harder to distinguish over the trials. The sequence of responses is expected to be of the form:

"... yes, yes, no, no ..."

A potential problem with ascending limits is **anticipation.** After observers become familiar with the method, they often realize that stimuli always become gradually more distinguishable over the trials. Consequently, observers will sometimes respond "yes" in anticipation even though they would have responded "no" if this particular stimulus had not been preceded by a series of hard conditions. An analogous problem with a descending series is **habituation,** where an observer will fall into the habit of saying "Yes" and continue doing so even when the stimuli are no longer distinguishable. Various procedures, such as intermixing ascending and descending series, have been proposed to deal with these problems.

One variation on the method of limits is to have the stimuli changed continuously instead of in discrete steps. Some of you have probably encountered this variation on the method of limits when you had an examination for hearing loss, as described in Box 2.2.

Box 2.2
Evaluations of Hearing Loss
Use Tracking Methods

A variation on the method of limits, called **tracking,** is often used to evaluate hearing loss. A stimulus, usually in the form of a tone, is allowed to change continuously in intensity. The patient is given a button and instructed to press the button while a sound is heard. During periods when the button is pressed, a descending series is in operation, so that the tone becomes gradually less intense. While the button is released, an ascending series goes into effect, such that the tone becomes gradually more intense.

The methods of limits have traditionally been used in conjunction with forms of yes-no psychophysical procedures to measure absolute thresholds. However, these methods are amenable to being used to measure difference thresholds as well as absolute ones and can be adapted to accommodate forced-choice as well as yes-no procedures.

Method of Constant Stimuli

When using the **psychophysical method of constant stimuli,** the experimenter determines ahead of time all of the conditions (ranging from easy to hard) that are going to be presented to the observer during an experiment. These conditions are then presented to the observer in random order. Usually this is done without replacement. In other words, if five conditions were going to be presented during an experimental session, each of the five conditions would initially be presented once in a randomized order. Then the same five conditions would be repeated for as many replications as desired for the experiment, each time in a new random order. This method eliminates the problems of anticipation and habituation.

Staircase Methods

One potential drawback of the methods of limits and constant stimuli is that a lot of time can be wasted collecting data that are far removed from the observer's threshold. The data are not of very much value in terms of evaluating the location of the threshold. Consider an observer tested with the method of limits whose responses on the first series of trials were

"no, no, no, no, no, no, no, no, no, no, no, no, no, yes, yes."

It seems obvious that the first ten or so conditions presented in the first series were well below threshold. When it comes time to repeat the series, these could probably be skipped in the interests of efficiency.

Staircase procedures, also called **adaptive procedures,** try to improve efficiency by applying rules based on an observer's recent performance to try to place future trials near her threshold. Following each series of trials (or in the limiting case as frequently as after each trial) the experimenter performs an analysis on the recent performance of the observer. If this analysis suggests the trials on which the observer is currently performing are below her threshold, conditions will be made easier on the next trial, and vice versa. Sophisticated staircase methods also adjust the step size. For example, if statistical analysis indicated an observer was currently operating 25 steps above his threshold, then a staircase procedure might immediately jump down 25 steps instead of wasting the next 25 trials moving the observer down one step at a time.

In research settings, efficiency is often not the highest priority. However, in a clinical setting, such as the office of an eye doctor, where only limited time is available to examine a patient, efficiency can be the deciding factor as to whether a reliable estimate of the patient's threshold is obtained. The most efficient staircase methods rely on use of complicated algorithms. Consequently, their use was limited historically, and staircase procedures were usually based only on simple rules. An example is the **two down, one up rule,** which states that whenever an observer gets two responses in a row correct the following trial should be made harder. After a single incorrect response, the following trial is made easier. With computers readily available in most research and clinical settings, use of more complicated staircase algorithms is now becoming more widespread.

Method of Adjustment

In the **psychophysical method of adjustment,** the experimenter simply asks the observer to find her own threshold. As a simple example, consider measurement of the absolute threshold for seeing a light. A quick and easy way to accomplish this would be to give the observer a dimmer switch and ask her to "adjust this dimmer switch until you can just barely see the light." The method of adjustment is efficient and simple to use in verbal subjects. However, the data obtained are subject to the limitations of Class B observations.

6 Application of the Theory of Signal Detection to Perception

Diagnostic Systems

A **diagnostic system** examines information it receives about its environment and then uses this information to decide (make a **diagnosis**) whether a statement about some specific state of affairs in the world is true or false. A few examples will make the concept of a diagnostic system more concrete.

- A weather forecaster examines the inputs from sensors on a weather balloon and must predict "rain tomorrow" or "no rain tomorrow."
- A radiologist examines an MRI scan from a patient and has to diagnose "cancer" or "no cancer."
- A psychiatrist considers results on a battery of psychological tests from a defendant in a criminal trial and must decide between "sane" and "insane."
- A perceiving organism examines a berry it has just picked from a plant and must decide whether it is "edible food that will provide nutrition" or "poison that will make me sick."
- An observer participating in an absolute threshold perceptual experiment using a yes-no psychophysical procedure must choose a response from the alternatives "yes, the stimulus was present" and "no, the stimulus was not present."

A theoretical framework, called **Signal Detection Theory (SDT),** was developed specifically to evaluate performance of diagnostic systems. SDT can be applied to any diagnostic system but in general is most valuable for analyzing diagnostic systems that must operate under conditions of uncertainty. These are conditions in which a diagnostic system is given insufficient information to be certain about the true state of affairs that exists, but must make a diagnosis nevertheless. SDT provides a rich, highly mathematical theoretical framework for addressing a number of methodological issues that arise when measuring the performance of any diagnostic system, including a perceiving observer.

False-Alarm Responses

The development of SDT was spurred in part by a historical problem encountered in perception studies that tried to measure thresholds. A control procedure commonly used with yes-no procedures has been to include **blank trials,** during which no stimulus is presented. If the observer responds Yes on a blank trial, this is referred to as a **false alarm** or a **false-positive response**. Blank trials are also sometimes referred to as *catch trials* because they are used to try to catch untrustworthy performance that might otherwise contaminate the scientific validity of the results.

However, when catch trials are included in studies in which observers are working near threshold, false alarm responses are encountered with moderate frequency, even from observers who would otherwise be considered reliable. So a methodological question arose about what an experimenter should do when confronted with false alarm responses.

Traditionally, two answers were given to this question, both of which are now known to be inappropriate, as will become apparent after we have learned more about SDT. The first was to eliminate false alarms by **punishing the observer** whenever they occurred. For example, the experimenter might charge into the room where the observer is participating in the study and shout in a loud voice something along the lines:

"I caught you! You just said 'yes' on that last trial and I had not even presented a stimulus! You are messing up the experiment! Start paying better attention!"

The second approach that has been traditionally proposed to deal with this problem is to leave the observer alone during the experiment but later subject the data to an analysis that adjusts the measured threshold by an amount that compensates for the contaminating effects of false alarms. Performing a data analysis of this type is commonly referred to as applying a **correction for guessing.**

The rationale for the correction-for-guessing data analysis procedures can be understood by working through how they would be applied in a specific example. Suppose an observer responded, "Yes,

I see it," on 75 out of 100 trials during which a stimulus was presented. Nominally, this indicates that the observer can detect this stimulus 75% of the time. However, suppose this same observer responded, "Yes, I see it," 50% of the time when blank trials were presented. The experimenter would not be able to trust that the observer really saw the stimulus 75% of the time, because the blank trials demonstrate that half of the time this observer "guesses yes" even when no stimulus is present. The correction-for-guessing procedure would readjust the percentage detected from (the measured) 75% to (a corrected) 50% based on the following rationale: The observer was presented with 100 stimulus trials and is assumed to have detected the stimulus on half (50 trials). On the remainder of the 50 trials that were not detected, the observer is assumed to have guessed yes on 50%, for an additional 25 "yes" trials. These two types of trials, actual detected trials plus guessing trials, added together account for the fact that the total number of measured "yes" responses was 75.

Correction-for-guessing procedures have enjoyed widespread usage in the perception literature. However, predictions made by these these correction procedures have been found to be wholly inadequate to account for how overall permormance changes when the numbers of false alarms are manipulated experimentally. The reason is that observers are not always simply guessing when they make false-alarm responses. A different underlying model that can accurately account for false alarms comes from SDT, as will be described shortly.

Can Observers Avoid Making Mistakes?

A second historical factor that led to development of SDT was a practical need to try to eliminate mistakes of observers performing vigilance tasks. Research on this topic was fueled by the escalation of the Cold War between the United States and the Soviet Union in the 1950s, shortly after the end of World War II. The United States Air Force employed radar technicians whose job was to observe the display of a radar screen and watch for approaching enemy airplanes or missiles. The radar screens exhibited "noisy" flicker whenever they were in operation. Suppose a radar technician were to see a small blip appear somewhere on the screen. The question confronting the radar operator would be whether that blip represented an incoming missile or just noisy flicker. Many of the earliest studies that eventually formed the basis for SDT

were geared towards the practical problem of trying to prevent radar operators from making mistakes while performing this vigilance task.

Outcomes of a Diagnosis

One of the contributions of SDT is a careful analysis of the potential outcomes that can occur whenever a diagnostic system is forced to make a binary decision about the presence or absence of some condition in the world. Technical terminology was developed to label these conditions. The absence of the relevant condition in the world is labeled as **noise,** and the presence of the condition is **signal plus noise,** commonly shortened to simply **Signal.** In the example of a radar operator, the signal condition would correspond to

"A blip on the screen is due to fact that a missile is present"

and the noise condition to

"No missile is present, and the blip is just due to a noisy screen."

The responses that diagnostic systems are allowed to make are also binary. The observer can diagnose **"noise"** or **"signal."** Thus, possible outcomes of conditions present in the world and diagnoses fall into four categories, labeled **hits** (or **true positives**), **false alarms** (or **false positives**), **misses** (or **false negatives**), and **correct rejections** (or **true negatives**), as summarized in Figure 2.17.

Possible Outcomes in Detection Tasks

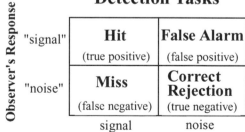

		signal	noise
Observer's Response	"signal"	**Hit** (true positive)	**False Alarm** (false positive)
	"noise"	**Miss** (false negative)	**Correct Rejection** (true negative)

Condition Present in World

FIGURE 2.17. The SDT provides a framework for evaluating the outcomes of any system that makes a diagnosis. There are two possible conditions present in the world, signal and noise, and two possible diagnoses, "signal" and "noise." This results in four possible joint outcomes, referred to as hits, false alarms, misses, and correct rejections.

FIGURE 2.18. The SDT conceptualizes the inputs to a diagnostic system as being noisy. The magnitude of the noise changes randomly over time.

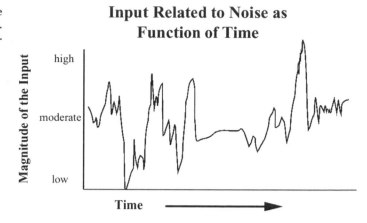

Input Related to Noise as Function of Time

Note that there are two ways that diagnostic systems can make mistakes: misses and false alarms. Thus, an analysis of a diagnostic system that examines only one of these types of mistakes will sometimes give a misleading impression about accuracy. This seems obvious when the outcomes of diagnoses are made explicit, as in Figure 2.17. However, this fact was not always fully appreciated prior to the development of SDT.

Signals Embedded in Noise

Signal detection theory assumes a particular model in terms of characteristics of the input information that is available to a diagnostic system. It is assumed that the information available to make a diagnosis is always embedded in noise. The sources of this noise can include the external environment that gives rise to the information and the internal processing mechanisms of the diagnostic system. Here it is sufficient for us to accept that, regardless of its source, information available to a diagnostic system is always embedded in noise. Furthermore, the level of noise is assumed to vary from moment to moment, as illustrated in Figure 2.18.

If one were to measure the amounts of noise present moment to moment over a long period, it would be possible to specify the amount likely to be present at any instant in terms of a probability distribution. This is illustrated in Figure 2.19.

This probability plot illustrates that the amount of noise most likely to be present at any moment is moderate but that there is a finite probability that the amount of noise is anywhere in the range from high to low.

Whenever a **signal** is present in the environment, some amount of information about that signal is potentially available as input to the diagnostic system. However, input related to the signal will always be intermixed with the ongoing input due to noise. Thus, the total amount of input to a diagnostic system will be the sum of the amount related to signal and that due to noise. This is illustrated for a case in which a signal of constant magnitude is present in Figure 2.20.

The dotted line along the bottom is reproduced from Figure 2.18, which showed the input due to noise. The vertical arrows represent the magnitude of input pertaining to the signal. This signal input is added to the noise. The smooth line at the top shows the sum of the inputs from signal plus noise. The total amount of input in the presence of a signal is, of necessity, larger than that for noise alone. Consequently, graphs of the signal (plus noise) probability distribution will always show it to the right of the noise distributions, as illustrated in Figure 2.21.

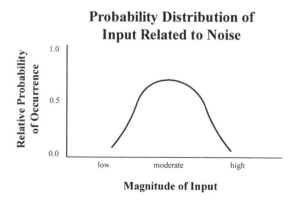

Probability Distribution of Input Related to Noise

FIGURE 2.19. When averaged over a long period of time, the magnitude of the input to a diagnostic system related to noise can be characterized statistically in terms of a probability distribution.

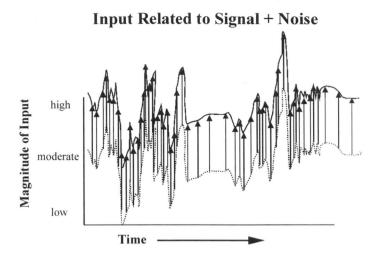

Input Related to Signal + Noise

FIGURE 2.20. Any signal that is input to a diagnostic system rides on top of the noise being input at the same moment. The bottom dotted line is reproduced from Figure 2.18 and shows the magnitude of input due to noise alone. The arrow symbols illustrate the magnitude of a signal being input. The upper solid line illustrates the overall magnitude of the input from signal plus noise that is being input over time.

Analysis of Probability Distributions Underlying Diagnostic Decisions

The last few figures have included scales on the horizontal or vertical axes labeled **magnitude of input.** That is a somewhat abstract theoretical construct used in SDT to explain what happens internally in a diagnostic system. However, as a heuristic device to help understand this concept, a hypothetical example will be used in which magnitude of input corresponds to rate of electrical firing by a neuron in the brain.

As will be described further in Chapter 5, there are **contrast-detecting neurons** in visual processing areas of the brain that respond to differences between dark and bright regions of an image. Suppose you were visiting Antarctica and went for a walk on an overcast day out into a region where there was nothing in your field of view except snow and a white haze in the background. The firing rate of a contrast-detecting neuron would be relatively low, because there is not very much contrast present in this snowfield. However, the firing rate would not be zero, because neurons respond with a spontaneous firing rate even in the absence of contrast. Furthermore, the firing rate would not be constant but would vary somewhat over time. The instantaneous firing rate might be 30 electrical impulses per second at one moment, 35 impulses per second a few moments later, and still later 20 impulses per second. If we were to plot the probability distribution of the firing rate, it would be normally distributed around some low level that corresponds to its average amount of spontaneous activity. This is illustrated in the left-hand probability distribution in Figure 2.22.

The figure also illustrates what would happen if a penguin suddenly appeared on the snowfield. The high contrast between the white and dark portions of the penguin's body would cause the neuron to fire at a higher rate. Again, the rate of firing would not necessarily be constant but would vary over time, and can be characterized in terms of the probability distribution shown on the right. Examination of the two probability distributions in the figure reveals that the neuron seldom exceeds moderate levels in the condition of snow alone but responds mostly with high levels in the presence of a penguin.

Next imagine what would happen if a snowstorm started to roll in. The perceived contrast of the black and white portions of the penguin

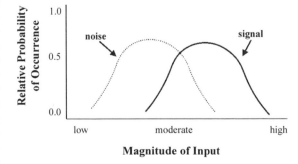

Probability Distributions Related to Signal and to Noise

FIGURE 2.21. The probability distribution due to noise is the same as that shown in Figure 2.19. The probability distribution due to an added signal is shifted to the right.

FIGURE 2.22. In this hypothetical example, the input to a diagnostic system is characterized as the firing rate of a neuron. The probability distribution of the firing rate in the presence of noise (only snowfield is present) is shown on the left. The probability distribution of the firing rate when a signal (penguin is standing in the snow) is also present is shown on the right.

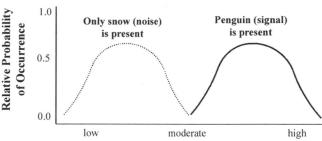

Probability Distributions Related to Presence of Penguin on Overcast Day

Only snow (noise) is present

Penguin (signal) is present

Relative Probability of Occurrence

1.0

0.5

0.0

low   moderate   high

Rate of Firing in Neuron

decreases as the density of the snowflakes swirling in the air increases (Figure 2.23).

Hypothetical changes in the probability distributions of activity in the contrast-detecting neuron under various densities of swirling snowflakes are illustrated in Figure 2.24.

The leftmost and rightmost conditions are reproduced from Figure 2.22. The probability distribution for the "penguin in a moderate snowstorm" condition is not shifted away from the "snowfield" (noise) condition as much as the probability distribution for the "penguin on an overcast day" condition, and that of the "penguin in a heavy snowstorm" condition even less.

Consider the **level of accuracy** that that can be attained by a diagnostic system that makes its decision about the presence or absence of a penguin on the basis of the firing rate of this neuron. The potential level of accuracy depends on the magnitude of the rightward shift of the penguin (signal) probability distribution from the snowfield

(noise) distribution. Thus, the diagnostic system will be able to make fewer mistakes when diagnosing the presence or absence of a penguin on an overcast day, when the response to the signal is strong relative to the response to noise, than under conditions such as a snowstorm, when the response to the signal is weaker. Note that mistakes will be impossible to avoid completely whenever the distributions due to noise and signal partially overlap.

Another theoretical construct provided by SDT, the **criterion,** allows a deeper understanding of the relationship between the two types of mistakes a diagnostic system can make. The criterion is a position along the horizontal axis of the probability distribution, established by the diagnostic system in response to factors that will be described shortly. Once the criterion is established, the diagnostic system uses the following rule to determine its diagnosis: Whenever magnitude of input exceeds the criterion, respond "signal;" otherwise, respond

FIGURE 2.23. The scene of a penguin standing in the snow changes as the weather changes from an overcast day to a snowstorm.

overcast day

moderate snowstorm

heavy snowstorm

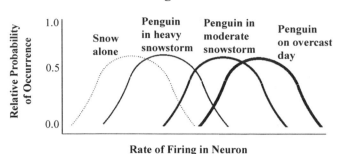

Probability Distributions of Penguin During Snowstorms

FIGURE 2.24. The probability distributions for the snowfield alone shown on the left and for a penguin standing on the snowfield on an overcast day are the same as those shown in Figure 2.22. The probability distributions shown in the middle illustrate the changes expected to be caused by the presence of a snowstorm.

"noise." Figure 2.25 illustrates this concept by showing a criterion that might be used by our hypothetical diagnostic system that is trying to detect the presence of a penguin based strictly on the firing rate of the neuron.

As illustrated in Figures 2.26 and 2.27, SDT makes very specific predictions about the proportions of various outcomes (hit, miss, correct rejection, and false alarm) that will occur for this diagnostic system, given these probability distributions and this criterion.

The **correct rejections** will correspond to the proportion of the **noise distribution to the left** and the **false alarms** to the proportion of the **noise distribution to the right** of the criterion. In the penguin detector example, false alarms are diagnoses of "penguin" when no penguin is present. A correct rejection is a diagnosis of "snowfield alone" when no penguin is present.

Similarly, **misses** correspond to the proportion of the **signal distribution to the left** and **hits** to the proportion of the **signal distribution to the right** of the criterion. In our example of a penguin detector, a hit corresponds to diagnosis of "penguin" when one is in fact present and a miss to a diagnosis of "snowfield alone" when it turns out a penguin was present.

In the examples just shown, the criterion of diagnostic system was placed in about the middle of the magnitude axis. However, in general, a diagnostic system can have the criterion placed anywhere along the horizontal axis. A diagnostic system in which the criterion is placed **towards the left end** of the input axis is said to be **liberal**. A liberal criterion will produce many false alarms but also lots of hits. An example is illustrated in Figure 2.28.

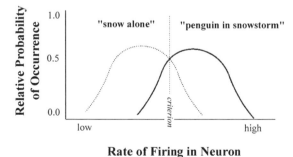

FIGURE 2.25. The SDT conceptualizes the decision made by a diagnostic system in terms of a criterion. This figure illustrates a specific criterion that a hypothetical diagnostic system could use to try to discriminate between the presence and the absence of a penguin in a snowstorm. Whenever the firing rate of a contrast detecting neuron is higher than the criterion, the diagnosis is "penguin is present." Otherwise, the diagnosis is "penguin is absent."

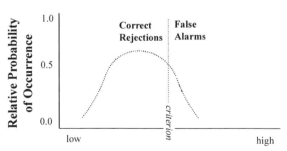

Rate of Firing in Neuron

FIGURE 2.26. The SDT predicts the number of false alarms that are expected to occur based on the area of the noise distribution that falls to the right of the criterion. The area of the noise distribution that falls to the left of the criterion is used to predict the number of correct rejections.

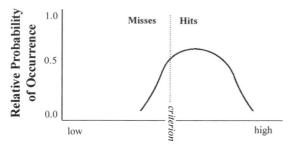

Rate of Firing in Neuron

FIGURE 2.27. The SDT predicts the number of hits that are expected to occur based on the area of the signal distribution that falls to the right of the criterion. The area of the signal distribution that falls to the left of the criterion is used to predict the number of misses.

We now have a deeper understanding of why an observer participating in a threshold experiment would be expected sometimes to give false alarm responses on catch trials. The trials on which a reliable observer says "yes" are not due to "guessing" but are simply trials where the observer is using a liberal criterion and the noise level on that trial happens to exceed the criterion.

A diagnostic system in which the criterion is placed **towards the right side of the input axis** is said to be **conservative.** A conservative criterion seldom results in false alarms. However, there is no free lunch. As illustrated in Figure 2.29, a conservative criterion also yields fewer hits.

When an observer shifts to a new criterion, SDT allows specific predictions to be made about how the proportions of hits, misses, false alarms, and correct rejections should change. These predictions are in some cases orders of magnitude different from those derived from the correction-for-guessing data analysis procedures described above. Numerous studies over the past 30 years have overwhelmingly confirmed that SDT predictions are more accurate.

Analyses of Receiver Operating Characteristics Curves

All of the figures shown above to illustrate underlying probability distributions are theoretical constructs about what is going on "inside the head" of the observer making the diagnosis. These distributions are not actually measured. What is measured in an experiment are the proportions of hits, misses, false alarms, and correct rejections. These empirical results are typically plotted in the form of a **receiver operating characteristics (ROC) curve.** In this section, we will first describe the form and interpretation of ROC curves. Then we will relate the shapes of these curves to the underlying theoretical probability distributions associated with signal and noise that we have been discussing.

Although there are four possible outcomes of a diagnosis (hit, miss, false alarm, correct rejection), there are only two degrees of freedom. For example, if the value for the hits is 80%, then it is known

Liberal Criterion

Rate of Firing in Neuron

FIGURE 2.28. The criterion is not fixed but can be set at different positions by the diagnostic system. When the criterion is set towards the left of where the probability distributions for noise and signal intersect, as illustrated in this hypothetical example, the diagnostic system is said to have a liberal criterion.

Conservative Criterion

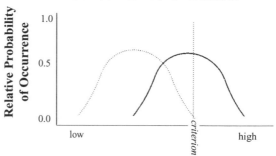

Rate of Firing in Neuron

FIGURE 2.29. When the criterion is set towards the right of where the probability distributions for noise and signal intersect, as illustrated in this hypothetical example, the diagnostic system is said to have a conservative criterion.

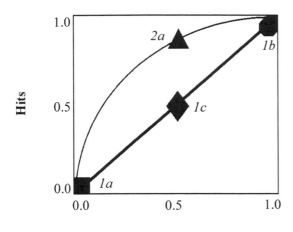

Hits

False Alarms

FIGURE 2.30. The performance of a diagnostic system is typically characterized in the form of a receiver operating characteristics (ROC) curve, as illustrated here. The proportion of hits is plotted as a function of the proportion of false alarms. The points labeled 1a, 1b, 1c, and 2a correspond to hypothetical performance under four different conditions described in the text.

(automatically) that the misses have to be 20% (100 − 80 = 20). Similarly, if the false alarm rate is known, the rate of correct rejections is also specified. Thus, an ROC curve needs to show values only for hits and false alarms, because the other two values can be easily derived from the same curve. The form of an ROC curve is shown in Figure 2.30.

The proportion of hits is plotted on the vertical axis and that of false alarms on the horizontal. Consider the data point labeled 1a. This data point represents the location on the ROC curve where performance would be expected to fall for a diagnostic system that was **simply guessing** but was **extremely conservative** and always responded "noise" instead of "signal." An example would be an observer participating in a yes-no absolute threshold experiment for detecting a light who simply closed her eyes and responded "no" on every trial. The ROC curve demonstrates that this observer never produces a false alarm but never has any hits either.

The data point labeled 1b represents the performance that would be expected for this same **guessing observer** if she changed her criterion to **extremely liberal** and responded "yes" on every trial. If one measured only the percentage of "yes" responses for this observer, it would appear that she can detect the light very well. She would report that she sees the light even when it is extremely dim.

Similarly, if one calculated the percentage of misses, the result would demonstrate that this observer did not miss a single trial. Based on this analysis alone, one might conclude that the observer was a very good diagnostic system for detecting the presence of the light. However, note that this observer would also exhibit 100% false alarms!

Next, consider what the performance would look like for this same **guessing observer** if she adopted an **unbiased criterion,** such as flipping a coin following each trial and responding "yes" for heads, "no" for tails. The data point labeled 1c shows that performance for the guessing observer under these conditions will be 50% hits but also 50% false alarms.

These three examples should make it obvious that neither percentage of "yes" responses nor percentage of hits reveals anything about diagnostic ability. A diagnostic system that is simply guessing can exhibit performance on either of these measures that varies anywhere from 0% to 100%.

Note that all three of the data points in the last figure that involve only guessing fall on the **rightward-leaning diagonal** of the ROC graph. This is a fundamental characteristic exhibited by data plotted in the form of an ROC curve. Any points that fall along the diagonal reflect **total lack of diagnostic ability,** regardless of the percentage of hits or false alarms.

Consider next the data point labeled 2a. This represents performance of a diagnostic system where false alarms are 50% and hits are 85%. This observer has **some amount of diagnostic capability,** because the hits are being generated at a higher rate than would be expected based on the false alarms.

The data point labeled 2a represents the responses of an observer using a criterion that is generating 50% false alarms. What would happen if this observer changed her criterion to give either more or fewer false alarms? The general answer is that her performance would shift along a smooth ROC curve like the one drawn through the data point 2a in the figure. A more **liberal criterion** will result in a **shift to the right along this curve.** In the extreme case, if the observer becomes so liberal that she says yes on every trial, her performance will overlap with 1b. Similarly, a more **conservative criterion** will produce a **shift to the left** and in the extreme will result in a performance that overlaps with 1a. Note that the smooth ROC curve characterizes the performance of this diagnostic system regardless of what criterion is used.

The ROC curve for this observer falls **above the diagonal.** This is another fundamental characteristic of data plotted in the form of an ROC

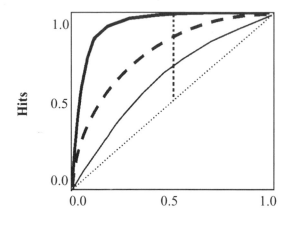

Hits (y-axis), **False Alarms** (x-axis)

FIGURE 2.31. The dotted, thin solid, dashed, and thick solid lines represent hypothetical ROC curves under conditions in which the signal can be discriminated from the noise with various levels of accuracy. The points where the dotted vertical line intersects with the ROC curves illustrate where observer's performance would fall when a criterion level is chosen that results in 50% false alarms.

curve. Whenever the ROC curve of a diagnostic system falls above the diagonal, that system is exhibiting a **diagnostic accuracy better than chance.** Parenthetically, it can be noted that a curve that falls below the diagonal is from a perverse diagnostic system, one that can distinguish signals from noise but always reports the opposite of its true diagnosis.

Indices of Diagnostic Accuracy

The higher the ROC curve falls above the diagonal, the higher the level of diagnostic accuracy. Three examples are shown in Figure 2.31.

The dotted curve along the diagonal reflects chance performance. The thin solid ROC curve indicates performance somewhat better than chance, the dashed curve even better, and the thick solid curve best of all. The vertical dashed line illustrates the hit rates that would be attained for each of these curves when using a criterion with a false-alarm rate of 50%. What do these shifts in the heights of the curve correspond to in terms of the probability distributions we discussed earlier? This is illustrated in Figure 2.32.

The vertical dashed line in this figure shows the criterion level corresponding to 50% false alarms, the same level as illustrated by the vertical dashed line intersecting the ROC curves in Figure 2.31. The three probability distributions are theoretical constructs derived from the empirical data plotted in the three ROC curves shown in Figure 2.31. These three distributions are shifted various distances to the right of the noise distribution. The arrows under the horizontal axis designate the amount of shift for each. In SDT, the technical term d' is used to refer to the magnitude of this rightward shift. The larger the value of d', the more accurate the diagnoses. The units of d' are standard deviations of the underlying probability distributions. On the ROC curve, d' can be related to the height of the curve above the diagonal. The d' measure serves as one **index of accuracy** of a diagnostic system.

Another index of accuracy, A, is calculated by measuring the percentage of the area of the box enclosing the ROC plot that falls under the ROC curve. The value of the A index is more convenient than the value of d' in terms of providing a straightforward specification of diagnostic accuracy. The value of the A index corresponds to the percentage of correct trials that would be obtained from the observer if tested on a two-alternative forced-choice psychophysical task. An intuitive understanding of this relationship between the A index and expected

FIGURE 2.32. Performances represented by the hypothetical ROC curves illustrated in Figure 2.31 are shown here in terms of their underlying probability distributions. The probability distributions are shifted to the right of the mean of the probability distribution associated with noise alone (dotted vertical line) by the amounts shown by the arrows at the bottom of the figure. The lengths of these arrows, scaled in units of standard deviations of the probability distributions, are referred to as d'.

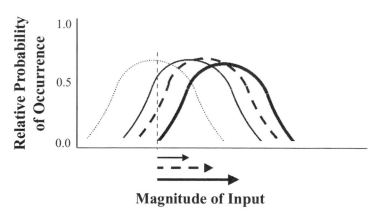

Relative Probability of Occurrence (y-axis)

Magnitude of Input

percentage correct performance on a two-alternation forced-choice task can be gleaned by noting that the area that falls under the rightward diagonal on the ROC plot (corresponding to total lack of diagnostic accuracy) will be 50% (corresponding to chance performance on a two-alternation forced-choice task).

Various other indices of accuracy have been developed for analyzing ROC curve results. The choice of the proper index to use when analyzing a particular perceptual diagnostic system falls beyond the scope of this book's purposes, and the interested reader should consult one of the texts from the reading list at the end of this chapter.

Recently, studies have been carried out with monkeys in which performance on a psychophysical detection task was related to the electrical firing rate of a single neuron in the monkey's brain. In these studies, the underlying probability distributions were not simply theoretical constructs but were calculated based on physiological recordings from the neuron, allowing a direct comparison between the measured ROC curve and the measured underlying probability distribution. Those studies are discussed in Chapter 10.

Indices of Bias

Other indices can be used to quantify the **bias,** liberal or conservative, of the criterion being used by a diagnostic system. The most commonly used index of bias is **beta,** the ratio of the height of the signal distribution to the height of the noise distribution. Beta can range from zero to infinity as the criterion changes from extremely liberal to extremely conservative. If the criterion is placed at the position where the noise and signal distributions intersect, then beta equals 1.0. An observer using this criterion will balance the numbers of the two types of mistakes (misses and false alarms). A value of beta less than 1.0 will result in more misses than false alarms, and vice versa.

Bias is a property of the diagnostic system that comes into play whenever a diagnostic system adopts a criterion. Bias might be, in part, an inherent property of the diagnostic system. For example, some observers are just intrinsically more conservative or liberal in their willingness to produce false alarms versus misses. However, bias can also be altered by external factors. Two in particular, **payoffs** and **a priori probabilities,** have been demonstrated to be able to influence the bias of a human observer. Payoffs have to do with rewards or punishments that are associated with the various outcomes of a diagnosis. The influence of payoffs on the criterion has been demonstrated in numerous laboratory studies in which payoffs associated with each type of outcome are explicitly controlled using monetary rewards. An example is presented in Box 2.3.

Box 2.3
Explicit Payoffs Using Money

Suppose you were in a perception experiment in which you were told that you would receive $1 for every hit but would be penalized by $20 for every miss. Following an observation, you think you might have seen a very dim light flash, but you are not sure. Are you going to respond "yes" or "no"? Now, suppose the rules are altered and you are to receive $1 for every hit and be penalized $20 for every false alarm. On the next trial, you see what you think may have been a very dim light flash, but you are not sure. Are you going to guess yes or no?

When actual experiments have been done along these lines, the results are clear. Observers in the first set of conditions approach an asymptotic pattern in which they always respond "yes" and those in the second one in which they always respond "no." These results make it obvious that one should not base conclusions about how well an observer can see based simply on how often the observer responds "yes" or "no."

When analyzed with TSD methods, these two sets of conditions produce identical values for indices of diagnostic accuracy such as d' or A. Only beta changes.

Most diagnostic systems do not operate under conditions in which exact monetary payoffs are given for each diagnosis. However, even in situations with no explicit payoffs, there are always implicit, perhaps social, payoffs.

Consider the example in the section on false-alarm responses, in which an experimenter yells at an observer for making a mistake (false alarm) during a catch trial. Observers will attempt to alter their subsequent performance to avoid this social punishment. Thus, a punished observer will simply move to a more conservative criterion that reduces the chance of false alarms. However, the total number of mistakes that will be made on future trials is set by d', over which the observer has no control. Thus, the only real effect of this treatment will be to reduce the number of mistakes in the form of false alarms while simultaneously increasing the number of mistakes in the form of misses. Some additional real-world examples in which social rewards influence criteria of diagnostic systems are presented in Box 2.4.

strong. Similarly, in a situation where observers have an expectation that signals are very likely to be present, a liberal criterion is adopted, and even slight evidence will be reported as signal. Box 2.5 illustrates the influence of a priori probabilities using the earlier example of a radar operator working for the Air Force.

Box 2.4
Implicit Payoffs

Suppose you are working for the Air Force as a radar operator. You have been informed that the consequences of missing an enemy missile could be catastrophic to the United States. You are instructed to adopt a liberal criterion that will eliminate any chance of a miss. SDT informs us that this strategy will inevitably also produce many false alarms. You are working in the middle of the night. You see a small blip on the screen that you think is probably just noise but that could be a missile. You call and wake up your commanding officer, but when she arrives, it is established that it was a false alarm. An hour later you see a similar blip on the screen and wake up the commanding officer a second time. Again, it turns out to be a false alarm. What are the likely consequences if you continue to repeatedly wake up your commanding officer every time you see a small blip on the screen regardless of how insignificant it appears? With those consequences in the back of your mind, what are you likely to report if a third blip shows up on the screen, similar to the first two, later that same evening?

Box 2.5
A Priori Probabilities

Imagine that you are a radar operator working for the Air Force. You have been in the Air Force for 364 days. You have been spending 8 hours per day, 40 hours per week, looking at radar screens for missiles. During this period, you have seen tens of thousands of small blips on the radar screen. None of these has ever been an enemy missile; they have all just been noisy blips. Today is your 365th day on the job, and you see a small blip on the screen. Are you more or less likely to report "missile" for this small blip today than the first time you saw a similar blip during your first day on the job? This example illustrates that bias about the likelihood that this blip represents a missile can be altered by expectations formed because of prior experiences. These influences pose serious problems for any diagnostic system performing vigilance tasks under conditions where the probability of a signal's being present is small. One solution is to perform frequent unannounced drills in which a signal's presence is simulated and payoffs in the form of rewards and punishments are given based on performance during the drills.

A second set of factors that influence the choice of a criterion, in addition to payoffs, are referred to as the **a priori probabilities.** These are the expectations an observer has about the likelihood that a signal might actually be present in the world. In situations where there is a strong expectation that only noise is likely to be present, observers will adopt a conservative criterion and will not report "signal" unless the input suggesting signal is very

All diagnostic systems operate with bias set to some value in the range between extremely conservative to extremely liberal. Whenever there are situations in which the relative proportions of the two kinds of mistakes diagnostic systems make (misses versus false alarms) need to be maintained at a particular value, this can be achieved by adjusting the payoffs and a priori probabilities. A neutral bias is usually not the desired state for most diagnostic systems. For example, in our criminal justice system it is usually considered more appropriate to operate with a bias "better 10 guilty should go free than one innocent should be found guilty." Each diagnostic system needs to be evaluated to determine the most appropriate bias at which it should operate and then have the payoffs and a priori probabilities adjusted accordingly.

7 Relating Behavioral Phenomena to Biological Processing

Any psychophysical experiment conducted to specify whether an observer has perceptual knowledge about some aspect of the immediately surrounding environment is bound to yield one of two outcomes: Either the observer can perceive "what is out there" or not. These two outcomes lead to different sets of questions regarding perceptual processing.

Whenever we obtain the first outcome, questions arise about **sensory coding.** How did neural processing allow this information to make it from the environment to the observer's perceptual knowledge base? Consider the simple example of an observer who can see a light. Information specifying the status of the light bulb must have been present at each step along the causal chain from the environment to the behavioral response. However, that information is coded in different forms at various stages along the causal chain. At the front of the eye, the information is coded in the structure of the light rays being transmitted into the eye. In the retina, the information is coded in terms of

voltages across membranes of neurons, and so on. In principle, it should be possible to specify the sensory coding at every stage of neural processing. We will deal extensively with issues of sensory coding in Chapter 6.

Consider next the situation in which an observer is not able to pick up some specific information about the immediately surrounding environment. When that happens, an important issue that must be addressed about processing has to do with **critical locus.** Where along the chain of sensory processing did the information get lost? Consider the example in which an observer reports that she cannot see a light. Would information that the light is present have been detected if instead of measuring the behavioral response we had measured the sensory code at the level of the visual cortex? Or of neurons in the retina? Or of the light rays forming the retinal image?

These two types of questions provide a common rationale for many studies of perception. First, a behavioral study is performed to find out whether an observer has access to a particular kind of perceptual information. If the answer is yes, then neuroscience methods are used to pursue questions about sensory coding. If the answer is no, then neuroscience methods are used to pursue questions about the location of the critical locus. This rationale is usually implicit and applied in an informal or ad hoc manner. However, the psychologist **Davida Teller** has tried to formalize this rationale in the context of **psychophysical linking hypotheses,** as described in Box 2.6.

Box 2.6
Psychophysical Linking Hypotheses

Since psychological and biological events come from logically separate universes of discourse, some have questioned whether statements linking these two realms of events, called **psychophysical linking hypotheses,** can ever attain the status of being analytically true. Psychophysical linking hypotheses come in four basic types, as systematized by Davida Teller.

1. **Biology implies psychology.** These are cases in which a biological phenomenon is observed and the observation is used to draw a conclusion about some psychological fact. An example would be: A "grandmother cell" identified in an observer's brain is observed

to be firing action potentials; one may conclude the observer is experiencing a percept of his grandmother.

2. **NOT psychology implies NOT biology.** If hypothesis 1 is accepted as true, then in general hypothesis 2 must be accepted also, because it is the logical contrapositive. An example would be: The observer is not experiencing a percept of his grandmother; one may conclude the grandmother cell in his brain is not firing action potentials.

3. **Psychology implies biology.** This is the logical converse of hypothesis 1. There is no logical necessity for it to be true even if hypothesis 1 and 2 are true. An example

would be: The observer is experiencing a percept of his grandmother; one may conclude the grandmother cell identified in his brain is firing action potentials.

4. **NOT biology implies NOT psychology.** This is the logical contrapositive converse and in general must be accepted if hypothesis 3 is true. An example would be: The "grandmother cell" identified in the observer's brain is not firing action potentials; one may conclude that the observer is not experiencing a percept of his grandmother.

Most linking hypotheses that fall into categories 3 and 4, *cannot* be accepted as generally true, because the causal relationships between brain cells and behavior are many-to-one rather than one-to-one. Thus, a particular psychological state is consistent with a number of possible biological states. Some linking hypotheses that fall into categories 1 and 2 have been accepted by many visual scientists as being universally true. The most general of these is one proposed by Brindley:

Whenever two stimuli cause physically indistinguishable signals to be sent from the sense organs to the brain, the [percepts] produced by these stimuli, as reported by the subject in words, symbols, or actions, must also be indistinguishable.

Brindley's psychophysical linking hypothesis is stated in the logical form of category 1 but is meant to be used in the form of its category 2 contrapositive. The observed phenomenon would be that percepts formed by two stimuli *are* distinguishable. Based on that psychological evidence alone, one is forced to the conclusion that the state of the brain when the observer is viewing one of these two stimuli is different from the state when viewing the other.

Summary

The scientific study of perception relies on measures of behavior as a probe to learn about properties of percepts. A number of specialized behavioral methodologies, called psychophysical methods, have been developed to collect behavioral phenomena that relate to each of the three basic aspects of percepts described in Chapter 1. Psychophysical methods of studying What-perception measure the perceptual information an observer has gained about his immediate surroundings and relate this to equivalent empirical information about the surroundings that can be obtained with instruments. Methods of studying How-perception measure perceptually guided actions of an observer and use these actions to infer properties of percepts that are deemed necessary to guide these actions. Methods of studying subjective perceptual experiences depend on phenomenological descriptions in the form of Class B observations.

Specialized psychophysical procedures have been developed for measuring perceptual thresholds. Absolute thresholds are the smallest amounts of some physical quantity that can be detected by an observer. Difference thresholds are the smallest differences between two physical quantities that can be discriminated.

Many aspects of perception can be conceptualized in the context of observers performing in the role of diagnostic systems. Observers must make a decision, i.e., a diagnosis, about "what is out there" based on uncertain information provided by the sense organs. The performance of observers under these conditions can be modeled by SDT, which assumes these diagnoses depend on internal processing of signals embedded in noise. SDT allows an observer's performance to be characterized according to two independent parameters: an index of sensory discriminability and an index of observer bias.

Psychological phenomena derived from psychophysical methods can also be used to draw conclusions about what kinds of biological processes must be occurring inside the head of an observer. The rationale for drawing these conclusions has been formalized within the context of psychophysical linking hypotheses.

Selected Reading List

Baird, J.C., and Noma, E.J. 1978. *Fundamentals of Scaling and Psychophysics*. New York: Wiley.

Brindley, G.S. 1970. Introduction to sensory experiments. In *Physiology of the Retina and Visual Pathway*, 2nd edition, pp. 132–138. Baltimore: Williams & Wilkins.

Fechner, G.T. (1801–1887), 1966. *Elements of Psychophysics*, trans. H.E. Adler, eds. D.H. Howes and E.G. Boring, with an introduction by E.G. Boring. New York: Holt, Rinehart and Winston.

Finney, D.J. 1971. *Probit Analysis*. New York: Cambridge University Press.

Gescheider, G.A. 1985. *Psychophysics: Method, Theory, and Application*, 2nd edition. Hillsdale, NJ: Erlbaum.

Green, D.M., and Swets, J.A. 1966. *Signal Detection Theory and Psychophysics*. New York: Wiley.

Lee, D.N., and Reddish, P.E. 1981. Plummeting gannets: A paradigm of ecological optics. *Nature* 293:293–294.

Levine, G., and Parkinson, S. 1994. Detection, discrimination, and the theory of signal detection. In *Experimental Methods in Psychology*, pp. 205–233. Hillsdale, NJ: Erlbaum.

Stevens, S.S. 1951. *Handbook of Experimental Psychology*. New York: Wiley.

Swets, J.A. 1988. Measuring the accuracy of diagnostic systems. *Science* 240:1285–1293.

Swets, J.A., and Pickett, R.M. 1982. *Evaluation of Diagnostic Systems: Methods from Signal Detection Theory*. New York: Academic Press.

Teller, D.Y. Linking propositions. 1984. *Vision Res.* 24:1233–1246.

Woodworth, R.S. 1972. *Woodworth & Schlosberg's Experimental Psychology*, Volume 1, Sensation and Perception. New York: Holt, Rinehart and Winston.

3
The Perceptual Environment: What Is Out There to Be Perceived?

Questions

After reading Chapter 3, you should be able to answer the following questions.

1. Define what is meant by the term "perceptual environment," and describe some of the ways a perceptual environment might differ from the physical environment.

2. Describe the wave properties and particle properties of light and explain why these properties are paradoxical.

3. What is the significance for perception of the difference between radiometric and photometric units?

4. Describe the basic photometric units used to measure amounts of light.

5. Describe how Snell's law confers image-forming properties that allow light to convey information to observers about the spatial structure of their surroundings.

6. Use examples to illustrate how laws of geometrical optics can provide simple graphical and analytical methods of determining where images will be formed by lenses having spherical surfaces.

7. Explain how spatial and temporal properties of the environment can be described from a special perspective called the frequency domain.

8. If you were an engineer employed by a company that manufactures dimmer switches, why would it be important for you to understand issues of perceptual scaling?

9. What are some differences between natural and artificial environments, and what implications do these differences have on perception?

10. Are environments meaningful?

1 Defining the Perceptual Environment

For purposes of this text, the **perceptual environment** will be defined simply as the **surroundings of observers that can be sampled with their own sense organs.** This definition is general and probably adequate for all of the exteroceptive sensory systems, although this text will deal only with the perceptual environment for vision. Questions about whether specific aspects of the perceptual environment are contributing information to What-perception, How-perception, or subjective perceptual experiences of an observer need to be addressed by using the methods described in Chapter 2. Definitions of the perceptual environment simply delineate what is potentially available to be perceived.

The perceptual environment extends out in all directions from the observer. Under some conditions, as when an observer views a distant star, the perceptual environment extends literally to the far ends of the universe (Figure 3.1).

Perceptual environments can be described from various perspectives, including physical properties, formal information, and functional significance from the point of view of a particular observer.

2 Light Energy

Light, a form of **electromagnetic energy,** has been present since the beginning of time, when a Big Bang created the universe. It is hard to imagine what life would be like in a dark universe in which

FIGURE 3.1. The perceptual environment includes the surroundings of an observer that can be sampled by the sense organs. In the case of vision, the perceptual environment can reach to the ends of the universe when viewing distant stars.

light energy did not exist. A few species of animals that live in caves and deep in the ocean have adapted to surviving in the absence of light. However, these are only rare exceptions to the general, pervasive rule that animals have evolved to live in an environment that is bathed in light energy.

All forms of electromagnetic energy, including light, are said to have a **dual nature.** Some properties make sense only if light is considered to exist in the form of **waves,** others only if it is considered to consist of small individual **particles.** The dual properties of light have relevance to understanding many aspects of visual perception, such as our abilities to reliably detect small amounts of light and to see color.

Dual Nature of Light

One of the properties of light that have led physicists to conclude it has a wave-like nature is illustrated in Figure 3.2.

The top panel illustrates **diffraction**, a property of light that causes it to spread out after it passes through a narrow slit. When the light that has passed through the slit falls onto a projection screen, it forms a bright blob, as indicated by the intensity profile shown on the right. The bottom panel illustrates **interference**, a phenomenon that can occur when light waves from a source are allowed to interact after passing through a pair of slits. Based on the intensity distribution of light

Light   Light passes   Intensity of
Source   through   light falling
               narrow slits   onto projection
                                   screen

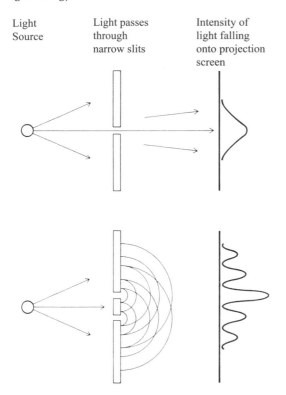

FIGURE 3.2. After light has passed through a narrow slit, it forms a diffraction pattern on a projection screen placed behind the slit, as illustrated in the top panel. An explanation of diffraction is that the light falling on the projection screen is not limited to the rays that pass straight through the slit but also includes light that has spread out after passing through the slit, as illustrated by the arrows in the top panel. The bottom panel illustrates interference, an uneven distribution of light falling on a projection screen after light has passed through two narrow slits. An explanation of interference is that light consists of waves that can interact with one another as illustrated by the overlapping semicircles in the bottom panel. Adapted from L. Kaufman, *Sight and Mind: An Introduction to Visual Perception,* © 1974, by permission of Oxford University Press, Inc.

from a single slit as shown in the top panel, one might expect to see two similar overlapping distributions in the bottom panel, one produced by each slit. Instead, a series of alternating dark and bright bars, called an **interference pattern,** is produced at the projection screen. Interference patterns can be understood by assuming that the light passing through each slit is a wave and that the waves from the two slits interfere with one another. For example, positions where peaks from each wave come together result in a bright region, and positions where two troughs interact result in a dark region.

Other properties of light energy have led physicists to consider it to be composed of discrete particles rather than of waves. For example, when measurements are made of light being emitted by a source or absorbed in a substance, the smallest amount of light that can exist is an individual packet called a **photon.** Furthermore, if more than one photon is present, then the amount has to be an integral number of photons. For example, the eye can absorb one photon or a packet of two photons or 10,000,000,003 photons but cannot absorb 0.5 or 1.3 photons. These discrete properties of light place some fundamental limits on the reliability of an observer's perceptual performance when small amounts of light are involved. These issues will be further covered in Chapter 6 in a discussion of perceptual processing conceptualized as a statistical process.

It is tempting to combine the concepts of waves and particles and think of photons as tiny bullets of light traveling through space and oscillating in a wavelike manner as they do so. However, physicists have demonstrated that the fundamental nature of light is neither that of a particle nor that of a wave but, paradoxically, both. A striking example comes from making measurements of the interference patterns formed by light passing through two slits, as illustrated in Figure 3.2, under the special conditions described in Box 3.1.

When considering light energy as waves, it is helpful to use an analogy to the waves that are formed on the surface of water when you throw a pebble into a pond (Figure 3.3).

A concentric ring of waves will travel out in all directions from where the pebble enters the water. Each wave in the concentric ring is surrounded on the outside by the wave that was produced prior to it and on the inside by the wave that was produced after it. Consider three basic measurements that could be made of the waves that are formed. First, the distance between the concentric rings of waves could be measured and referred to as their **wavelength.** Second, the speed, or **velocity,** at which each wave is traveling over the water could be measured. Third, if a small stick were floating in the water, it would bob up and then down – one **cycle** – as the first wave passed by and then repeat this cycle for every succeeding wave. The number of cycles completed in one second could be counted and referred to as the **frequency** of oscillation.

Light can be characterized in the same ways. When light energy is emitted from a source, it can be conceptualized as waves that travel out in all directions through three-dimensional space. The outermost wave will form a wave front that can be

Box 3.1
Demonstration of Light Properties That Appear Paradoxical

Interference can be demonstrated using the double-slit arrangement shown in the bottom of Figure 3.2. A small probe can be used to measure light intensity at each location on the projection screen. The device will identify regions where few photons are detected and regions where many photons are detected, corresponding to the dark and bright areas of the interference pattern that is visible to the eye. A surprising result is obtained when these measurements are made under light conditions so dim that only one photon is present to pass through the slits at any given moment in time. The regions of high and low intensity corresponding to areas of the interference patterns can still be demonstrated in terms of the probability distributions of absorption within each region over many trials. These results demonstrate that, in some real sense, each single photon must pass through both slits and interfere with itself.

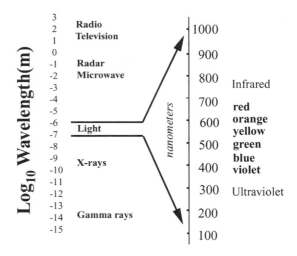

FIGURE 3.4. Electromagnetic radiation extends from very short wavelengths, as in gamma rays, to very long wavelengths, as in radio waves. "Light" is the name given to electromagnetic radiation that has wavelengths falling between about 400 and 700 nm. When human observers look at light they report seeing the colors red, orange, yellow, green, blue, and violet for various portions of the spectrum, as illustrated on the right side.

FIGURE 3.3. When a pebble is thrown into a pond of water, concentric rings of waves are formed on the surface of the water around the point of impact. The emission of light from a point source is analogous, except that waves are emitted in three-dimensional space instead of along a two-dimensional surface.

visualized as an expanding sphere. The second wave will form another expanding wave front just inside the first, and similarly for subsequent waves. We can specify the **wavelength** of light energy in terms of the distance from the crest of one wave to that of the next. Similarly, we can measure the **frequency** of light energy in terms of the number of wave oscillations per second. The **velocity** of light is the speed at which each wave travels through the medium.

The Visible Spectrum

Specific bands of wavelengths that make up the electromagnetic spectrum are given different names, as shown in Figure 3.4.

Towards the short end of the spectrum are gamma rays and X-rays, which have wavelengths less than one billionth of a meter. At the long end are radio and television waves, which have wavelengths of several kilometers. It is important to keep in mind that gamma rays and radio waves, and all of the other forms of electromagnetic energy in between, are not qualitatively different from one another in terms of their fundamental properties. They just differ quantitatively in terms of their wavelengths. This applies to light as well.

Light is the name given to a band of wavelengths in the electromagnetic spectrum extending from about 400 to 700 nm. The reason this particular band is singled out for a special name is that these wavelengths **can be seen by the human eye.** Note that light is a **psychophysical** entity in that it is defined with reference to both physical and psychological terms. It is a biological and a psychological fact that when electromagnetic radiation with wavelengths within this range stimulates the eye, a physiological response is produced in the brain and humans see light.

Normal human observers looking at wavelengths of light report a sensation of **color.** Wavelengths of light shorter than 490 nm are typically reported to look violet on blue, above 490 and up to 570 nm green, up to 600 nm yellow or orange, and above 600 nm red. However, these facts tell us relatively little about the operation of color perception, because in our ordinary environment we are usually confronted with mixtures of many wavelengths. Under these conditions, the relationships between wavelength composition of light and perceived color are more complicated. We will discuss color vision in more detail in Chapter 7. Wavelengths just shorter than the visible spectrum are called **ultraviolet** and those just longer **infrared.** Some animals have eyes that are adapted to see these portions of the spectrum, although they are invisible to primates. Too much ultraviolet light can cause cataracts to form in human eyes, and for this reason it is prudent to wear sunglasses that block out ultraviolet rays when spending time in very bright outdoor conditions, such as on the beach or skislopes.

Specification of How Much Light Is Present

When studying visual perception it is often important to know how much light energy passes into or through a given medium during a given time period – for example, how much light energy per second is emitted by a light bulb and what percentage of that light enters an observer's eye. The units, measures, and systems used to specify magnitude of light are somewhat esoteric, and the details need to be mastered only by lighting engineers and visual science specialists. Here, the goal is limited to providing a brief overview of some of the major concepts involved in understanding how light is measured. Standard reference texts can be consulted for more detailed information.

Transmittance of Light Through a Medium

For many questions about how much light passes through a medium, absolute amounts are not important, and it is necessary to specify only the relative amounts that are involved. **Transmittance** refers to the proportion of the light energy that makes it through a medium. It can range from 1.0 when all of the light is transmitted to 0.0 for an opaque medium. A common convention is to specify transmittance in units of **density,** defined as the base 10 logarithm of the reciprocal of transmittance. Density becomes greater as the percentage of transmitted light becomes smaller. For example, a density of zero corresponds to a transmittance of 1.0. A density of 3.0 corresponds to a transmittance of only 0.001.

If a medium transmits all wavelengths of light equally, it is said to have a **neutral density.** If a medium transmits some wavelengths more than others, it is said to exhibit **selective spectral transmission** and takes on a colored appearance. For example, air transmits more short than long wavelengths to the eye and this gives the sky its bluish appearance. The effects of selective spectral transmission can be characterized by a **spectral transmittance curve** that shows the proportion of light energy that makes it through the medium at each wavelength. Light that does not make it through a medium is either reflected or absorbed. For many media, the primary reason light fails to be transmitted is because of absorption. These media are often characterized by a **spectral absorbency curve,** which is simply the reciprocal of the spectral transmittance curve.

Radiometric and Photometric Units

Physicists most commonly use **radiometric units** to quantify absolute amounts of light that are present. Radiometric units are based on the amount of energy that is present in light. However, radiometric units are not very convenient for specifying the amount of light involved in human perception. Single photons of different wavelengths have different energies, but our eyes essentially ignore that fact and simply count the number of photons absorbed. In other words, every absorbed photon is treated the same as any other by the eye, regardless of its energy level. Thus, the most appropriate units for specifying light in relation to the act of perception are those that can be related to number of photons present rather than to overall amount of energy. Further complicating the issue is the fact that photons of different wavelengths have differential probabilities of being absorbed by the eye.

Consequently, the most appropriate units for perception need to be expressed in terms of the number of photons absorbed within the eye rather than of the total number that enter the eye. **Photometric units** meet this requirement. A fixed magnitude of light, as specified in photometric units, will result in a fixed number of photons absorbed in the eye. The basic photometric unit used for measurement of amount of light is one **lumen**. The technical specification of a lumen is provided in Box 3.2.

Box 3.2
Conversions Between Photometric and Radiometric Units

Radiometric units measure light energy in units of ergs per second. It is possible to convert back and forth between photometric units of lumens and radiometric units of energy, but only if the wavelengths that are involved are known. The official conversion is made at a wavelength of 555 nm, where 10^7 ergs per second of radiant energy are taken as equal to 680 lumens. At this wavelength, one lumen corresponds to 4.12×10^{15} photons per second. The numbers of photons that make up one lumen at other wavelengths vary, and the calculations involved go beyond the scope of this book. The interested reader should consult a reference text on lighting and photometry.

Intensity and Luminance from Light Sources

Two specialized photometric terms are used to specify the number of lumens emitted by a light source, depending upon whether its surface is small (point source) or occupies an extended area. For **point sources,** the amount of light emitted is defined in terms of **intensity,** as illustrated in Figure 3.5.

Intensity is simply the **flux forming a cone of unit solid angle (steradian) in a given direction from the source,** specified in **candelas,** where **one candela is equal to one lumen per steradian.** The cone spreads the lumens out over an increasingly wide area with increasing distance from the source. However, since intensity is specified in terms of

Intensity of a Point Source

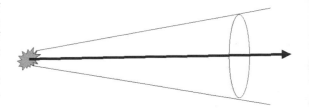

FIGURE 3.5. The amount of light emitted by a point source can be specified in terms of the number of lumens in a cone of a standard size (steradian) that extends in a given direction from the source. An intensity of one lumen per steradian is called a candela.

lumens within the cone, it is independent of the distance from which it is measured. This fact makes intensity a property of the source.

Intensity does not have to be the same in all directions. Consider an automobile headlight seen from a distance. The intensity of the light radiating straight ahead is very high, but the intensity radiating to the side is low.

If light is being emitted from an extended rather than a point source, then its magnitude is measured in units of **luminance** rather than intensity. An intuitive understanding of luminance can be attained by conceptualizing an extended source as being made up of a very large number of point sources, as illustrated in Figure 3.6.

Like intensity, **luminance** is specified in a given direction from a source, depicted by the thick arrow in Figure 3.6. It is possible to conceptualize this arrow, and the corresponding cone, as emanating from a single point source on the surface. If only

Luminance of an Extended Source

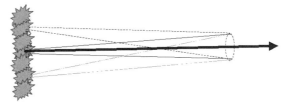

FIGURE 3.6. The luminance emitted by an extended source of light can be conceptualized in a manner similar to the intensity of a point source as described in Figure 3.5. The extended surface is simply treated as though composed of a large number of individual point sources. The amount of light emitted in a cone that extends in a given direction is adjusted for the number of contributing point sources by specifying luminance in units of candelas per square meter of the emitting surface.

Projected Area of Source

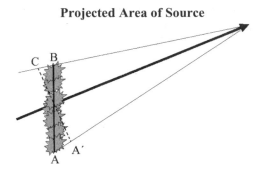

FIGURE 3.7. The projected area of the source shown in this figure in the direction of the thick arrow is A'C rather than AB. A measurement of the projected area adjusts for foreshortening when the luminance is measured in a direction other than normal to the emitting surface.

that one point source were present, the lumens within the cone would correspond to the intensity emitted by that point. However, an extended surface provides additional lumens emitted from the surrounding area that also enter the same cone. The surrounding area can be conceptualized as being made up of a large number of individual point sources. The dotted cones in the figure illustrate the contribution of two of these. Measures of luminance adjust for the contribution of the other point sources on the surface by dividing the intensity value that is measured by the area of the emitting surface. Thus, luminance can be thought of, roughly speaking, as the mean of the intensities of all of the individual point sources that contribute lumens to a cone in a given direction. The technical specification of luminance for an emitting surface is **candelas per square meter of the emitting surface.** Like intensity, luminance is a property of a source and does not vary with distance.

A technicality involved in the measurement of luminance is that the area of the emitting surface is its **projected area,** as illustrated in Figure 3.7.

True point sources do not exist, because all sources that emit light have some surface area. However, as a practical matter, whenever the longest dimension of the surface of an emitting source is less than 1/20 of the distance from which light is being measured, it is usually acceptable to treat it as a point source.

Illuminance Falling on Surfaces

A different set of photometric units is used when specifying the amount of light falling on a surface. In this case, units of **illuminance** are used,

defined as the number of **lumens falling on each square millimeter of the surface.** The amount of illuminance produced on a surface by a point source varies with the square of the distance from the source, a relationship referred to as the **inverse square law.** The basis for an intuitive understanding of this relationship is provided in Box 3.3.

Box 3.3
Inverse Square Law for Illuminance

Figure 1 shows lumens emitted from a point source being measured with a probe having a surface of fixed diameter at two distances, 1 m and 4 m. At the closer distance, the surface of the probe just fills the end of the light cone. For purposes of this example, this probe is assumed to measure 1 lumen per second. When the same probe is moved to a distance of 4 m, it only fills 1/16 of the cone. This is because the radius of the circle at the end of the cone is proportional to its distance from the surface, and the area of a circle increases with the square of its radius ($4^2 = 16$). Thus, the total flux illuminating the surface of the probe at this greater distance is only 1/16 lumen per second.

Lumens Falling on Surfaces from a Point Source

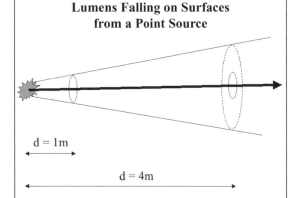

Figure 1. Lumens from a point source are being measured by a probe depicted by the small circles at two distances, 1 m and 4 m: the probe fills the entire cone of light when placed at the 1 m distance, but fills only $^1/_{16}$ of the cone when placed at the 4 m distance.

Luminance from a Surface Due to Reflection

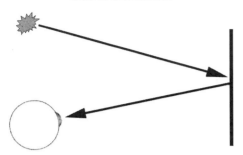

FIGURE 3.8. The light entering the eye of this hypothetical observer does not come directly from a light source but is reflected from a surface. The amount of light entering the eye depends on both the illuminance falling on the surface and the proportion of light reflected from the surface in the direction of the eye.

Luminance Reflected from Surfaces

In artificial environments, observers often view luminous sources such as video displays that really emit light. However, we seldom view luminous sources directly in the natural environment. Instead, the light that enters our eyes from surfaces is usually reflected from other sources, as illustrated in Figure 3.8.

The luminance emanating from these surfaces depends on two factors: the illuminance falling on the surface, as just defined, and **reflectance,** the proportion of the incident photons reflected by the surface in the given direction. A surface with a reflectance of 1.0 is one from which all incident photons are reflected, and one with a reflectance of 0.0 is one from which none are reflected. **Luminance from a reflecting surface** is simply its **illuminance multiplied by its reflectance.**

Luminances of Typical Sources

The absolute range of luminances confronted in daily activities is an astounding 15 log units. Some examples of luminances of typical sources and highly reflective surfaces are shown in the Figure 3.9.

Retinal Illuminance

In studies of perception, it is often desirable to be able to specify the illuminance falling on the retina of an observer. The only light rays that can reach the retina are those that pass through the pupil at the front of the eye. Thus, retinal illuminance depends upon both the luminance of the surface being viewed and the area of the pupil. Units of **trolands** are typically used to specify retinal illuminance, where one troland corresponds to the illuminance produced by **viewing a surface having a**

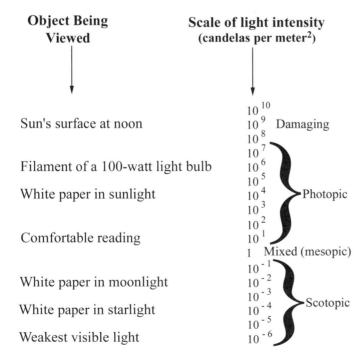

FIGURE 3.9. The human eye responds to light intensities that range over 15 log units, from amounts that are just visible (10^{-6} candelas per square meter) to amounts that are damaging to eye tissue (10^9 candelas per square meter). Adapted from R. Sekuler and R. Blake, *Perception*, Third Edition, © 1994, by permission of McGraw-Hill.

luminance of one candela per square meter through a pupil having an area of one square millimeter.

Photometric Units Applied to Nonstandard Conditions

There is one additional complication involved in use of photometric units that we have ignored in this discussion. Our eyes have two different types of photoreceptors. One type operates under daytime, **photopic** conditions and the other under dim, **scotopic** conditions. The relationships between wavelength and proportion of photons that are absorbed are somewhat different for the two types of receptors. The photometric quantities we have discussed here are defined in terms of photopic vision. A second system of **scotopic photometric units** is available to specify the effects of light on human observers more accurately under dim illumination. The technical details of how these two types of photometric units differ go beyond the scope of this book, but exact specifications can be found in the publication from the Commission Internationale de L'Eclairage in the list of suggested readings at the end of this chapter. However, the general rule is that a system of light measurement and specification for any class of observer, be it human, monkey, or some other type of observer, is meaningful only if it is based on the effects of light energy on the eye of that observer. In the remainder of this text, unless otherwise stated, it can be assumed that whenever issues involving magnitude of light are discussed, specifications are in photopic photometric units.

The perceptual experience most closely related to photometric intensity or luminance is **brightness.** In general, the more photons absorbed in the retina, the brighter the percept. However, the relationship is not as straightforward as one might expect, for two reasons. One involves issues of perceptual scaling, a topic covered later in this chapter. The other involves complex interactions involving absolute and relative amounts of luminance in various parts of a scene. These issues are covered in Chapter 7.

3 Formation of Images by Interactions of Light with Surfaces

The light energy from the environment that impinges on the eye of an observer is not uniform. Instead, the amount of illumination varies with direction.

Furthermore, the distribution of illumination is not random but is structured by the surrounding environment. James J. Gibson coined the term **optic array** to refer to the structured nature of light illuminating a single point in space where an observer's eye might potentially be located. The structural optic array produced by a particular environment for an eye located at a particular location is illustrated in Figure 3.10.

It is the presence of the optic array that makes it possible for an eye to pick up information about the locations and two-dimensional shapes of objects in the surrounding environment. In order to understand how that is accomplished, we need to understand some details about how images are formed.

As light travels through the environment, it interacts with surfaces that separate media. When light encounters a surface between one medium and another, as, for example, when light passes from air into water, one of two things can happen. Sometimes light is reflected back into the original medium (**reflection**). Other times light passes through the surface and into the new medium (**refraction**). In both cases, the direction of light travel is altered in specific ways. It is quite remarkable that these simple interactions with surfaces allow light to form images that specify the locations and shapes of the objects that it has encountered during its travels through space (Figure 3.11).

The discipline concerned with formation of images by the interactions of light and surfaces is **geometrical optics.** This discipline is less concerned with physical properties of light than with formal

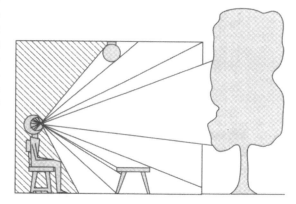

FIGURE 3.10. The spatial distribution of the light entering the eye of an observer is structured by the locations of the light sources and of the surfaces that reflect light from those sources into the eye. Adapted from J.J. Gibson, *The Ecological Approach to Visual Perception,* © 1986, by permission of Lawrence Erlbaum Publishers.

Light Interacts with Surfaces to Form Images of Objects

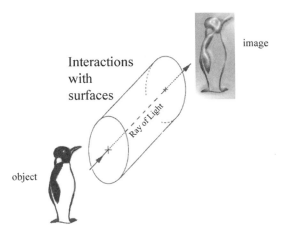

FIGURE 3.11. Light rays that travel through the environment reflect off objects, such as the body of a penguin. When light rays subsequently interact with certain types of surfaces that separate media, they form images that capture the spatial structure of the objects from which they were reflected.

concepts that apply to image formation. A central concept is a **ray of light,** used to characterize the path taken by light between various locations in the environment. Although not physically correct, a ray can be thought of as a path that might be taken by a photon or as the normal to a wave front.

Reflections from Surfaces

The relationship between relative directions of the incident and reflected light rays from a surface depends upon whether the surface is specular or diffuse. **Specular** reflection is the basis for the formation of images by a mirror or the smooth surface of a pond. In the case of **specular** surfaces, the angle of reflection is the same as the angle of incidence but opposite in sign, with an angle normal to the surface taken as 0°. An illustration of reflection from a specular surface is shown in Figure 3.12.

Each reflected ray can be seen only by an eye positioned along its trajectory. In Figure 3.12, eye X can see only the ray emanating from location A and reflected at 70 degrees and eye Y only the ray reflected at −70 degrees. Other rays (not shown) emanating from location A would be seen only by eyes at other locations. Note that as an eye moves, for example from the position occupied by eye X to the position occupied by eye Y, the rays reflected from the object at point A all appear to be coming

from a stationary point A′ behind the surface, called its **virtual image.**

Diffuse surfaces also reflect the incoming rays, but in a disordered manner. The reflections along the direction of a given ray scatter in many directions. Thus, unlike the case with a mirror, viewers at any location in front of a diffuse surface are able to view reflections from the same ray, as illustrated in Figure 3.13.

Diffuse surfaces do not form images from the reflected light. However, if there is an image projected onto the plane of the diffuse surface, this image can be seen by observers viewing the surface. This property allows diffuse surfaces to be used for practical applications, such as projection screens used in movie theatres (Figure 3.14).

In this example, a projector is used to form an image on the surface of the screen. The diffusing properties of the screen allow the members of the audience to view rays from all portions of the image regardless of where they are sitting in the auditorium.

Snell's Law

When rays pass through a surface between two media, they are bent, or **refracted,** in a manner that depends on the refractive indices of the two media.

Reflections from Specular Surface

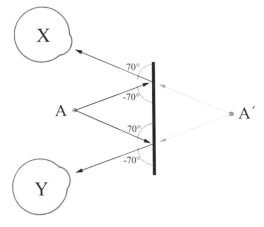

FIGURE 3.12. The figure illustrates two rays coming from object A. One ray is incident to the surface at an angle of 70° and is reflected at an angle of minus 70°. To an observer positioned at location Y, this reflected ray appears to be coming from the direction of a point A′ that lies behind the surface. The second ray is incident at minus 70° and reflected at 70°. To an observer positioned at location X, this reflected ray also appears to be coming from the direction of point A′.

Reflections from Diffuse Surface

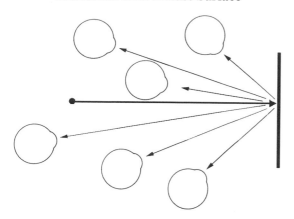

FIGURE 3.13. Reflections of light along a single ray are scattered in many directions at a diffuse surface and can be viewed from many different locations.

The **refractive index** of a medium is a number that quantifies the reduction in the velocity of light traveling through it compared to the velocity of light in a vacuum.

The exact amount of bending of light that takes place during refraction is described by **Snell's law:**

$$N' \sin r = N \sin i$$

where i is the angle of the incident and r that of the refracted ray and N and N' are the refractive indices of their respective media. In other words, the magnitude of the sine of the angle of the light leaving the surface will become equal to that of the angle of the light approaching the surface when they are multiplied by the refractive indices of the two media. Another way of thinking about the operation of Snell's law is presented in Box 3.4.

Box 3.4
Snell's Law

The rules that govern the path taken by rays of light as they pass from one medium to another can be understood with an analogy. Suppose you are home and need to get to the supermarket. You have two choices for your route. One is to use local streets for a distance of 10 miles. This choice takes about 30 minutes due to congested traffic conditions. A second choice is to take a freeway route that is 15 miles but only takes 20 minutes. The street route is the shortest distance, but the freeway route is the shortest time. Light rays always follow the route that is fastest rather than shortest. Snell's law is simply a mathematical formula that can be used to determine what this fastest route will be.

Refraction at a Spherical Surface

A **spherical surface** is a surface with a shape corresponding to a portion of the surface of a sphere. **Convex spherical surfaces** that transmit light have a special property that allows light rays that pass

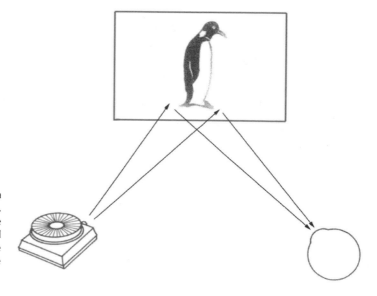

FIGURE 3.14. A projector is forming an image at the plane of a diffusing screen. Light rays from each point on the screen are reflected in many directions, as illustrated in Figure 3.13. Thus, observers can view the image from any location in front of the screen.

through them to form **real images** of objects. When we state that a real image is formed, we mean that all of the rays that emanate from the object physically converge at its image. **Concave spherical surfaces** form **virtual images,** similar to those described for specular reflecting surfaces, in which the rays do not physically converge but appear to be emanating from the location of the virtual image. We will deal here only with convex surfaces and real images. A more general treatment of spherical surfaces can be found in any geometrical optics textbook.

The image that will be formed by a spherical surface can be calculated based on **ray tracing** and Snell's law. In these calculations, the deviation of each ray is determined at each surface. However, ray tracing methods are somewhat cumbersome and can become very complicated, especially when applied to images formed by complex optical systems composed of more than one surface. A generalized system of equations and rules, **geometrical optics,** provides a more convenient set of procedures for characterizing images formed by complicated systems, such as eyeballs and optical sensors, made up of several surfaces. Here we describe some of the basic rules of geometrical optics that apply to spherical optical systems.

The simplest case is a single spherical surface between two media having different refractive indices. An example is the front surface of the eye that separates a medium of air on the outside from a fluid medium on the inside. The refraction of light by a single spherical surface is illustrated in Figure 3.15.

This diagram shows a **convex surface** that converges light rays that pass through it from left to right. The surface separates two media, one, on the

Image Formation by a Spherical Surface (graphical method)

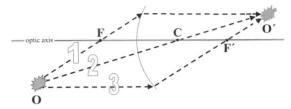

FIGURE 3.16. The location of the image point that is formed by a spherical surface for a given object point can be determined graphically. Symbols and details are explained in the text.

anterior (left) side, that has a refractive index of N, and a second, on the **posterior** (right) side, with a refractive index of N'. The most anterior portion of the spherical surface, called its **vertex,** is at **A,** and its **center of curvature**, the location where the center of the sphere defined by this surface is located, is at position **C**. The straight line that passes through the vertex and the center of curvature defines the **optical axis.**

All spherical surfaces have two **focal points,** which fall on the optical axis. As illustrated in Figure 3.15 by the lower dashed line, light rays that are parallel with the optical axis approaching from the anterior side become bent at the surface such that they converge at the **posterior focal point, F'**. As illustrated by the appear dashed line, light emanating from the **anterior focal point, F**, will form rays that become bent at the surface such that they run parallel to the optical axis. Any rays of light that pass through the center of curvature, **C**, pass through the spherical surface undeviated, as illustrated for one representative ray in Figure 3.15 by the thick dotted line. A major principle in geometrical optics is that all relationships are reversible. Thus, if light were approaching from the right side, the same relationships would hold, and the arrowheads on the dotted and dashed lines would simply be turned around.

The three types of rays illustrated in Figure 3.15 are sufficient to determine the **image point** formed by a spherical lens for any **object point** based on graphical methods. An example is illustrated in Figure 3.16.

From the object point **O** one ray can be drawn that passes through the anterior focal point F. It is known that this ray will be bent to run parallel to the optical axis. A second ray can be drawn from the object through the center of curvature **C**. It is

Properties of Spherical Surfaces

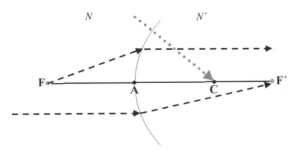

FIGURE 3.15. This schematic illustrates how three specific rays are altered as they pass through a spherical surface that separates two media. Symbols and details are explained in the text.

known that this line will continue across the surface undeviated. A third line can be drawn through the object that runs parallel to the optical axis. This line will be bent such that it passes through the posterior focal point F'. All three of these lines are constrained to cross at the point **O'** where the image will be formed.

Graphical methods based on geometrical optics are a valuable tool for understanding images formed by spherical surfaces. However, for many applications it is necessary to analyze properties of images with more quantitative methods. Geometrical optics provides simple formulas for solving many of these problems. The simplest case involves determining where an image will be formed by a spherical surface for an object that falls along the optical axis, as illustrated in Figure 3.17.

This diagram illustrates a spherical surface that forms an image point **O'**, of an object point, **O**. Three representative rays are shown in the diagram. The distance from the object point to the vertex **A**, is l and that from the vertex to the image point l'. When using the analytical formulas based on geometrical optics, many distances need to be characterized by a parameter called **reduced distance**, in which distance as measured by a ruler is divided by the refractive index of the media. Expressed as reduced distances, object distance becomes l/N and image distance l'/N.

Another important convention in geometrical optics is that distances are usually expressed in units of **diopters (D)**, which are the reciprocal of distance in meters. For example, a distance of 0.5 m would be specified as 2.0 D, 10 m as 0.1 D. Specify-

ing distance in diopters has certain advantages, as elaborated in Box 3.5.

> ## Box 3.5
> ## Diopters
>
> Specifying the power of lenses in diopters has a practical advantage. Power in diopters is additive. This means a 5 D lens can be replaced with a combination of two lenses with powers of 2 D and 3 D, which will yield the same effect. Specifying the power of a lens in terms of some other parameter, such as its focal length, would not allow this property to be expressed in additive terms.

Combining the concepts of reduced distance and diopters should make it apparent that object and image distances can also be expressed in diopters as (N/l)D and (N'/l')D.

A fundamental property of any spherical refracting surfaces is its **dioptric power**, P, expressed in units of diopters. Dioptric power can be related to the object and image distances specified in the units just defined:

$$P = N'/l' - N/l$$

Another way of specifying the power of a spherical surface is with respect to its radius of curvature, defined as the distance from the vertex to the center of curvature. If we designate that distance, as measured with a ruler, as r, then:

$$P = (N' - N)/r$$

The way these formulas are typically used in practice is that values for two of the three parameters of interest (dioptric power, object distance, image distance) are known. Then the third can be solved based on algebraic addition and subtraction. For example, if the dioptric power of a spherical surface and the object distance are known, the image distance can be calculated.

The two formulas just given are sufficient to determine the position of any image formed from an object point that falls on the optical axis. We need just one additional formula for applying these same analytical methods to image formation above and below the optical axis. An example of this situation is illustrated in Figure 3.18.

Image Formation by a Spherical Surface (analytic method)

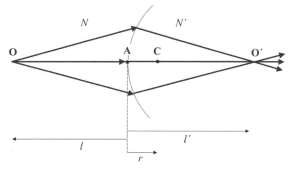

FIGURE 3.17. The location of the image point O' that is formed by a spherical surface for a given object point O that falls on the optical axis can be determined with analytical methods based on the parameters illustrated. Symbols and details are explained in the text.

Magnification of an Image by a Spherical Surface

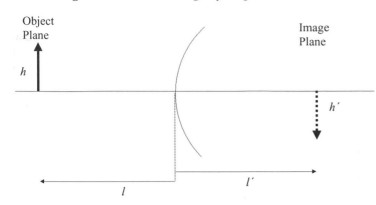

FIGURE 3.18. Specifying the location of the image point that is formed by a spherical surface for a given object point that falls off the optical axis is more complicated. The distances from the vertex of the object plane, *l*, and image plane, *l'*, can he determined the same as for points on the optical axis as illustrated in Figure 3.17. However, the heights above or below the optic axis of the object point, *h*, and image point, *h'*, must also be determined.

The relationship of the object distance, *l*, and image distance, *l'*, is specified by the formulas just described. All that is required in addition is to determine the height *h'* of the image above or below the optical axis, based on the height *h* of the object. That relationship is determined by:

$$M = h'/h = l'/l,$$

where *M* is **magnification,** defined as the ratio of image to object heights. The formula states that the ratio between image and object heights is the same as the ratio between image and object distances.

Refractions by Thin Lenses

Our discussion to this point has involved a spherical surface that separates two media having different refractive indices. Another common situation involves **lenses** in air, where the refractive indices anterior and posterior to the surface are identical. For many applications involving lenses, it is sufficiently accurate to ignore the thickness of the lens and simply treat the lens as an infinitesimally thin spherical surface having a designated power. When dealing with a thin lens in air, the general formula for dioptric power can be simplified:

$$P = 1/l' - 1/l = 1/r,$$

where *P* is power of the lens, *l* and *l'* are the object and image distances, and *r* is the radius of curvature. Distance *r* is sometimes referred to as the focal distance of the lens.

Some general relationships can be noted between object and image distances. It should be apparent from examination of the above formulas that there has to be a reciprocal relationship if the power of the lens is kept constant. As an object gets closer to a lens, its image moves farther away, and vice versa. This is illustrated in Figure 3.19.

There are limits to this reciprocal relationship. When an object has moved as far away from a lens as it could possibly get, this distance is called **optical infinity.** Theoretically, optical infinity is a very long distance, but as a practical matter, any distance beyond about 6 meters can be treated as

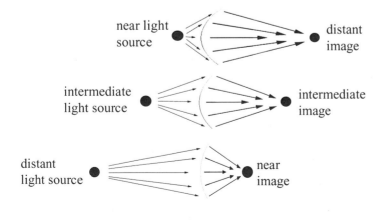

FIGURE 3.19. For a spherical lens of fixed power there is a reciprocal relationship between the distance of an object, in these examples a point source of light, and the distance of the image that is formed.

weak convex lens

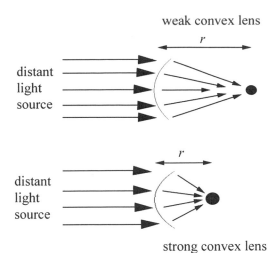

distant light source

r

distant light source

r

strong convex lens

FIGURE 3.20. When an image is formed of an object a long distance away from a lens, the image is formed at the focal distance of the lens, which is equal to its radius of curvature *r*.

being at optical infinity for the types of issues that concern us here. When the object is at optical infinity, object distance specified in diopters is zero $(1/\infty)$. The formula for thin lenses just discussed reveals that under these conditions the reciprocal of image distance $(1/l')$ is equal to the reciprocal of focal distance $(1/r)$. It follows that the image distance measured with a ruler, l', is equal to the focal distance of the lens, r.

When object distance is held constant, lenses with a higher power will form images at a closer distance. This general rule is illustrated in Figure 3.20 for the condition in which the object is at optical infinity.

Refractions by Thick Lenses

The use of the thin-lens equations ignores the influence of the thickness of the lens on image formation. For some applications, this is not sufficiently accurate. Geometrical optics provides **thick-lens approximations** that can be helpful in these situations. When using the thick-lens approximations, the relationships described above for thin lenses are applied in a two-part manner. All of the relationship that would occur anterior to a single spherical surface are shown anterior to one plane, called the **anterior principal plane.** Relationships that would take place posterior to a single spherical surface are shown posterior to a second, **posterior principal plane.**

These thick-lens approximations can even be applied to more complex image-forming systems, such as eyeballs, that are composed of a number of spherical surfaces separated by different media. Further discussion of this issue will be deferred until Chapter 4, where optical properties of eyeballs are discussed.

4 Formal Descriptions of Spatial and Temporal Properties

The optic array that gives rise to the retinal image is structured in time as well as space as illustrated in Figure 3.21.

A prerequisite step before many questions about perception can be addressed is to have appropriate methods for characterizing spatial and temporal properties of environmental stimulation in a formal manner.

It often turns out to be convenient to describe these properties from a specialized perspective referred to as being in the **frequency domain.** This perspective is somewhat nonintuitive and depends on understanding some technical concepts that may at first seem foreign and esoteric if you have not previously encountered them. However, a grasp of these concepts is a prerequisite for understanding the explanations of a number of perceptual phenomena described later in this book. The advan-

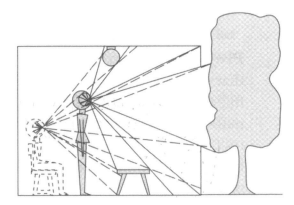

FIGURE 3.21. As illustrated in Figure 3.11, the light entering the eye of an observer is structured by the spatial locations of the light sources and surfaces that reflect light from those sources into the eye. Whenever the relative locations of an observer and objects in the environment change, the structure also changes. Thus, the light rays entering the eye are structured in both space and time. Adapted from J.J. Gibson, *The Ecological Approach to Visual Perception*, © 1986, by permission of Lawrence Erlbaum Publishers.

FIGURE 3.22. This figure shows a grating formed by alternating black and white vertical bars of equal width.

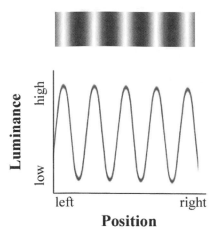

Position

FIGURE 3.24. The top panel shows a sine wave grating. It is similar to the square wave grating shown in Figure 3.22 except that the edges between the dark and bright bars appear fuzzy instead of abrupt. The reason for this is apparent from the luminance profile shown in the bottom panel. The luminance is modulated gradually from bright to dark according to a mathematical sine wave function.

tages of using descriptions in the frequency domain will be highlighted in Chapter 8, where perception of two-dimensional patterns and shapes are discussed.

Spatial Sine Wave Gratings

In optics, a **grating** is a series of alternating parallel bright and dark bars (Figure 3.22).

A grating consists of a repeating basic pattern called one **cycle,** one bright bar next to one dark bar. Gratings can be described in a formal manner. Figure 3.23 provides one example of a description of the grating shown in the previous Figure 3.22.

This description is in the form of a **luminance profile across space** and is said to be in the **spatial domain,** because it involves specifying the light intensity at each spatial position along the grating. It could be measured by moving a small photodetector probe horizontally across the grating. It is important to keep in mind that the "description" of

a grating is not the same thing as the "grating" itself. Figure 3.22 is an actual grating and Figure 3.23 only its description.

The grating in Figure 3.22 is called a **square wave grating,** based on its luminance profile. Note that the transition between the dark and the bright bars is abrupt, such that the luminance profile has square corners at the borders between each pair of bars. Another type of grating, called a **sine wave grating,** and its luminance profile are illustrated in Figure 3.24.

The top panel shows an actual sine wave grating. Like the square wave grating, it has a repeating pattern. In this case, a cycle consists of a fuzzy-looking bright region next to a fuzzy-looking dark region. There is no sharp border between the bright and the dark bars. The luminance profile changes gradually from bright to dark and then bright again, forming a sine wave, as illustrated in the bottom panel; hence the name "sine wave grating."

These descriptions in the spatial domain, as illustrated in Figure 3.24, provide one way of specifying a sine wave grating. However, it turns out that there is another way of describing sine wave gratings that is more general. One can simply provide the values of four specific parameters, discussed below: 1) **spatial frequency,** 2) **contrast** (also called **amplitude**), 3) **orientation,** and 4) **spatial phase.** If the

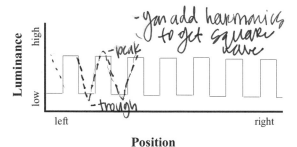

Position

FIGURE 3.23. A grating of the type shown in Figure 3.22 can be described in terms of its luminance profile. This luminance profile could be produced by moving a small photocell along a horizontal path from the left to the right side of the grating.

Spatial Frequency of Sine Wave Gratings

FIGURE 3.25. The top and bottom panels show sine wave gratings with different spatial frequencies. The grating in the top panel has a higher spatial frequency than the grating in the bottom panel.

values of these four parameters are known, then the spatial properties of the sine wave grating are completely specified. Describing a sine wave grating by providing values for these four parameters provides a description in the **spatial frequency domain**.

Spatial Frequency

"Spatial frequency" refers to the frequency with which the grating changes from dark to bright per unit distance across the stripes. An example of two sine wave gratings that differ only in spatial frequency is shown in Figure 3.25.

Each sine wave grating is being viewed through a separate window. The two windows are the same width. The top grating has 4.5 cycles displayed within the window. The bottom grating has only 1.5 cycles displayed in the window. The top grating is said to have a higher spatial frequency than the bottom grating, since it has more "cycles per window."

Instead of cycles per window, the convention in studies of perception is to specify spatial frequencies in units of **cycles per degree of visual angle** (abbreviated as **cy/deg** or **cpd**). Additional information about these units is provided in Box 3.6.

Contrast

The terms "contrast" and "amplitude" refer to the magnitude of the difference between the luminances of the bright and the dark bars. Figure 3.26 illustrates two sine wave gratings that have the same spatial frequency but differ in contrast.

The mean luminance across the entire grating is the same for both stimuli. However, the contrast between the dark and bright bars is higher for the sine wave grating on the left than for the one on the right. In the luminance profile, the amount of contrast is reflected in the difference in height between the peaks and troughs. The usual convention in perception studies is to specify the contrast of a grating in units of **Michelson contrast**, defined as:

$$Michelson\ Contrast = (L_{max} - L_{min})/(L_{max} + L_{min}),$$

where L_{max} is the luminance at the peak of a bright bar and L_{min} is the luminance at the trough of a dark

Contrast of Sine Wave Gratings

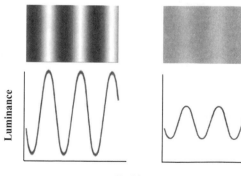

FIGURE 3.26. The top panel shows two sine wave gratings and the bottom panel their luminance profiles. The gratings are identical except that the grating on the left has a higher amplitude of luminance modulation, or contrast, than the grating on the right.

Box 3.6
Specification of Spatial Frequency in Cycles Per Degree of Visual Angle

In general, cycles per window is not a very convenient way to specify spatial frequencies for perception studies. The size of the window that is used to display the grating is arbitrary, and whenever we change the size of the window, the number of cycles per window will change.

Another way to specify spatial frequency is in units of cycles per millimeter (cy/mm). This specification can be made in terms of the distal stimulus or the proximal stimulus. Two distal stimuli having the same spatial frequency will produce proximal stimuli that vary in spatial frequency with distance, as illustrated in Figure 1. In the example shown in the top panel, the observer is looking at a grating from a particular distance. If the image of the window formed on the retina is 3 mm wide, since there are 3 cycles of grating present in the window, it is easy to calculate that the grating has a spatial frequency of 1 cy/mm on the retina. In the example shown in the bottom panel, the observer is looking at the same grating in the same window. However, since the observer is farther away, the size of the retinal image is smaller. For example, if the image of the window has

been reduced to 1 mm in width, the spatial frequency of the grating is now 3 cy/mm on the retina.

For issues having to do with perceptual processing of spatial frequency information, it is generally assumed that the brain has access only to the proximal stimulus. Thus, units of spatial frequency specified for the proximal stimulus are generally more convenient than units specified for the distal stimulus.

There is an obvious technical problem with measuring gratings in cy/mm on the retina. How can a ruler be put into the back of the eye of the observer to measure the retinal image? Fortunately, there is a way to approximate retinal image size with units that can be easily measured in a noninvasive manner from the front of the eye. Straight lines drawn from the two sides of an object to the corresponding image will cross near the front of the eye to form a **visual angle,** as illustrated in Figure 1. To a first approximation, the angle formed by the object at the front of the eye is proportional to the size of the retinal image formed. For perception studies, spatial frequency is typically specified in cy/deg.

Cycles per Degree of Visual Angle

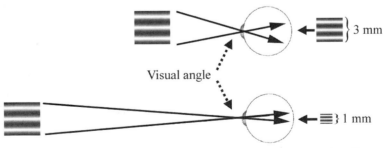

FIGURE 1. The top and bottom panels illustrate an eye viewing the same sine wave grating from two different distances. The size of the retinal image formed is smaller for the greater viewing distance. The visual angle formed at the front of the eye is roughly proportional to the size of the retinal image.

Spatial Amplitude Spectrum Plot

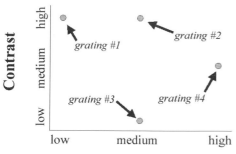

FIGURE 3.27. Gratings having a particular spatial frequency and contrast can each be represented as a single point in a spatial amplitude spectrum plot.

bar. Note that this is a relative measure. Michelson contrast specifies the amount of contrast that is present relative to the maximum possible contrast available given the mean luminance. Other definition of contrast are used occasionally in perception research, but whenever we refer to contrast in this text, it can be assumed we are referring to Michelson contrast.

A convenient way to represent both spatial frequency and contrast of one or more sine wave gratings is with a **spatial amplitude spectrum plot,** as illustrated in Figure 3.27.

On this type of plot, each sine wave grating is represented by a single point. This diagram specifies the spatial frequencies and amplitudes for four sine wave gratings. The first has low spatial frequency and high contrast, a second medium frequency and high contrast, a third medium frequency and low contrast, and a fourth high frequency and medium contrast.

Orientation

"Orientation" refers to the directions in which the bars of a grating point. All of the gratings illustrated so far have had a vertical orientation, in which the bars point straight up and down. However, this does not have to be the case. The bars can point in any direction, including horizontal or oblique (Figure 3.28).

A convenient graphical representation that simultaneously specifies spatial frequency, contrast, and orientation is called a **spatial amplitude polar plot,** illustrated in the Figure 3.29.

In this type of plot, each spot represents a single sine wave grating. The origin of the plot, where the horizontal and vertical axes cross, designates a spatial frequency of zero, and higher spatial frequencies are represented at proportionately greater distances away from the center. Contrast is indicted by the size of the spot, with large spots designating high contrast and small spots designating low contrast. Orientation is indicted by the angle between the horizontal axis and a line from the origin to the center of the point.

To be technically accurate, each orientation would have to be represented by two points falling along two lines passing out of the origin in opposite directions. For example, a 45° rightward leaning oblique grating would be represented along the diagonal passing upward and to the right of the origin (as shown for grating #3) and simultaneously along the diagonal passing downward and to the left (not shown). The representation in the upper right quadrant has positive values and the lower left quadrant has negative values. The component of the representation with negative values is hard to conceptualize and has to do with technical details of the mathematical formulas used to

Orientation of Sine Wave Gratings

FIGURE 3.28. This figure illustrates a vertical grating on the left and the same grating oriented horizontally on the right.

Vertical   **Horizontal**

Spatial Amplitude Polar Plot

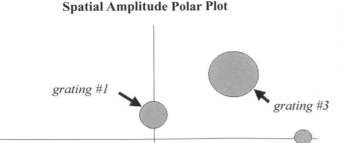

grating #1

grating #3

grating #2

FIGURE 3.29. Gratings having particular spatial frequencies, contrasts, and orientations can each be represented with a single symbol on a spatial amplitude polar plot. The location of each symbol specifies the spatial frequency and orientation and the size of the symbol the contrast. Three separate sine wave gratings are represented on this plot. The first is a vertical grating of low spatial frequency and medium contrast. The second is a horizontal grating of low contrast and high spatial frequency. The third is an oblique grating of high contrast and medium spatial frequency.

describe gratings in the frequency domain. These details go beyond the scope of our interests, and in the spatial amplitude polar plots in the text we will be concerned only with the positive components of the representation.

with respect to some other grating or some other landmark.

Spatial phase can be shown graphically along with the other spatial parameters, frequency, contrast, and orientation. However, it is difficult to

Spatial Phase

"Spatial phase" refers to the position of the grating relative to some fixed location, such as the window through which it is being viewed. This is illustrated in Figure 3.30, which shows two gratings with identical spatial frequency, contrast, and orientation but different phases.

The grating on the bottom has been shifted slightly to the right compared to the grating on the top. Thus, these two gratings have different positions relative to the window. Shifts in spatial phase can be specified in degrees based on the relationship that one cycle of sine wave corresponds to 360°. For example, if a grating's phase is shifted by plus or minus 180°, the center of each dark bar is positioned at the location previously occupied by a bright bar. In the example shown in the figure, the grating in the bottom panel is shifted by 90° with respect to the grating in the top panel. While there is no need to be concerned here with additional technical details about how magnitudes of phase are specified, it is always possible to specify the phase of any grating with respect to the window through which it is being viewed or

**Spatial Phase of
Sine Wave Gratings**

FIGURE 3.30. The two sine wave gratings shown are identical except that the position of the grating in the bottom panel has been shifted in position, or phase, relative to the grating in the top panel.

FIGURE 3.31. A slide of a penguin in the projector produces an image of a penguin on the screen.

represent four dimensions at once on the same two-dimensional plot. For the descriptions of sine wave gratings in the remainder of this text, we will usually show simplified graphical representations in the form of spatial amplitude polar plots (three parameters) or spatial amplitude spectrum plots (2 parameters). These simplified representations are convenient as heuristic devices when our primary purpose is to illustrate the fact that parameters are being specified for one or more sine wave gratings. Simply keep in mind that these simplified graphical representations would have to be expanded to a four-dimensional plot in order to give a complete specification of the sine waves that are being described.

Fourier's Theorem

Up to this point, it may not be apparent that any of the preceding discussion of sine wave gratings has relevance to perception. As we look around in our environment, we seldom, if ever, see sine wave gratings. Instead, we see objects having complex shapes. The key to relating sine wave gratings to complex shapes is provided by **Fourier's theorem,** which states *any* spatial pattern can be formally characterized in terms of a **superposition of sine wave gratings.** Fourier's theorem can be illustrated with a hypothetical example. Suppose a slide of a penguin were put into a slide projector and its

image projected onto a screen. Figure 3.31 illustrates what will happen.

Next, consider what would happen if a slide of a sine wave grating were put into the same projector. This is illustrated in Figure 3.32.

The outcome of these two examples is not surprising. When a slide of a penguin is placed into the slide projector, an image of the penguin is seen on the screen. When a slide of a sine wave is put into the slide projector, an image of a sine wave is seen on the screen. Suppose next that several projectors are all pointed at the same screen. There are also several slides, but each slide is simply a sine wave grating. The slides differ from one another because the various slides have different spatial frequencies, contrasts, phases, and orientations. If the appropriate slides have been picked, the image shown in Figure 3.33 will be formed on the screen when all of these projectors are turned on simultaneously, superimposing the images of the sine wave gratings.

This example illustrates what Fourier's theorem means when it states that it is possible to produce a complex pattern by simply superimposing sine wave gratings.

Obviously, this image of the penguin would not appear for just any combination of sine wave gratings that happened to be superimposed. One would need to know precisely which sine wave gratings to put into the projectors to produce an image of a penguin. How would one figure that out? The answer is provided by a mathematical procedure

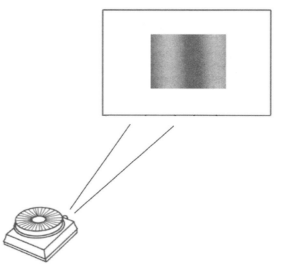

FIGURE 3.32. A slide of a sine wave grating in the projector produces an image of a sine wave grating on the screen.

FIGURE 3.33. Several slides of sine wave gratings all superimposed on the screen at the same time produce an image of a penguin.

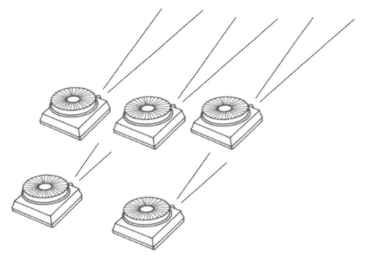

called **Fourier analysis.** Given any shape described in the spatial domain, Fourier analysis can provide a list of the specific sine wave gratings that would need to be superimposed to produce that pattern.

Recall that any single sine wave grating can be specified by simply providing values for each of four parameters: spatial frequency, contrast, orientation, and spatial phase. Thus, *any* complex spatial pattern can be described in a standard format in a table with four columns, as illustrated in Figure 3.34.

Each row in this table specifies one sine wave grating that needs to be superimposed to produce the given pattern. The sine wave grating in each row is characterized by four numbers, one in each column, that provide the specifications for the parameters of this grating. The information provided in

Description of Complex Pattern in Spatial Frequency Domain

	Frequency	Contrast	Phase	Orientation
1				
2				
3				
.				
.				
.				
N				

FIGURE 3.34. Any complex shape can be described in the frequency domain in a standardized format consisting of a table with rows and columns. Each row in the table has four columns that list the four parameters needed to specify one sine wave grating.

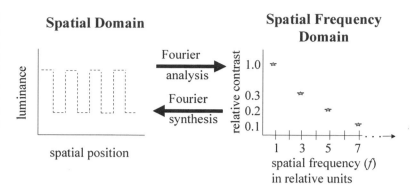

FIGURE 3.35. The plot on the left provides a description for a square wave grating in the spatial domain. The plot on the right shows a description of the same grating in the spatial frequency domain. Fourier analysis and Fourier synthesis can be used to move back and forth between these two descriptions.

this table is a formal description of a complex shape in the spatial frequency domain.

There are thus two equivalent ways that can be used to specify the shape of any complex two-dimensional object, such as the image of a penguin. The first is a description in the spatial domain, in which the intensity of every spatial location in the image is specified. The second is a description in the frequency domain, in which parameters are specified for superimposing sine wave gratings. These two descriptions are equivalent, in the sense that either can be used to specify the spatial shape of the same complex two-dimensional pattern.

Fourier analysis is a tool that can be used when one already has a description of a complex shape in the spatial domain and wants to translate this description into the spatial frequency domain. A complementary tool is **Fourier synthesis.** Given a description of any shape in the spatial frequency domain, Fourier synthesis provides the corresponding description of the same shape in the spatial domain.

An intuitive understanding of the relationships between these two domains can be attained by considering a simple example of how one would superimpose sine wave gratings to produce a square wave grating. A square wave grating, like any other two-dimensional shape, can be described in the spatial frequency domain. In other words, it is possible to superimpose some number of sine wave gratings to produce a square wave grating. The specific sine wave gratings that have to be superimposed to produce a square wave grating are illustrated in Figure 3.35.

The left of Figure 3.35 shows a description of a square wave in the spatial domain in terms of its luminance profile. The sine wave gratings that will need to be superimposed to produce this square grating are shown on the right side in the spatial frequency domain in the simplified form of a spatial amplitude plot. Using relative units, the lowest

spatial frequency sine wave grating that will be needed, labelled 1 in the figure, is called the fundamental (F). Superimposed on the fundamental is a sine wave grating having a spatial frequency three times that of the fundamental, called the 3rd harmonic ($3F$) next, the 5th harmonic ($5F$) is superimposed, followed by the 7th harmonic ($7F$), and continuing in this same pattern with additional **odd-harmonic** sine wave gratings. The degree to which the superimposed sine wave gratings resemble a square wave grating will continue to improve with addition of odd-harmonics until the discrepancy becomes infinitesimally small.

The contrasts of the gratings need to be adjusted in a pattern based on the reciprocals of their spatial frequencies relative to F. In other words, the contrast of $F2$ must be adjusted to be $^1/_2$ that of F, $F3$ to $^1/_3$ that of F, and so on.

Figure 3.36 illustrates the luminance profiles for the square wave and the first three sine waves (F, $3F$, and $5F$) that would be superimposed to approximate the square wave. This figure has been drawn such that relative phase relationships can be appreciated as well as contrasts and spatial frequencies.

The phase of F is adjusted such that the peaks of its bright bars and the troughs of its dark bars correspond to the bright and dark bars of the square wave. The contrast of F is adjusted until the peaks of its bright bars are just brighter and the troughs of its dark bars just darker than those of the square wave. Note that the profile of this single sine wave approximates the square wave, except that its peaks and troughs are rounded and those of the square wave are flat. The middle panel illustrates how the position of $3F$ can be adjusted relative to the fundamental until their phases are in an arrangement called peaks-subtract. As a result, the superimposed profile of F and $3F$ will have peaks and troughs that are somewhat flattened out, approximating more closely the luminance profile of the square wave. The bottom panel illustrates the effect of super-

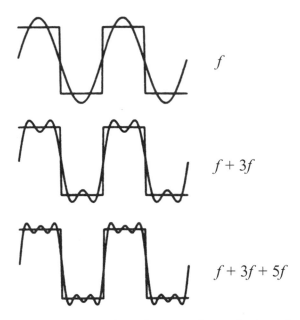

f

$f + 3f$

$f + 3f + 5f$

FIGURE 3.36. This figure illustrates the amplitudes and spatial phase relationships of the first three sine wave gratings that would need to be superimposed to produce a particular square wave grating.

imposing 5F in a similar peaks-subtract phase relationship with 3F, further flattening out the peak and trough regions. With the superposition of each additional odd harmonic, the peaks and troughs become more flattened and the overall profile more like a perfect square wave. Additional harmonics can be added until the resulting profile approximates a square wave with any desired level of accuracy.

Temporal and Temporal Frequency Domains

The major concepts involved in understanding descriptions of changes that take place over time in the **temporal domain** and the **temporal frequency domain** are directly analogous to those described in the previous section for the spatial and spatial frequency domains. Thus, a brief description will suffice here.

Consider what happens to a small region at the very center of a television screen while a program is on. The intensity of this small portion of the screen varies over time in a complex manner. These changes over time could be specified by placing a

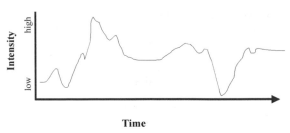

FIGURE 3.37. Intensity of a small spot of light can change in an arbitrary manner over time.

photocell directly over the center of the screen. The output of the photocell might look something like the example shown in Figure 3.37.

The output of this photocell provides a description of the changes in intensity as a function of time, a description in the **temporal domain.** These changes can also be characterized in the **temporal frequency domain,** analogously to the way spatial patterns are characterized in the spatial frequency domain. Fourier's theorem informs us that complex patterns of changes of intensity over time, as illustrated in Figure 3.37, are equivalent to superposition of some number of intensities, each simply modulated sinusoidally over time. Figure 3.38 illustrates the profile of a single light intensity being modulated sinusoidally over time.

Recall from our discussion of spatial properties that four parameters can specify an individual sine wave grating. Characterization of **sinusoidal modulation in time** is even simpler; it can be accomplished with only three parameters, **tempo-**

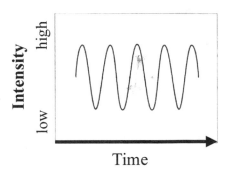

FIGURE 3.38. This figure illustrates sinusoidal modulation of light intensity over time.

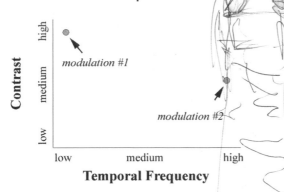

FIGURE 3.39. Illustration of a temporal amplitude spectrum plot.

ral frequency, amplitude, and phase. There is no need to specify orientation when referring to a sinusoidal modulation in time because time spreads out in only one dimension, unlike a 2-dimensional image in space.

"Temporal frequency" is typically specified in Herty (Hz) where 1 Hz refers to the condition where one cycle of the sine wave is completed each second. "Amplitude" means the difference between the intensity at the times when intensity is at its peak and the times when intensity is at a trough. "Phase," when used in reference to time, refers to the time when the sine wave profile has reached a specified part of its 360° cycle, e.g., 0°, with respect to some landmark time. For example, if two sinusoidal modulations of the same frequency are plus or minus 180° out of phase with respect to one another, then the moment when one is at its maximum intensity will occur at the same moment when the second is at its minimum intensity.

The changes in intensity over time can be completely characterized in the temporal frequency domain in terms of a table that has three columns. Each row describes a sinusoidal modulation of intensity over time that has a frequency, amplitude, and phase as specified in the three columns. Simplified descriptions can also be provided in graphical form, ignoring phase, in terms of a **temporal amplitude spectrum plot,** as illustrated in Figure 3.39.

Two sinusoidal modulations are illustrated in this plot. One is being modulated sinusoidally slowly over a wide contrast range, the second fast over a medium contrast range. When we use these simplified temporal amplitude spectrum plots, it is with the understanding that a full description of temporal change would require phase to be specified as well.

5 Viewer-Centered Descriptions of the Environment

The previous sections of this chapter discuss descriptions of the environment that are objective in the sense of not being tied to the point of view of any one observer. This final section describes some properties of the environment that are tied more closely to a viewer-centered point of view.

Perceptual Scaling

When measuring environments, some **metrics,** or systems of measurement, have more utility than others. Consider the example of a light bulb attached to a dimmer switch. For purposes of this example, assume that the light bulb generates a photometric intensity of 100 candelas with no attenuation and 0 candelas when fully attenuated. Photometric intensity can be represented as illustrated in Figure 3.40.

This scale is convenient for measuring photometric intensity because equivalent distances as measured with a ruler anywhere along this scale correspond to equivalent changes in intensity. This attribute can be stated in another manner in relation to a dimmer switch. Suppose each 1/4 turn of the dimmer switch knob changed the intensity by an amount designated by one tick on this scale. In that case, the intensity would change by 10 candelas for every 1/4 turn regardless of whether the knob was turned from the off setting, the fully on setting, or anywhere in between. We will refer to a dimmer switch with these properties as one in which there is a **linear relationship** between rotation of the knob and photometric intensity.

However, units of photometric intensity refer simply to the effectiveness of light in terms of stim-

Intensity of light bulb (candelas)

0   50   100

FIGURE 3.40. Photometric intensity represented on a scale with ticks spaced at equal distances.

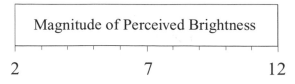

FIGURE 3.41. Brightness represented on a scale with ticks spaced at equal distances.

ulating the eye. They do not directly measure the magnitude of the subjective perceptual experience produced by light, called **brightness**. For example, we do not know whether stimulating the eye with 100 units of photometric intensity will be perceived to have the same brightness as the combination of two other lights each having 50 units of photometric intensity.

Brightness can be quantified for a given observer by using the Class B psychophysical method called magnitude estimation that was described in Chapter 2. Recall that when using this procedure, an observer is simply asked to look at a series of lights and report a number for each that represents the magnitude of its brightness. Suppose these methods reveal that the brightness has a value of 2 when the light bulb is turned off and 12 when the dimmer switch is fully on. The exact values used here for these two ends of the scale are arbitrary, and unimportant for the purposes of this discussion. All we really care about here are the brightness values that fall between these two extremes. They can be represented on a scale where ticks corresponding to equal increments in brightness are spaced at equal distances (Figure 3.41).

A dimmer switch knob that was adjusted so that each 1/4 turn changed the brightness by an amount designated by one tick on this scale would change brightness by an equivalent amount regardless of the initial starting point. A switch having these properties is one in which there is a linear relationship between rotations of the knob and brightness.

What is the best way to scale attenuation of the dimmer switch? Is it more important to design the dimmer switch to guarantee that 1/4 turns produce equal changes in photometric intensity or equal changes in brightness? A switch based on photometry would be most appropriate if the purpose is to change intensity by amounts that can be specified in an objective manner. However, for a company trying to sell dimmer switches to humans to use in controlling the brightness of their environment, the objective standard is of less importance than the functional utility of the dimmer switch. Most customers would be unsatisfied with a dimmer switch

in which sometimes the knob has to be turned several revolutions before any change in brightness is noticed and other times only a tiny rotation causes large changes. These unhappy customers would probably not be satisfied with an assurance that the switch was designed so that each small turn results in a comparable change in photometric intensity.

This example illustrates a general principle of perceptual environments. Physical measures of the environment obtained with devices such as rulers or photometers do not necessarily provide the most convenient metric for specifying magnitudes in relationship to an observer. A more appropriate metric for many purposes is one in which physical magnitudes have been scaled in such a way that equivalent distances along the scale correspond to equivalent changes in perceived magnitude. Achieving such a metric is the goal of **perceptual scaling.**

In order to achieve perceptual scaling, we need to know the relationship between the scale of physical magnitude that is being employed and the scale of perceived magnitude. This relationship can be seen by plotting the values of one scale against those of the other. Hypothetical data for the dimmer switch example are illustrated in Figure 3.42.

Examination of this plot reveals that hypothetical observers experience a brightness of 2 when looking at a light bulb generating 0 candelas, a brightness of 6 for 10 candelas, etc. Notice that the relationship between intensity and brightness is curved rather than forming a straight line. Thus, a dimmer switch that is constructed so that its rotations have a linear relationship with photometric intensity will produce large changes in brightness when starting from a low setting but only tiny changes when starting from a high setting. In other

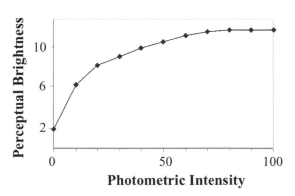

FIGURE 3.42. Plot illustrating relationship between photometric intensity and perceptual brightness.

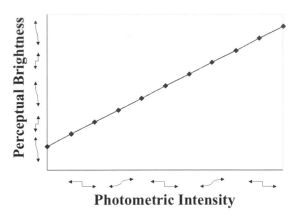

FIGURE 3.43. Plot of same data shown in Figure 3.42, but with the axes stretched and distorted in a manner that makes the relations fall along a straight line.

words, the relationship between rotations of this dimmer switch and the bulb's brightness is **nonlinear.**

This nonlinear relationship exists when the results are scaled as illustrated in Figure 3.42. However, imagine that the graph was drawn on a sheet of rubber that could be stretched. It would be possible, in principle, to turn the relationship into a straight line by stretching and distorting the sheet of rubber in the horizontal and vertical directions, as illustrated in Figure 3.43.

The relationship between intensity and brightness is now a straight line when specified in relation to the axes on this stretched piece of rubber. A fixed distance as measured with a ruler anywhere along the horizontal direction always results in an equivalent change in vertical distance. An engineer could, in principle, build a dimmer switch in which rotations of the knob are linearly related to perceived magnitude by working in units that correspond to equal distances along these distorted and stretched axes.

The solution achieved by stretching a piece of rubber is merely an illustration. An engineer needing to apply equivalent units of voltage attenuation to a dimmer switch will use a more practical, general solution, in the form of a mathematical transformation of the axes, to achieve a functional result.

Steven's Law

Two specific mathematical operations have been proposed as general solutions to perceptual scaling. These solutions are applicable not only to brightness but more generally to any perceptual scaling

between physical intensity and perceived magnitude. One, **Steven's law,** states that a linear relationship can be graphed between physical intensity and perceptual magnitude if the graph is distorted in a specific manner that can be described mathematically as plotting both the horizontal and vertical axes on logarithmic axes, as illustrated in Figure 3.44.

Stated mathematically, this relationship is a **power function,** in which the perceived magnitude P is equal to the magnitude of the physical intensity I raised to some power n:

$$P = I^n.$$

The parameter n is the slope of the straight line when plotted on log-log coordinates. As illustrated in Figure 3.44, the value of this parameter varies with stimulus conditions and sensory modality. Stevens' law has been demonstrated to provide a reasonably accurate description of perceptual scaling for several type of physical stimulation.

Weber-Fechner Law

A different solution to perceptual scaling is the **Weber-Fechner law,** proposed by Fechner. The rationale for the Weber-Fechner law is derived from a relationship involving perceptual thresholds called **Weber's law.**

Weber's law describes a very general relationship that holds between difference thresholds measured

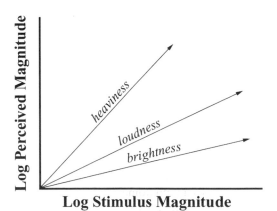

Log Stimulus Magnitude

FIGURE 3.44. Steven's law refers to the fact that perceived magnitude across many conditions and sense modalities falls along a straight line when the logarithm of perceived magnitude is plotted against the logarithm of stimulus magnitude. Three examples are shown here, perceived heaviness and weight of an object, perceived loudness and amplitude of a sound wave, and perceived brightness and intensity of a light source.

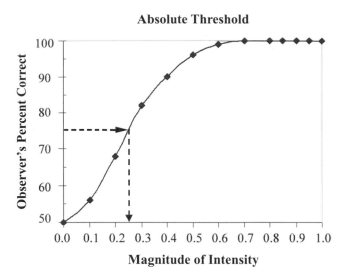

FIGURE 3.45. Plot of a hypothetical psycho-metric function such as might be obtained when measuring an absolute threshold using the methods described in Chapter 2.

form standards having different intensities. Recall the discussion of measuring absolute and difference thresholds in Chapter 2. In a hypothetical example, it was determined that the absolute threshold for seeing a light was 0.25. In other words, as the physical intensity of a comparison stimulus was increased from 0 to 0.25, the observer's ability to distinguish the comparison from a standard of 0 improved from 50% (chance) to 75% correct (Figure 3.45).

Suppose we were to measure another threshold, this time a difference threshold from a standard with the value obtained from this first measurement, i.e., from a standard having an intensity of 0.25. This procedure can be repeated in an iterative manner for several more cycles, adjusting the standard on each iteration to be at the difference threshold of the previous iteration. Hypothetical results

obtained from this procedure are illustrated in Figure 3.46.

Examination of this graph reveals that the size of a just-noticeable-difference (JND) threshold is not constant but increases as the standard is increased. Another way of stating this finding is that the size of JNDs as measured with a ruler varies along the horizontal axis. Weber's law asserts that JNDs can all be made the same size by a logarithmic transformation of the horizontal axis. This is illustrated graphically in Figure 3.47.

Weber's law can also be stated in the form of an equation:

$$K = JND/S,$$

where K is a constant called the **Weber fraction** and S is the intensity of the standard. Weber's law stated

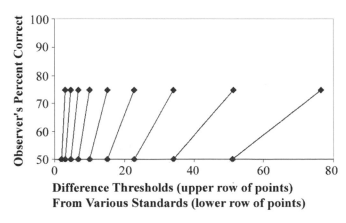

FIGURE 3.46. Plot illustrating hypothetical difference thresholds that might be obtained using different standards.

FIGURE 3.47. Same data as shown in Figure 3.46, but with the horizontal axis transformed by logarithmic scaling.

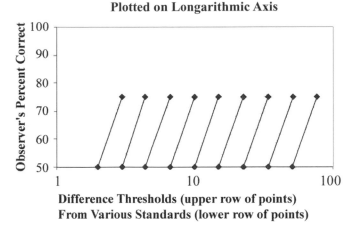

Difference Thresholds Plotted on Longarithmic Axis

Difference Thresholds (upper row of points)
From Various Standards (lower row of points)

in words is that the ratio between the JND and its standard is constant.

Fechner attempted to apply Weber's law to suprathreshold perception by using the following rationale: When a stimulus moves from an intensity of 0 to an intensity that corresponds to one JND above the absolute threshold, the magnitude of the perceptual change is 1 JND unit. When a stimulus moves from an intensity that corresponds to one difference threshold above the absolute threshold to one that is 2 JNDs above that threshold, this also results in a perceptual change of 1 JND unit, and similarly with an increase from an intensity of 2 to 3 JNDs, from 3 to 4 JNDs, etc.

Fechner argued that all of these changes should be perceptually equivalent, since each corresponds to one unit of perceptual magnitude. Additionally, Fechner argued that JNDs should be additive under suprathreshold conditions. For example, adding together four lights that are each 1 JND above the threshold should result in the same brightness as a single light that is 4 JNDs above the threshold. These assumptions, combined with Weber's law, lead to the Weber-Fechner law. Stated mathematically, the Weber-Fechner law states that perceived magnitude should be proportional to the logarithm of physical intensity:

$$P = k \log S,$$

where P is the perceived magnitude, S is the physical intensity of the stimulus, and k is the Weber fraction.

It is now known that the assumptions of the Weber-Fechner law do not hold, at least not under all conditions. In addition, studies performed over the past few decades have established that Stevens'

law is more accurate for perceptual scaling than the Weber-Fechner law under a variety of modalities and conditions. However, as a first-order approximation, results of the two laws are often indistinguishable if one restricts their use to a middle range of conditions, avoiding the extremes. Furthermore, the Weber-Fechner law is easier to implement as an engineering solution to problems of perceptual scaling. For this reason, the Weber-Fechner law remains in wide use by both scientists and engineers. For example, a company that is making dimmer switches is likely to adjust the relationship between turns of the knob and changes in photometric intensity to be approximately logarithmic.

Ecological Descriptions

Environments can be described from an **ecological perspective**, as emphasized by Gibson (see Chapter 1). This level of description is closest to our everyday notions of what is meant by the term "environment." Gibson argued that when we look around we do not see photons, rays, or spatiotemporal frequencies. Instead, we perceive trees, grass, the sky above, and the ground below.

Natural environments have some special properties. Consider the two stimuli shown in Figure 3.48 that might be used as stimuli in a study of perception.

Historically, observers in perceptual experiments have been asked to view artificial stimuli, as shown on the left, much more often than natural scenes, as shown on the right. The usual assumption has been that what was learned about perception of artificial stimuli could be generalized to perception of

FIGURE 3.48. The left side of the figure illustrates an artificial stimulus that might be used in an experiment on perception performed in a laboratory. The photograph on the right illustrates a natural scene that might be viewed by a primate.

natural scenes. However, there are some differences between artificial and natural environments even in some very basic properties, such as spatiotemporal frequency content. For example, natural environments are more regular than would be expected if the elements that make up environments were assembled randomly. They tend to have smooth transitions in space and time and less abrupt changes than are present in many artificial stimuli. Described in the frequency domain, natural images have a characteristic signature in which the amplitudes decrease as the spatial and temporal frequencies (f) increase as illustrated in Figure 3.49. Stated quantitatively the amplitudes decrease with $1/f^2$.

Such special properties of natural environments have been exploited by biological perceivers to reduce the computational power that is needed to extract perceptual information. Ecological perception has an advantage over general purpose perception because it does not have to process all information about the environment in one way. When confronted with complex spatiotemporal patterns of perceptual information that are ambiguous about "what is out there," biological organisms do not perform an exhaustive set of inferences about every possible object that might be present. Instead, biological perceptual processing is specialized to be efficient at picking up only certain types of information that reliably specify states of affairs

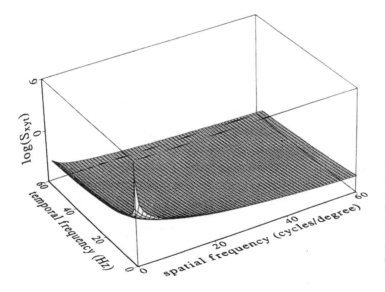

FIGURE 3.49. This figure illustrates the characteristic signature of natural environments when characterized in the spatial frequency and temporal frequency domains. Adapted from J.H. van Hateren, Spatiotemporal contrast sensitivity of early vision. *Vision Res.* 33:257–267, © 1993, by permission of Elsevier Science.

that matter to the organism. The philosopher **Andy Clark** calls this **niche-dependent perception;** it is defined by the parameters that matter to an animal with a specific way of life in its natural environment.

Engineering designs of perceiving robots have also had some success in exploiting niche-dependent perception, as described in Box 3.7.

Box 3.7
Herbert the Robot Perceives in a Special Niche

Herbert is a robot designed to function only within a coke-can-littered environment in the MIT Mobile Robot Laboratory. Herbert's computational load is substantially reduced by exploiting reliable facts that pertain to this niche: An entire world populated only by obstacles and tables with surfaces on which cans are sometimes found. Herbert does not have to perform representations of other objects. Also, Herbert can rely on niche-idiosyncratic facts, such as knowing cans sit on tabletops and ordinarily do not change their positions. The information built into Herbert about this niche comes from the roboticists at MIT who designed this robot. Biological observers also exploit built-in information that codes certain features of the environment that have been constant over evolutionary time. Chapter 11 discusses issues pertaining to how much perceptual information is built in and how much is based on experience.

Meaningful Environments

The descriptions of ecological environments in the last section illustrate an important general principle of perception that was emphasized by Gibson. An essential element of perception is that perceiving organisms stand in meaningful relationships to an external world. The act of perception is always carried out by an observer who engages in causal relationships with the environment in which perception takes place. For this reason, biological observers usually have little interest in evaluating all of the logical possibilities about "what might be

out there" based on the information provided by their sense organs. Instead, biological perceptual systems zero in on trying to answer questions about the environment that are meaningful: "Where is food?" "Who is a potential mate?" "When should I duck to protect my head from being hit by a flying object?"

When I look around my natural environment, I see

- the floor, a surface on which I know I can walk;
- a chair, an object on which I know I can sit;
- an apple, an object I know I can eat;
- a ledge outside the window of my 12th-floor apartment, a place on which I know it would be dangerous to stand;
- a friend I know will smile back at me.

These kinds of meaningful aspects of biological perception will need to be taken seriously by any engineer attempting to build a machine that perceives in a manner similar to the way humans perceive. It will not be sufficient simply to provide a machine with a computer program that operates on abstract entities. In order to "perceive," a minimal requirement will be that a processing system needs to be implemented in a physical system that has causal relationships with the world and thus must include sensors that can pick up meaningful information and motor outputs that can in turn act upon it.

Summary

Perception never occurs in isolation. Perception occurs in an environment that an observer samples with sense organs. The perceptual environment can be described from several perspectives. One level of description involves physical properties, which in the case of visual perception involves light, a form of electromagnetic radiation that exists throughout the universe. Another important level of description is derived from geometrical optics. Descriptions on this level characterize interactions between light rays and surfaces such as mirrors and lenses. A convenient way to describe spatial and temporal properties of environments in relation to many perceptual phenomena is from a specialized perspective called the frequency domain. Finally, environments can be described from an observer-centered perspective that takes into account properties of the environment that matter to a particular observer. These include descriptions in which magnitudes are scaled into convenient units based on perceptual scaling. Descriptions of an environment can sometimes be greatly simplified if restricted to a niche within which an observer has built-in knowledge about idiosyncratic properties. Biological perceivers are always engaged in meaningful causal relationships with their environments, and this aspect of perception will need to be taken into account in attempts to design and build perceiving machines.

Selected Reading List

Clark, A. 1997. *Being There: Putting Brain, Body, and World Together Again*. Cambridge, MA: MIT Press.

Commission Internationale de L'Eclairage. 1983. The Bases of Physical Photometry. Publication CIE 18.2 (TC-1.2). Commission Internationale de L'Eclairage, Paris.

Cornsweet, T.N. 1970. *Visual Perception*. New York: Academic Press.

Fechner, G.T. (1801–1887), 1966. *Elements of Psychophysics*, trans. H.E. Adler, ed. D.H. Howes and E.G. Boring, with an introduction by E.G. Boring. New York: Holt, Rinehart and Winston.

Gibson, J.J. 1979. *The Ecological Approach to Visual Perception*. Boston: Houghton Mifflin.

Perkowitz, S. 1996. *Empire of Light*. New York: Henry Holt.

Smith, G., and Atchison, D. 1997. *The Eye and Visual Optical Instruments*. Cambridge, UK: Cambridge University Press.

Southall, J.P.C. 1943. *Mirrors, Prisms and Lenses: A Text-book of Geometrical Optics*. New York: Dover.

Stevens, S.S. 1951. *Handbook of Experimental Psychology*. New York: Wiley.

Wandell, B. 1995. *Foundations of Vision*. Sunderland, MA: Sinauer Associates.

4
Sensing the Environment: What Mechanisms Are Used to Sample, Image, and Transduce Physical Stimulation from the Environment?

Questions

After reading Chapter 4, you should be able to answer the following questions.

1. What are the potential consequences if perceptual databases are not updated and maintained?
2. What is the function of perceptual filtering?
3. Describe the relationship between the individual samples of the environment being processed by a visual system at any moment and the observer's percepts.
4. Explain the function of eye scanning patterns and describe some of the smart processes involved in their guidance and control.
5. Describe how the neural eye movement control systems operate to allow us to move our eyes.
6. Describe the perceptual phenomena called entoptic images and explain how they can be accounted for by the gross structure of the eye.

7. How do visual disorders such as being near- or farsighted relate to the optical properties of the eye?
8. How does the distribution of photoreceptors across the back surface of the eye affect perception?
9. How is photopigment in the eye replenished and recycled?
10. Why do people have trouble seeing when they first walk into a movie theater on a sunny afternoon?
11. Describe the cascade of events in the photoreceptors of the eye that transduce light energy into electrical activity.
12. What is the smallest amount of light humans can see?

1 Maintaining a Perceptual Database

As we discussed in Chapter 3, the light energy that bombards our bodies carries with it abundant information about the properties of the surrounding environment. In this chapter, we will describe some of the mechanisms an observer uses for the initial sampling, imaging, and transducing of selected portions of that environment. These mechanisms provide the inputs that are used for perceptual processing. As we discuss these mechanisms, it is helpful to keep in mind the larger goal they serve. They allow an observer to keep her perceptual knowledge base of information about properties of the immediately surrounding environment up to date (Figure 4.1).

As captured by the colloquial adage used by computer programmers, "garbage in, garbage out,"

the quality of the output in any information processing system depends heavily on the properties of its inputs. A perceptual database is no different. In order to maintain fidelity, it must be updated whenever important changes occur in the immediately surrounding environment. Otherwise, it will soon become hopelessly out of date.

However, a huge amount of processing resources would be needed to track all changes in the surrounding environment in real time in an environment that changes rapidly. Perceptual systems do not have access to unlimited processing resources, and for this reason an attempt to monitor all changes in the environment simultaneously could easily lead to a deterioration in performance due to information overload. Consider, for example, the situation your visual system is confronted with while driving an automobile at a high rate of speed on a curving mountain road.

Filtering, a process by which only selected parts of the environment are sampled, is one mechanism that perceptual systems employ to help prevent information overload. Selective filtering allows limited processing resources to concentrate on the kinds of perceptual information that are deemed most important in the current situation. In the automobile example, the most important information probably has to do with the direction the road curves immediately in front of the automobile rather than, for example, the colors of individual flowers on the side of the road. Considered more generally, in the case of biological perceivers, the information that is deemed most important is that discovered over evolutionary time to be most relevant to survival. Selective filtering based on knowledge gained over evolutionary history is one important aspect of perceptual processing in biological organisms, including humans, that differen-

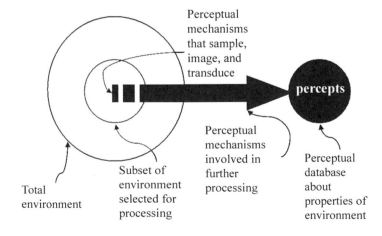

FIGURE 4.1. Perceptual systems are constantly sampling selected portions of the immediately surrounding environment. In vision, the rays of light that are selected for sampling are imaged onto the back of the eye and transduced into electrical signals that are processed by the nervous system. Ultimately, these signals are used to form percepts.

tiates it from the way it might be performed in a general-purpose perceptual machine.

Some portions of a person's immediately surrounding environment, like the regions behind the back, are sampled only occasionally, when he turns his head around. This accounts for the surprise when, for example, someone sneaks up from behind and says "Boo!" The surprise is due to the fact that the person's perceptual database about the properties of the surroundings behind his back had become out of date. Sampling mechanisms involved in perceptual processing need to try to avoid this kind of surprise. This involves making decisions about which information in the perceptual database is most likely to be out of date and thus have the highest priority for being updated.

Once this decision has been made, perceptual systems direct movements of the head and eyes so that the needed information can be sampled.

The highest bandwidth for visual processing is associated with the **direction of gaze,** also called the **line of sight,** the portion of the environment located in the direction the eyes are pointed. As we will discuss further below, objects in the environment that fall along the line of sight are imaged onto a specialized anatomical region at the back of the eye called the **fovea.** Psychologically, this portion of the environment is perceived as being **straight ahead** in the direction one is looking.

The bandwidth of visual processing decreases with eccentricity in all directions from straight ahead as illustrated in Box 4.1.

Box 4.1
Bandwidth of Visual Processing is Highest for Straight Ahead

Direct your gaze to the center dot in the left-hand panel of Figure 1. Without moving your gaze from this position, see how many letters you can identify. The letters located near the center of gaze are easy to identify, but those farther away become harder to resolve. Notice that there is a strong tendency to want to move your eyes away from the center when trying to resolve letters away from the direction of gaze. This is because the natural inclination of the visual system is to direct the eyes so that they point towards the intended object of regard. The right panel gives an indication of the quantitative decline in resolution as one moves outward from the direction of gaze. The letters in this panel have been scaled so that they are about equally resolvable. This panel illustrates that the bandwidth of the visual system allows detailed information to enter from the direction of gaze but only coarse information from the periphery.

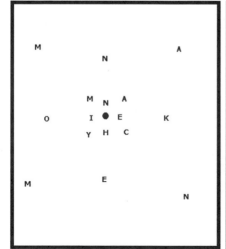 

FIGURE 1. This figure illustrates the decrease of processing bandwidth as a function of eccentricity.

The visual system continuously monitors the low-resolution input coming in from the periphery of the visual field and makes decisions about which portions of the environment need to be sampled most intensively at the next instant. Orienting and scanning activities are then programmed to reposition the eyes so that the higher-bandwidth processing of the fovea can be brought to bear on those portions of the visual field.

Orienting and Scanning Activities

Orienting and scanning movements of the head and eyes serve to bring the image of the intended object of regard onto the fovea. It is interesting as a his-

---

Box 4.2
Perceptual Learning and the Law of Effect

Classical psychology theories of learning were based on the law of effect. Behavior causes changes to the environment, which in turn lead to effects in the form of positive or negative reinforcement. Behaviors that lead to positive reinforcement are supposed to increase in frequency, and those leading to negative reinforcement should drop out. However, the law of effect cannot operate if behavior does not cause a change in the environment. Thus, theories of learning based on the law of effect had trouble accounting for empirical findings demonstrating that learning can take place when animals are simply allowed to explore. Furthermore, animals will work in order to be allowed to explore, even when the exploration has no apparent rewarding consequences. These kinds of facts led some psychologists to propose that perception is intrinsically motivating. Issues of perceptual learning go beyond the scope of our interests here, but it is worth noting that most biological perceivers appear to have a natural propensity to want to explore/perceive as well as manipulate their surroundings. This aspect of biological perception will not be mimicked by any machine, or model of perception, that simply passively processes incoming sensory inputs.

---

torical note that classical psychological theories of learning found it difficult to explain why these **perceptual activities** should occur. This issue is discussed in Box 4.2.

Exploratory activities go on continually during ordinary perception, usually without conscious effort or awareness. They can be measured with **scan patterns** that superimpose a record of where an observer's eyes are pointed onto the scene being observed, as illustrated in Figure 4.2.

An observer was instructed to look at the photograph on the left. The observer's direction of gaze was measured continuously and plotted with a black ink trace, as shown on the right. This scan pattern reveals that the environmental stimulus being examined is not scanned uniformly. Certain portions, such as the eyes, nose, and mouth, are sampled extensively, while others are sampled only coarsely or not at all.

Relationships of Samples to Percepts

These facts about orienting and scanning activities have strong implications regarding the relationships between samples being processed by the visual system at any moment and percepts. Consider the hypothetical room shown schematically in the left panel of Figure 4.3.

Suppose you were to walk into this room and look around. If we were to examine your scan patterns, we would discover that at any one instance only a small portion of this scene was being processed in detail by your visual system. At one moment, the direction of gaze might be directed towards the bottom of the chair. The high-bandwidth information traveling from the fovea to the central brain processing areas at that moment would concern the details illustrated in the top panel on the right. This would likely last for only a fraction of a second. Then your eye would jump to a new location, perhaps to the top of the chair, as illustrated in the second panel on the right, and so on. On average, human observers take in about three samples per second while scanning a scene.

Now consider what your percept of the room would be like. It would have little in common with the fleeting images taken in during each short fixation period. Instead, your perception would be of a stable room. This example highlights the homunculus fallacy. Even if there were a homunculus sitting in your brain looking at the image being processed at any one instant, the percept of the homunculus

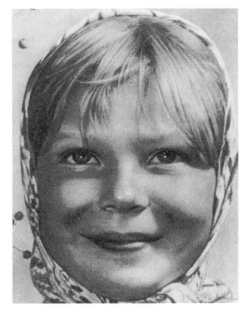
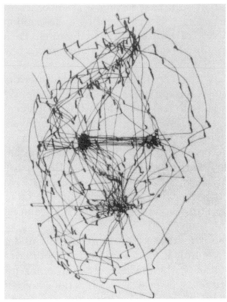

FIGURE 4.2. The panel on the right shows a scan pattern of an observer viewing the face in the left panel. Adapted from A.L. Yarbus, 1967, *Eye Movements and Vision* (translated by B. Haigh), © 1976, by permission of Plenum Press.

would be something like the sequential panels shown in the right side of the figure. These percepts of the homunculus have little or no explanatory value regarding the observer's percept, which is of a stable external environment consisting of a continuous, structured room. Similarly, it is a fallacy of some perceptual theories to describe the sampled images' being processed, whether from the level of description of information processing or that of neuroscience, as though these were sufficient to provide an explanation for percepts.

How Smart Processes Guide Scans

Observers looking at real world scenes are generally unaware that they are taking in information sequentially via scanning patterns. Instead, they usually have the impression that they see the entire scene in detail. However, experimental studies reveal that this is not the case. The perception of a detailed world is partly an illusion supported by the ability to revisit any part of the scene. The gist of a scene can be determined within

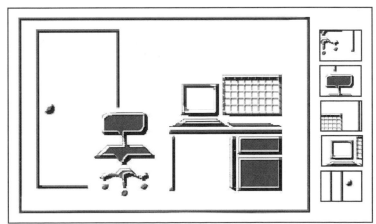

FIGURE 4.3. The left panel shows a room that contains a door, a chair, a desk with a computer sitting on it, and a calendar on the wall. The right panel shows five successive images formed on the fovea of an observer looking around in this room.

a single short sampling lasting on the order of 0.1 sec. However, this gist includes a description of only a few of the most salient features. Each subsequent sample takes in, or updates, only a few additional features.

If this low-capacity sampling mechanism were guided by dumb processes, it would seriously limit our ability to maintain a high-fidelity perceptual database. However, in biological perceivers this sampling mechanism is not as limiting as it may seem, because it is guided by smart mechanisms. Attention is normally drawn to those parts of the scene most relevant to the task at hand. Two kinds of processes, one **top down** and the other **bottom up,** are responsible for guiding scanning patterns.

The top-down process is goal directed and under voluntary control. It allows us to direct our eyes to portions of the scene that are important for specific purposes, even when such actions are not automatically triggered by bottom-up processing of stimulus features. This top-down process may be responsible for certain kinds of perceptual learning that are involved in specialized skills. An example might be the way a professional race car driver learns to focus attention exclusively on the features of the environment that are critical for steering. Top-down effects can be very powerful. For example, studies have demonstrated that under some conditions in which observers give complete attention to particular objects or events in a scene, they become essentially blind to irrelevant objects.

The bottom-up process is stimulus-driven by properties of the scene. Some stimulus properties automatically grab the attention of biological observers. For example, some small mammals respond automatically to the silhouette of a hawk passing overhead. These kinds of mechanisms evolved in biological organisms to deal with conditions in the environment that have important consequences if missed. These specialized bottom-up processing mechanisms allow observers to engage in **niche-dependent perception** as described in Chapter 3. A general-purpose perceiving machine unable to make the same assumptions about "what is likely to be out there that is important to me" cannot gain the benefits of niche-dependent perception and must carry out exhaustive item-by-item scans of the entire scene to glean the same information.

One low-level feature that is highly salient for most biological perceivers, including humans, is a sudden change in the environment. For example, the appearance of a new object in the scene will usually elicit an eye movement in its direction even in the absence of voluntary control or awareness. This scanning eye movement allows the perceptual database to be updated about the change. The fact that this happens so automatically, and without conscious awareness, reinforces the impression that ones sees everything in the scene, rather than just the information that grabs the attention and leads to scanning eye movements. However, experimental manipulations demonstrate that this impression is mistaken. Even a major change in the immediate surroundings will fail to be noticed if it takes place at a location away from straight ahead and does not trigger focused attention near the time it occurs.

One experimental paradigm used to demonstrate this phenomenon in the laboratory involves production of flicker by introducing brief blanking fields between successive frames of a changing scene. Figure 4.4 illustrates the use of this procedure.

Each frame presents a scene that lasts for 240 ms. Frames are separated by blank periods that last 80 ms. Thus, observers are presented with a new sample of the scene every 320 ms, at about the same frequency as occurs when viewing natural scenes with scanning patterns. After every two frames, a change is made in the scene. In the example shown in the figure, a statue is positioned in front of a wall in the first two frames, but the wall is removed in the subsequent two frames. This change in the scene is obvious to observers when they become aware of it or if they are instructed ahead of time where to focus their attention. However, when observers are instructed simply to view the scene and report any changes, most do not detect the change until after many alternations – some not until the sequence has been repeated more than 80 times over a period of 50 seconds. On the other hand, when the stimulus frames are not separated by blanks, the change is usually detected immediately on the first alternation. The explanation for these laboratory findings is that the flicker from the blanks swamps the stimulus-driven bottom-up process, which evolved to operate during ordinary conditions, under which only a small number of temporal changes typically occur in a scene at the same time.

These experimental results demonstrate that the processes used by biological organisms to detect changes in the environment are smart but not infallible. Biological predators have also perhaps learned to try to exploit weaknesses in these processes, as described in Box 4.3.

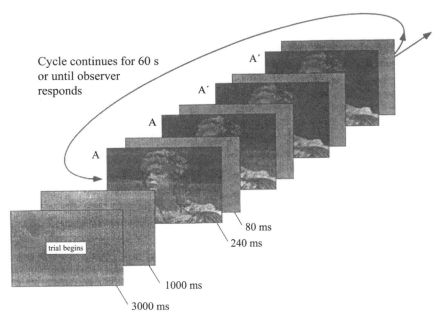

Cycle continues for 60 s
or until observer
responds

80 ms

240 ms

1000 ms

3000 ms

FIGURE 4.4. This figure illustrates a procedure to introduce blanking fields while an observer views a scene.
Details are described in the text. Adapted from R.A.

Rensink, et al, To see or not to see: The need for attention
to perceive changes in scenes, *Psychol. Sci.* 8:368–373,
© 1997, by permission of Blackwell Publishers.

Box 4.3
How Does a Cat Catch a Bird?

Take the next opportunity you have to watch
a cat stalking a squirrel or a bird. Note how
it approaches the prey. The cat makes quick
movements of a few steps and then holds
perfectly still, sometimes for many seconds.
If the potential prey notices the cat during
the quick movements, as usually happens, it
flees. However, if the perceptual database of
the prey is not updated during the quick
movement, the prey will continue whatever
it is doing, seemingly unaware that something dangerous is now part of its immediately surrounding environment. In order to
capture the prey, the cat simply needs to
play the statistical odds of obtaining enough
sequential episodes of undetected movement
to get within striking distance.

Perceptual Versus Memory Databases

An unresolved issue is that of where perception
ends and memory begins. This issue has to do with

whether the information in the perceptual database was obtained an instant ago, a few
seconds ago, an hour ago, or perhaps 20 years
earlier. Consider what happens when you close
your eyes. The perceptual experience of your
immediately surrounding visual environment stops
quickly, but the perceptual/memory database
about the immediate surroundings persists for a
longer period – it is simply no longer being
updated.

An extreme conservative approach taken by
some perceptual theories is to limit the term "perceptual database" to only the information causing
one's immediate perceptual experiences. This conservative approach would argue that another term,
such as "memory database," should be used when
referring to any information that was obtained
prior to the current moment.

An extreme liberal approach taken by other perceptual theories is that the time elapsed since the
information was entered into the database is irrelevant and that all information entered via the sense
organs is perceptual. This extreme liberal position
is illustrated with an example in Box 4.4.

Over the past 30 years or so, cognitive scientists
have argued for a plethora of storage mechanisms,
with names like "iconic storage" and "short-term
visual store," that fall somewhere in between pure
perception and pure memory. Those distinctions go

Box 4.4
Did Sacagawea Perceive or Remember the Route to the Northwest Territories?

When the historical explorers Lewis and Clark went on their expedition to explore the Northwest Territories, they took along a Native American guide, Sacagawea, because "she knew the rote." Sacagawea had not been in the Northwest Territories since childhood. The extreme liberal position would argue that Sacagawea had knowledge about the route based on a perceptual database last updated during childhood. The fact that the database happened to be several years out of date presumably did not matter much because the major features Sacagawea used to identify the route had not changed appreciably in the intervening years.

2 Eye Movements

Eye movement control systems play a critical role in the ability to carry out scanning patterns and orienting movements that allow observers to sample the environment efficiently. Usually eye movements are performed in coordination with movements of the head and trunk. However, muscles attached to the eye also allow primates to sample the environment without moving the head or body.

Extraocular Muscles Rotate the Eyeball Within the Orbit

Each eyeball sits in a cavity in the skull called the **orbit.** It is sheathed in connective tissue and is able to rotate in the orbit like a ball in a ball-and-socket joint. Six extraocular muscles attach between the eyeball and the orbit, as illustrated in Figure 4.5.

The extraocular muscles are organized into three pairs, with each member of a pair having roughly opposing actions: **lateral and medial rectus, superior and inferior rectus,** and **superior and inferior oblique.** A cut portion of the **levator muscle** is also shown. This muscle passes over the top of the eyeball and attaches to the eyelid (not shown).

The place where each muscle attaches to the skull inside the orbit is called its **origin.** Its point

beyond the scope of this book. A pragmatic approach is to use the general term "perceptual database" to refer to any information about the immediately surrounding environment that has been updated by sensory input within an interval on the order of seconds.

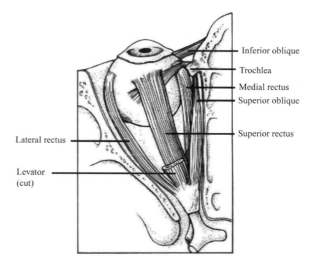

FIGURE 4.5. View of the eyeball and attached muscles in the orbit. The left panel shows the left eye as seen from the left side of the head with the lateral wall of the orbit removed. The right panel shows the same eye as viewed from the top with the roof of the orbit removed. Adapted from E.R. Kandel, et al, *Principles of Neural Science,* Third Edition. Norwalk, CT: Appleton & Lange, © 1992, by permission of McGraw-Hill Companies.

of attachment to the eyeball is its **insertion**. All of the extraocular muscles except the inferior oblique originate at the back of the orbit. The inferior oblique originates at the front of the orbit floor on the nasal side. All of the muscles follow an essentially straight course from the origin to the insertion except the superior oblique. It passes through a structure attached to the medial wall of the orbit called the **trochlea** that acts as a pulley.

By examining the origin and insertion of each muscle, it is possible to discern what effects that muscle is likely to exert on the movements of the eye. There are three principal axes of eye rotation (Figure 4.6).

Lateral movements consist of **abduction,** away from the nose, and **adduction,** towards the nose. **Vertical movements** are either **elevation,** upward, or **depression,** downward. Finally, **torsional** movements allow the eye to rotate over small angles around the direction of gaze. **Intorsion** involves rolling towards the nose and **extorsion** away from the nose.

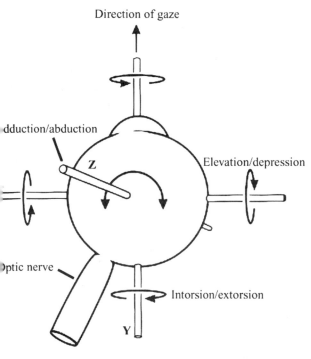

FIGURE 4.6. Eye movements can be characterized in terms of rotation around three axes, designated in this figure as x, y, and z. Adapted from E.R. Kandel, et al, *Principles of Neural Science,* Third Edition. Norwalk, CT: Appleton & Lange, © 1992, by permission of McGraw-Hill Companies.

The lateral rectus attaches to the outside of the eyeball and is primarily responsible for abduction. The medial rectus attaches to the inside of the eyeball and is primarily responsible for adduction. The superior rectus produces primarily elevation and the inferior rectus primarily depression. The inferior oblique and superior oblique assist with these lateral and vertical movements and also help stabilize the eye with respect to torsional movements.

The actions of the muscles just described are overly simplified, because they apply only when initiated while the eye is aiming straight ahead. More generally, their true actions vary depending on the position of the eye in the orbit. The computational problem confronting the brain in figuring out what contractions of specific eye muscles are needed to accomplish moving the eye from one position in the orbit to another is nontrivial. The interested reader seeking more detailed information is referred to Von Noorden's text that appears in the reading list at end of the chapter. Ophthalmologists need to understand these details for practical reasons, to perform eye muscle surgery to compensate for neurological defects.

There is a minor difference between the extraocular muscles of monkeys and those of humans that may have functional implications, as described in Box 4.5.

Brainstem Nuclei Control Extraocular Muscles via Cranial Nerves

One motor neuron axon innervates about 10 extraocular muscle fibers, compared to about 140 muscle fibers in skeletal muscles. This heavy innervation facilitates precise neural control of eye muscles. Innervation of the extraocular muscles comes from axons carried by three cranial nerves that originate from the **oculomotor, trochlear, and abducens nuclei** in the brainstem (Figure 4.7).

The **abducens nerve (VI)** innervates the lateral rectus, the **trochlear nerve (IV)** the superior oblique, and the **oculomotor nerve (III)** the other extraocular muscles.

The Four Major Eye Movement Control Systems

In order to take in visual information from the environment, the brain needs to organize the actions of the eye muscles. When a person looks around or performs scanning patterns, the brain has to rotate

Box 4.5
Why Are Humans So Susceptible to Having Crossed Eyes?

About 3% of the human population is afflicted with a clinical ophthalmologic disorder called **strabismus,** in which the two eyes do not point in the same direction. Strabismus is seldom seen in monkeys, and one possible explanation for the difference in prevalence between humans and monkeys has to do with differences in eye muscles. Figure 1 illustrates an extraocular muscle called the accessory lateral rectus that is present in monkeys but has been lost over evolutionary time in humans.

This muscle is thought to be too small and weak to have much effect on monkey eye movements, but it could play a role in stabilizing the eye in the orbit. Based on its origin and insertion, its major effect is expected to be counteracting drifts of the eye towards the nose. An interesting observation in this regard is that in the vast majority of cases of human strabismus with their onset in infancy, the child is cross-eyed rather than wall-eyed.

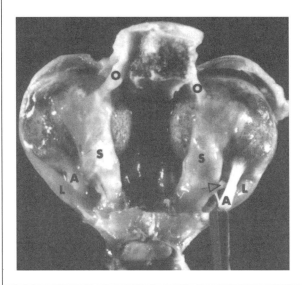

FIGURE 1. This photograph shows the orbits of a monkey cadaver head in which the top of the skull and brain were removed to expose the eyeballs and extraocular tissue. Muscles that can be seen in this view from above are the superior obliques (O), superior recti (S), and lateral recti (L) and the accessory lateral recti (A), which are not present in humans. The arrowhead points to the optic nerve of the right eye. Reproduced from R.G. Boothe, et al, Accessory lateral rectus orbital geometry in normal and naturally strabismic monkeys. *Invest. Ophthalmol. Vis. Sci.* 31:1168–1174, © 1990, by permission of the Association for Research in Vision and Ophthalmology.

the eye quickly from one location to the next to allow the fovea to take samples. While a sample is being taken, however, its image needs to be kept on the fovea and stabilized, even when the object or the head moves. These functions are accomplished by four interacting eye movement control systems organized by higher-order brain centers that are functionally and to some extent anatomically separate (Figure 4.8).

These four centers operate, at their final common output, on a common pool of motoneurons in the oculomotor, trochlear, and abducens nuclei in the brainstem. The **saccadic system** serves to generate eye movements called **saccades** that quickly rotate the eye. It can be used to move the direction of gaze rapidly from one location to another. The **fixation and smooth pursuit system** allows one to maintain the image of a desired object of regard on the fovea,

even in the presence of drift created by slow movements of the object, the body, or both. The **retinal image stabilization system** is primarily reflexive in nature and assists with maintaining a stable image when the head moves. The **vergence control system** coordinates binocular movements of the eyes when the gaze moves between far and near locations. Discussion of the vergence system is deferred until Chapter 9, which covers issues of binocular vision.

Saccadic System

The saccadic system generates all fast eye movements, including those that move **fixation,** defined as a willfull pointing of the direction of gaze towards an intended object of regard, from one

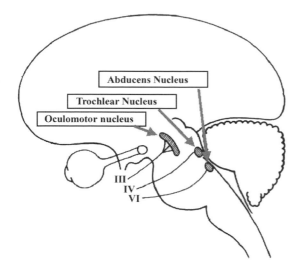

FIGURE 4.7. Locations in the brain of the nuclei (abducens, trochlear, and oculomotor) and cranial nerves (III, IV, and VI) that provide innervation to the extraocular eye muscles. Adapted form J. Lang, *Strabismus*, Slack Incorporated, 1984, by permission from J. Lang.

The saccadic system involves a number of different neural structures in the brain (Figure 4.9).

The **superior colliculus** in the midbrain can generate spontaneous saccades, i.e., in the absence of volition. The **posterior parietal cortex** participates in visually guided intentional saccades. The **frontal eye fields,** in conjunction with the **supplementary eye fields** and **dorsolateral prefrontal cortex,** appear to be associated with top-down volitional control of saccades independent of stimulus properties. These cortical areas feed down to a brainstem saccadic generator, directly in the case of the frontal eye fields and also indirectly, via the superior colliculus.

It takes a human about 200 ms to generate a refixation saccade in response to a new stimulus. The velocity of saccades depends on their size, increasing proportionately with distance up to a maximum of about 800°/sec. The neurological system that controls saccades operates continuously. However, as each individual saccade is initiated, it typically operates in a **ballistic** manner, meaning that it continues to completion without alteration. If the neurological control system determines that the size of a completed saccade, or of a saccade in progress, is in error, it programs a new, "corrective" saccade.

In a perceptual phenomenon called **saccadic suppression,** visual sensitivity for seeing during a saccade is lowered. Some of this decreased visual sensitivity can be accounted for on the basis of the blurring and masking of the retinal image as it moves rapidly across the fovea. However, there is also evidence for a neural suppression mechanism that acts to decrease sensitivity further during saccades. The presumptive function of neural saccadic suppression is to prevent the blurry, moving retinal image present during the saccade

location to another during scanning patterns. The saccadic system is also responsible for fast phases of eye movements associated with some forms of **nystagmus.** Nystagmus is an involuntary oscillatory motion of the eye with a fast component going in one direction and a slow component in the opposite direction. Nystagmus is present in some neurological disorders and can also be triggered in normal subjects under certain environmental conditions. The saccadic system is also responsible for generating the **rapid eye movement (REM)** characteristic of the dream cycle of sleeping.

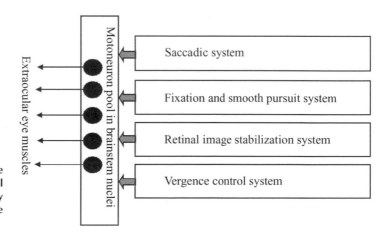

FIGURE 4.8. Schematic illustrating the four separate eye movement control systems, which operate by selectively activating a pool of motor neurons in the brainstem nuclei.

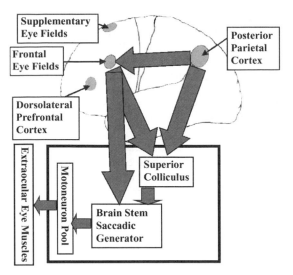

FIGURE 4.9. Schematic illustrating some of the major brain regions and pathways involved in regulating saccadic eye movements. Cortical areas in the parietal lobe (posterior parietal cortex) and the frontol lobe (supplemental eye fields, frontal eye fields, and dorsolateral prefrontal cortex) interact with one another in planning volitional saccades. The superior colliculus in the midbrain helps carry out volitional saccades and is also involved in the planning of automatic saccades. Execution of saccades is coordinated by a diffuse set of neural circuits in the brainstem, referred to collectively as the brain stem saccadic generator. The final neural pathway from the brain to the extraocular muscles passes through a motoneuron pool that is distributed across the abducens, trochlear, and oculomotor nuclei.

from interfering with processing of samples of the environment obtained during the fixation periods between saccades. Because saccades typically last only 20–50 ms, vision is affected for only a short time compared to the 200–300-ms intersaccadic fixation period. Issues of saccadic suppression are discussed further in Chapter 5, in relation to neural circuits in the thalamus that process visual information.

Fixation and Smooth Pursuit System

The fixation and smooth pursuit system is used to maintain a fixed relationship between the fovea and an intended object of regard in the environment. When the object of regard is stationary in the environment, the typical movements produced are called **fixation movements**. When the object is moving in the environment, these tracking move-

ments are referred to as **smooth pursuit movements**. The neural circuits involved in controlling fixation and smooth pursuit differ in some of their details, but they largely overlap and will be treated as one system here.

Feedback regarding slip of the object of regard's image from the fovea is monitored continuously. This feedback is used to match eye and target motion to maintain the image on the fovea. Pursuit control can respond to this feedback with a latency of about 125 ms. Smooth pursuit tracking movements can reach velocities up to 100°/sec but have a linear relationship target velocity only up to about 40°/sec. In addition to feedback, the smooth pursuit system uses a smart, predictive mechanism to improve tracking accuracy.

When the velocity capabilities of smooth pursuit are exceeded, or when the image slips off the fovea for other reasons, saccades intrude to place the target back on the fovea. Then smooth pursuit continues. Many physiological conditions, such as fatigue, alcohol, and other drugs, can degrade the quality of smooth pursuit movements.

Smooth pursuit is subject to voluntary control in terms of deciding which object of regard, if any, to track. However, it is generally not possible to generate smooth pursuit in the absence of a target to track, except as described in Box 4.6.

Box 4.6
Voluntary Smooth Pursuit as a Parlor Trick

While standing in front of a mirror, hold your finger in front of your eye and slowly move it back and forth while looking at it. You will be able to see your eyes smoothly moving while tracking the finger. Now remove your finger and try to move your eyes in the same motion as you had previously used to track your finger. You will most likely discover that it is impossible. The eyes, with no target to follow, will now move in a typical scanning pattern involving saccades interspersed with short fixations. However, with enough practice some individuals can learn to generate voluntary smooth pursuit. Those individuals have in their arsenal an amazing parlor trick, similar to the ability to wiggle one's ears.

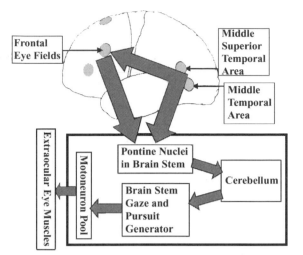

FIGURE 4.10. Schematic illustrating the major brain regions involved in regulating smooth pursuit movements. Extrastriate cortical areas (middle temporal area and middle superior temporal area) and the frontal eye fields located in the frontal lobe interact with one another in planning and controlling volitional and predictive smooth pursuit movements. Execution of smooth pursuit involves the cerebellum and brain stem areas including the pontine nuclei and a diffuse set of neural circuits referred to collectively as the brain stem gaze and pursuit generator. The final neural pathway to the extraocular muscles passes through a motoneuron pool distributed across the abducens, trochlear, and oculomotor nuclei.

Neural control of predictive and voluntary smooth pursuit involves cortical circuits located in the frontal eye fields and two visual processing areas, called **middle temporal (MT)** and **middle superior temporal (MST),** that are heavily involved in visual motion processing as well as in eye movement control (Figure 4.10). The kinds of motion processing that takes place in these cortical areas are discussed in more detail in Chapter 10.

The cortical areas project to pontine nuclei in the brainstem, which in turn project to nuclei in the cerebellum. The cerebellum interacts with brainstem nuclei to generate stable gaze during fixation or pursuit movements.

Retinal Image Stabilization System

Additional eye movement control systems assist with stabilization of the retinal image. These systems are reflexive, not under voluntary control, and they operate continuously whether or not the observer is attempting to fixate or engage

in smooth pursuit. The feedback that drives these systems includes nonvisual signals as well as retinal slip.

Nonvisual signals come from the **semicircular canals** and the **otoliths** of the inner ears. The semicircular canals respond to angular acceleration of the head and the otoliths to linear acceleration. The stabilizing eye movements driven by inputs from these structures result from the **vestibuloocular reflex.** This reflex is very effective at stabilizing drift during brief head movements but cannot compensate for sustained constant-velocity motion. Its latency in responding to a head movement is very fast, on the order of 15 ms.

The neural circuitry associated with the vestibuloocular reflex has to generate two kinds of signals. The first is an **eye-velocity signal** that can compensate for head motion. The second signal that must be generated is just as important but not as apparent. Once the eyes have moved to any eccentric position in the orbit, they have a natural tendency to drift back to look straight ahead, the physiologic position of rest. To keep the eyes in the new position generated by the eye-velocity signal, a **position signal** must also be generated to hold the eye in the new position. Computationally, the magnitude of the position signal that is needed can be obtained by performing mathematical **integration of the velocity signal.** Neural network models have been developed to demonstrate how neural circuits might be able to solve both of these computational problems, as described in Box 4.7.

The neural network described in the box is able to generate a combined velocity and position signal that minimizes eye position error when it is sent to the oculomotor plant. This is an example of a dynamic neural network model, in which the input consists of time-varying signals and the output needs to be coded in patterns that can be used to produce a desired movement.

Neural networks are dealt with further in Chapter 6 and specifically with dynamic models in Chapter 10.

The stabilizing eye movements driven by visual inputs are controlled by the **optokinetic reflex.** This reflex is organized by brainstem structures that receive direct input from the retina and additional information from cortical visual processing areas of the brain. It has a longer latency in responding to head movement than the vestibuloocular reflex but responds in a sustained manner even during prolonged constant-velocity head motion. The longer latency is due to the time it takes for the information that the retinal

Box 4.7
Using a Neural Network to Model the Vestibuloocular Reflex

To permit the vestibuloocular reflex to operate, the brain must transform the input signals that come from the vestibular system into the patterns of innervations that need to be sent to the oculomotor neurons. A neural network model that simulates these activities, labeled according to biologically relevant operations, is illustrated in Figure 1.

The bottom trace shows the horizontal head position H in degrees as a function of time. The desired output, shown at the top of the diagram, is the horizontal eye position E in degrees as a function of time that compensates for the head rotation. The input to the network is a signal from the left, L, and right, R, semicircular canals that encodes shorterm changes in head velocity by changes in the rate of electrical firing, ΔRv_1, around a background rate R_0. This is shown in the second row from the bottom. Note that the magnitude of the change in firing rate is equivalent to the derivative of the change in head position, $\dot{H}$. The neural output needs to be in the form of a modulation ΔRm around the background rate, R_0, of the electrical signals sent to the motoneurons

of the extraocular muscles, that is proportional to the sum of desired change in eye velocity, $\dot{E}$, and new position, E. The velocity signal is easy to compute, since it is just the reversal of the input signal. It can be computed by a three-neuron loop modeled after known anatomical connections from the semicircular canal to the **medial vestibular nucleus (MVN)** to motoneurons in the abducens nucleus (**6**). This three-neuron circuit from the right circular canal to motoneurons innervating the lateral rectus muscle (L) is illustrated by the thick line in the Figure. A signal of the same magnitude, but reversed in sign, is transmitted from the left canal to the medial rectus muscle (M) via the oculomotor nucleus (3). The eye position command is more complicated, and must be derived from the eye-velocity signal by integration with respect to time. Figure 1 illustrates the connections of a neural network model that has been demonstrated to be able to accomplish this function. These connections are modeled after known anatomical connections in the MVN and in the **nuclei propositus hypoglossi (NPH)** in the monkey.

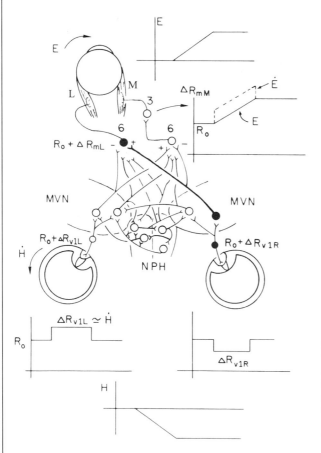

FIGURE 1. A neural network that has been demonstrated to be able to perform the computations needed for the vestibuloocular reflex. Adapted from D.A. Robinson and D.B. Arnold, Creating pattern generators with learning neural networks. The Society for Neuroscience Short Course 3 Syllabus, *Neural Computation*, St. Louis, MO, 1990, by permission of the Society for Neuroscience.

image has moved to travel from the eye up the optic nerve to central brain processing areas. Box 4.8 describes a way to demonstrate the fact that the vestibuloocular reflex has a shorter latency of responding to sudden motion.

Box 4.8
The Vestibuloocular Reflex Responds Faster Than the Optokinetic Reflex

Rapidly move this page up and down while you attempt to read. The text is blurred because the optokinetic reflex, responding only to retinal slip, takes time to respond. Now hold the text still and at the same speed as previously, move your head up and down. Reading is now easy. The image stabilization system can compensate better in the latter case because it now has access to short-latency input about head movement from the vestibular system.

Obviously, the optokinetic reflex response cannot continue compensating continuously over sustained periods. Consider what happens to an observer seated in a swivel chair who starts to rotate in one direction. The optokinetic and vestibuloocular reflexes both start to operate due to the initial acceleration.

As the observer continues to rotate at a constant velocity, the vestibuloocular response abates, but the optokinetic response continues trying to compensate for the retinal slip being produced by the rotation of the head. Soon, however, the eye has rotated as far as it can go in the orbit. As a result of this physical limitation, the optokinetic nystagmic reflex consists of two alternating phases. In one, the eye rotates smoothly in the orbit to compensate for retinal slip. When the eye has rotated as far as it can move comfortably, a phase of fast movement in the opposite direction, under the control of the saccadic eye movement system, returns the eye towards the center of the orbit. The rhythmic alternation between these two phases creates a characteristic nystagmic response.

People do not ordinarily appreciate the functional significance of the retinal image stabilization systems, because they operate effortlessly without conscious effort. Only when they fail to function properly, as in patients with neurological disorders, is their functional utility made apparent. Patients with these disorders can have difficulty reading signs or even recognizing people while engaged in ordinary activities such as walking or riding in an automobile over rough terrain. The functional significance of stabilizing image motion on a sensor is also appreciated by engineers designing physical devices, as illustrated in Box 4.9.

Box 4.9
Image Stabilization in Machines

Perhaps you have watched home movies taken with an old-fashioned camera that did not include a mechanism to compensate for image motion. The jittery moving picture captured with these devices can be unpleasant to watch unless the person operating the camera made a special effort to avoid sudden movements. Modern technology has now borrowed from biological vision, and most current video cameras include a mechanism that samples the motion of the image. Whenever transient motion of the entire image is detected, this is assumed to reflect movement of the camera rather than of the world. The output of the camera is the result of algorithms that compensate for some of this motion in order to reduce jitter in the image.

---

3 The Eyeball

The eyeball is a critical accessory structure for vision that enhances the kinds of information about the environment that can be sampled and processed.

Gross Structure

Human and monkey eyeballs are virtually identical except for size. Figure 4.11 illustrates some of their main structural components.

A right eye is illustrated as it would be seen if sliced in half along the horizontal meridian and

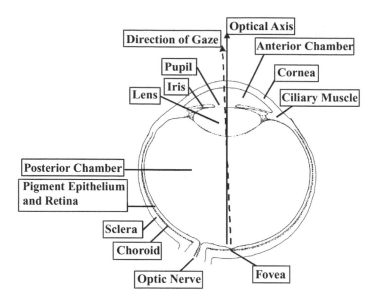

FIGURE 4.11. Illustration of the major components of the eyeball.

the top removed. The view is from the top into the remaining eye cup. The front of the eye is at the top of the figure; the side facing towards the nose, called the **nasal side,** is at the left; and the side towards the ear, called the **temporal side,** is at the right.

The outer white covering of the eye is the **sclera.** The **choroid** layer beneath the sclera is opaque. Its function is to help reduce glare, both by reducing the amount of extraneous light passing into the eye through the sclera and by reducing internal reflection of light that enters through the pupil. Some animals with eyes specialized for operating under dim light conditions have an additional specialization to deal with reflected light inside the eye, as described in Box 4.10.

The **pigment epithelium** layer lies between the choroid and the retina and performs certain metabolic functions for the retina. In addition, cells of the pigment epithelium are filled with a black pigment, **melanin,** which helps to reduce light scatter further inside the eye. The neural **retina** forms the innermost layer of tissue that surrounds the **posterior chamber.**

Light rays enter the front of the eye through the **cornea** and then the **anterior chamber,** which is filled with a liquid called the **aqueous fluid.** The back of the anterior chamber is covered by the **iris,** which contains a pigment responsible for the characteristic color of an eye, such as blue, brown, or hazel. The iris is a smooth sphincter muscle with a small circular opening called the **pupil** in its center. The iris acts as a diaphragm to control the amount

Box 4.10
Why Do Some Animals' Eyes Shine in the Dark?

You have perhaps had the experience while driving along a country road at night with your headlights on and seeing an animal on or alongside the road. The most prominent feature you can see is the eyes, which seem to shine in the dark. Animals that see well at night have eyes in which a tradeoff has been made; some resolution has been sacrificed in order to achieve maximum sensitivity to dim light. These animals have a specialized structure, called a **tapetum,** located behind the retina that reflects light. This decreases resolution of the eye because it is impossible for the brain to determine whether light that is detected came directly from the pupil or was reflected within the eye first. However, this mechanism increases sensitivity compared to eyes like ours, in which extraneous light is mostly absorbed by the choroid. Any light not detected when it first reaches the retina has additional chances of being detected after being reflected by the tapetum. Some of the light that is reflected by the tapetum passes back out of the eye through the pupil, and this is what causes the eyes to shine in the dark.

of light that passes through the pupil to the interior of the eye. Neural control of the pupil comes from neurons in the **ciliary ganglion,** which are in turn innervated by **parasympathetic** components of **cranial nerve III.** Brainstem structures such as the **Edinger-Westphal nucleus** send inputs to the ciliary ganglion and allow the **pupillary reflex** to control the amount of light that enters the eye based on ambient lighting conditions.

The iris also receives input from neural circuits involved in the **near reflex,** which respond differentialty depending on whether one is looking at a faraway or a near object. This input involves both parasympathetic and sympathetic influences, and because of this, the pupil responds to psychological conditions such as state of arousal as well as to ambient light levels.

Immediately behind the iris are the **lens** and the **ciliary muscle,** to which the lens is attached by ligaments called **zonules.** Together, the cornea, anterior chamber, and lens act as a compound optical system that bends the entering light rays to form an image on the retina.

After light rays leave the lens, they pass through a thick viscous substance called the **vitreous fluid** in the posterior chamber and are brought to focus on the retina. The **fovea** is a specialized portion of the retina at the back of the eye where rays are directed for objects in the **direction of gaze.** Surprisingly, the anatomical location of the fovea is a couple of degrees off the optical axis of the eye. The neural tissue in the retina other than photoreceptors is thinned out in the foveal region. This forms a depression called the **foveal pit** that minimizes the distortion caused when light passes through neural tissue. The nerve fibers of the retina pass out of the eye at the back, where they form the **optic nerve.**

Fundus

Under ordinary conditions, it is impossible to see inside an eye by looking into the pupil. However, it is possible to view the back of the eye with an instrument invented by Helmholtz, called an **ophthalmoscope.** The photograph in Figure 4.12 shows what is seen when looking into an eye with an ophthalmoscope.

The back of the eye as seen with an ophthalmoscope is called the **fundus.** Many people have experienced having an eye doctor use an ophthalmoscope to view the fundus when they go in for a routine exam. The **optic disk,** the place where the optic nerve leaves the eye, can be easily identified when looking at the fundus. It is the bright circular region on the right in Figure 4.12. Blood vessels enter and leave the eye at the optic disk and can be seen streaming out in all directions in a treelike pattern. To the left in Figure 4.12 is a small, intensely dark region devoid of blood vessels. This is the fovea. It is surrounded by a larger circular region, called the **macula,** that is also darker than the surrounding retina. The macular region looks

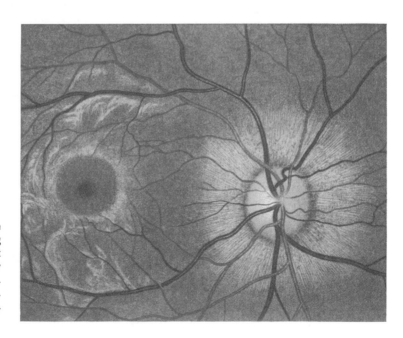

FIGURE 4.12. An early photograph of the fundus of the eye taken using an ophthalmoscope, a device that had only recently been invented by Helmholtz. Reproduced from G.L. Johnson, Philos. Trans. Roy. Soc. London, ser. B, vol. 194, Pl. 1, 1901, by permission of the Royal Society.

darker than the rest of the fundus because it contains a yellow pigment. A schematic of what a fundus looks like when an observer is viewing an object is shown in Figure 4.13.

A hypothetical observer is looking at the clock in the bell tower depicted in the top panel, and the image of the scene is superimposed onto the fundus in the bottom panel. The region of the fundus where the image of the clock falls is the fovea. This image on the fundus is the **proximal stimulus** for vision.

Entoptic Images

A number of perceptual phenomena, referred to collectively as **entoptic images,** are related to properties of the eyeball. Shadows of the blood vessels can be visualized using a flashlight. Hold a small penlight to the side of one eye so that it is touching the sclera and aiming toward the center of the eye. In spite of the fact that the choroid is relatively opaque, enough light will enter the eye to form shadows. Slowly aim the light in different directions, and you will experience a complex shape somewhat like tree branches. This perceptual experience is caused by the shadows of blood vessels at the back of the eye. When observers are asked to draw a picture of the shape experienced, the details of the picture agree with the complex shape of the blood vessels seen in the fundus with an ophthalmoscope.

When looking at a homogeneous surface such as the sky or a white ceiling, one can sometimes see small spots that move along a fixed trajectory. This percept is likely to be caused by shadows formed by individual blood cells passing through small capillaries in the eye. This can be confirmed by taking one's own pulse and correlating the movements of the spots with the heartbeat.

Again, while looking at a homogeneous surface, diffuse percepts, called floaters, can sometimes be seen. Unlike the small spots caused by blood cells, floaters are relatively stable rather than moving along a fixed trajectory. Floaters are most likely the result of shadows formed by impurities floating in the vitreous fluid. Floaters are often not visible when the eye is still but can be seen for a short period following a saccade, presumable because they jiggle from the acceleration and deceleration of the eye movement.

If a bright white surface is viewed while looking alternately every few seconds through a blue and a yellow filter, a dark ring is visualized around the point of fixation. This is caused by differences in

FIGURE 4.13. Schematic view of the image formed at the back of the eye (bottom panel) while viewing the scene shown in the top panel. Adapted from J. Lang, *Strabismus,* Slack Incorporated, 1984, by permission from J. Lang.

absorption of these wavelengths by the yellow pigment in the macular region of the retina compared to that in the surrounding retina.

4 Optical Properties of the Eyeball

Accommodation

The length of the eyeball is constant, but optical power requirements vary as the plane of fixation is changed from far to near. The first component of the

emmetropia

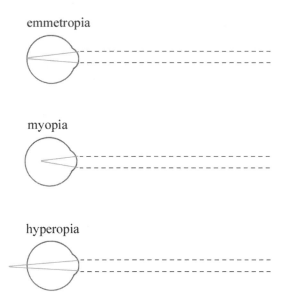

myopia

hyperopia

FIGURE 4.14. Illustration of the relationships between the length of the eye and the power of its optical components for eyes that are emmetropic, myopic, and hyperopic.

• cornea has a fixed power
○ lens - can be weak.

eye's compound optical system, the cornea, has a fixed power. However, the power of the second optical component, the lens, can vary from weak to strong because it is made of an elastic material that can change shape when stretched.

The stretching imposed on the lens, called **accommodation,** is controlled by the interactions between the zonule fibers and the ciliary muscle. When accommodation is completely relaxed, the zonule fibers stretch the lens so that it flattens and goes to its minimum power. When accommodation is activated, the ciliary muscle acts in opposition to the zonule fibers, allowing the lens to thicken, and its power increases. Regulation of accommodation comes from the brainstem via fibers of the third cranial nerve that descend to the ciliary ganglion. Neurons in the ciliary ganglion in turn innervate the ciliary muscle.

The **refractive state** of an eye is specified based on how well the power of the optical components is matched to the length of the eye. Since power varies with accommodation, refractive state must be specified with reference to a particular accommodative state. Typically, this is done with accommodation in its most relaxed state. This state can be achieved with drugs called **cycloplegics.** Drops of a cycloplegic drug placed topically on the front of the eye temporarily block parasympathetic input to the ciliary muscles, causing them to relax and

allowing the lens to go to its resting (minimum power) state.

In normal eyes, images of distant objects are focused at the plane of the retina while in a cycloplegic state. The technical term for this state is **emmetropia.** Eyes that are too long, so that rays of light from distant objects are focused in front of the retina under cycloplegic conditions, are said to be **myopic.** Eyes too short, so that the images of distant objects form beyond the retina, are said to be **hyperopic.**

These relationships are illustrated in Figure 4.14. They apply only to eyes in a relaxed state while viewing distant targets. When the eyes are allowed to adjust their accommodative state and view targets at various distances, the situation is more complicated.

Myopia

The consequences of viewing objects at various distances with a myopic eye, with and without an external corrective lens, are illustrated in Figure 4.15.

When a myopic observer views distant objects with accommodation relaxed, the image falls in front of the retina. Accommodation will only make things worse, because it increases the power of the lens and moves the image even farther in front of the retina. One way a myopic person can bring objects into good focus is by bringing them close to the eye. This is the reason humans with myopia are sometimes said to be **nearsighted.**

Myopic eye viewing near target

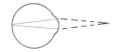

Myopic eye viewing distant target

Myopic eye viewing distant target with optical correction

FIGURE 4.15. Illustration of the images formed in a myopic eye under three viewing conditions.

The only way for a myopic eye to see distant objects clearly is with the aid of concave lenses, which have negative power. Recall from Chapter 3 that the powers of combinations of lenses are additive. Thus, if an external lens with negative power, in the form of glasses or a contact lens, is placed in front of the eye, whose optical components have too much position power, the total power of the optical system will be reduced. This has the effect of moving the image towards the retina. By a suitable choice of power for the external lens, a myopic eye can be allowed to focus images of distant objects onto the retina. A myopic viewer wearing glasses will have to use accommodation, the same as an emmetropic individual will to bring near objects into focus.

Hyperopia

Hyperopic eyes have the opposite problem of those with myopia; the optical system does not have enough power. The consequences of having a hyperopic eye are illustrated in Figure 4.16.

An individual with hyperopia can bring distant targets into focus by accommodating. However, this can lead to eyestrain, since the ciliary muscles must remain in a continuous state of increased tension, whereas normal eyes can remain in a state of rest while viewing distant targets. Furthermore, if the degree of hyperopia is high, then the degree

Hyperopic eye with no accommodation

Hyperopic eye during accommodative effort

Hyperopic eye with no accommodation but wearing optical correction

FIGURE 4.16. Illustration of the images formed in a hyperopic eye under three viewing conditions.

of accommodation needed will be too great to maintain when viewing near targets. This is the reason individuals with hyperopia are sometimes said to be **farsighted.** They can use moderate amounts of accommodation to see clearly at a distance but have to strain to see near targets. Since the hyperopic eye's optical system does not have enough power, it can be corrected with an external convex lens that provides sufficient additional power to focus distant targets onto the retina while the eye is in an accommodative state of rest.

Presbyopia

Accommodation works only because the lens in the eye is elastic and changes its shape when the ciliary muscles act on it. In children, the lenses are very elastic and accommodation can easily change their shape sufficiently to bring into focus objects at very near distances. The lenses become less elastic with age. The functional effect of the loss of lens elasticity is called **presbyopia.**

Persons with presbyopia have difficulty bringing close objects into focus. Figure 4.17 shows how the **near point,** the nearest distance which objects can be brought into good focus by accommodating, degrades as a function of age in humans.

At birth, most humans can focus to a near distance less than 10 cm. By college age the near point has moved out to about 10 cm, and by about the age of fifty to a point beyond arm's length. This is the age when the effects of presbyopia become severe enough that many humans need reading glasses, even when the cycloplegic refractive state of the eye remains emmetropic. In the elderly, the lens has lost almost all of its elasticity.

One way to eliminate some of the problems associated with presbyopia is to wear **bifocal lenses.** The power of the top part of a bifocal lens is different from the power of the bottom part. There is a natural tendency for individuals to look through the top of the lens when viewing distant objects and through the bottom when viewing near objects, especially when reading. With a little practice, most individuals can learn to exaggerate this natural tendency and view objects at various distances through bifocals. A modern variation on this, called **progressive lenses,** has power that varies gradually from top to bottom. These solutions to presbyopia are often ineffective for individuals who must perform tasks such as viewing a computer display monitor. Individuals trying to view these displays through the bottom of their lenses frequently complain that the unusal

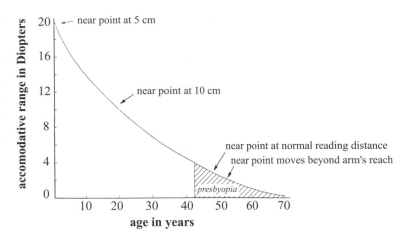

FIGURE 4.17. This figure shows population norms illustrating how the maximum accommodative power that can be generated in a human eye decreases with age. Adapted from A. Duane, Studies in monocular and binocular accommodation with their clinical applications. *Am. J. Ophthalmol.* 5:865–877, © 1922, by permission of Elsevier Science.

position leads to neck pain. "Computer glasses" with a fixed power adjusted for the distance of the monitor sometimes provides a better solution for these individuals.

Astigmatism

Astigmatism is a defect in the ability to bring into focus lines of certain orientations due to irregularities in the shape of the lens. A normal lens has a spherical shape, like a basketball. This allows the lens to bring lines of all orientations into good focus. Some eyes have lenses with an abnormal shape, more like a football. These lenses have more curvature, and thus more power, in some orientations than in others. Figure 4.18 illustrates a test that can be used to check for astigmatism. With a normal lens, all of the lines in the diagram should look equally crisp. With astigmatism, lines of some orientations (for example, the vertical) will look crisper than others.

FIGURE 4.18. Lines of all orientations can be seen equally well by a normal eye, but some orientations look crisper than others to an astigmatic eye.

Astigmatism can be corrected with glasses. How this works can be imagined based on the football example. If the lens in an eye is shaped like a football set vertically, this defect can be corrected by glasses in which the lens is shaped like a football set horizontally. The defect in the lens of the eye and the shape of the lens in the glasses cancel out, allowing all orientations to be seen clearly.

Reduced Eyes and Schematic Eyes

Even though the eye is actually a compound optical system, it is adequate for many purposes to treat it as though it were a single lens. Depending on the level of accuracy that is needed, this approximation can be based on a thin lens model (**reduced eye**) or on a thick lens model (**schematic eye**).

An analysis of image formation in a reduced eye is identical to the procedures described in Chapter 3 for analyzing thin lenses. It simply requires specifying the indices of refraction in front of and behind the lens and the locations, relative to the vertex (A), of the anterior focal point (F), posterior focal point (F'), and center of curvature (C). Typical values for a reduced eye in a human and a monkey are presented in Box 4.11.

In cases where somewhat more accuracy is needed than is provided by the reduced eye, a **thick lens approximation** can be applied that takes into account the fact that the actual physical components making up lenses or systems of lenses are not infinitesimally thin. The primary concept one needs to understand in order to extend the thin lens analyses described in Chapter 3 to an analysis based on thick lenses is **nodal distance**. Recall that a ray through the center of curvature of a

Box 4.11
Reduced Eyes of Human and Monkey

In the human and monkey eyes, the index of refraction in front of the eye, N, can be approximated as 1.0 and the index of refraction within the eye, N', as 4/3. The posterior focal length, A to F', of a human eye is about 22 mm, corresponding to a power of about 60 D (4/3 ÷ 0.022). The posterior focal length in a monkey eye is only about 18 mm, corresponding to a power of 74 D. The anterior focal length, F to A, is 16.7 mm (1/60 D) in the human eye and 13.5 mm (1/74 D) in the monkey eye. The radius of curvature, A to C, is 5.6 mm ([4/3 − 1]/60 D) in the human eye and 4.5 ([4/3 − 1]/74 D) in the monkey eye.

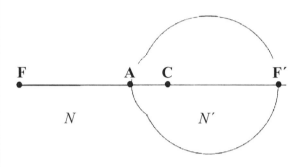

FIGURE 1. Schematic of a reduced eye.

thin lens remains undeviated. In a thick lens approximation, that relationship is redefined into one that is slightly more complicated, as illustrated in Figure 4.19.

A pair of **nodal points,** N and N', replace the single center of curvature, C, used with a thin lens. The trajectory of a ray from an object that passes through the center of curvature is modeled for a thick lens system by drawing a straight line segment from the object to the first nodal point. Then a second straight line segment parallel to the first is drawn continuing from the second nodal point.

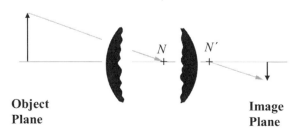

Object Plane

Image Plane

FIGURE **4.19.** Illustration of the concept of separation of nodal points. See text for details.

The distance separating the first and second nodal points is used to redefine the relationships between the vertex and the principal points that were defined for thin lenses. These relationships are now applied in a two-part process, as illustrated in Figure 4.20.

Two parameters, the first and second principal points, P and P', replace the single parameter, A, the vertex of a lens, that is used in thin lens approximations. Recall that in a thin lens the anterior and posterior focal lengths are both measured from A. In an analysis based on a thick lens approximation, the anterior focal length, f, is measured from F to the **anterior principal point,** P. The posterior focal length, f', is measured from the **posterior principal point,** P', to F'. Similarly, object distance, l, and image distance, l', are measured with reference to the principal points rather than the vertex. Other relationships in the figure are as described for a thin lens model in Chapter 3.

The concept of a schematic eye uses the thick lens model approach to define the eyeball based on these six parameters, called the **cardinal points:** two nodal, two principal, and two focal points. A number of schematic eyes, each based on slightly different assumptions, have been derived for humans. Examples of the values of the cardinal points in typical schematic eyes

FIGURE 4.20. Illustration of image formation by a thick lens. See text for details.

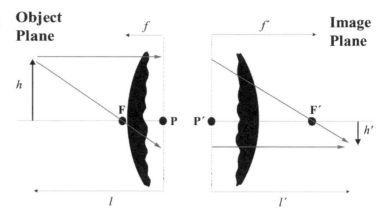

applied to humans and monkeys are illustrated in Figure 4.21.

Additional details about schematic eyes can be found in any standard physiological optics text.

5 Photoreceptors

Distribution Across the Retina

There are about 120 million rods and 6 million cones in the human retina, together forming about 70% of all of the exteroceptor cells of the body. However, they are not distributed equally over the surface of the retina. The highest density of cones is in the fovea, and this accounts for the higher reso-

lution of processing in central vision. Figure 4.22 shows the density of cones in the left eye of a human.

The fovea is at the center of the figure. The gray scale illustrates that the density of cones decreases in all directions from the fovea, with somewhat more density along the horizontal than the vertical axis. The exception is a small island with no cones present to the left of the fovea in the figure. This island corresponds to the optic disc, where the axons that form the optic nerve leave the eye. The distribution of rods is quite different, as illustrated in Figure 4.23.

The fovea has no rod receptors. The area of highest density is a donut shape about 20° eccentric from the fovea. The quantitative differences between the distributions of rod and cone receptors

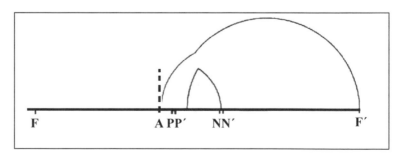

FIGURE 4.21. Application of a schematic eye model to human and monkey eyes. Distances of the six cardinal points in millimeters from the vertex of the eye are shown for both a human and a monkey eye.

Distance from Vertex (A)	Human	Monkey
Anterior focal point (**F**)	-14.98	-13.48
Posterior focal point (**F′**)	23.90	18.01
Anterior principal point (**P**)	1.55	1.03
Posterior principal point (**P′**)	1.85	1.22
Anterior nodal point (**N**)	7.06	5.56
Posterior nodal point (**N′**)	7.36	5.75

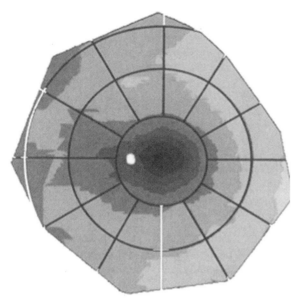

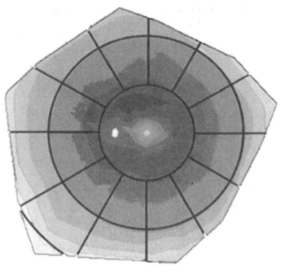

FIGURE 4.22. Cone density in the human retina represented with a gray scale, with the darkest regions corresponding to the highest densities. Reproduced from C.A. Curcio, et al, Human photoreceptor topography. *J. Comp. Neurol.* 292:497–523, © 1990, by permission of Wiley-Liss, Inc., a subsidiary of John Wiley & Sons.

FIGURE 4.23. Rod density in the human retina represented with a gray scale, with the darkest regions corresponding to the highest densities. Reproduced from C.A. Curcio, et al, Human photoreceptor topography. *J. Comp. Neurol.* 292:497–523, © 1990, by permission of Wiley-Liss, Inc., a subsidiary of John Wiley & Sons.

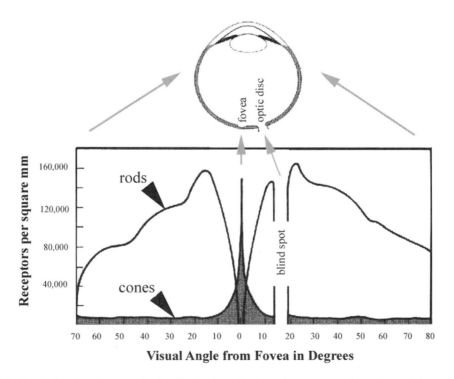

FIGURE 4.24. Graph showing the quantitative distribution of rods and cones over the surface of the retina. Adapted from results originally published by M.H. Pirenne, *Vision and the Eye,* © 1948, by permission of The Pilot Press, London.

in the fovea are even more dramatic than might be inferred from this gray-scale representation. A better appreciation of the quantitative relationships can be gleaned from Figure 4.24. This graph shows rod and cone densities plotted as a function of eccentricity along the horizontal meridian.

This plot illustrates the distribution for a right eye, highlighting the fact that cone receptors are concentrated in the center of the fovea. Rods are most numerous in the periphery. No photoreceptors are present where the optic nerve and blood vessels pass through the retina at the optic disc, and this has obvious perceptual consequences that give rise to its other name, the **blind spot.** The blind spot can be demonstrated as shown in Box 4.12.

The differences in distributions of rods and cones correspond to differences in function, a perceptual finding referred to as **duplicity theory.** The rods operate best under dim illumination conditions and give the most sensitive thresholds for detecting the presence of light. This accounts for the fact that humans can actually detect very dimly illuminated objects best by looking off to the side instead of directly at the object. Astronomers are aware of this fact and will look to the side of, instead of directly at, a very dim star when viewing it through a tele-scope, causing the image to fall onto the periphery of the retina instead of onto the fovea. However, rods do not allow very good acuity or color vision. These functions are provided by the cones. This is the reason one looks directly at an object if one needs to scrutinize its detailed shape or color properties. The differences between rod and cone function are also due to factors besides their distributions on the retina, an issue discussed further in Chapter 5.

The retina of the eye appears to be built "backwards," as illustrated in Figure 4.25.

When light rays reach the back of the eye, they have to first pass through the neural tissue before they reach the photoreceptors, where light is absorbed. This is not necessarily as serious a problem as it may first appear, because nerve tissue in the retina is relatively transparent. Thus, the amount of scatter and absorption is limited enough not to have an appreciable effect on vision except where it is most fine, in the straight-ahead direction. The anatomical specialization called the fovea appears to have evolved to minimize this problem for straight-ahead vision. The **foveal pit** is a depression in the retina where the cell bodies have been shifted to each side to allow light to reach the receptors without passing through very much retinal tissue.

Box 4.12
The Blind Spot Can Be Used to Perform a Disappearing Magic Trick

Close your left eye and look at the white tip of the magic wand at the left side of the figure. Slowly move the page closer to and farther away from your face while continuing to fixate on the magic wand. At a distance of about 1/3 m, the rabbit disappears from vision. At closer or farther distances, it reappears. The explanation is that the image of the rabbit falls on the optic disc at the distance at which it disappears (Figure 1).

FIGURE 1. This figure allows a demonstration of the blind spot where viewed as described in the text.

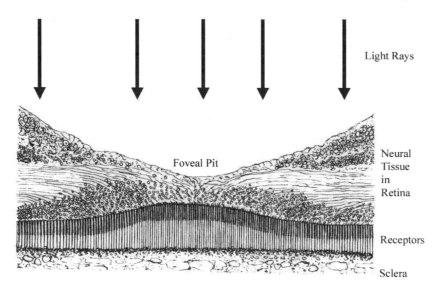

FIGURE 4.25. Illustration of relationships between light rays, neural tissue in the retina, and receptors. Adapted from S. Polyak, *The Vertebrate Visual System,* © 1957, by permission of the University of Chicago Press, Chicago.

Basic Morphology

The basic morphological features of individual rod receptors are illustrated in Figure 4.26.

The **outer segment (OS)** contains the photopigment, which absorbs light. The outer segment is joined to the **inner segment (IS)** by a thin connecting **cilium (CC)**. The IS contains the **nucleus (N)** of the cell as well as metabolic machinery such as **mitochondria (M)**. Connections with neurons in the retina are made at the **synaptic terminals (ST)**.

The outer segments have a morphological specialization in the form of an elaborate system of stacked membranous **disks** that dramatically increase the surface area of the membrane. This enables the outer segments to be densely packed with visual pigment and thus highly effective at absorbing light. The disks are stacked such that light that escapes being absorbed as it passes through one disk has additional chances to be absorbed as it passes through subsequent disks.

The basic morphology of cones is similar, but the widths of the disks in their outer segments is not uniform. In cones, the disks are widest near the cilium and become progressively thinner towards the top of the outer segment, giving a "cone-like" appearences.

The disks are constantly being renewed, but the mechanisms used to achieve this in rods and cones are somewhat different due to differences in their morphology. In rods, the disks are formed at the cilium. As each new disk is formed, it pinches off from the plasma membrane and becomes a separate disk. The individual disks migrate outward in the outer segment until they reach the tip. Then they are discarded and taken up by the pigment epithelial cells as **phagocyte material (P** in Figure 4.26). This entire process is regulated as a circadian rhythm, with disk shedding occurring each morning. A functional implication is that we have the most rod disks available in the nighttime hours, prior to disk shedding. In cones, the disks do not pinch off and renewal takes place throughout the outer segment by mechanisms that are less well understood.

Absorption of Light by Photopigments

This section emphasizes details that apply specifically to rods, the receptor type that has been studied most extensively. The events in cones are similar but involve different photopigments.

The photopigment in human and monkey rods, **rhodopsin,** is located within the membrane of the outer segment disks. Each molecule of rhodopsin is formed from **opsin,** a large transmembrane protein that spans the disk membrane seven times, and **retinal,** a vitamin A aldehyde that attaches to the opsin within the membrane. These relationships are illustrated in Figure 4.27.

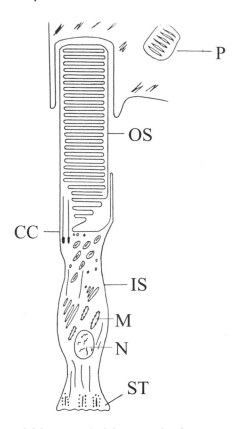

FIGURE 4.26. Anatomical features of rod receptors. See text for details. Adapted from PM Iuvone, Cell biology and metabolic activity of photoreceptor cells: Light-evoked and circadian regulation. In *Neruobiology and Clinical Aspects of the Outer Retina,* edited by M.B.A. Djamgoz, S.N. Archer, and S. Vallerga, pp. 25–55, London: Chapman & Hall, © 1995, by permission of Kluwer Academic Publishers.

Retinal can exist in two different geometric shapes, called the **11-cis** and the **all-trans** isomers. The 11-cis shape can fit into the opsin molecule attachment site within the disk membrane. When the 11-cis form of retinal binds to the opsin protein, it forms rhodopsin, a photolabile compound that can undergo a chemical reaction in response to absorbtion of light energy. When a photon is absorbed, a covalent chemical bond is broken, resulting in isomerization of retinal back to the all-trans configuration. This is the only light-dependent step in vision. The all-trans form of retinal no longer fits into the attachment site, and this leads to a cascade of events, described in the next section, that eventually leads to vision.

Rhodopsin does not absorb all wavelengths equally. A plot of the relative numbers of incident photons absorbed as a function of their wavelength, called a **spectral absorption plot,** is illustrated by the filled circle symbols in Figure 4.28.

Photons with a wavelength near 510nm are absorbed most efficiently, with dropoffs in sensitivity at both shorter and longer wavelengths. The spectral absorption of rhodopsin can be related to a behavioral **scotopic spectral sensitivity curve.** Absolute thresholds are measured, using the methods described in Chapter 2, for a number of lights having wavelengths across the visible spectrum. These threshold values can be corrected to account for absorption of photons in tissues of the eye prior to reaching the photoreceptors. The open circle symbols are the reciprocals of these thresholds, and these define the scotopic spectral sensitivity curve. The scotopic spectral sensitivity curve

FIGURE 4.27. This figure depicts the fact that the photopigment rhodopsin is located in the disk membranes of the outer segments of rods. Each molecule is formed from opsin, a protein that spans the membrane seven times, and attached retinal, as illustrated in the inset.

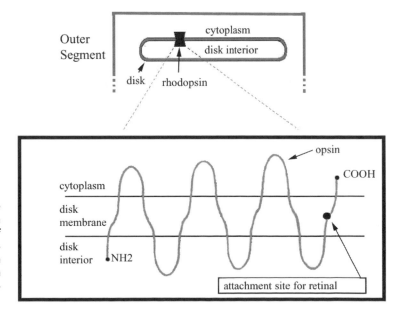

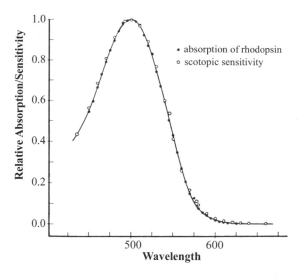

FIGURE 4.28. Spectral absorption of rhodopsin compared to scotopic spectral sensitivity of a human observer. Reproduced from G. Wald and P.K. Brown, *Science* 127:222–226, © 1958, by permission of American Association for the Advancement of Science.

can be essentially superimposed on the spectral absorption plot for rhodopsin. This is illustrated by the smooth curve plotted in Figure 4.28. Thus, the photochemistry of rhodopsin provides the explana-

tion for the shape of the behavioral scotopic spectral sensitivity curve.

The cone photopigments are also composed of an opsin and 11-cis retinal, but the opsin in cones is slightly different from that found in rhodopsin, causing a shift in the spectral absorption curve. There are three types of cones in humans and monkeys, each of which contains a different type of opsin and has a slightly different spectral absorption curve. The properties of the individual cone types are dealt with in Chapter 7, which discusses color vision. For the purposes of the current discussion, it is sufficient to consider a single average **cone spectral absorption curve**. The cone absorption curve is less sensitive than that of rods but is also shifted towards longer wavelengths, as illustrated in Figure 4.29.

When the behavioral spectral sensitivity curve is measured under daylight conditions in which the cones are detecting light, it is called the **photopic spectral sensitivity curve**. This behavioral function can be essentially superimposed on the average cone spectral absorption curve.

Differences in the spectral absorption curves of rods and cones can account for a perceptual phenomenon called the **Purkinje shift**, named after the Bohemian Physiologist Jan Evangelista **Purkinje**, who first described it. You can observe the Purkinje shift yourself, as described in Box 4.13.

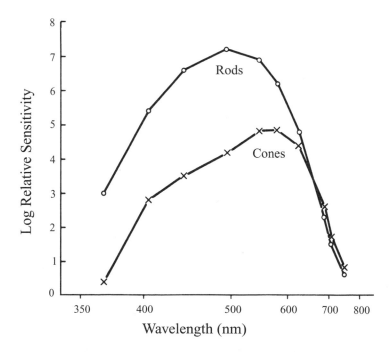

FIGURE 4.29. Comparison of relative spectral absorption curves for rods and cones. From G. Wald, *Science* 101:653–658, © 1945, by permission of American Association for the Advancement of Science.

Box 4.13
Brightness of Flowers in the Garden at Sunset

Go sit in a flower garden that has yellow and green flowers late in the afternoon when the sun is about to set. Pay close attention to the relative brightness of the yellow and green flowers. If you continue looking at the same flowers during sunset, the flowers will lose their colored appearance as the eye switches over from detecting light with cones to using rods. However, the flowers that had been green during the daylight will now appear relatively brighter and those that had been yellow relatively darker. This change in relative brightness for different colors as one goes from photopic to scotopic conditions is called the **Purkinje shift.** Its explanation is found in the differences in the spectral absorption curves of rods and cones.

The flowers that look yellow under daylight conditions predominantly reflect wavelengths near 570 nm. These wavelengths are absorbed efficiently in cones. Thus, under daylight conditions the yellow flowers produce many photon absorptions in the eye and appear bright. The flowers that look green under daylight conditions reflect shorter wavelengths, near 510 nm, that are not absorbed as efficiently by cones. Thus, these flowers produce fewer photon absorptions and appear less bright.

During sunset, the cones stop functioning and the rods become active. Now the shorter wavelengths, near 510 nm, are absorbed more efficiently than the longer wavelengths, near 570 nm, causing a reversal in their relative brightness compared to daylight conditions.

Recycling of Photopigment

Following absorption of a photon, the linkage between opsin and retinal is broken, and the all-trans retinal diffuses away. The all-trans isomer must be converted back to 11-cis before it can be recycled and attached again to the opsin. This recycling takes place in the pigment epithelium. A form of vitamin A that cannot be synthesized by humans is needed for this recycling, and this accounts for the fact that a nutritional deficiency in vitamin A can lead to blindness.

It can take 30 minutes or more to restore all of the rhodopsin molecules following exposure to a bright light. This fact accounts for the time course of a perceptual phenomenon called **dark adaptation.** Following exposure to a bright light, sensitivity for detecting a dim light is decreased and returns only gradually. This can be measured in the form of a dark adaptation curve. An observer is initially exposed to a bright light. Then the bright light is turned off and the individual sits in darkness. An absolute threshold for detecting a small dim light is measured quickly, and the time elapsed since the adapting light was turned off is noted. A short time later the absolute threshold is measured again and the elapsed time noted. This is done repeatedly for 30 minutes or more. When the process is completed, the results can be plotted in the form of a **dark adaptation curve** in which sensitivity, defined as the reciprocal of absolute threshold, is plotted as a function of time in the dark. The shape of a hypothetical dark adaptation curve is shown in Figure 4.30.

This simple shape is obtained if conditions are chosen such that only rods are operating. Under conditions in which cones are allowed to operate as well, the dark adaptation curve takes on a more complex shape, the initial portion of which is due to cones and the later portion,

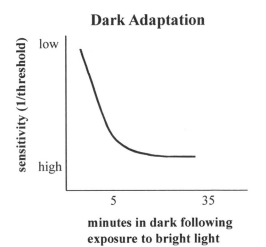

FIGURE 4.30. Sensitivity for seeing light is decreased following exposure to a bright light and then increases gradually over several minutes. The time course over which this happens is called a dark adaptation curve.

corresponding to what is shown in Figure 4.30, to rods.

Dark adaptation sensitivity continues to improve for about 30 minutes. The functional consequences of dark adaptation are apparent in situations like that described in Box 4.14.

Box 4.14
Why Can't I Find My Seat in the Movie Theater on a Bright Afternoon?

You have perhaps encountered the following problem if you have gone to a movie in the afternoon on a bright day. When you first enter the dark movie theater, you have trouble finding your seat, perhaps stumbling over people already sitting in their seats. However, if you look around after you have been sitting in your seat for several minutes, you discover that you can see much more clearly, and wonder why you had so much trouble earlier. The explanation is related to the fact that it takes time for biological eyes to restore rhodopsin molecules after they have been bleached by light.

Phototransduction

The general term **transduction** refers to the conversion of the physical energy that impinges on one of the sense organs into an electrophysiological signal that can be processed by the brain. In the case of vision, the term **phototransduction** is used to refer specifically to the conversion of light energy into a neural signal by the photoreceptors. Phototransduction takes place in the outer segments of the rods and cones. The absorption of each photon by a molecule of photopigment triggers a cascade of events that eventually leads to changes in electrical potential across the photoreceptor membrane. In order to understand the details of this process, it is necessary to first understand some of the basic membrane physiology of receptors.

Electrical current, sometimes called **dark current**, is constantly flowing through the receptor membrane in the dark, as illustrated in Figure 4.31.

The current enters the outer segment through **cGMP-gated Na$^+$ channels** located in the outer

segment and flows back out through **non-gated K$^+$ channels** located primarily in the inner segment. A high density of **pumps** in the inner segment serves to maintain steady intracellular concentrations of Na$^+$ and K$^+$ in the face of these large fluxes. As a result of the dark current, the inside of the receptor is maintained at around $-40\,mV$ relative to the surrounding extracellular space. This voltage is called the **photoreceptor membrane potential.** The absorption of a photon of light has its effect by altering the flow of dark current, which in turn alters the membrane potential.

When retinal changes from the 11-cis to the all-trans configuration (Rh*) following absorption of a photon (hv), it drifts away from its attachment site with opsin. As a result, opsin is able to interact with a G protein called **transducin.** This event activates **phosphodiesterase,** which in turn lowers the concentration of **cGMP.** Ionic Na$^+$ channels along the rod's outer segment membrane then close, causing a reduction in the inward flow of dark current and hyperpolarization of the voltage across the membrane.

Amplification occurs during this process because absorption of a single photon allows opsin to continue interacting with transducin molecules until another molecule, **arrestin,** binds with opsin. A typical absorption results in interactions with about 500 transducin molecules before the process is stopped. Once the process is stopped, the opsin

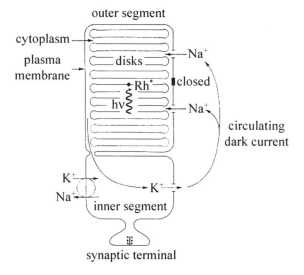

FIGURE 4.31. Photoreceptors generate dark current. Symbols and details are presented in text. Reproduced from T.D. Lamb, Transduction in vertebrate photoreceptors: The roles of cyclic GMP and calcium. *Trends Neurosci.* 9:224–228, © 1986, by permission of Elsevier Science.

FIGURE 4.32. The effect of absorbing a single photon on the voltage across the membrane of a rod and cone photoreceptor. Reproduced from K.-W. Yau, Phototransduction mechanisms in retinal rods and cones: The Friedenwald Lecture. *Invest. Ophthalmol. Vis. Sci.* 35:9–32, © 1994, by permission of the Association for Research in Vision & Ophthalmology.

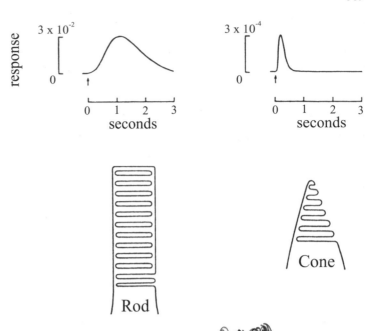

molecule is put into a quiescent state in which it is unable to interact with light again until it encounters new 11-cis retinal.

As a result of the amplification, absorption of a single photon leads to a change in membrane current of about 1 pA and a consequent change in voltage of about 1 mV. Rods respond slowly; the effect on membrane voltage of a single photon absorption lasts over a second, as shown on the left side of Figure 4.32.

If additional photons are absorbed by the same photoreceptor while its membrane is still responding to the first, the electrical responses will be superimposed. This results in **temporal summation**, as illustrated in Figure 4.33.

The response to the first absorbed photon hyperpolarizes the membrane to −41 mV, and then the voltage begins to return towards its baseline level. If no other absorptions occurred the membrane would continue depilarizing to −40 mV. However, in the example shown, a second photon is absorbed before the membrane has returned to its baseline level. As a result, the membrane voltage is hyperpolarized to a magnitude greater than −41 mV. The amount of temporal summation in this example is only partial, because the response to the absorption of the first proton had already started to abate before the second photon was absorbed. However, if two or more absorptions happen near enough in time to one another, the responses summate completely so that absorptions of two photons produce twice the effect of either alone. In a rod, **complete temporal summation** occurs when photons are absorbed within about 100 ms of each other.

The period of complete temporal summation in rods can be related to **Bloch's law,** a psychophysical relationship that holds between intensity and duration when measuring absolute thresholds for flashes of light. Bloch's law states that the number of photons per second and the total duration of a flash can be traded off as long as the total number of photons in the flash remains constant. Bloch's law holds up to about 0.1 second.

There is a tradeoff between temporal summation and **temporal resolution,** the ability to differentiate

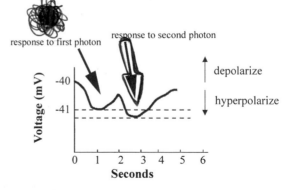

FIGURE 4.33. The effect of absorption of a photon on membrane voltage is influenced by the recent history of the membrane. Thus, two absorption events produce a greater effect when they occur close in time than when they occur separately.

small changes in absorption over time. Rods, having a relatively long period of temporal summation, do not allow very good temporal resolution. Rod-mediated (scotopic) vision cannot resolve light flickering on and off faster than about 12 Hz. The temporal response of cones is faster as shown on the right side of Figure 4.32, and this allows cone-mediated (photopic) vision to detect flicker up to almost 60 Hz. Of course there is a tradeoff for this higher temporal resolution. Cones are not as sensitive as rods when it comes to detecting small amounts of light because they cannot take as much advantage of temporal summation.

Photoreceptor voltage increases in a graded manner with additional numbers of photons absorbed only up to a limit, as illustrated schematically in Figure 4.34.

Under dark conditions, the resting membrane voltage is −40 mV because of the dark light current. An intense light that causes simultaneous absorption of about 100 photons is sufficient to close all of the channels and drive the membrane potential to its saturation level of −70 mV. A very intense light that causes absorption above this limit has no further effect on amplitude. For light intensities at intermediate levels, the membrane voltage will fall between the limits of −40 and −70 mV.

Recall from Chapter 3 that this limited dynamic range must be used to signal small changes in light over a range of ambient light levels of about 15 log units, a factor of more than 1 trillion to 1. This is accomplished by the mechanism of **photoreceptor light adaptation,** whereby the photoreceptor largely ignores the prevailing ambient light level and responds only to perturbations around that ambient level. When a photoreceptor is suddenly exposed to an intense light, a surge of cGMP release quickly closes all of the gated channels, and the electrical potential across the membrane becomes saturated at −70 mV. However, over a period of seconds, **calcium** alters the effectiveness of cGMP such that the new ambient light level results in a steady-state voltage of only about −40 mV. In other words, the voltage returns to its baseline and is once again ready to respond to small changes around the current ambient level. The dependence of this adaptation on calcium can be demonstrated experimentally. When specialized experimental methods are employed that allow the calcium in a photoreceptor to be clamped at a constant level, the photoreceptor still responds to an initial increase in light level but will not readapt its baseline response to this new level. Light adaptation is rapid, being completes in a minute or so, in contrast to dark adaptation, which can take over 30 minutes.

This mechanism of photoreceptor light adaptation can be related to **perceptual light adaptation.** When exposed to an intense light, as when one steps outside from a dark room into bright sunlight, one is dazzled by the light and initially has difficulty seeing. This corresponds to the period when the photoreceptor response is saturated. However, over a period of several seconds, one is able to see more clearly. Perceived intensity, i.e., brightness, of an intense unchanging stimulus declines by as much as 90% over several seconds. This corresponds to the period when the photoreceptors are adapting to the new baseline level.

6 How Much Light Does It Take for Us to See?

S. Hecht, S. Schlaer, and M.H. Pirenne carried out a psychophysical study in the 1940s to answer the question: How much light has to be present in order to be seen? They wanted to determine the smallest amount of light that was needed under optimum conditions. They adjusted several parameters to accomplish this. For example, they presented the light 15° to 20° eccentric from the fovea, the region with the highest density of rod photoreceptors. They used light with a wavelength near 510 nm, the

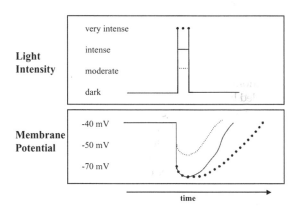

FIGURE 4.34. This figure illustrates that increasing the number of photons that are absorbed simultaneously by a photoreceptor causes a corresponding increase in hyperpolarization of the membrane voltage, but only up to limit of about −70 mV. Response to a hypothetical light flash of only moderate intensity is depicted by the fine dotted line, to an intense flash that hyperpolarizes the membrane potential to its limit by the smooth line, and to an even more intense flash by the coarse dotted line.

portion of the visible spectrum absorbed most efficiently by rhodopsin. Each observer sat in the dark for over 30 minutes prior to participating in the experiment so that all of the rhodopsin molecules were replenished. The light was presented in a brief flash of about 0.1 second so that complete temporal summation would take place. Finally, the light was imaged onto a small spot so that the photons would be absorbed in photoreceptors situated near one another and all signaling to a common pool of retinal neurons.

Under these optimal conditions, the observer reported whether or not each flash of light could be seen, and the intensity was adjusted until it was detected on half the trials. When the experiment was over, the experimenters placed a photodetector at the location where the observer's eye had been and measured the number of photons in the flash. These measurements revealed that in order for an observer to detect a flash of light, it must contain about **90 photons when it reaches the front of the eye.** Not all of these 90 photons are absorbed in the photoreceptors, because some of them are reflected or absorbed by ocular media other than photoreceptors. For example, some photons are reflected back out of the eye from the cornea or the front of the lens, and others are absorbed in the aqueous and vitreous fluids in the eye. Calculations reveal that approximately half of the photons that reach the front of the eye make it to the retina. Thus, in order for an observer to detect a light, it must contain about **45 photons that reach the retina.** About 80% of the photons that reach the retina are absorbed in retinal tissue other than receptor outer segments or pass between receptors to be absorbed in the pigment epithelium and choroid. Thus, **fewer than 10 photons were absorbed by rhodopsin molecules in photoreceptor outer segments.** The photons that were absorbed were scattered over a region of the retina that contained perhaps 300 to 500 individual rods, so the odds of any one receptor's absorbing more than one photon under these conditions are nil. Thus, **a single photoreceptor is sensitive enough to register absorption of a single photon.**

This is an extraordinary degree of sensitivity. As a standard of reference, consider that an ordinary flashlight bulb emits about 20 trillion photons each tenth of a second. Note also that the exact number of photons that get absorbed in the photoreceptors from trial to trial will not be identical even when the intensity of the distal light flash is the same. The number of photons absorbed depends ultimately on statistical probabilities. These **photon statistics** pose a fundamental limit on our ability to reliably detect dim lights, as will be discussed further in Chapter 6, which covers perception modeled as a statistical process.

Summary

The act of perception involves constantly sampling our immediate surroundings in order to keep our perceptual database up to date. The neural processes responsible for perceptual processing do not have enough bandwidth to sample the entire surroundings continuously. Biological systems minimize this problem by selective filtering based on smart mechanisms. This allows perceptual processing to concentrate on those aspects of the perceptual database that have the highest priority for updating. The initial filtering is accomplished by scanning movements in which the eyes are repositioned, at a rate of about three times per second, in the direction of the environment to be sampled in the next instant.

Scanning eye movements are regulated by four oculomotor neural control systems that coordinate movements of the eyes. Each of these neural systems operates at a final common pathway on pools of motoneurons in brainstem oculomotor nuclei. Axons from these motoneurons innervate the extraocular muscles of the eye via cranial nerves.

During each fixation, a reflex called accommodation attempts to adjust the optics of the eye appropriately so that a clear image is formed on the retina at the back of the eye. If the power of the optics of the eye is not matched to the length of the eyeball, clear images cannot be obtained under all viewing conditions. The optical properties of the eye can be modeled based on thin lens approximations or with somewhat more sophisticated models based on thick lens approximations.

The retinal image is sampled by photoreceptors located across the surface of the retina. The highest density of sampling under normal daylight conditions occurs at the fovea and under dim light conditions about 20° eccentric to it. The light energy that is focused onto each receptor is transduced into neural signals via a complex cascade of molecular events. Photoreceptors adjust to the prevailing light levels via processes called light and dark adaptation.

A photoreceptor is sensitive enough to register a single photon, and under optimal conditions, a human or monkey observer can detect as few as 10 absorbed photons.

Selected Reading List

Bennett, A.G., and Rabbetts, R.B. 1984. *Clinical Visual Optics*. London: Butterworths.

Berkley, M.A., and Stebbins, W.C. 1990. *Comparative Perception*, Vol. 1, Basic Mechanisms. New York: Wiley.

Biederman, I., Mezzanotte, R.J., and Rabinowitz, J.C. 1982. Scene perception: Detecting and judging objects undergoing relational violation. *Cog. Psychol.* 14:143–177.

Crawford, M.L.J. 1977. Central vision of man and macaque: Cone and rod sensitivity. *Brain Res.* 119:345–356.

Hecht, S., Schlaer, S., and Pirenne, M.H. 1942. Energy, quanta, and vision. *J. Gen. Physiol.* 25:819–840.

Jonides, J., Irwin, D.E., and Yantis, S. 1982. Integrating visual information from successive fixations. *Science* 215:192–194.

MacLeish, P.R., Shepherd, G.M., Kinnamon, S.C., and Santos-Sacchi, J. Sensory Transduction. 1999. In *Fundamental Neuroscience*, ed. M.J. Zigmond, F.E. Bloom, S.C. Landis, J.L. Roberts, and L.R. Squire. pp. 671–718. San Diego, CA: Academic Press.

Neisser, U. 1967. *Cognitive Psychology*. New York: Appleton-Century-Crofts.

Ogle, K.N. 1976. *Optics*, 2nd edition, 3rd printing. Springfield, IL: Charles C. Thomas.

Packer, O., Hendrickson, A.E., and Curcio, C.A. 1989. Photoreceptor topography of the retina in the adult pigtail macaque *(Macaca nemestrina)*. *J. Comp. Neurol.* 288:165–183.

Polyak, S. 1957. *The Vertebrate Visual System*. Chicago: Chicago University Press.

Sperling, G. 1960. The information available in brief visual presentations. *Psychol. Monogr.* 74:1–29.

Von Noorden, G.K. 1983. *Atlas of Strabismus*, 4th edition. St. Louis, MO: Mosby.

Wald, G. 1968. The molecular basis of visual excitation. *Nature* 219:800–807.

Walls, G. 1963. *The Vertebrate Eye and Its Adaptive Radiation*. New York: Hafner.

Yarbus, A.L. 1967. *Eye Movements and Vision*, trans. B. Haigh. New York: Plenum.

Yau, K.-W. 1994. Phototransduction mechanisms in retinal rods and cones: The Friedenwald Lecture. *Invest. Ophthalmol. Vis. Sci.* 35:9–32.

5
Perceptual Processing I. Biological Hardware: What Properties of Neural Tissues Support Perceptual Processing in Humans and Monkeys?

Questions

After reading Chapter 5, you should be able to answer the following questions.

1. Describe the overall functional architecture of the visual system from the levels of analysis of neurons, micronetworks, hypercolumns, brain nuclei and areas, and streams of processing.
2. Describe the essential anatomical and physiological characteristics of a neuron that allow it to serve as the basic unit for processing perceptual information.
3. Give some examples of micronetworks that can solve elementary perceptual functions.
4. What is a hypercolumn?

5. Explain what it means to state that hypercolumns are organized in a topographical fashion into brain nuclei and areas.
6. Describe the basic functional organization of the retina.
7. How do the left and right portions of the visual field map onto structures in the left and right halves of the brain?
8. What filtering functions are performed as perceptual information passes through the thalamic relay station called the lateral geniculate nucleus?
9. Describe the geniculostriate pathway, including the neural structures that are involved and the kinds of perceptual processing that take place in these structures.

10. Illustrate with some examples how the visual system is organized into functional streams of processing.

The human brain, weighing only about 1.5 kilograms, is the most complex structure in the known universe. It is composed of over 100 billion individual components connected together by thousands of kilometers of cabling that form hundreds of trillions of connections. This book does not make any attempt to describe the overall biological properties of the brain. The interested reader is referred to standard textbooks of neuroanatomy, neurophysiology, and neuroscience. This chapter focuses only on those aspects of **functional architecture** of the brain that allow it to accomplish perceptual processing of information derived from the eyes.

1 Levels of Analysis

Functional architecture can be described from a number of different levels of analysis. The basic level that well be emphasized here is that of a single **neuron.** Small groups of interconnected neurons that perform some elementary function are called **micronetworks.** Groups of micronetworks that process and analyze information about one small region of the visual scene form a functional unit

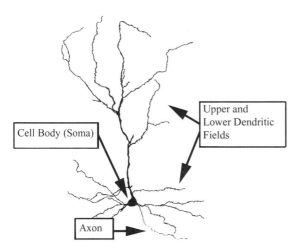

FIGURE 5.1. This tracing was made from a pyramidal neuron from monkey visual cortex as seen under a microscope. Adapted from J.S. Lund and R.G. Boothe, Interlaminar connections and pyramidal neuron organisation in the visual cortex, area 17, of the macaque monkey. *J. Comp. Neurol.* 159:305–334, © 1975, by permission of Wiley-Liss, Inc., a subsidiary of John Wiley & Sons.

called a **hypercolumn.** Groups of hypercolumns, one for each part of the visual scene, ·are organized in a topographic fashion in brain **nuclei** or **areas.**

Neurons

The basic functional biological units in the brain that process perceptual information are neurons. A prototypical example of a visual processing neuron, called a **phramidal cell,** found in the visual cortex of a monkey brain is shown in Figure 5.1.

Neurons come in a variety of shapes, but all have three morphologically specialized regions that can be identified when neural tissue is viewed with a microscope: **dendrites,** a **cell body,** and an **axon.** Communication from one neuron to the next takes place at specialized sites called **synapses** located at the **axon terminals.** In the pyramidal neuron shown in Figure 5.1, the axon is relatively long and passes out of the bottom of the figure so that its axon terminal cannot be seen. Information flow from one neuron to the next is polarized, typically passing from the axon of one neuron to the dendrites or cell body of the next, as illustrated in Box 5.1.

The dendrites often split into a number of treelike branches to form **dendritic fields** over which inputs can be received. The shape of a neuron's dendritic fields has a major influence on the sources of inputs it can receive. Similarly, an axon can branch to form one or more **axon terminal fields,** allowing the neuron to have a sphere of influence at particular sites.

The cell body integrates and processes the incoming signals and then generates a new signal that is propagated down its axon. It accomplishes this with electrophysiological mechanisms originally described by the British biologists Alon **Hodgkin** and Andrew **Huxley** in the 1940s and summarized in Box 5.2.

When the generated electrical signals reach an axon terminal, they cause release of vesicles across the synapse to the next neuron.

Every neuron that processes visual information has a **receptive field,** defined as the specific locations on the retina, or equivalently in the visual field, where stimulation by light can modulate its electrical state. Electrical activity is typically measured by positioning an electrode close enough to the neuron to record its action potentials using methods described in Box 5.3.

Box 5.1
Communication from One Neuron to the Next Occurs at a Synapse

The direction of information flow is polarized, passing from the axon terminals of one neuron to the dendrites or cell body of the next (Figure 1). The neuron that transmits the information is said to be **presynaptic,** and the neuron that receives the information is said to be **postsynaptic.** Transmission of information within an axon is electri-cal, but transmission across the synapse is chemical. The axon terminal of the presynaptic neuron contains **vesicles** filled with a **neurotransmitter** that cross the synaptic cleft to stimulate the postsynaptic neuron. Some types of neurotransmitters have an excitatory effect on the postsynaptic cell, while others are inhibitory.

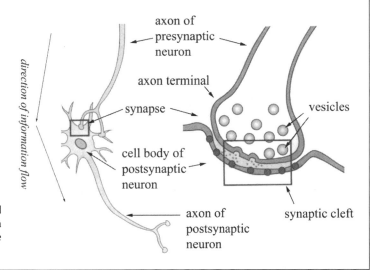

FIGURE 1. Signals are transmitted from the axon of one neuron across a synapse to the cell body or dendrite of a second neuron.

Box 5.2
Sensory Coding of Signals in Terms of Voltages

Signals are coded in neurons as voltages maintained across their membranes. At its resting state, the voltage inside a neuron cell body is about $-70\,mV$ with respect to the surrounding brain tissue. This **resting potential** is maintained by neural channels and pumps that regulate the flow of ions across the cell membrane. When an **excitatory input** is received at a synapse, the membrane is **depolarized** slightly, becoming less negative. When an **inhibitory input** is received, the membrane further **hyperpolarizes,** becoming even more negative than at the resting potential. The membrane potential at any moment reflects the sum of all of the recent inputs, weighted by factors such as how many vesicles were released at each synapse and the distance of each synapse from the cell body.

In most neurons, whenever the membrane potential becomes depolarized to a critical level of about $-40\,mV$, called its **threshold potential**, voltage-sensitive Na^+ **channels** open, resulting in a positive impulse. The impulse, called an **action potential,** or **spike,** is propagated down the axon to the terminals. The exceptions are a few classes of neurons that have very short axons, in which the membrane potential propagates to the terminal directly, without generating a spike. Most of the neurons in the retina fall into this category and communicate with one another via graded membrane potentials rather than with spikes. However, the output neurons in the retina whose axons travel up the optic nerve to central brain areas all generate spikes, as do most neurons in central visual processing brain regions.

Box 5.3
Measuring Action Potentials in Monkeys

A guide tube is placed at the appropriate location on the skull during a surgical procedure performed on the anesthetized animal. Following the monkey's recovery from surgery, a microelectrode can be positioned in the appropriate brain area by passing it to the appropriate depth through the guide tube. This does not cause any pain, because there are no pain receptors in brain tissue; similar procedures are sometimes performed on humans prior to brain surgery. The microelectrode tip can be used to record action potentials from single neurons while the retina is stimulated with appropriate visual stimuli. Typically, the action potentials are amplified and displayed in real time on an oscilloscope during the experiment as well as stored for later analyses (Figure 1).

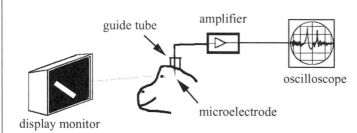

FIGURE 1. Methods used to record electrical activity of single neurons in the brain.

Micronetworks

Neurons do not function as isolated entities. Instead, small groups of neurons (on the order of 10^1 to 10^3) interconnect to form **micronetworks** that perform elementary processing functions. Identifying these micronetworks and discovering the way they operate is a major goal of neuroscience. The existence of micronetworks is usually not apparent based on simply looking at neural tissue, because neurons are so numerous and tightly packed together. This is illustrated in Figure 5.2, which shows a photograph of a thin slice of brain tissue from visual cortex of a monkey.

This slice of brain tissue has been treated with **Nissl's stain,** which makes the cytoplasm in the cell body of each neuron turn dark so that it can be visualized as a small dark spot. The dendrites and axons of the neurons do not stain. If they were visible, the space would be so packed with tissue that it would be impossible to resolve the individual neurons. The connections between neurons are also not apparent from tissue stained in this manner. On average, each neuron in the brain makes about 5,000 synaptic connections with its nearby neighbors.

In order to try to identify micronetworks, neuroanatomists have developed a number of chemical procedures that stain only selected portions of nervous tissue. One staining method that has been widely used historically to learn about

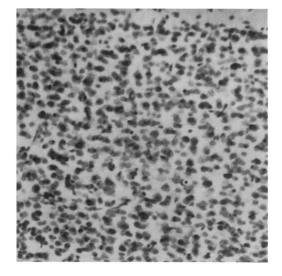

FIGURE 5.2. Photograph of a Nissl-stained thin slice of brain tissue from the visual cortex of a monkey.

micronetworks was developed by **Camillo Golgi** in the 1800s and exploited by the neuroanatomist **Ramón y Cajal** near the turn of the 20th century to prepare detailed atlases of the shapes of dendritic fields and local axonal projections of neurons that are present in various brain regions. Golgi and Cajal shared the Nobel Prize in 1906. Further information about use of **Golgi's method** is presented in Box 5.4.

Anatomical observations of stained neural tissue are often used to infer the presence of micronet-works. An example illustrating the basic rationale of this approach is shown in Figure 5.3.

The two neurons illustrated schematically in this figure represent types that were discovered based on examining Golgi-stained material from visual cortex of monkey brain. The dendrites of the neurons are drawn with bumps that represent **spines,** specialized appendages where most of the excitatory synaptic connections are made. The thin smooth lines denote axons. As will be elaborated on

---

Box 5.4
The Golgi Method of Staining Brain Tissue

Golgi's method does not actually allow one to see the connections that are formed between neurons. However, the method has two convenient properties. First, it stains most of the neuron, including the cell body, dendrites, and at least portions of the axon. Second, and just as important, the method stains only a small proportion of the neurons that are present, thus allowing those neurons that are stained to be visualized without being obscured by the dense background of neighboring neurons. Figure 1 shows a drawing made by Cajal of some morphological types of neurons he visualized in the visual cortex of a human brain based on examining tissue prepared with the Golgi method.

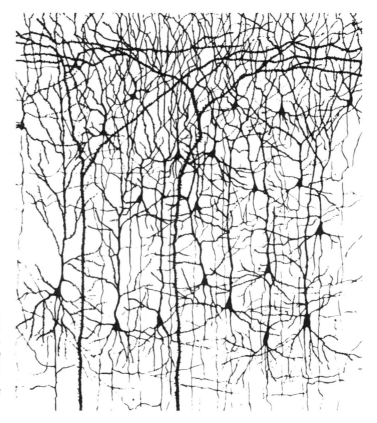

FIGURE 1. **Drawing made by Cajal of neuronal tissue in visual cortex that had been prepared with Golgi's method. Reproduced from** *Cajal on the Cerebral Cortex,* **edited and translated by J. DeFelipe and E.G. Jones, © 1988, by permission of Oxford University Press.**

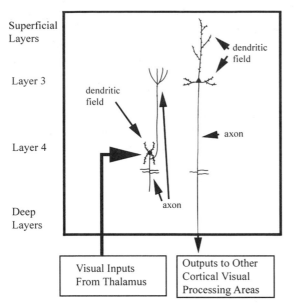

The important thing to note here is that the two neurons represented here appear to be specialized for "listening in" to information in certain layers and for "talking to" other layers. The neuron on the left is in a good position to receive visual input signals from the thalamus and to communicate with cell bodies and dendrites located in layer 3. The neuron on the right listens in to signals entering layer 3 and sends a projection to other cortical processing areas. Based on these anatomical observations alone, one might infer the presence of a micronetwork that receives visual inputs from the thalamus, processes this information in some way, and then passes the results on to another cortical area.

Many micronetworks involved in visual processing have been discovered, as illustrated in this example, by making inferences from anatomy. In other cases, micronetworks have been discovered based on inferences from physiological function. The sections that follow provide examples of both.

FIGURE 5.3. Schematic illustrating presumed connectivity of neurons in monkey visual cortex inferred from examination of Golgi-stained tissue. Adapted from R.G. Boothe, et al, A quantitative investigation of spine and dendrite development of neurons in visual cortex (area 17) of *Macaca nemestrina* monkeys. *J. Comp. Neurol.* 186: 473–490, © 1979, by permission of Wiley-Liss, Inc., a subsidiary of John Wiley & Sons.

Hypercolumns

In humans and monkeys, groups of perhaps 10^1 to 10^2 micronetworks are organized into a **hypercolumn,** a small slab of neural tissue responsible for processing the information from one small portion of the visual scene (Figure 5.4).

The term hypercolumn was coined by Nobel laureates **David Hubel** and **Torsten Wiesel** in the 1970s to describe a basic anatomical organization of visual cortex. They proposed that primary visual cortex is organized into functional units, which they

later in this chapter, the visual cortex receives visual inputs from the thalamus and sends outputs to other cortical processing areas. Other details of this figure are not of concern now.

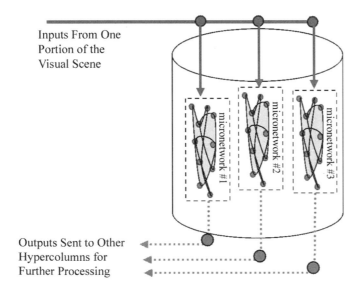

FIGURE 5.4. A hypercolumn is a small slab of tissue containing micronetworks that process various kinds of information from one small portion of the visual field.

called hypercolumns, each of which is responsible for processing information from one location in the visual scene. We will apply this concept more generally to other parts of the visual system in addition to the visual cortex. The term "hypercolumn" will be used to denote a **functional unit of neural machinery located in a visual processing area of the eye or brain organized to process information from a single small portion of the visual scene.** Within a hypercolumn, information might be processed by a number of different micronetworks. For example, some micronetworks might extract information concerning shapes of objects, others their color, and still others their motion. Neighboring hypercolumns do the same processing for nearby portions of the visual field.

Brain Areas and Nuclei

Neurons in the brain are not distributed uniformly. Instead, cell bodies are organized into densely packed clusters. Neuroanatomists have identified and given names to a large number of these densely packed brain regions based strictly on **cytoarchitecture,** which refers to distirctive characteristics of cell structure that can be seen under the microscope. Identifiable clusters of cell bodies in the cortex of the brain are generally called **areas** and those in other parts of the brain, such as the brainstem and thalamus, **nuclei.** In the early 1900s the German neuroanatomist Korbinian **Brodmann** proposed a mapping of the human cerebral cortex into fifty-two discrete areas based on cytoarchitecture. Brodmann's nomenclature is still in widespread usage.

Other divisions of the brain into separate areas have been made based on other criteria, such as their physiological properties, and, especially for sensory processing areas, on their organization into **topographically organized mappings. A topographic mapping** refers to the fact that a visual processing area of the brain is organized such that regions of the visual scene that are near one another are processed by hypercolumns that are also near one another. Topographic mappings of brain regions can be related to either the proximal stimulus (**retinal topography**) or the distal stimulus (**field topography**).

Topographic mapping for a hypothetical visual processing area in the left hemisphere of the brain is illustrated in Figure 5.5.

The left panel shows the original scene, a standing penguin, with nine specific locations marked. The observer is fixating at location 5. The middle panel shows the inverted image of the scene formed on the retina. The right panel shows the surface of a hypothetical visual processing area. Circles represent hypercolumns organized in **topographical fashion** across the surface of this brain area. Note, for example, that the region of the visual field identified as "6" in the original scene falls between the regions labeled "5" and "7." This relationship is preserved in the inverted retinal image and remains present in the organization of the hypercolumns. However, note that the topographic mapping across the hypercolumns is distorted and that the representations of some parts of the visual scene are magnified relative to others. For example, the distance from "5" to "6" is the same as the distance from "6" to "7" in both the original scene and the retinal image but not in the topographical mapping. In addition, the mapping that takes place in one hemisphere of the brain includes only the **contralateral hemifield,** the half of the visual field labeled "5" through "9," which falls in front of the opposite side of the head. The **ipsilateral hemifield,** labeled "1" through "5," is mapped onto the corresponding visual processing area located in the other

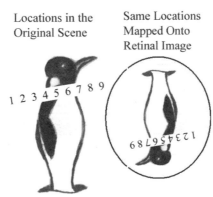

FIGURE 5.5. In topographic mapping, locations in a scene (left panel) are mapped onto the retinal image (middle panel) and onto hypercolumns in visual processing areas of the brain (right panel).

Locations in the Original Scene

Same Locations Mapped Onto Retinal Image

Processing of These Locations by Hypercolumns in Hypothetical Visual Processing Areas in Left Hemisphere of Brain

hemisphere of the brain. The properties of topographical mapping illustrated for the hypothetical brain area in Figure 5.5 are typical of many actual brain areas involved in visual processing.

A number of visual processing areas of the brain can be identified based on strictly cytoarchitectural criteria. For example, **Brodmann's areas** 17, 18, and 19 are all involved in visual processing. Visual processing areas have also been defined based on topographic mappings, and these have been given names such as V1, V2, and V3. In some cases, brain regions defined based on these two different criteria are isomorphic. For example, Brodmann's Area 17 corresponds to V1 and Area 18 to V2. However, in other cases the two methods of defining brain regions do not correspond. For example, Brodmann's Area 19 encompasses V3, V4, and V5. In addition, many brain areas have one or more "common" names. For example, Area 17 (V1) is frequently referred to as either the **primary visual cortex** or the **striate cortex.** Similarly, V5 is commonly called the **middle temporal area,** typically shortened to **MT.** This plethora of names can easily lead to confusion for the nonspecialist (and indeed for the specialist as well). The remainder of this text generally uses a single name or two for each structure to minimize the confusion. However, when reading the original scientific literature on this topic, it should be kept in mind that the terminology will not always correspond exactly to what is used in this text.

Neurons form connections with one another over various distances. The connections between neurons that are confined to a single brain area are referred to as **intrinsic** connections. The most numerous intrinsic connections are **short-range,** between neurons within one micronetwork. Moderate numbers of **medium-range** connections also pass between nearby micronetworks within a single hypercolumn, and some **long-range** intrinsic connections go between hypercolumns within the same brain area. Fiber bundles of axons that connect one brain area with another are called **extrinsic** connections.

2 Brain Areas Involved in Visual Processing

Many of the basic brain areas that are involved in processing visual information are the same in human and monkey. Some of the major extrinsic connectivity between these brain areas is illustrated Figure 5.6.

One major projection from the **retina** goes to the **lateral geniculate nucleus** in the thalamus. Brainstem nuclei that receive direct retinal input include the **superior colliculus, pretectum,** and **accessory optic nuclei.** There is also a minor projection to the **suprachiasmatic nucleus** in the hypothalamus that is responsible for coordinating circadian rhythms between the eye and the hypothalamus based on the light-dark cycle of a 24-hour day.

Secondary and higher-order processing of visual information takes place in the **inferior pulvinar,** the **striate visual cortex,** and a number of higher-order cortical areas referred to collectively as the **extrastriate cortex.**

The extrinsic connections formed between the eye and these nuclei travel in the **optic nerve** and **optic tract.** The projection to the lateral geniculate necleus in the thalamus dominates visual processing in primates. It is phylogenetically newer than the pathways projecting to the brainstem and hypo-

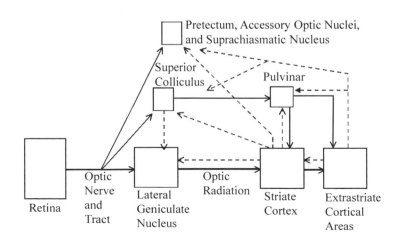

FIGURE 5.6. **Major brain nuclei and areas, and their interconnections, that are involved in processing visual information in humans and monkeys.**

FIGURE 5.7. View of brain from below, illustrating major components involved in the geniculostriate pathway. Adapted from D.H. Hubel, The visual cortex of normal and deprived monkeys. *American Scientist* 67:532–543, © 1979, by permission of D.H. Hubel.

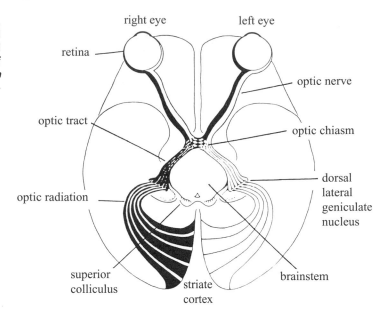

thalamus. Neurons in the lateral geniculate nucleus project to the striate cortex via a major extrinsic fiber tract called the **optic radiation.** The name **geniculostriate pathway** is sometimes used to refer to the major pathway from retina to lateral geniculate to striate cortex. The basic organization of the geniculostriate pathway is illustrated schematically in Figure 5.7.

The extrastriate cortical areas receive most of their visual inputs from the striate cortex. However, a smaller projection reaches extrastriate areas via the brainstem. One brainstem pathway carries information from the eye to the superior colliculus in the midbrain to the pulvinar in the thalamus, to the extrastriate cortex. Over thirty extrastriate cortical areas play a role in processing visual information, and their interconnections with one another and with striate cortex are complex. There are also feedback pathways from striate and many extrastriate areas to the thalamic and brainstem nuclei.

Retina

Basic Anatomical Organization

Viewed under the microscope, the retinas of humans and monkeys are virtually identical. Monkey retinal tissue that has been stained to show neuron cell bodies is shown in Figure 5.8.

The retinal tissue is organized into layers of cell bodies and synaptic connections. Layers are referred to as "outer" when they are closer to the sclera than to the vitreous. A diagrammatic view of the retina that illustrates the shapes of the various types of neurons that are present and their interconnections is shown in Figure 5.9.

The nuclei of the rods and cones form the **outer nuclear layer.** Cell bodies of four classes of interneurons, **bipolar, interplexiform, horizontal,** and **amacrine** cells, form the **inner nuclear layer. Ganglion cells,** whose axons travel up the optic nerve to central visual processing areas in the brain, form the **ganglion cell layer.** Connections of the receptors with horizontal and bipolar cells are made in the **outer synaptic layer.** Connections between bipolar, amacrine, and ganglion cells form in the **inner synaptic layer.** The interplexiform cells connect horizontal and amacrine cells.

One functional pathway in the retina reflects a **forward flow** of information. This pathway passes predominantly from receptors to bipolar cells to ganglion cells. Another set of functional pathways involves **lateral flow** of information. This function is subserved primarily by connections formed by horizontal cells, amacrine cells, and their interactions via interplexiform cells. Lateral flow allows information from distant receptors to also have an influence on forward information flow.

Neural processing of visual information begins in the retina after photoreceptors transduce light energy into an electrical signal, as described in Chapter 4. Recall that the effect of absorption of light on the photoreceptors is to hyperpolarize the membrane voltage. Thus, the regulation of neurotransmitter release from rods and cones is nonintuitive, being released continuously in the dark and

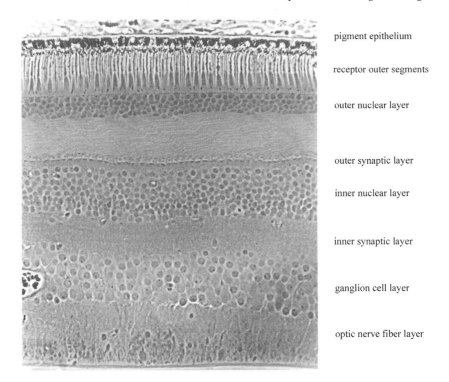

pigment epithelium

receptor outer segments

outer nuclear layer

outer synaptic layer

inner nuclear layer

inner synaptic layer

ganglion cell layer

optic nerve fiber layer

FIGURE 5.8. Photograph of tissue from monkey retina treated with Nissl's stain to highlight the organization of cell bodies into layers. Adapted from B.B. Boycott and J.E. Dowling, Organization of the primate retina: Light microscopy. *Philos. Trans. R. Soc. Lond. B* 255:109–194, © 1969, by permission of the Royal Society.

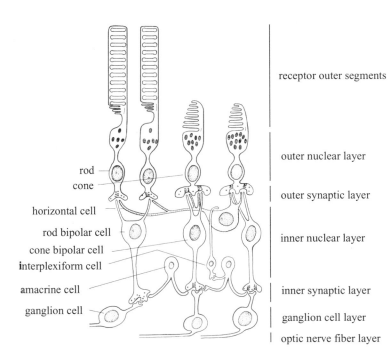

receptor outer segments

outer nuclear layer

rod
cone

outer synaptic layer

horizontal cell

rod bipolar cell
cone bipolar cell
interplexiform cell

inner nuclear layer

amacrine cell
ganglion cell

inner synaptic layer

ganglion cell layer

optic nerve fiber layer

FIGURE 5.9. Schematic drawing of cell types present in the various layers of the retina. Adapted from J.E. Dowling and B.B. Boycott, Organization of the primate retina: Electron microscopy. *Proc. R. Soc. Lond. B* 166:80–111, © 1966, by permission of the Royal Society.

FIGURE 5.10. Photoreceptors have the nonintuitive property that light causes a decrease in neurotransmitter release. Reproduced from K.-W. Yau, Phototransduction mechanisms in retinal rods and cones: The Friedenwald Lecture. *Invest. Ophthalmol. Vis. Sci.* 35:9–32, © 1994, by permission of the Association for Research in Vision & Ophthalmology.

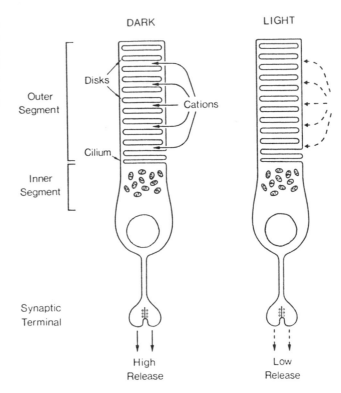

at decreasing rates with increasing light intensities (Figure 5.10).

Transmitter release in the absence of light stimulation is sometimes called the **dark light signal.** Ganglion cells in the retina typically have a **spontaneous discharge rate** of twenty to sixty spikes per second when the eye is in the dark. Thus, the effect of light on visual processing, at least in the early sensory pathways, is to modulate an ongoing signal rather than to create activity out of a quiescent state. Fluctuations in the spontaneous activity associated with the dark light signal may be responsible for the reports sometimes obtained from observers in a completely dark room that they experience a vague sensation of light stimulation.

On- and Off-Center Processing

Bipolar cells have receptive field shapes that come in two types, referred to as **on-center and off-center**, as illustrated in Figure 5.11.

Circular, Concentric, Antagonistic Center-Surround Receptive Field

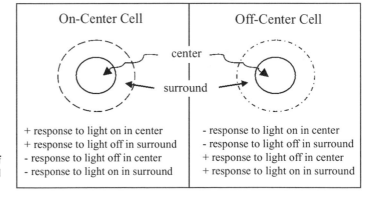

FIGURE 5.11. Schematic illustration of the receptive field organization of retinal bipolar cells.

The receptive fields are characterized by a **circular, concentric, antagonistic center-surround** organization. The term "circular" refers to the fact that the region on the retina where light can influence a bipolar cell is essentially circular in shape. The overall receptive field can be demarcated into two separate regions, a circular **center region** and a **concentric surround region**. The responses elicited by a light being turned on or off in the center of the visual field, whether hyperpolarizing (–) or depolarizing (+), are always **antagonistic** to those elicited by the same stimulus in its surround. On-cells are excited (depolarizing response) when light is turned on in the center and off-cells when light is turned off in the center.

In general, the function of center-surround receptive fields is detection of contrast between an object and its surround. The most effective overall stimulus for an on-center neuron is a bright spot that just fills the center of the receptive fill in the presence of a dark surround, as, for example, the full moon against a dark sky. The most effective overall stimulus for an off-center neuron is a dark spot that just fills the center in the presence of a bright surround, as, for example, a black widow spider on a white wall. Both types of neurons respond only weakly to uniform illumination and are not very much affected by changes in ambient light intensity.

On- and off-center bipolar neurons are present in roughly equal numbers. Every photoreceptor sends

---

Box 5.5
The Neurotransmitter Glutamate Produces Opposite Effects in On- and Off-Center Bipolar Neurons

The responses of the two types of bipolar cells are in opposite directions, and this is somewhat surprising, since they are both responding to release by the photoreceptors of the same neurotransmitter, **glutamate.** In off-center bipolar cells, glutamate causes an excitatory (depolarizing) response via a ligand-gated channel. In on-center bipolar cells, glutamate causes an inhibitory (hyperpolarizing) response

via a second messenger system. The photoreceptor is depolarized in the dark, causing a high rate of release of glutamate. When light is absorbed, the receptor hyperpolarizes, decreasing the rate of glutamate release. Since the on-center bipolar hyperpolarizes in response to glutamate, its response to light is to depolarize. The off-center bipolar cell responds in the opposite manner (Figure 1).

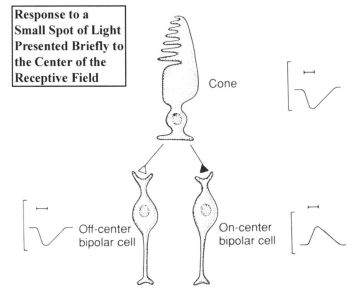

Response to a Small Spot of Light Presented Briefly to the Center of the Receptive Field

Cone

Off-center bipolar cell

On-center bipolar cell

FIGURE 1. Schematic illustrating the responses of cone photoreceptors and off-center and on-center bipolar neurons to a small spot of light presented briefly in the center of the receptive fields of the neurons. Conventions for the graphs are the same as in Figure 5.12. Adapted from E.R. Kandel, et al, *Principles of Neural Science,* Third Edition. Norwalk, CT: Appleton & Lange, © 1991, by permission of McGraw-Hill Companies.

outputs to both types. The on-center and the off-center bipolar cells can be differentiated anatomically by the fact that their connections with amacrine cells are made in different sublayers of the inner synaptic layer. The response elicited by light in the center of the receptive field of both on- and off-center bipolars is carried by feedforward pathways involving connections from receptors in the center of the receptive field to bipolar neurons, as described in more detail in Box 5.5.

The influences of the receptive field's surround are carried by lateral connections in the retina from receptors in the surround to horizontal cells and then to receptors and bipolar cells in the center. Some of the details of the circuitry that is involved are described in Box 5.6.

On- and Off-Center ganglion cells respond to light stimulation in the same manner as the corresponding bipolar cells except that their output consists of spikes superimposed on the graded potentials (Figure 5.12).

This makes sense, since their axons travel up the optic nerve, a distance too great to signal reliably with graded membrane potentials.

Box 5.6
Micronetworks Operate in the Retina to Create the Surround Response

The retina is simply an extension of the brain and contains micronetworks that perform processing on visual information before signals are sent up the optic nerve. An example of the kinds of processing that are performed is illustrated in Figure 1, which shows the connections responsible for the surround response of an on-center bipolar neuron.

Horizontal cells are excited (depolarized) by glutamate. Thus, reduction in glutamate from light in the surround causes the horizontal cell to be inhibited (hyperpolarized). The horizontal cell releases a gabanergic neurotransmitter onto the receptor terminals, causing inhibition (hyperpolarization). Thus, the hyperpolarizing response of the horizontal cell leads to disinhibition (depolarization) of the receptor in the center of the receptive field. Consequently, the on-center bipolar cell is inhibited (hyperpolarized).

Lateral interactions similar to those illustrated for horizontal cells in the outer synaptic layer are facilitated by amacrine cells in the inner synaptic layer. Amacrine cells are particularly involved in processing fast temporal changes, whereas horizontal cells are more concerned with steady-state lateral interactions. Some types of amacrine cells produce action potentials superimposed on top of their graded potentials, presumably because they need to register temporal changes quickly.

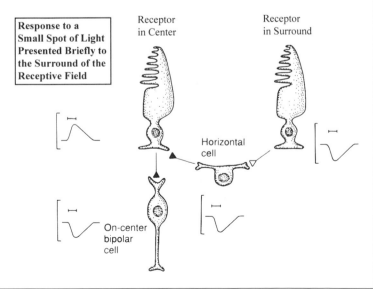

FIGURE 1. Schematic illustrating the responses of cone photoreceptors and an on-center bipolar neuron to light presented briefly in the surround region. Conventions for the graphs are the same as in Figure 5.12. Adapted from E.R. Kandel, et al, *Principles of Neural Science*, Third Edition. Norwalk, CT: Appleton & Lange, © 1991, by permission of McGraw-Hill Companies.

Response to a Small Spot of Light Presented Briefly to the Surround of the Receptive Field

Receptor in Center

Receptor in Surround

Horizontal cell

On-center bipolar cell

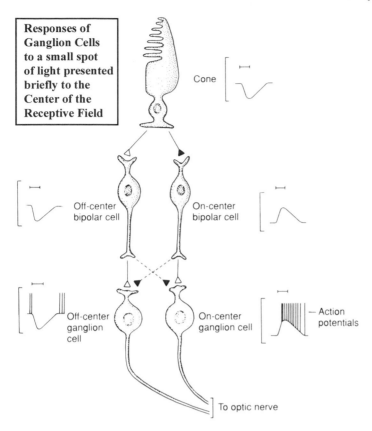

Responses of Ganglion Cells to a small spot of light presented briefly to the Center of the Receptive Field

FIGURE 5.12. Schematic illustrating responses of cones, bipolar cells, and ganglion cells to a small spot of light presented briefly in the center of the receptive fields of the neurons. The graphs next to each type of neuron show the changes in membrane voltage in response to a brief presentation of a small spot of light. Hyperpolarizing responses are depicted by downward deflections in the graphs and depolarizing responses by upward deflections. The horizontal lines demarcate the time during which the light stimulus was presented. On- and off-center ganglion cells respond similarly to the corresponding types of bipolar cells that provide their input except that they produce action potentials that can be seen as vertical spikes superimposed on the changes in membrane potential. Adapted from E.R. Kandel, et al, *Principles of Neural Science*, Third Edition. Norwalk, CT: Appleton & Lange, © 1991, by permission of McGraw-Hill Companies.

Rod and Cone Signal Processing

In addition to the organization of the retina into pathways that process information based on "on" and "off" responses, information flow is also organized according to whether it is derived from rod or cone receptors. Recall from Chapter 4 that **duplicity theory** refers to the general idea that the kinds of information picked up by rods and cones are processed differently and used to subserve different visual functions. Some differences in the properties of the information derived from rods and from cones are due to different properties of the receptors themselves, as discussed in Chapter 4. Other differences in function are conferred primarily by differences in the patterns of the connections rods and cones make with bipolar and ganglion cells.

The degree of spatial convergence is less for cones than for rods. A single bipolar or ganglion cell receives connections from many rods, allowing signals from many rod receptors to be pooled. However, this also reduces spatial resolution, because differences in photon absorptions by neighboring rods are averaged out by the pooling. Only a few cones converge onto each bipolar or ganglion

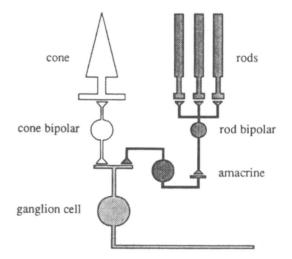

FIGURE 5.13. Feedforward pathways from cones pass directly to bipolars and then to ganglion cells, whose axons leave the retina. All signals derived from rods must pass through amacrine cells before being relayed to ganglion cells for transmission out of the eye. Reproduced from P.H. Schiller, On the specificity of neurons and visual areas. *Behav. Brain Res.* 76:21–35, © 1996, by permission of Elsevier Science.

cell, allowing higher spatial resolution, attained at a cost of poorer sensitivity.

Specialized **rod bipolar** and **cone bipolar** cells transmit the rod and cone signals. All rod bipolar cells are of the on-center type. All of the ganglion cells are functionally connected to either only cones or a mixture of rods and cones. Rod signals are relayed from bipolar neurons to ganglion cells via amacrine neurons and have no private pathway to get out of the retina (Figure 5.13).

Under scotopic lighting conditions, the ganglion cells that are functionally connected to a mixture of rods and cones carry signals only from rods, because the cones are not sensitive enough to respond. Under photopic conditions, these ganglion cells carry signals only from cones, because the response of the rod receptors is saturated and no longer modulated by lighting conditions. Under **mesopic** (intermediate) conditions, in which both rod and cone receptors are active, inhibitory interactions in the retina allow only one type of signal to gain access to the ganglion cells for transmission to higher brain centers.

Alpha, Beta, and Gamma Ganglion Cells

Ganglion cells in humans and monkeys can be classified into three anatomical types, called **alpha,** **beta,** and **gamma** based on their sizes and shapes. Examples of an alpha, also called **parasol,** and a beta, also called a **midget,** ganglion cell from the peripheral human retina are shown in Figure 5.14.

At any given retinal eccentricity, alpha cells have relatively larger cell bodies and dendritic fields. This enables them to have larger receptive fields. They also have large-diameter axons, allowing spikes to travel from the retina to central processing areas with short latency. Beta cells have medium-sized cell bodies and small dendritic and receptive fields. Their axons have a medium diameter, giving rise to spikes that reach central processing areas with medium latencies. Gamma cells have not yet been as well characterized as the other two classes and may be a heterogeneous class. They appear to have small cell bodies, dendrites of variable sizes, and small-diameter axons. Functionally, many gamma cells have unusual receptive field properties, making them hard to classify.

Optic Nerve, Optic Chiasm, and Optic Tract

The axons of the roughly 1 million ganglion cells from each eye leave the retina at the optic disc and form the **optic nerve** for that eye. Fibers from gan-

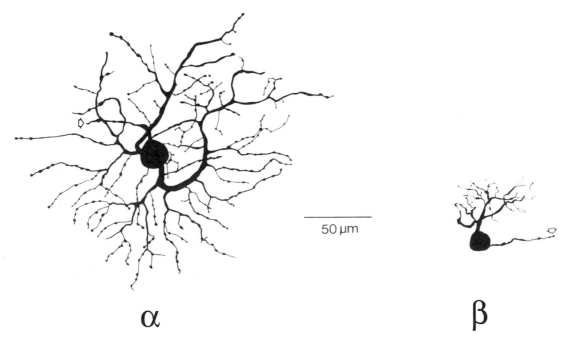

50 μm

α   β

FIGURE 5.14. Tracings of parasol (alpha) and midget (beta) ganglion cell from a human retina as seen under a microscope. Arrowheads point to where each axon leaves the plane of the section. Adapted from R.W. Rodieck, et al, Parasol and midget ganglion cells of the human retina. *J. Comp. Neurol.* 233:115–132, © 1985, by permission of Wiley-Liss, Inc., a subsidiary of John Wiley & Sons.

glion cells **decussate** in the **optic chiasm.** This means that ganglion cells located in the **nasal half of each retina cross** to the contralateral hemisphere of the brain, whereas ganglion cell axons from the **temporal retina remain uncrossed.** Together, the crossed fibers from the contralateral eye and the uncrossed fibers from the ipsilateral eye, form the **optic tract** in each hemisphere that transmits information from the contralateral hemifield to central visual processing brain structures. As illustrated in Figure 5.15, the effect of decussation is that information about the right visual field is processed by the left hemisphere of the brain and that about the left visual field by the right.

Neurologists make use of this decussation pattern to help localize brain damage in patients with visual field defects, as illustrated in Box 5.7.

Brainstem Nuclei

Ganglion cell axons in the optic tract that enter the brainstem terminate in the pretectum, accessory optic nuclei, and superior colliculus. These fibers come primarily, perhaps exclusively, from alpha and gamma cells, although this issue has not been studied extensively. Some of these are probably collateral fibers of ganglion cells that also project to the lateral geniculate nucleus.

Neurons that project to the pretectum are involved in the pupillary light reflex. The pretectum projects to the Edinger-Westphal nucleus, which projects in turn to the ciliary ganglion and finally to the sphincter muscles of the iris. The projection from the pretectum is bilateral, leading to both a **direct response** of the pupil in the eye stimul-

---

Box 5.7
Inferring Location of Neurological Damage Based on Behavior

Neurologists and neuroophthalmologists must often try to infer the location of brain damage caused by conditions such as a stroke or brain tumor. Figure 1 illustrates how these inferences are made based on certain perceptual deficits.

Consider a patient who can see normally in all parts of the visual field in the left eye but is blind in the right eye, with no apparent damage to the right retina (A in Figure 1). It can be inferred that the damage must be in the optic nerve in front of the optic chiasm (A in Figure 5.15).

Similarly, consider a patient who has a normal nasal field in each eye but is blind in the temporal field in each eye (B in Figure 1). The clinical term for this condition is **bitemporal hemianopia.** It is the classic visual field defect of chiasm lesions (B in Figure 5.15) and would immediately lead to concern about conditions such as a tumor of the pituitary gland that sits below the chiasm.

Homonymous hemianopia is a condition in which half of the visual field is blind in each eye (hemianopia) and the defect is either on the left side or on the right side in both eyes (homonymous). This type of field defect indicates damage to only one hemisphere somewhere beyond the optic chiasm. Defects in left fields (C

in Figure 1) point to damage in the right hemisphere (C in Figure 5.15) and those in right fields to left hemisphere damage.

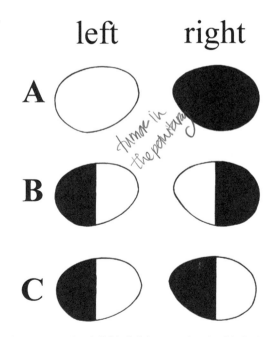

FIGURE 1. Visual field deficits associated with brain damage at locations labeled •A, •B, and •C in Figure 5.15.

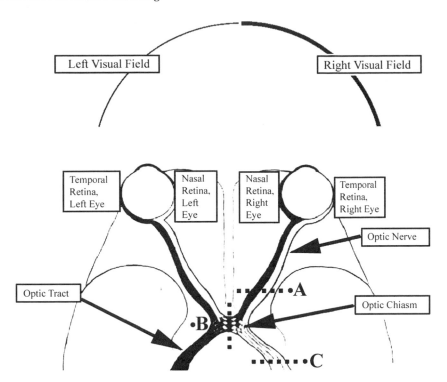

FIGURE 5.15. Decussation at the optic chiasm has the functional effect that neural processing in the left brain is primarily concerned with information derived from the right visual field and vice versa. Information about the right visual field is processed along the neural pathways shown with thick, bold lines, the left visual field with thin lines. Brain damage at locations A, B, or C will lead to specific visual field deficits that are discussed in Box 5.7.

ated by light and a **consensual response** in the fellow eye.

The projections to the accessory optic nuclei are involved in the basic reflex functions that help stabilize the retinal image during movements of the head, as discussed in Chapter 4.

The largest projection to the brainstem is to the superior colliculus, in the midbrain, called the **optic tectum** in nonmammalian vertebrates such as amphibians and birds, in which it is the principal brain structure involved in processing information from the eyes. The neurons in the superior colliculus are organized into dorsal and ventral layers. The dorsal layers receive visual information from the retina and via feedback from the visual cortex. The ventral layers receive information from other sensory modalities.

There is a topographic organization across the surface of the superior colliculus such that adjacent regions of space map onto adjacent hypercolumns. The hypercolumns processing visual information project down through all three dorsal layers. In other words, a hypercolumn penetrating straight down from the surface and passing through the dorsal layers would encounter neurons that process information from the same region of the visual field.

The same relationship holds up roughly across the ventral layers as well. For example, if an object is located upward and to the lift of the head, the neurons in the superior colliculus responding to the sight (dorsal layers) and sound (ventral layers) of this object will lie above and below one another.

Because of this organization, one major functional role of the superior colliculus is thought to be combining sensory information from the different sensory modalities. This information is then used to help orient the eyes and head to salient stimuli in the environment. Extrinsic connections from the superior colliculus project mostly to parts of the brain that control eye movements, described in Chapter 4, and head and neck movements.

In addition, the superficial layers of the superior colliculus also send information onto the pulvinar in the thalamus. These nuclei in turn have extensive extrinsic connections to extrastriate visual areas. The functional role of these anatomical projections has not been heavily studied, but they may play a role in directing visual attention, in coordination with eye movements, towards relevant portions of the visual scene.

Lateral Geniculate Nucleus

The lateral geniculate nucleus of the thalamus consists of two anatomical structures. The **ventral lateral geniculate nucleus** receives sparse retinal input, and its functional role in vision is not well defined. Discussion here will be limited to the **dorsal lateral geniculate nucleus (dLGN)**, a major thalamic relay station that transmits retinal information along the geniculostriate pathway.

The dLGN is organized into six layers, as illustrated in Figure 5.16, which shows a Nissl-stained section from a monkey brain as seen under the microscope.

Each layer contains a complete topographical representation of the contralateral visual field but

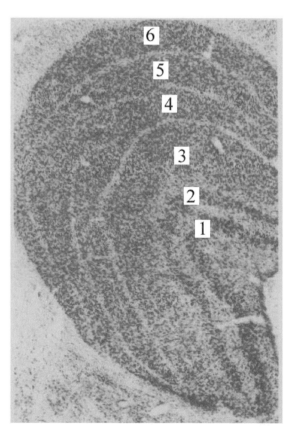

FIGURE **5.16. Photograph of a section through the dorsal lateral geniculate nucleus of a monkey brain. Reproduced from J.R. Wilson, et al, Effects of aphakia on the geniculostriate system of infant rhesus monkeys. *Acta Anat.* 142:193–203, © 1991, by permission of S. Karger A.G., Basel.**

receives this representation via fibers that come exclusively from either the ipsilateral or the contralateral retina. Layers 1, 4, and 6 receive from the contralateral eye and layers 2, 3, and 5 from the ipsilateral. The visual representations in the various layers are in spatial register and this accounts for the fact that a column of dLGN tissue that penetrates from layer 6 at the surface through all of the layers to 1 forms a functional hypercolumn.

The two ventral layers, 1 and 2, are called **magnocellular** because they are composed of large cell bodies. The four dorsal **parvocellular layers** are composed of neurons with small cell bodies. There is also a separate diffuse group of cells that lie ventral to layer 1 and also between the other layers, referred to collectively as the **koniocellular cells.**

The three types of retinal ganglion cells, alpha, beta, and gamma, described above have specific projection patterns onto the three types of dLGN layers. The alpha cells whose axons project to the lateral geniculate nucleus terminate in the magnocellular layers. The beta cells project predominantly, perhaps exclusively, to the parvocellular layers. Some gamma cells project to koniocellular neurons. It is uncertain whether the axons that terminate on koniocellular neurons are collateral branches of gamma cell axons that also terminate in brainstem nuclei or come from a separate population of gamma cells that project only to the dLGN.

The receptive field properties of individual neurons in the dLGN appear to be essentially the same as those of the retinal ganglion cells providing their input. This has led to the frequent characterization of the function of the dLGN as merely to serve as a **relay station** between the retina and the visual cortex. A better characterization is that the dLGN acts as a **filtering device.** Recall from Chapter 4 that biological perceptual systems need to limit the amount and type of information that is subjected to extensive processing in order to avoid information overload. The dLGN plays a role in several kinds of filtering functions. It does this with micronetworks that receive inputs from the brainstem and cortex as well as from the eye, as illustrated in Box 5.8.

A number of filtering functions have been proposed for the dLGN. One is to allow the dLGN to perform a gating function with regard to saccadic suppression, as described in Chapter 4. This function may play a role in perceptual organization of the inputs obtained from scanning patterns, as illustrated in Box 5.9.

Another proposed filtering function of dLGN micronetworks emphasizes the role of the extensive extrinsic projections, from the brainstem, pre-

Box 5.8
Micronetworks in the dLGN Receive Inputs from the Brainstem and Cortex in Addition to the Eyes

Information that passes from the retina through the dLGN is processed by micronetworks that receive inputs from the brainstem and visual cortex (Figure 1).

This diagram illustrates a relay neuron that receives several intrinsic and extrinsic inputs. In addition to the retina, extensive inputs are shown coming from the cortex, pregenicu-

late nucleus, and other brainstem areas. Intrinsic connections with interneurons come from axons and specialized synapses formed by presynaptic densities. Excitatory synaptic connections are depicted by filled triangles and squares and inhibitory connections by open symbols. Output from the relay neuron passes to the visual cortex.

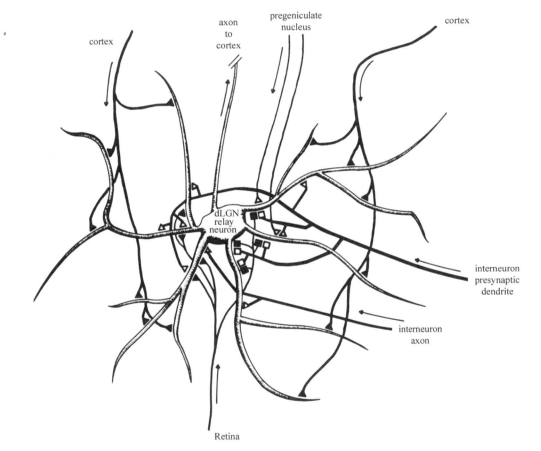

FIGURE 1. Schematic illustrating micronetworks that have been inferred based on observed anatomical connections within the dorsal lateral geniculate nucleus. Adapted from J.R. Wilson, Synaptic organization of individual neurons in the macaque lateral geniculate nucleus. *J. Neurosci.* 9:2931–2953, © 1989, by permission of Society for Neuroscience.

Box 5.9
The dLGN May Gate Flow of Information During Scanning Patterns

Recall from the discussion in Chapter 4 that the visual system needs to process information obtained during sampling periods while the eye is fixating. These periods of fixation are separated by saccadic eye movements in an alternating sequence, as illustrated in the top row of Figure 1.

If the cortex responded to the information received during the fixation periods in a sustained manner, the result would be as depicted in the second row. The information being processed during one fixation would continue flowing into the cortex during subsequent fixations. Some type of gating mechanism is needed, perhaps along the lines depicted in the third row, to allow the information obtained during each fixation to be segregated as illustrated in the fourth row. The transient response triggered by each saccade is used to turn off the flow of

information coming in from the previous fixation as designated by the negative sign arrows passing from the transient response to the sustained response. As will be elaborated on in later sections discussing P-streams and M-streams of neural processing, the detailed information transmitted during each fixation period as a sustained response is routed primarily through the parvocellular layers of the dLGN and the transient response triggered by each saccade primarily through the magnocellular layers.

The exact cellular mechanisms that regulate this gating process are not well understood. However, it is known that **acetylcholine** inputs to the dLGN from the **pedunculopontine nucleus** and **gabanergic** inputs from the **pretectum** carry signals that are related to the onset and termination of saccades and may be involved in regulating these processes.

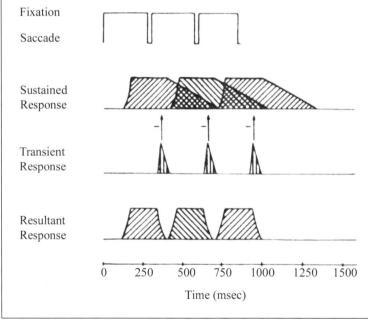

FIGURE 1. Proposed sustained and transient responses to information obtained during scanning patterns and the resultant signals that would be sent to the cortex following processing by micronetworks in the dLGN. Adapted from B.G. Breitmeyer, Unmasking visual masking: A look at the 'why' behind the veil of the 'how.' *Psychol. Rev.* 82:52–69, © 1980, by permission of American Psychological Association.

geniculate nucleus, and cortex as illustrated in Box 5.10.

Another proposed filtering function of the dLGN is to gate transmission of sensory information to the cortex according to the wake/sleep cycle, as described in Box 5.11.

Striate Cortex

Basic Gross Anatomy

The **striate cortex** is so named because myelinated axons form a prominent striation in its middle

Box 5.10
Extrinsic Pathways to dLGN May Allow Cognitive Enhancement of Processing

Extrinsic pathways from the brainstem, pregeniculate nucleus, and the cortex may feed information into dLGN micronetworks that are involved in directing or focusing attention on specific areas of the visual scene that have been determined by cortical processing to deserve further scrutiny. This kind of filtering could be used, for example, to amplify incoming visual signals in ways that complete sensory input patterns under conditions when afferent stimulation is weak (Figure 1).

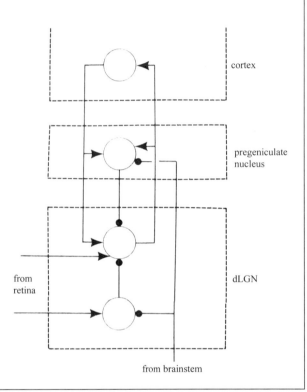

FIGURE 1. Schematic illustrating some interconnections of the dLGN with the retina, brainstem, pregeniculate nucleus (PGN), and cortex. Reproduced from E. Harth, et al, The inversion of sensory processing by feedback pathways: A model of visual cognitive functions. *Science* 237:184–187, © 1987, by permission of the American Association for the Advancement of Science.

layers, termed the **stripe of Gennari**, so distinct that it can be seen at low magnification in unstained dissected tissue. A number of the other names are also used to refer to this anatomical structure. It corresponds to Brodmann's cytoarchitectural Area 17. It is also often called the **primary visual cortex** or the first cortical visual area, V1, based on connectivity and topographic mapping. The gross anatomical location of this structure in the cerebral cortex is roughly similar in humans and monkeys in the sense that it lies at the back of the brain near the posterior pole of the occipital lobe. However, its anatomical location is slightly different in humans and monkeys, because there are differences in the details of the way the cortex bends and folds to form elevated convolutions called **gyri** and grooves called **fissures** or **sulci.**

In humans, most of the striate cortex is buried in a fissure on the medial surface of the occipital lobe, with only a small portion extending onto the lateral surface, as illustrated in Figure 5.17.

In monkeys, the striate cortex is located mostly on the dorsolateral surface of the occipital cortex, called the **operculum,** that lies posterior to the **lunate sulcus,** as illustrated in Figure 5.18.

The top panel is a view of a monkey brain from behind. The position of the lunate sulcus on the left hemisphere is indicated by the dotted line. (The X

Box 5.11
The dLGN May Gate Sensory Information from the Eyes During Sleep and dreaming

The dLGN is involved in the wake/sleep cycle and in dreaming, and may function to regulate the amount of information that is allowed to pass from the eye to the cortex during the various stages of sleep and dreaming. The neural inputs to the dLGN that could potentially mediate in these functions include a **histaminergic** input from the **hypothalamus** associated with wakefulness and a **serotonergic** input from the **raphe nuclei** associated with sleep. These inputs tend to be diffuse **neuromodulatory** projections rather than conventional synaptic contacts with individual neurons. Why the brain needs to regulate flow of sensory information in conjunction with the wake/sleep cycle and during dreaming is an interesting question in itself but one that goes beyond the scope of this book.

and arrow are discussed in the next section.) The striate cortex covers the dorsal and lateral surface of the posterior pole of the brain and wraps around to the medial surface. If followed to the medial surface between the two hemispheres, the striate cortex continues into a buried fissure, as in the human brain.

A block of tissue has been cut out of the right hemisphere of this brain. The bottom panel shows a view of what would be seen looking to the left from within the groove where the block was removed. The portion of striate cortex that falls on the dorsolateral surface is at the top (A), and the portion in the buried fissure lies below (B). The lunate sulcus is at the upper right of the photograph, and the border of the striate cortex, just behind the lunate, is indicated by the arrow.

Topographical Mapping

The contralateral hemifield is topographically mapped onto the striate cortex in each hemisphere in both man and monkey. The X symbol in the upper panel of Figure 5.18 marks the projection of the fovea in the monkey brain. A charge in anatomical position along the arrow corresponds to movement in the topographic projection away from the fovea along the horizontal meridian. Movement along the dotted line parallel to the lunate sulcos

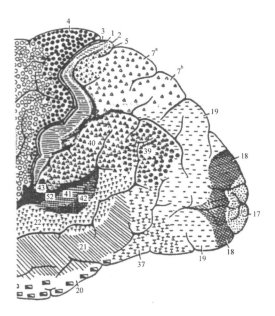

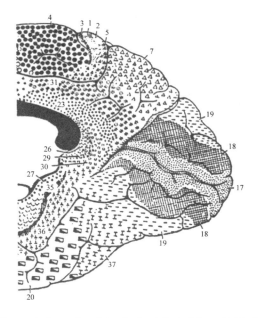

FIGURE 5.17. Location of Brodmann's areas on the lateral (left panel) and medial (right panel) surface of the posterior part of the human brain. Brodmann's area 17, the striate visual cortex, is located in the occipital lobe at the posterior pole of the brain. Adapted from R. Lindenberg, et al, *Neuropathology of Vision*. Philadelphia: Lea & Febiger, © 1973, by permission of Lippincott Williams & Wilkins.

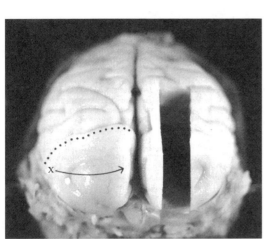

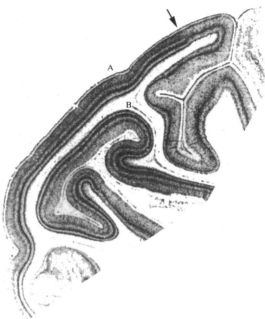

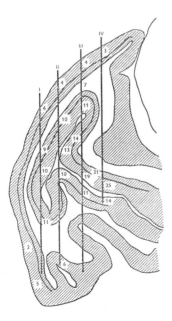

FIGURE 5.18. Location of the striate visual cortex, Brodmann's area 17, on the dorsolateral surface of the occipital lobe of the monkey brain. A block of tissue has been removed from the right occipital lobe. Top panel: view of left and right occipital lobes as seen from back of brain. Bottom panel: higher magnification view of what would be seen from within hole from which block was removed. See text for details. Adapted from D.H. Hubel, The visual cortex of normal and deprived monkeys. *American Scientist* 67:532–543, © 1979, by permission of D.H. Hubel.

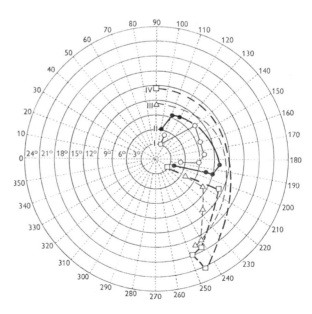

FIGURE 5.19. Mapping of the topographical projection of the visual field onto the striate cortex. The left panel shows recording sites from a monkey's striate cortex and the right panel shows the location of receptive fields corresponding to each recording site. Reproduced from P.M. Daniel and D. Whitteridge, The representation of the visual field on the cerebral cortex in monkeys. *J. Physiol. (Lond.)* 159:203–221, © 1961, by permission of Physiological Society.

corresponds to downward movement along the vertical meridian. This topographical projection was first mapped by Talbot and Marshall in the 1940s for the central visual field by recording electrical potentials in the cortex while stimulating the eye with small spots of light. This mapping was extended to the peripheral visual field by Daniel and Whitteridge in the 1960s; Figure 5.19 is reproduced from their study.

The left panel illustrates a section from a monkey brain from the same view as in Figure 5.18. Recordings were made with a small needle electrode. The vertical lines indicate tracts formed during four separate penetrations of the electrode through the cortical tissue. The numbers next to the electrode tracts designate the eccentricity from straight ahead, in degrees, where electrical responses could be elicited by light flashes presented in the visual field. The exact locations within the visual field where light evoked responses were elicited during each electrode tract are plotted in the right panel. Based on a detailed analysis of raw data of the kind shown in this figure, the entire topographic projection onto the striate cortex has been mapped out. More recently, the topographic mapping has been demonstrated by exposing anesthetized monkeys to a target-shaped visual stimulus after injection of radioactively labeled 2-deoxyglucose that is taken up by active brain cells. The stimulus is shown at the top of Figure 5.20.

The bottom photograph is of a flattened section of the brain cut parallel to the surface. The dark regions reflect a high degree of radiolabeling. The center of the target can be seen projected onto the left side of the tissue, with mappings of successive rings moving successively to the right. Similar topographic mappings have been performed recently for human subjects by using noninvasive brain imaging methods such as positron emission tomography (PET) and functional magnetic resonance imaging (fMRI).

Cortical Magnification

More striate cortex tissue is devoted to processing information from the central visual field than from the periphery. This accounts for the distortion in the topographic mapping apparent in Figure 5.20. The term **cortical magnification factor** refers to the length in millimeters of the part of the cortex that is involved in processing information about each degree of the visual field, measured radially from the fixation point. Magnification is about 13 mm per degree in the foveal region and decreases gradually to less than 0.1 mm per degree in the far periphery.

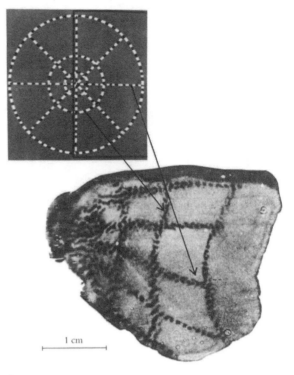

FIGURE 5.20. Topographic mapping of monkey striate visual cortex as assessed with radioactive labeling of active areas of the brain (bottom panel) obtained while viewing a radial stimulus (top panel). Reproduced from R.B. Tootell, et al, Deoxyglucose analysis of retinotopic organization in primate striate cortex. *Science* 218:902–904, © 1982, by permission of Association for the Advancement of Science.

The gradual reduction in cortical magnification with eccentricity can be related to decreases in density of cones and of ganglion cells, as illustrated for monkeys in Figure 5.21.

As described in Chapter 4, cone density is highest in the fovea and decreases monotonically with eccentricity. Ganglion cell density shows a similar distribution except that it is low in the fovea because ganglion cells have been displaced outward to form the foveal pit. However, even though the ganglion cells are physically displaced out of the fovea, they remain connected to cones in the fovea. Thus, if one were to plot the positions of their receptive fields instead of the physical locations of their cell bodies, as is shown Figure 5.21, the ganglion cell distribution would look the same as that of the cones.

Consider the cortical machinery that makes up one hypercolumn, as illustrated schematically in Figure 5.22.

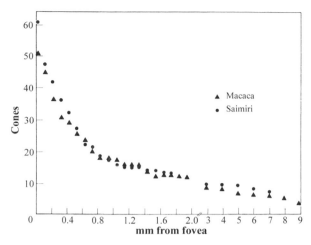

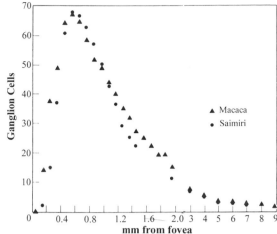

FIGURE 5.21. Density of cones (top panel) and ganglion cells (bottom panel) as a function of retinal eccentricity for two genera of monkeys (*Macaca* and *Saimiri*). Reproduced with permission from E.T. Rolls and A. Cowey, Topography of the retina and striate cortex and its relationship to visual acuity in rhesus monkeys and squirrel monkeys. *Exp. Brain Res.* 10:298–310, 1970.

Every hypercolumn is of similar size, has similar micronetwork circuitry, and receives input from a roughly constant number of ganglion cells. However, the spacing of that constant number of ganglion cells on the retina is not uniform. The ganglion cells are some five neurons deep near the fovea but widely spaced in a single layer in the far periphery. Imagine building up the cortical surface by beginning at the foveal projection with a single hypercolumn built up from inputs from a set number of ganglion cells. Then imagine what would happen if one added additional hyper-columns by using, successively, inputs from the same number of ganglion cells. The expected result is illustrated in Figure 5.23.

Even though the number of ganglion cells per hypercolumn remains constant (two per hypercolumn in this hypothetical example), the area of the visual field that is sampled by successive hypercolumns is increasingly spread out. As illustrated in Figure 5.23, this is simply due to the fact that the ganglion cells are more spread out in the peripheral retina. This schematic is oversimplified, because there are actually two steps involved in this process: magnification first from ganglion cells to dLGN cells and then from dLGN cells to cortical hypercolumns.

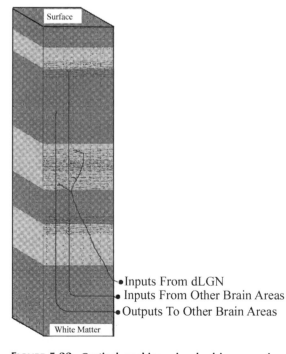

FIGURE 5.22. Cortical machinery involved in processing information within one cortical hypercolumn. Adapted from J.P. Frisby, *Seeing: Illusion, Brain and Mind,* Oxford University Press, © 1979, by permission of J.P. Frisby.

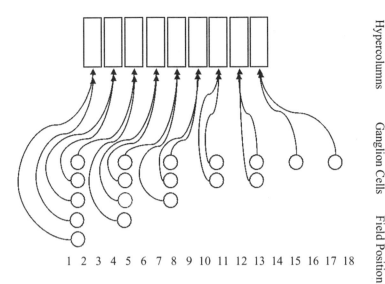

Hypercolumns

Ganglion Cells

Field Position

1 2 3 4 5 6 7 8 9 10 11 12 13 14 15 16 17 18

FIGURE 5.23. Schematic illustrating how cortical magnification might arise automatically out of uneven distribution of ganglion cells in retinal tissue.

The variation of sensitivity for some visual functions depending on eccentricity can be understood with reference to cortical magnification. The sensitivity of many visual functions is limited by the amount of cortical machinery available to process information. For these functions, sensitivity measured at various eccentricities can be made similar if scaled according to magnification so that the same amount of cortical tissue is involved in processing the stimuli located at each eccentricity.

An example of the effects of scaling based on cortical magnification is given in Chapter 4. Revisit the demonstration shown in Figure 1 in Box 4.1 and see if you now have a more complete understanding of why the letters that have been scaled to different sizes at different eccentricities have similar visibility.

Intrinsic Connectivity

The historical convention in neuroanatomy has been to divide all neocortical areas of the brain into six principal layers. In cases where this does not provide a good fit to what is actually seen with the microscope, neuroanatomists have tended to subdivide some of the layers into additional sublaminae rather than using a numbering scheme with more than six layers. This has led to somewhat arcane numbering systems for

some cortical areas, and the striate cortex is no exception. The layering convention that has been applied to the striate cortex can be appreciated in sections of tissue prepared with Nissl's stain that show the locations of the cell bodies, as shown in Figure 5.24.

The top panel shows a section of striate cortex tissue as seen at low power, as in previous figures. The piece of tissue within the box includes parts of the striate cortex from both the dorsolateral surface and the buried fissure. The bottom panel shows the striate cortex at higher magnification and demonstrates the numbering of the layers of the striate cortex. The layer nearest the surface is designated 1, and the deepest layer before entering the white matter (w) is designated 6. Layers in between are fit into this six layer scheme by using as many sublayers as necessary to accommodate the anatomically distinct layers (e.g. 4A, 4B, 4C).

The micronetworks that are present in the striate cortex can be better appreciated by looking at the shapes of the axonal and dendritic fields of the neurons that are present in the various layers, as summarized in Figure 5.25.

The inputs from the dLGN terminate in specific sublaminae rather than diffusely, as illustrated in the left side of Figure 5.25. The three subpopulations of dLGN neurons (parvocellular, magnocellular, and koniocellular) remain largely separated. Parvocellular layers terminate primarily in sublayers 4Cβ and 4A, perhaps with small projections to 1 and 6. There are two subpopulations of magnocellular inputs, one projecting to the entire depth of 4Cα and the other only to the superficial portion. Both also have a small projection to layer 6. Koniocellular cells terminate in specialized anatomical structures called **blobs** in layer 3.

The right side of Figure 5.25 reveals that within layer 4C, the primary recipient zone for dLGN inputs, neurons of a distinctive type, called **stellate cells,** are the major postsynaptic targets. Stellate cells come in two vaFrieties, with and without spines on the dendrites, and send only local axonal projections.

Another distinctive type of neuron, called a **pyramidal cell,** has an apical dendrite that extends vertically over several layers, giving rise to dendritic fields within restricted sublayers. Pyramidal neurons often send out long axons that make extrinsic connections with other cortical areas and nuclei. Neurons in upper layers of the striate cortex tend to project to extrastriate cortical areas. Some pyramidal neurons with cell bodies located in layer 5 project to the superior colliculus, accessory optic nuclei, and pulvinar. Large pyramidal neurons with cell bodies in layer 6 have projections to the dLGN.

Some cortical neurons in the sublayers of 4 that receive input from the dLGN have concentric on- or off-center receptive fields similar to those seen at earlier stages of processing in the retina and dLGN. However, as soon as information is transmitted to second- or higher-order cortical neurons, new receptive field properties are seen that are not present at earlier stages of processing. Most of these neurons respond best to lines or bars of a particular orientation moving in a particular direction. These properties of striate cortex neurons, referred to as **orientation selectivity** and **direction selectivity**, were first discovered by Hubel and Wiesel in electrophysiological studies performed on anesthetized animals in the 1960s and 1970s. A schematic illustrating the mapping of the receptive field for an orientation- and direction-selective neuron is shown in Figure 5.26.

The left side of the figure illustrates the configuration of the stimuli relative to the receptive field of the neuron. An elongated ban of light was moved orthogonal to its orientation in both directions on each trial and the orientation of the ban was varied between trials. The electrical responses recorded from a microelectrode posi-

tioned close to the neuron are shown on the right side of the figure. Each time the neuron fired an action potential, a spike of electrical activity was recorded that can be seen in the figure as a thin vertical line extending above and below the background electrical activity being generated in the surrounding cortical tissue. This particular neuron gives the strongest response (most spikes/record) to a left oblique oriented ban moving upward and to the right across the receptive field as shown in the fourth row in the figure.

Hubel and Wiesel devised a classification scheme for neurons according to receptive field type based on examining changes in firing rate with various stimulus geometries. They referred to neurons of the type shown in Figure 5.26 as **simple cells.** The reception fields of single cells are characterized by easily defined spatially discrete, on- and off-responsive regions and by selectivity for contours of specific orientations and directions of movement.

We will describe micronetwork models that try to account for directional selectivity in Chapter 10. Hubel and Wiesel proposed that a simple micronetwork based strictly on connectivity between the dLGN and cortical neurons might account for the emergence of orientation selectivity in striate cortex. Their scheme is illustrated in Figure 5.27.

A cortical unit that responds best to horizontal orientations is assumed to receive selective inputs from dLGN neurons whose receptive fields are aligned in a horizontal row. Similarly, the inputs to a cortical neuron that responds to vertical orientations are presumed to be aligned in a vertical row.

Columnar Organization

Hubel and Wiesel made another groundbreaking observation about the organization of orientation selectivity in cortical neurons. They discovered an ordered arrangement of orientation preferences across neighboring neurons. They noted that when a vertical penetration was made with the recording electrode straight down through all the layers of the cortex, all of the neurons tended to have similar orientation preferences. However, as the electrode was moved laterally across the cortex, the preferred orientation gradually shifted. Hubel and Wiesel used the term **ordered orientation columns** to refer to this columnar organization, illustrated in Figure 5.28.

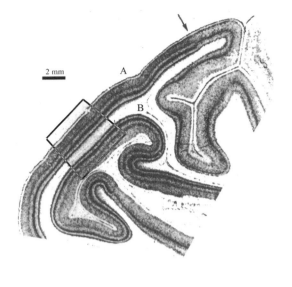

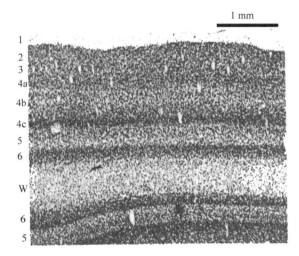

FIGURE 5.24. Photographs of a low-magnification view of a slice of the striate cortex of a monkey as shown previously in Figure 5.18 (top panel) and, at a higher magnification (bottom panel), of the portion of the tissue within the box in the top panel. Adapted from D.H. Hubel, The visual cortex of normal and deprived monkeys. *American Scientist* 67:532–543, © 1979, by permission of D.H. Hubel.

Each small slab, or column, of cortical tissue appears to be organized to analyze the inputs coming into layer 4 for a single orientation. Adjacent columns span all orientations systematically within a short distance – about 1 mm, the approximate width of a single hypercolumn. Thus,

there appear to be sufficient micronetworks within each hypercolumn to analyze the portion of the visual scene mapped to that hypercolumn for all orientations.

Hubel and Wiesel also discovered a columnar organization of cortical tissue with respect to a receptive field property called **ocular dominance.** This property reflects the extent to which the receptive field properties of a neuron are related to stimulation of the left eye, the right eye, or some combination of the two. The degree of ocular dominance can be characterized by using an **ocular dominance rating scale.** For example, a neuron might be categorized as 1 if it responds only to the left eye, 2 if to both eyes but more strongly to the left, 3 if equally to both eyes, 4 if more strongly to the right, on 5 if only to the right eye.

When these kinds of ratings are made for neurons in the dLGN, all the cells have ratings of either 1 or 5, depending on their layer. In other words, all dLGN neurons are **monocular.** There are no **binocular** cells, with ratings of 2, 3, or 4. This total lack of binocular neurons is maintained in layer 4 in the striate cortex, where most dLGN axons terminate. Binocular neurons, with ocular dominance categories 2, 3, or 4, are first encountered in striate cortex higher-order neurons that lie in layers above and below the dLGN recipient sublaminae.

When recording from single neurons in striate cortex, Hubel and Wiesel discovered a columnar organization for **ocular dominance** similar to that for orientation. When a vertical penetration was made with the recording electrode, the binocular neurons recorded from all exhibited a preference for the same eye. However, as the electrode was moved laterally across the cortex, the preferred eye shifted in a regular fashion from left to right and back to left every 0.5 mm. The physical layout of ocular dominance and orientation columns was conceptualized by Hubel and Wiesel with an **ice cube tray model,** as illustrated in Figure 5.29.

Each "ice cube" corresponds to a slab of cortical tissue that processes information about one orientation from one portion of the visual scene as seen from a point of view dominated by one eye. An entire "tray of ice cubes" corresponds to one hypercolumn and processes information about one portion of the visual scene regarding all orientations as seen by both eyes.

When first described by Hubel and Wiesel, hypercolumns were simply conceptual models based on functional relationships noted while recording from single neurons. However, an

FIGURE 5.25. Drawings of inputs from the lateral geniculate nucleus to the striate cortex of the monkey, along with several neuronal types that are present and their outputs. Adapted from E.R. Kandel, et al, *Principles of Neural Science*, Third Edition. Norwalk, CT: Appleton & Lange, © 1991, by permission of McGraw-Hill Companies.

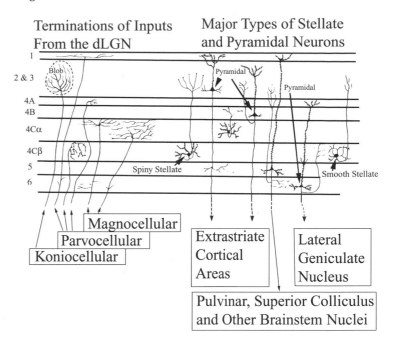

FIGURE 5.26. Illustration of the electrical responses elicited from a neuron in the striate cortex that responds to bars of particular orientations that more across its receptive field in a given direction. Reproduced from D.H. Hubel and T.N. Wiesel, Receptive fields and functional architecture of monkey striate cortex. *J. Physiol.* 195:215–243, © 1968, by permission of Cambridge University Press.

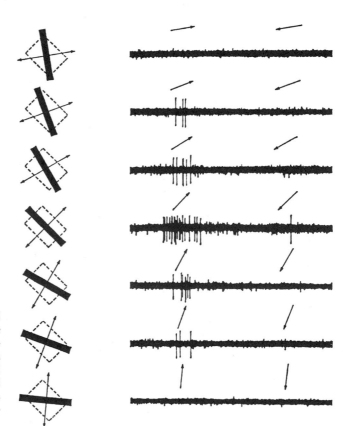

Cortical neuron receiving input from three dLGN neurons to achieve vertical orientation selectivity

Cortical neuron receiving input from three dLGN neurons to achieve horizontal orientation selectivity

FIGURE 5.27. Illustration of how cortical neurons with orientation selectivity could be formed based on selective connectivity with dLGN neurons.

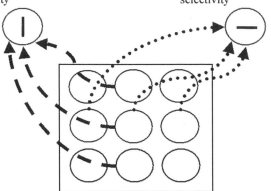

Locations of circular receptive fields of nine individual dLGN neurons

anatomical marker that shows up when striate cortex tissue is stained for the mitochondrial enzyme, **cytochrome oxidase,** provides a potential anatomical marker for an organization based on hypercolumns. Figure 5.30 shows a thin slice of neural tissue from the striate cortex cut in a plane approximately tangential to the surface and stained for this enzyme.

The slice of tissue passes obliquely through several layers, starting with layer 1 at the surface on the right side and ending where layer 6 passes into the white matter on the left. Layer 4C, where the dLGN axons terminate, is in the middle. The tissue exhibits a series of stripes **in layer 4C** that correspond to the **ocular dominance columns** associated with the left and right eyes. In addition, the blobs can be visualized in layers 2 and 3. Although not easy to see in this single slice of tissue, when information from several slices is reconstructed in three dimensions, it can be seen that the blobs are aligned

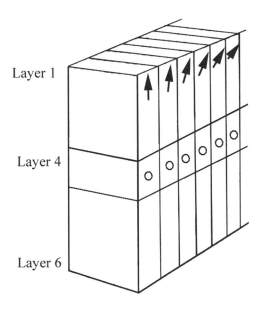

FIGURE 5.28. Orientation columns in the striate cortex.

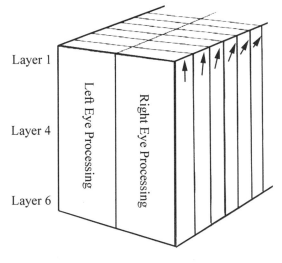

FIGURE 5.29. The ice cube tray model proposed by Hubel and Wiesel as one possible arrangement of orientation and ocular dominance columns in the striate cortex.

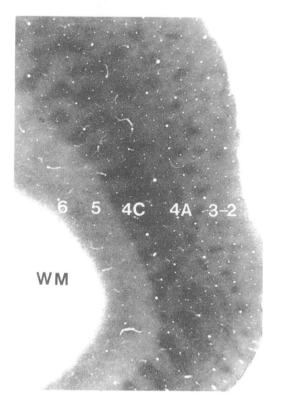

FIGURE 5.30. Photograph of a slice of striate cortex tissue from a monkey brain that has been stained for the mitochondrial enzyme cytochrome oxidase. Layer 1 is at the far right at the surface of the brain. The slice is at an oblique angle through the cortical tissue and passes into the white matter (WM) at the far left. The approximate locations of layers 2–3, 4A, 4C, 5, and 6 are labeled. Reproduced from M. Tigges, et al, Competition between an aphakic and an occluded eye for territory in striate cortex of developing rhesus monkeys: Cytochrome oxidase histochemistry in layer 4C. *J. Comp. Neurol.* 316:173–186, © 1992, by permission of Wiley-Liss, Inc., a subsidiary of John Wiley & Sons.

and motion, that will be described in subsequent chapters. There are about 3,500 hypercolumns in the striate cortex.

Long-range horizontal connections that extend a millimeter or more interconnect hypercolumns. They appear to selectively connect neural modules associated with similar kinds of processing. For example, an orientation column associated with one hypercolumn receives input from neurons in nearby hypercolumns that have similar orientations specificity.

3 Extrastriate Pathways

Once information leaves the striate cortex, it fans out to reach extrastriate cortical regions located in the occipital, temporal, and parietal lobes. The second cortical visual processing area, V2, corresponds to Brodmann's cytoarchitectural Area 18. It gets major input from V1, along with a very sparse projection from the dLGN and from the pulvinar. Cytochrome oxidase staining reveals an organization based on repeating patterns of a **thick stripe** and a **thin stripe,** separated by **interstripe regions,** as illustrated in Figure 5.32.

Neurons in the layer 3 blobs of the striate cortex project to the thin stripes of V2. Layer 3 neurons in the interblob regions project to the pale interstripes. Neurons from layer 4B project to the thick stripes.

over each ocular dominance column at intervals of about 0.5 mm. This finding has led to a conceptualization of each hypercolumn as being organized around four blob structures, as shown schematically in Figure 5.31.

A hypercolumn in striate cortex is a slab of tissue that occupies about 1 mm² on the surface and penetrates through all six layers. It processes various kinds of information for a single location in the visual scene. A hypercolumn contains a pair of ocular dominance columns (one for each eye) and a complete set of orientation columns. It also contains other microcircuits involved in processing other types of visual information, such as color

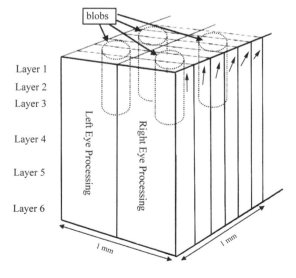

FIGURE 5.31. Blobs in relation to ice-cube model.

interblob to   blob to
interstripe   thin stripe

Area 18
(V2)

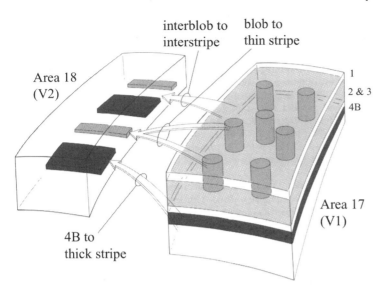

1
2 & 3
4B

Area 17
(V1)

4B to
thick stripe

FIGURE 5.32. The thick, thin, and inter-stripe regions of V2 and their inputs from V1. Adapted from J.H. Martin, *Neuroanatomy. Text and Atlas*, Second Edition. Stamford, CT: Appleton & Lange, © 1996, by permission of McGraw-Hill Companies.

Each thick-pale-thin-pale cycle spans about 4 mm and receives inputs from four to five hypercolumns. Thus, the modular organization in V2 is more complex than that in the striate cortex. In the striate cortex, a square millimetre of cortex represents a hypercolumn that processes information about a specific locus in space. In V2, there appears to be a multiple discontinuous mapping of the visual field, with each area mapped in triplicate.

This organization of V2 illustrates a general rule that while each extrastriate area appears to be highly organized, the geometry of the organization is not necessarily the same from area to area. The changes from V1 to V2 suggest that the organization has shifted from one dominated by lower-level features, such as ocular dominance, orientation, and point-to-point mapping, to one concerned with more integrated features as reflected in percepts. These changes in processing described at the level of neural tissue can be mapped onto the schemes that have been developed by psychologists to account for the transformations of sensations into percepts as summarized in Chapter 1 (see Figure 1.12).

Outputs from V1 and V2 continue on to other extrastriate pathways. With progress through the system, topographic precision declines and sometimes disappears, and functional properties of neurons become more abstract. The receptive field properties of individual neurons are not as obviously tied to specific sensory stimuli but have relationships that are more abstract.

There are over a billion cortical neurons in the striate and extrastriate cortical areas. This gives a ratio of about 600 visual processing cortical neurons for each input line from the dLGN. Over thirty separate extrastriate areas have been identified in the monkey, defined by a combination of cytoarchitecture, patterns of connectivity, and functional properties of the neurons. These areas are interconnected with a complex set of over 300 identified extrinsic interconnections, illustrated in the schematic in Figure 5.33.

This diagram suggests a **hierarchical organization** of visual processing brain areas into about ten levels. A hierarchical organization means that the processing performed in some areas precedes that performed in others. Thus, the outputs at one stage can be considered to form the inputs needed at the next stage. Sometimes this hierarchical organization is obvious. For example, the retina projects to the dLGN, which in turn projects to the striate cortex. However, the organization is usually more complex than a strictly hierarchical relationship. For example, cortical area V1 projects to V2, which in turn projects to V3. However, a direct projection from V1 to V3 bypasses V2.

Also, many connections between brain areas involve **feedback** as well as **feedforward,** because they have reciprocal connections that go in both directions. Feedforward connections tend to terminate in layer 4 of the receiving area. Feedback connections tend to originate in the upper layers and terminate in the upper and lower layers. The fact that the origins and targets are in different laminae means the feedforward and feedback pathways are not strictly reciprocal but reflect a more complicated **recurrent architecture**.

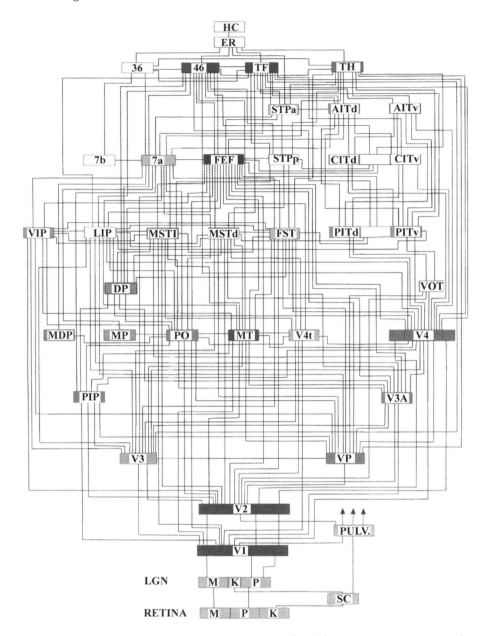

FIGURE 5.33. Connectivity of extrastriate cortical areas. Adapted from D.C. Van Essen, et al, Information processing in the primate visual system: An integrated systems perspective. *Science* 255:419–423, © 1992, by permission of D.C. Van Essen.

It is at the extrastriate stages of neural processing that some characteristics of human and monkey visual processing probably start to diverge significantly. However, some extrastriate areas appear to be roughly homologous in humans and monkeys, and it is on those areas that the remainder of this chapter and the chapters that follow concentrate.

4 Streams of Processing

As information flows from the retina through the visual processing brain areas, it is organized into **parallel processing streams.** This organization allows information that pertains to one portion of the visual scene to be processed simultaneously along several distinct routes.

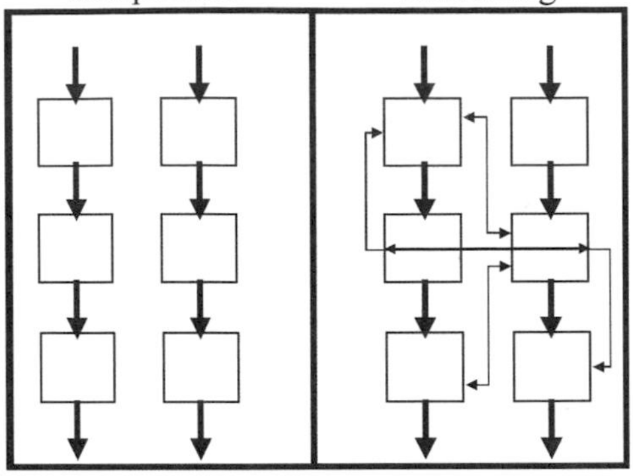

Parallel Streams

Independent   Interacting

FIGURE 5.34. Parallel stream organization in which streams operate in a truly independent fashion (left panel) and in which the streams interact at certain stages (right panel).

Perceptual scientists who adopt an evolutionary approach have speculated that parallel processing resulted from constraints due to evolution. Evolution can tinker with properties of brains only if many of their subsystems are unessential for the survival of the organism. Otherwise, any small change is more likely to lead to extinction than to evolution. Consequently, many evolved brain circuits act in parallel and are somewhat redundant.

In reality, the term "parallel" in reference to the information processing performed in brain circuits is something of a misnomer. Parallel streams of processing are only semi-independent, because they always interact with one another at certain stages, as illustrated schematically in Figure 5.34.

Processing streams differ in terms of their scope or extent. In some cases, a single stream of processing diverges into two or more parallel streams that persist for no more than a few synapses and then collapse back into a single stream again. Other processing streams continue to run in parallel over a span of several brain areas.

An example of parallel processing streams restricted to a single area already discussed is **rod and cone processing streams** in the retina. These two parallel streams form at the photoreceptor layer of the retina, continue through the bipolar cells, and merge back into a single anatomical processing stream at the ganglion cells.

Another example of parallel processing streams that extend over more than one brain processing area is the way separate on- and off-center receptive fields form at the level of bipolar neurons in the retina and continue to similarly organized ganglion cells, dLGN neurons, and nonoriented neurons in the dLGN recipient layers of the striate cortex. This organization into separate **on- and off-steams** of processing is illustrated in Figure 5.35.

Beyond layer 4 of the striate cortex, the on- and off-streams dissipate. However, many striate and extrastriate cortical neurons continue to reflect the influences of inputs from the on- and off-streams, because their receptive fields can be separated into separate "on" and "off" regions.

The on- and off-streams are both very ancient systems phylogenetically and must play a very important role related to survival. One possible function is described in Box 5.12.

Box 5.12
Why Do We Need Both On- and Off-Streams?

It seems as though Mother Nature went to a lot of work to give us both on- and off-streams of processing. Couldn't the brain just figure out that a light has gone off when a neuron with an on-center receptive field stops firing? One possible reason both on- and off-streams evolved early in biological perceivers may have to do with the rapidity with which information about both "on" and "off" needs to be signaled. For example, a fish in the ocean must be able to respond quickly to a predator rising from the dark depths and appearing bright because it is illuminated from above, as well as to a seabird diving from above, signaled by a dark shadow against the bright sky.

Peter Schiller and colleagues tested the functional significance of the on- and off-streams in monkeys by using a pharmacological agent that rendered neurons in the on-stream inoperative without significantly altering neurons in the off-stream. Monkeys were impaired in responding to increases but not decreases in intensity for small spots of light, while other aspects of visual processing, such as orientation and direction selectivity, remained unimpaired.

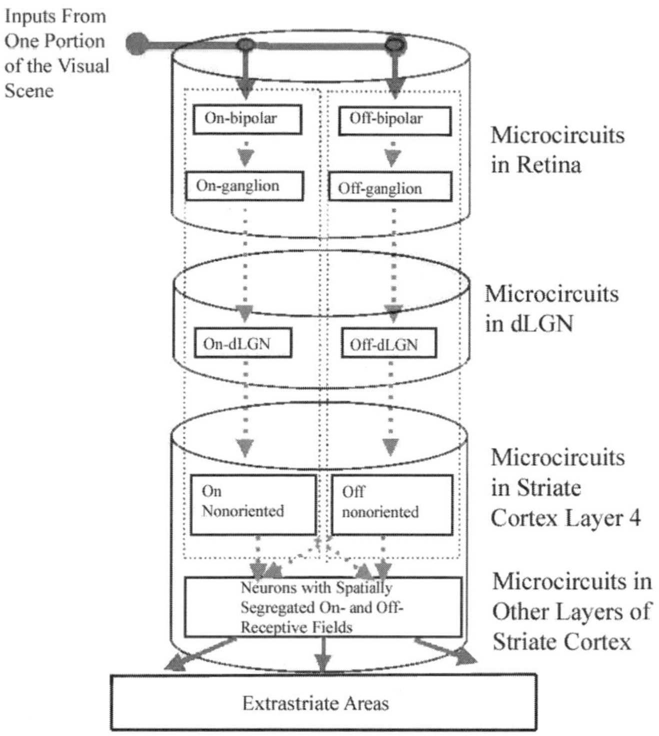

FIGURE 5.35. Functional organization of on- and off-parallel processing streams.

Inputs From One Portion of the Visual Scene

On-bipolar | Off-bipolar

On-ganglion | Off-ganglion

Microcircuits in Retina

On-dLGN | Off-dLGN

Microcircuits in dLGN

On Nonoriented | Off nonoriented

Microcircuits in Striate Cortex Layer 4

Neurons with Spatially Segregated On- and Off-Receptive Fields

Microcircuits in Other Layers of Striate Cortex

Extrastriate Areas

M-, P-, and K-Streams

The three types of retinal ganglion cells, **alpha, beta,** and **gamma** form the beginnings of **M-, P-,** and **K-streams** of processing, respectively. These streams share many of the same neurons as the on- and off-streams, but the information carried in these two sets of streams is essentially orthogonal. Both alpha and beta neurons come in both the on- and off-center varieties. Gamma neurons often have unusual receptive field properties, not easily classified as either On or Off center.

The M-, P-, and K-streams derive their names from their route through the dLGN. M-stream information passes through neurons in the magnocellular layers, P-stream information through the parvocellular layers, and K-stream information through the koniocellular neurons, located primarily in interlaminar zones. These streams continue to the striate cortex, where they become reorganized and feed into extrastriate processing streams.

The P-stream is the largest in terms of numbers of neurons that are involved. Approximately 80% of the fibers traveling up the geniculostriate pathway carry P-stream information along the anatomical route shown schematically in Figure 5.36.

The P-stream derives predominantly, perhaps exclusively, from retinal beta ganglion cells and

passes through the parvocellular layers of the dLGN and then primarily to sublaminae 4Cβ and 4A in the striate cortex. Interneurons in the striate cortex continue the information flow from layer 4 to the blob and interblob regions of layers 2 and 3. Finally, these feed the extrastriate processing streams described in the next section.

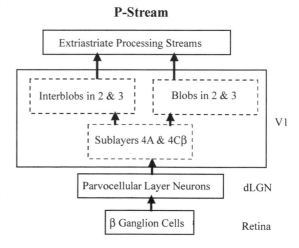

P-Stream

Extrastriate Processing Streams

Interblobs in 2 & 3 | Blobs in 2 & 3

Sublayers 4A & 4Cβ

V1

Parvocellular Layer Neurons — dLGN

β Ganglion Cells — Retina

FIGURE 5.36. Functional organization of the P-stream of visual processing.

M-Stream

```
┌─────────────────────────────────────┐
│     Extrastriate Processing Stream   │
└─────────────────────────────────────┘
                  ▲
         ┌─────────────────┐
         │   Sublayer 4B   │
         │        ▲        │      V 1
         │  Sublayer 4Cα   │
         └─────────────────┘
                  ▲
┌─────────────────────────────────────┐
│     Magnocellular Layer Neurons      │    dLGN
└─────────────────────────────────────┘
                  ▲
      ┌───────────────────────┐
      │    α Ganglion Cells    │      retina
      └───────────────────────┘
```

FIGURE 5.37. **Functional organization of the M-stream of visual processing.**

The M-stream makes up about 10% of the population of axons passing up the geniculostriate pathway. The anatomical pathway for the M-stream is shown schematically in Figure 5.37.

The M-stream begins with α ganglion cells, travels through the magnocellular layers of the dLGN, and then terminates primarily in sublamina 4Cα in striate cortex. Interneurons transmit information to layer 4B and then to extrastriate processing streams.

The K-stream is not well defined, and relatively little is known about its properties. It passes from the retina through the koniocellular cells of the dLGN and then on to the striate and extrastriate cortices, as shown schematically in Figure 5.38. The K-stream is most likely derived from the γ type of ganglion cells, which are in actuality a heterogeneous population of several ill-defined subtypes. As described above, most, perhaps all, of these ganglion cells have projections to the koniocellular neurons in the dLGN, although some of them probably have collaterals that project to other nuclei as well, (not shown in Figure 5.38) such as the superior colliculus. The dLGN koniocellular neurons are probably also a heterogeneous group. They are found primarily in the interlaminar zones between the M- and P-layers, and in a layer of small cells called the S-layers that lie ventral to the M-layers. At least one subsystem of the K-stream projects from the dLGN to the blobs in layers 2 and 3 of the striate cortex. A few neurons in the dLGN are also known to project directly to the extrastriate cortex, bypassing the striate cortex, and these may be part of the K-stream.

The designation of the term "K-stream" to refer to this heterogeneous functional pathway has as its

primary role reminding us that there are functional subsystems involved in visual processing in addition to the M- and P-streams. These other functional systems make up about 10% of the geniculostriate pathway, which, it may be noted, is as large as the M-stream.

Behavioral studies carried out in the laboratories of **William Merigan** and **Peter Schiller** have tried to differentiate functions subserved by the M- and P-streams by placing lesions in either the parvocellular or the magnocellular portion of the dLGN. Examples of two such lesions are illustrated in Figure 5.39.

A complication of these studies is that the lesions produce indeterminate effects on neurons from the heterogeneous K-stream as well. Nevertheless, these studies have been widely interpreted as demonstrating functional differences between processing performed by the M- and by the P-stream. A summary of some of the primary functional differences that have been reported following lesions to the two systems is shown in Figure 5.40.

Lesions affecting the P-stream have their primary effects on color vision; fine discriminations involving textures, patterns, and shapes; contrast sensitivity and stereopsis when the stimulus information is carried primarily by high spatial frequencies. The primary deficits seen following lesions affecting the M-stream are in perception of motion and of flicker. Additional details of these deficits will be discussed in subsequent chapters.

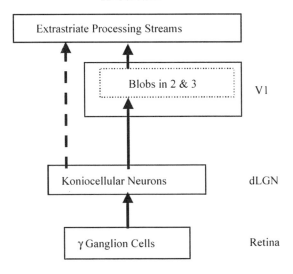

FIGURE 5.38. **Functional organization of the K-stream of visual processing.**

 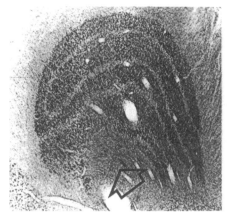

Parvocellular lesion   0.5 mm   Magnocellular Lesion

FIGURE 5.39. Photograph of a monkey brain in which lesions were produced within the parvocellular (left panel) and magnocellular (right panel) layers. Reproduced from P.H. Schiller, et al, Role of the color-opponent and broadband channels in vision. *Visual Neurosci.* 5:321–346, © 1990, by permission of Cambridge University Press.

More recent studies have demonstrated that except for color vision, both systems are capable of processing all of the visual attributes studied, but they do so for different ranges of spatial and temporal frequencies (discussed in Chapter 3). Both systems process low spatial and temporal frequen-cies and can contribute to stereovision (discussed in Chapter 9). The P-stream dominates acuity, as well as contrast detection at low temporal and high spatial frequencies. The M-stream mediates contrast detection at higher temporal and lower spatial frequencies.

Deficit magnitude following lesions

VISUAL CAPACITY		PLGN	MLGN
color vision		severe	none
texture perception		severe	none
pattern perception		severe	none
shape perception	fine	severe	none
	coarse	mild	none
brightness perception		none	none
coarse scotopic vision		none	none
contrast sensitivity	fine	severe	none
	coarse	mild	none
stereopsis	fine	severe	none
	coarse	pronounced	none
motion perception		none	moderate
flicker perception		none	severe

FIGURE 5.40. Summary of deficits found in monkeys in which either the parvocellular layers of the dLGN (PLGN) or the magnocellular layers (MLGN) were lesioned. Reproduced from P.H. Schiller, On the specificity of neurons and visual areas. *Behav. Brain Res.* 76:21–35, © 1996, by permission of Elsevier Science.

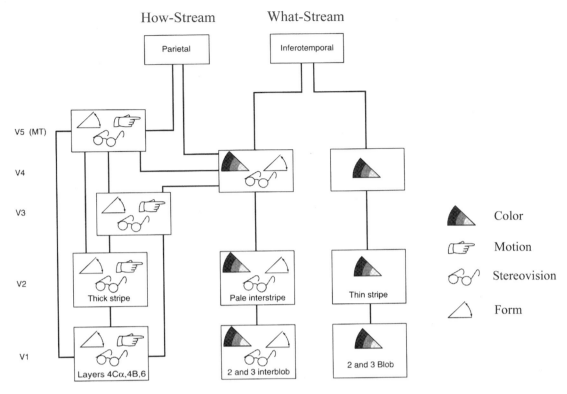

FIGURE 5.41. Schematic summarizing three major extrastriate streams of processing. Adapted from E.R. Kandel, J.H. Schwartz, and T.M. Jessell, *Principles of Neural Science*, Third Edition. Norwalk, CT: Appleton & Lange, © 1991, by permission of McGraw-Hill Companies.

Extrastriate Processing Streams

Information that leaves the striate cortex for extrastriate processing areas is organized into at least three functional streams that specialize in different kinds of perceptual information, as illustrated in Figure 5.41.

A **blob stream** originates from the blobs in layers 2 and 3 of the striate cortex, passes through the thin stripes of V2, and continues on to other extrastriate areas. It appears to be heavily involved in processing of color information and will be described further in Chapter 7.

An **interblob stream** originates from the interblob regions of layers 2 and 3 of the striate cortex, passes to the interstripe regions of V2, and then goes on to other extrastriate areas. It appears to be heavily involved in processing information about two-dimensional shape, described further in Chapter 8, and stereovision, described further in Chapter 9. Some color processing also takes place within this stream.

Information derived from the M-streams terminates in layers 4A, 4B, and 6 of the striate cortex.

Much of this information is funneled through layer 4B, which forms the beginning of an extrastriate **4B stream.** Neurons from 4B project to the thick stripes of V2 and appear to be involved in form, stereovision, and motion processing. They will be described further in Chapters 8, 9, and 10.

As the blob, interblob, and 4B streams continue through extrastriate cortical areas, they merge, diverge, and become reorganized in complex ways that are not well understood. However, a general organizational scheme has been noted that involves a division of labor between **ventral streams,** which continue primarily into the **inferotemporal lobe,** and **dorsal streams,** which continue into the **parietal lobe.** The inferotemporal lobe receives a dominant input from the blob and interblob streams and the parietal lobe from the 4B stream.

Functional differences between the dorsal and ventral streams having to do with object vision and spatial vision were originally proposed by **Mortemei Mishkin, Leslie Ungerleider,** and colleagues. Since then, various other naming schemes have been applied to this basic division, including "Where" and "What" streams. We will adopt a

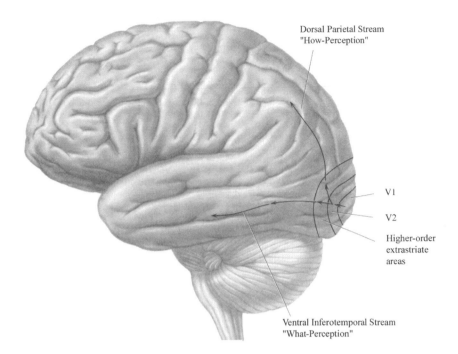

Dorsal Parietal Stream
"How-Perception"

V1

V2

Higher-order
extrastriate
areas

Ventral Inferotemporal Stream
"What-Perception"

FIGURE 5.42. Extrastriate cortical processing continues into a ventral stream that passes into the inferotemporal cortex, specialized for What-perception, and into a dorsal stream continuing into the posterior parietal cortex, specialized for How-perception. Adapted from J.H. Martin, *Neuroanatomy*, second Edition. Stamford, CT: Appleton & Lange, © 1996, by permission of McGraw-Hill.

naming scheme that maps onto the two kinds of perceptual outputs described in Chapter 1, What-perception and How-perception. As will be elaborated on in later chapters, the ventral stream appears to specialize in What-perception, while the dorsal stream is more specialized for How-perception (Figure 5.42).

Summary

The biological tissue that processes visual information is exceedingly complex, involving billions of neurons and trillions of connections. This tissue can be described from a number of levels of analysis. Groups of neurons are organized into micronetworks, then hypercolumns, and finally nuclei and areas.

The first structure involved in processing visual information is the retina, composed of several layers. The retina has a number of neuronal types organized into functional on- and off-center pathways involving forward and lateral flow of information. Three functional types of ganglion cells have axons that leave the retina and pass up the optic nerve.

Half of the axons traveling up the optic nerve decussate, causing a mapping of the contralateral visual field in each hemisphere. Some axons terminate in the brainstem in the superior colliculus, the pretectum, and the accessory optic nucleus. However, the dominant retinal projection in primates is along the geniculostriate pathway, first to the dorsal lateral geniculate nucleus (dLGN) in the thalamus and then to the primary visual cortex (V1) near the posterior pole of the brain. The dLGN acts as a filtering device for visual information.

Information from the retina is routed along P-, M-, and K-streams of processing that pass through different layers of the dLGN and terminate in different sublaminae of V1. Hypercolumns within V1 analyze the visual information for low-level features, such as oriented lines, and then pass processed information along to extrastriate cortical areas for further processing.

As information leaves the striate cortex and enters extrastriate areas, it becomes organized into a number of parallel streams that emphasize different functions. These extrastriate streams feed into two major higher-order visual processing streams. A dorsal stream routes information to the parietal lobe and is specialized for Howpereception. A ventral stream routes information to the inferotemporal lobe and is specialized for What-perception.

Selected Reading List

Daniel, P.M., and Whitteridge, D. 1961. The representation of the visual field on the cerebral cortex in monkeys. *J. Physiol. (Lond.)* 159:203–221.

DeFelipe, J., and Jones, E.G. 1988. *Cajal on the Cerebral Cortex.* New York: Oxford University Press.

Hubel, D.H. 1988. *Eye, Brain, and Vision.* New York: Scientific American Library.

Hubel, D.H., and Wiesel, T.N. 1968. Receptive fields and functional architecture of monkey striate cortex. *J. Physiol.* 195:215–243.

Hubel, D.H., and Wiesel, T.N. 1974. Uniformity of monkey striate cortex: A parallel relationship between field size, scatter, and magnification factor. *J. Comp. Neurol.* 158:295–306.

Kuffler, S.W. 1953. Discharge patterns and functional organization of mammalian retina. *J. Neurophysiol.* 16:37–68.

Leventhal, A.G., Rodieck, R.W., and Dreher, B. 1981. Retinal ganglion cell classes in the Old World monkey: Morphology and central projections. *Science* 213:1139–1142.

Lund, J.S. 1988. Anatomical organization of macaque monkey striate visual cortex. *Ann. Rev. Neurosci.* 11:253–288.

Merigan, W.H., Katz, L.M., and Maunsell, J.H.R. 1991. The effects of parvocellular lateral geniculate lesions on the acuity and contrast sensitivity of macaque monkeys. *J. Neurosci.* 11:994–1001.

Mishkin, M., Ungerleider, L.G., and Macko, K.A. 1983. Object vision and spatial vision: Two cortical pathways. *Trends Neurosci.* 6:414–417.

Perry, V.H., and Cowey, A. 1984. Retinal ganglion cells that project to the superior colliculus and pretectum in the macaque monkey. *Neuroscience* 12:1125–1137.

Perry, V.H., Oehler, R., and Cowey, A. 1984. Retinal ganglion cells that project to the dorsal lateral geniculate nucleus in the macaque monkey. *Neuroscience* 12:1101–1123.

Rodieck, R.W., and Watanabe, M. 1993. Survey of the morphology of macaque retinal ganglion cells that project to the pretectum, superior colliculus, and parvicellular laminae of the lateral geniculate nucleus. *J. Comp. Neurol.* 338:289–303.

Saini, K.D., and Garey, L.J. 1981. Morphology of neurons in the lateral geniculate nucleus of the monkey: A Golgi study. *Exp. Brain Res.* 42:235–248.

Schiller, P.H. 1992. The ON and OFF channels of the visual system. *Trends Neurosci.* 15:86–92.

Schiller, P.H. 1996. On the specificity of neurons and visual areas. *Behav. Brain Res.* 76:21–35.

Talbot, S.A., and Marshall, W.H. 1941. Physiological studies on neural mechanisms of visual localization and discrimination. *Am. J. Ophthalmol.* 24:1255–1264.

Tootell, R.B., Silverman, M.S., Switkes, E., and De Valois, R.L. 1982. Deoxyglucose analysis of retinotopic organization in primate striate cortex. *Science* 218:902–904.

Van Essen, D.C., Anderson, C.H., and Felleman, D.J. 1992. Information processing in the primate visual system: An integrated systems perspective. *Science* 255:419–423.

Wilson, J.R. 1989. Synaptic organization of individual neurons in the macaque lateral geniculate nucleus. *J. Neurosci.* 9:2931–2953.

6
Perceptual Processing II. Abstractions: How Can Perceptual Processing Be Characterized and Modeled as Flow of Abstract Information?

Questions

After reading Chapter 6, you should be able to answer the following questions:

1. Compare and contrast the advantages of general-purpose and special-purpose perceptual processing.
2. Describe digital coding over a labeled line.

3. Explain some of the essential differences among digital, scalar, tuned filter, and temporal pattern coding.
4. Compare and contrast local with across-fiber coding schemes.
5. What is one bit of Shannon information?
6. What are formal neurons?
7. Describe the basic architectures used by neural network models.
8. How can vectors be used to represent perceptual coding and processing?
9. What is an algorithm, and what fundamental issues are involved in modeling human perceptual processing as carrying out an algorithm?
10. Elaborate about what it means to conceptualize perceptual processing as being essentially a statistical procedure.
11. Describe the conditions under which small amounts of noise can be exploited to allow detection of signals that are otherwise too weak to be detected.
12. Differentiate between the forward-flow problems and the inverse problems that are involved in perceptual processing.

Chapter 5 discussed some of the biological hardware involved in perceptual processing. In a change of focus, this chapter considers perceptual processing from the point of view of information and the operations performed on that information (Figure 6.1).

Information processing models are based on **formal abstractions.** These abstractions can include

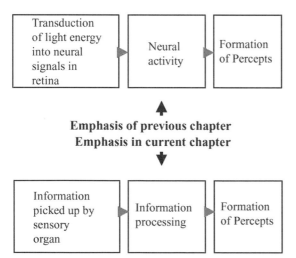

FIGURE 6.1. The current chapter emphasizes processing in terms of information flow, in contrast to the emphasis on neural hardware in the previous chapter.

issues related to how **sensory coding** takes place in biological tissue but can also be approached from a broader perspective less closely tied to biological details. Abstract models that attempt to incorporate many neuroscience facts are said to be **biologically realistic.** A potential limitation of these models is that if they become too complex, they can be just as opaque to understand as what they model. **Simplifying models** try to provide a conceptual framework for isolating some basic principles about sensory coding and processing without being overly concerned about biological details such as neurotransmitters or voltages.

Models at these various levels of abstraction can be **strictly conceptual,** as in an artificial intelligence–based model simulated on a computer. However, models can also be **implemented as a physical system,** as in a robotic machine.

1 General-Purpose Versus Specialized Processing

A recurring theme in the discussions of perception in biological organisms in previous chapters has been that processing is **specialized** rather than general purpose. Biological processing often provides ad hoc rather than general-purpose solutions to a particular perceptual problem, and these solutions work only because biological organisms are engaged in niche-dependent perception, as described in Chapter 3. In niche-dependent perception, perceptual processing mechanisms can take advantage of special knowledge about properties of the environment that have been built in over evolutionary time. Furthermore, the solutions to niche-dependent perceptual problems are often embodied in the actual physics of the way components of the biological hardware operate.

Specialized processing can be contrasted with **general-purpose processing** performed with digital devices as in a modern digital computer. An advantage of a general approach is that once a software program has been designed to solve a problem on one computer, the same program can be translated to run on other digital hardware. A solution achieved with special-purpose hardware often cannot be transferred to other hardware. However, general-purpose solutions pay a price in terms of loss of efficiency. For example, the computational neuroscientist **Christof Koch** has estimated that the brain performs 10^{16} operations per second while consuming less power than an electric light bulb. He points out that this level of performance would require the output of an entire power station if implemented with a digital computer.

The individual neurons in the brain operate about 1 million times slower than the individual components in electronic devices, but they make up for this by being massively parallel as well as taking advantage of solutions built into their hardware. An example is the fact that the biological membrane of a neuron automatically reacts differently to its current input depending on its recent history. This allows every neuron in the brain to perform current computations based on recent memories. Simulating the computations that are being performed automatically by this simple property of neuron membranes would be extremely difficult on a general-purpose digital computer. The scientist **Carver Mead** has documented a number of similar examples of the inefficiencies of general purpose digital computers stemming from the fact that they are not designed to take advantage of solutions built into their hardware. One example is elaborated in Box 6.1.

**Box 6.1
Digital Computers Fail to Take Advantage of Built-In Physical Properties**

Performing arithmetic operations such as addition and multiplication is an example of a task that is easy to accomplish with a general-purpose digital computer but difficult for biological brains. However, even here inefficiencies are built into these operations, due to the fact digital computers are designed to be general purpose. **Carver Mead** points out that a transistor, if left alone, can compute an exponential all by itself based on the physics of how it operates. However, as used in digital computers, all of the physics that is built into transistors is mashed down into a 1 or 0. Then algorithms are used to painfully build back up a system that can accomplish addition and multiplication, and a bunch of multiplication and addition operations are strung together to get an exponential. We pay a price of about a factor of 10^4 for using transistors in this general-purpose way instead of using the physics of the transistors directly.

These differences in efficiency account for the fact that biological organisms outperform even computers built with very fast microprocessors when performing on specialized tasks such as sensing and responding to the environment in real time.

Neuromorphics is an emerging new field in which researchers are trying to capture in devices constructed from silicon some of the essence of specialized processing that takes place in biological tissue. One point of emphasis in this field is to try to take advantage of the physics of the processing device in defining the elementary operations. In neurons, information is coded by creating voltages across membranes. These voltages vary in an analog fashion. In transistors, voltages are created across a gap between silicon and silicon dioxide. As used traditionally in standard integrated circuits, transistors are restricted to flipping between two voltage states. However, if allowed to, voltages across silicon can operate in an analog manner that mimics the physics of biological membranes. Labs around the world are currently exploiting **analog very large scale integrated (AVLSI)** circuit technology to design artificial noses, ears, and eyes. For example, Carver Mead has developed an artificial retina that contains thousands of individual sensing elements. Each element is made up of a sensor and adaptive circuitry that act as a photoreceptor and additional circuitry that mimics some of the feedforward functions reformed by bipolar cells and the lateral interactions performed by horizontal and amacrine cells.

The following sections will evaluate information coding and processing as it occurs in biological organisms and as simulated in models. The main theme will be trying to understand the essential aspects of perceptual information processing as it takes place in biological tissue. However, another issue to think about is the implications of these findings for this question: If perceptual processing could be completely abstracted and understood, could it then be implemented on nonbiological hardware to build not only a sense organ but ultimately a perceiving machine?

2 Sensory Coding

Coding Based on Spatial Mapping and Topography

A general property of the brain regions that are involved in early stages of processing perceptual information from the eyes is that they contain hypercolumns. This organization facilitates use of short- and middle-range intrinsic connections to compare information about various features that are present at the same location in the scene. Without this spatial organization, the length of the connections that would be required to carry out these kinds of analyses would have to be much longer on average.

Furthermore, hypercolumns are organized into topographical mappings. Recall from Chapter 5 that "topographical mapping" refers to the fact that a visual processing area of the brain is organized such that regions of the visual scene that are near one another are processed by hypercolumns that are also near one another. This allows longer-range intrinsic connections that pass between nearby hypercolumns to make direct comparisons of similar kinds of information, e.g., color, in several nearby areas of the scene.

Topographical mapping is distorted, as was discussed in dealing with the cortical magnification factor in Chapter 5. Some perceptual scientists such as **Eric Schwartz** have argued that this distortion performs useful coding functions. An example is illustrated in Figure 6.2.

The top panel illustrates how two stimuli on the retina, one small square and one large square, would be deformed by a smooth deformation of the retinal surface in a manner that approximates a mathematical function called **complex logarithmic mapping**. The shapes of the images in the final distorted map have some interesting properties. For example, they are now identical in size and shape, even though they were different sizes on the retina. This representation might be a convenient one to use if one wanted to do perceptual processing to analyze whether two objects have the same shape, regardless of their size. Consider trying to decide whether a letter is an "A" or a "B." The image of the letter on the retina will change depending on the distance from which it is viewed. However, its representation after undergoing complex logarithmic mapping would be invariant with changes in size or distance. The distortion that is present in the topographical mapping in the striate cortex resembles a complex logarithmic mapping function for stimuli that are imaged onto the fovea, as illustrated in the bottom panel of Figure 6.2.

Coding Based on Action Potentials

Within the biological retina, information is transmitted from one neuron to the next via graded electrical voltages. However, starting at the level

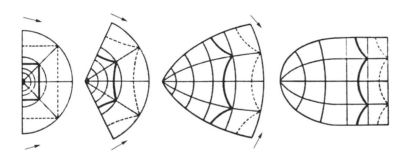

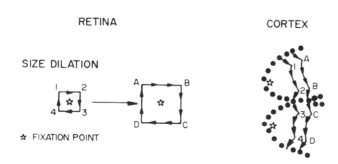

FIGURE 6.2. The top panel illustrates the effects of smoothly deforming a surface in a specific manner that approximates a mathematical function called complex logarithmic mapping. The bottom panel illustrates that topographical mapping in the striate cortex may be performing a similar function. Adapted from E.L. Schwartz, Computational anatomy and functional architecture of striate cortex: A spatial mapping approach to perceptual coding. Vision Research 20:645–669, © 1980, by permission of Elsevier Science.

of ganglion cells and continuing throughout the rest of the brain, information is transmitted down the axon in the form of **action potentials**, as described in Chapter 5. We know visual information has to be carried by action potentials based on behavioral measures of reaction times. No other physiological response is fast enough to account for the reaction times with which humans can respond to visual stimuli. However, recall that an action potential, or spike, occurs in an all-or-none fashion and has a stereotypical form and magnitude that is the same for all neurons in the brain. This raises the question of how qualitative differences in perceptual qualities can arise. **Johannes Müller proposed one answer,** discussed in Box 6.2.

Box 6.2
Why Do We "See" With Our Eyes But "Hear" With Our Ears?

Müller formulated his **doctrine of specific energies of nerve fibers** in the early 1800s to try to deal with the question of how identical physiological signals could produce different qualities of perceptual experiences. He proposed that differences in central terminations of nerves were responsible. We "see" because neurons from the retina send action potentials to visual-processing portions of the brain. A neuron that transmitted the exact same pattern of action potentials from the ear would still result in hearing, because it terminates in an auditory processing part of the brain.

An implication of the stereotypical nature of action potentials is that any information must be coded in some aspect of the **spike train**, the timing of the spikes that are produced, rather than in variations in their size or shape.

Most neuroscience studies of visual processing have concentrated on only a single code: **rate of firing** over some period during or immediately following presentation of a stimulus to the receptive field. The simplest form of a rate code uses digital coding over a labeled line.

Digital Binary Coding Over a Labeled Line

In a digitized representation of information, the input is divided into some number of discrete states, called **bins**. For all of the input values within a particular bin, the corresponding output is given a single value. In **digital binary coding** the input is divided into two bins, as illustrated in Figure 6.3.

The output corresponding to any input within the first bin is given a value of 0 or a label such as "off." The second bin is given a value of 1 or a label such as "on." Digital binary coding involves an irreversible loss of information. In the example shown in Figure 6.3, an infinity of possible input states is reduced to only two possible output states.

In **labeled line coding,** the output of a processing device is digitized into one of two states, called **active** and **inactive**. The interpretation of what it means when an output line is active comes from reading a label attached to that line. An example of labeled line coding in a different context is described in Box 6.3.

Helmholtz emphasized the role of labeled lines for many kinds of sensory coding. Consider information that specifies the location of a stimulus in terms of its direction in front of the eye. Information about that stimulus will be imaged to a particular retinal location based on topographical

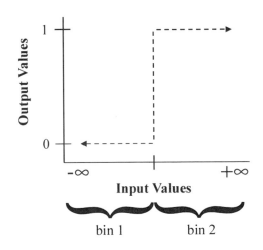

FIGURE 6.3. An example of binary digital coding. Values for all possible input states, ranging from −∞ to +∞, have been sorted into two bins, and a single output value given to each bin.

The basic idea is that a neuron fires at a high rate only when a particular feature is present. Thus, an active state of the neuron, defined as a firing rate that exceeds some set threshold level, can be interpreted as detection of that feature. Models based on feature detectors are typically formulated within a conceptual framework that assumes processing takes place along a hierarchy in which coding of features becomes increasingly complex at later stages. A neuron that serves as a detector for a low-level feature such as an edge would be expected to operate at a relatively early stage of processing, perhaps even as early as the retina. A detector for a complex feature such as grandmother would be expected to operate at a relatively late stage of processing, perhaps in a cortical area within the What-perception stream.

Scalar Coding

Sensory coding does not have to be binary, i.e., restricted to only two output states. Another coding scheme that could be used would sort the input into bins corresponding to three output states such as: 1. The feature is not present; 2. The feature might be present; 3. The feature is present. Similarly, a coding scheme could use four output states, or five, or six, and so on. The general term used when coding is based on classifying the inputs into some specific number of discrete output states is **digital scalar coding.** As the number of output states increases, a limiting case is reached where every input value is mapped onto a unique output state, and the relationship between input and output becomes a continuous monotonic function. This is referred to as **analog scalar coding.**

Analog scalar coding takes place automatically in the membranes of neurons, before an action potential is generated. However, the concept of analog

mapping and transmitted to the brain along a particular set of nerve fibers that have receptive fields at that location. When one of these nerve fibers is active, that fact carries information to the brain about the location of the stimulus. The information about location is conveyed not because any special code is transmitted, but simply because this fiber automatically specifies information about this particular location *whenever it is active.*

In general, labeled lines can act as **feature detectors.** In the example just presented, the feature was simply the location of the stimulus. However, a variety of stimulus features could be coded, in principle, by labeled lines, including more complex features, such as an "edge" or even "grandmother" (Figure 6.4).

Edge Detector   ### Grandmother Detector

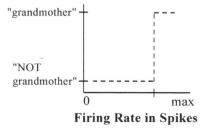

FIGURE 6.4. **This figure illustrates the use of a binary digital code along labeled lines to function as feature detectors. The labeled line on the left codes a low-level feature, an edge, and that on the right a more complex feature, grandmother.**

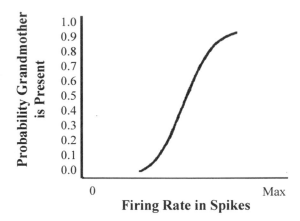

FIGURE 6.5. A hypothetical neuron carrying out scalar coding for the presence of grandmother.

scalar coding makes no sense with respect to a single action potential, since every spike generated is an all-or-none event. Thus, for neurons that generate action potentials, analog scalar codes apply only to higher-order variables such as average firing rate in spikes per second.

Whether analog or digital, on advantage of scalar codes over binary codes is that they can be used to signal a probability, rather than a deterministic decision, that some feature is present. For example, the higher the firing rate, the more confidence can be attributed to the inference that a particular feature is present in the environment. Figure 6.5 shows a hypothetical feature detector for grandmother based on a scalar coding scheme. The output can take on many different values, each of which signals a probability that grandmother is present. The more general issue of the role of statistical inferences in perceptual processing will be discussed below in the section on processing as the application of statistical procedures.

Coding Based on Tuned Filters

Most neurons that have been studied in biological organisms do not have properties that appear ideally suited for either digital binary coding over a labeled line or scalar coding. Most neurons fire maximally for some range of inputs, with decrements for input levels either above or below this range. The responses of these neurons are best characterized as those of **tuned filters** along some stimulus dimension(s). Consider a neuron in the cortex that responds differentially to lines of different orientations as illustrated in Figure 6.6.

The neuron is most highly tuned to the orientation that produces the highest firing rate. Thus, tuned filters provide another potential means to achieve either deterministic or probabilistic feature coding for the particular feature to which the neuron is most highly tuned. Whenever the neuron shown in Figure 6.6 fires at its maximum rate, this signals a high probability that a stimulus having the orientation to which it is tuned is present.

Empirical data regarding the frequency of occurrence of neurons with different types of tuning have often been used by perceptual scientists to infer the function of processing within a given brain area. For example, **Hubel** and **Wiesel** noted that many neurons in the visual cortex have highly tuned orientation selectivity, while neurons active at earlier stages of processing in the retina and dLGN do not. This has been used to argue that the visual cortex is the first processing area heavily involved in processing information about oriented contours.

A potential pitfall of coding schemes that depend on using tuning curves as feature detectors is that most neurons are tuned along some (unknown) number of stimulus dimensions. For example, a given cell might be tuned to a particular range of orientations but tuned to motion over a certain range of velocities as well. It is not possible to test an individual neuron along every possible stimulus dimension. Thus, an experimenter who tested a neuron along the dimension of stimulus velocity might conclude that the function of the neuron was to signal the perceived

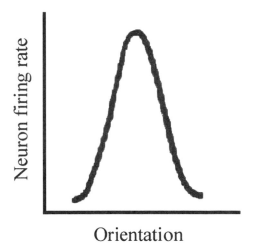

FIGURE 6.6. A hypothetical neuron acting as a tuned filter for orientation.

velocity. Another investigator testing the same neuron with orientations might conclude that the function of the neuron was to signal perceived orientation. In actuality, the neuron might be a detector for some complex feature that requires simultaneous tuning along a number of stimulus dimensions including, but not limited to, orientation and velocity.

Codes Based on Temporal Patterns

Each of the spike train coding schemes considered so far has been based simply on **firing rate.** However, neurons might also code information based on some more complicated aspect of the **temporal pattern** of the responses. One reason for thinking the brain might employ codes based on temporal patterns is that animals are often forced to make a behavioral response to a dynamically changing stimulus within a fraction of a second. An **instantaneous firing rate** can be computed quickly based on simply taking the reciprocal of the interval between the last two spikes received. However, this instantaneous firing rate is not very reliable because it typically varies over a huge range from one spike to the next. When decisions need to be made quickly, visual processing systems do not have the luxury of evaluating properties of the spike train over extended periods to establish a reliable **mean firing rate.** Furthermore, as we will demonstrate shortly, a much larger amount of information can potentially be transmitted within a very short period if the code is based on more sophisticated temporal properties than a simple average firing rate.

Information transmitted by means of action potentials can be completely characterized by the list of the times, relative to stimulus onset, when each spike occurs. Although time is a continuous variable, computational models of temporal coding often use a representation in which elapsed time is digitized into discrete bins. Consider the example of how time is measured with a stopwatch. A stopwatch with a hand that rotates around its face measures time continuously according to the position of its hard at each instant. A digital stopwatch only displays the current bin within which current time falls. For example the watch might display the digit 1 for a while, indicating current time falls within the first tenth of a second, then two, and so on. Using this type of digital representation, the temporal response pattern of a neuron can be represented as follows:

$$N_1, N_2, \ldots N_n,$$

where N_i is the number of spikes in the ith time bin and n is the total number of time bins in the analysis interval.

The duration of the individual time bins can be made as short as desired to satisfy the temporal resolution being modeled. For example, the time bins can be made so short that no more than one spike ever occurs within one bin. In this case, temporal coding can be modeled as a binary code with each time bin containing simply a 1 or 0 designating the presence or absence of a spike.

The number of features a neuron can potentially specify based on a temporal pattern code depends on the length of time allowed for transmission. Consider an example in which a binary code has to be sent within two consecutive time bins. Four features can potentially be coded, as illustrated in Figure 6.7.

If three time bins can be used, then eight possible features (2^3) can be coded; four bins yield sixteen (2^4); and so on. Thus, information that differentiates among an astronomical number of features, 2^{100}, could potentially be transmitted within 0.1 sec by using a simple binary temporal code based on bins that last about 1 msec.

There is accumulating evidence that spike trains of biological neurons sometimes utilize temporal pattern codes. Computational approaches have been used to try to address issues of how much information might be transmitted, in principle, by various codes. Two approaches that have proven particularly valuable for analyzing sensory codes are based on **Shannon information** and **ideal observers.**

	Bin 1	Bin 2
Feature 1	0	0
Feature 2	0	1
Feature 3	1	0
Feature 4	1	1

FIGURE 6.7. A temporal pattern code that must be transmitted within two time bin intervals can potentially specify up to four features.

Shannon Information and Ideal Observers

In trying to quantify information involved in perceptual processing, perception researchers have often utilized models along the lines illustrated in Figure 6.8.

Pioneering work on information processing models was done by Claude E. **Shannon** in the 1940s within the context of trying to specify how much information can be communicated over a transmission line. Within Shannon's model, the term **information** is used to quantify the extent to which an individual who is receiving information is **informed** in the sense of having **uncertainty reduced**. Consider the following two examples. In the first, a person is informed of the outcome of a coin toss. In the second, he is informed of the winner of a state lottery. He received more information in the second example, because the degree of uncertainty was reduced by a larger factor. In the first example, the person was able to reduce his uncertainty only from two possible alternatives (heads or tails) down to one. In the second, the uncertainty was reduced from perhaps millions of individuals (everyone who bought one or more lottery tickets) down to one.

Shannon used the concept of reducing uncertainty to quantify the amount of information that is communicated to a receiver. The total amount of uncertainty that is potentially available to be eliminated can be described formally as

$$U = \log_2 (P),$$

where U is the amount of uncertainty, specified in units called **bits**, and P is the number of possible events about which there was uncertainty before the information was received. If only one event is possible, then $P = 1$ and solving the above formula reveals that $U = 0$, indicating that there is no uncertainty to be reduced. If $P = 2$, then $U = 1$, indicating that transmission of 1 bit of information will be sufficient to eliminate all uncertainty. If $P = 4$, then $U = 2$, indicating that 2 bits will be needed, and so on.

Shannon's approach can be applied to perception in general by considering the environment as the information source, the observer as the receiver of the information, and perceptual processing as encoding, transmission, and decoding of the information. Shannon's approach can also be applied at a finer level of analysis to the neural processing of perceptual information at various stages of the brain. Consider the information source to be the eye and the receiver of the information to be the brain. The transmission lines are axons of ganglion cells traveling up the optic nerve. Figure 6.9 illustrates an early attempt by **R. Fitzhugh** in the 1950s to use this approach to quantify transmission of information in the spike train of a retinal ganglion cell.

The top panel shows the underlying model. Information is supposed to be present in the environment about whether or not a flash of light is present. That information is encoded by neural processing in the eye, transmitted to the brain by axons traveling up the optic nerve, and finally decoded by the brain and turned into a behavioral response such as "I see the light." Consider an observer performing on a two-alternative forced-choice task in which both of the alternatives are equally probable. In this situation, the upper limit of information that can be transmitted per stimulus presentation is one bit, because one bit is sufficient to reduce all of the uncertainty present in this situation. However, the amount transmitted might be less than one bit, for example, if there is noise or loss of information in the transmission line.

The second panel in the figure shows the measured amount of information transmitted about the presence of a flash on a single trial by one ganglion cell as a function of light intensity. Little information is transmitted for low intensities, but the amount of information increases to the maximum possible, one bit, by the time the flash is about three times more intense than its threshold.

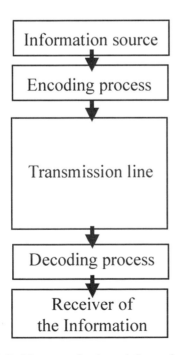

FIGURE 6.8. **Diagrammatic view of the model of information transmission developed by Shannon.**

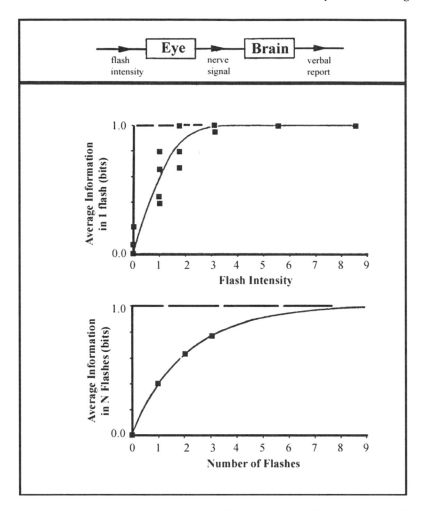

FIGURE 6.9. An early attempt to apply the concept of Shannon information to the spike train of a single neuron. Details are described in the text. Reproduced from R. Fitzhugh, the statistical detection of threshold signals in the retina. *J. Gen. Physiol.* **40**:925–949, © 1957, by permission of Rockefeller University Press.

The maximum amount of information that can be sent over a channel, a single ganglion cell axon in this example, is determined by the frequency of transmission multiplied by the total time spent transmitting. Thus, there are two ways to increase the amount of information that is transmitted: Either the bandwidth of the channel can be increased to allow a higher frequency of transmission, or transmission can continue for a longer time. The bottom panel in the figure illustrates how the amount of information transmitted by the ganglion cell to the brain increases as the number of repeated presentations (*n*) of a weak flash is increased. The uncertainty is reduced by only a small amount during each stimulus presentation, but by taking account all of the repeated presentations, information is gradually increased.

The concept of an **ideal observer** is derived from signal detection theory (SDT), discussed in Chapter 2. An ideal observer analysis calculates the expected performance of a hypothetical observer who makes optimal use of all available information of a certain type to make a decision. This approach can be used to model perceptual decisions made by whole organisms but can also be applied more generally to situations such as analyzing the information coded in spike trains of neurons.

The information carried in a given spike train is quantified in an ideal observer approach based on conditional probabilities. Consider the situation where two stimuli, α and β, are presented and the responses of a single neuron examined. It is possible, in principle, to estimate or measure the relative probabilities that any particular measure derived from a spike train(s) was present given the presence of α and then given the presence of β. For example, it might be established that the neuron fires at a mean rate of 100 spikes per second or more 90% of the time when a light is on (α) but responds with a rate this high only 10% of the time when it is dark (β). Once those probabilities have been calculated, the ideal observer confronted with a new spike train of 100 spikes per second by the same neuron simply calculates the relative likelihood that the measured response resulted from stimulus α or from β. This calculation can be described by the following equation:

$$L = P(S_\alpha)/P(S_\beta),$$

where L is the **likelihood ratio** that α rather than β is present. The value of this likelihood ratio is based on the conditional probability P that the measure S derived from the spike train would be present given α divided by the conditional probability that S would be present given β. Values of L greater than 1 designate a greater likelihood that α is present and less than 1 a greater likelihood that β is present. In our hypothetical example, L will have a value of 9 (.9/.1), which signals a strong likelihood for α.

The SDT provides tools that allow one to calculate how an ideal observer using this likelihood ratio will perform as described in Chapter 2. An ideal observer chooses a criterion value for L such that the response is α if the criterion is exceeded, otherwise the response is β. A more detailed example of how the signal detection theory approach can be applied to individual neurons is provided in Chapter 10.

Across-Fiber Vector Coding Based on an Ensemble of Filters

Up until this point, this chapter has considered only information that can be coded and transmitted by a single neuron. Another term that is sometimes used to refer to these kinds of processing schemes is **local coding,** meaning that each perceived feature is derived from the response of a single (local) cell. Local coding uses only one neuron, or perhaps a few for the sake of redundancy, to code for each single feature.

Next, the question of what kinds of information can be coded in the brain in terms of the relative activities of two or more neurons is addressed. These schemes are referred to as **across-fiber coding,** also called **vector coding,** for reasons discussed below.

Across-fiber coding is much more efficient than local coding. The potential amount of local coding scales only linearly with number of units, whereas vector coding scales exponentially. Consider a simple example in which four neurons are available to code for features. Local coding, if each neuron uses binary coding, allows only 8 (2 output states times 4 neurons) features to be coded for. Even if the single neurons are using digital scalar coding with five different possible output levels, only twenty features can be coded for (5 output states times 4 neurons). In contrast, across-fiber coding with four neurons and five possible levels of activation can produce $5^4 = 625$ combinations. The reason across-fiber coding gives these greater efficiencies is because any given output state of one neuron can be used to code many different features depending on the simultaneous output states of the other neurons.

The reason across-pattern coding is called vector coding is because the actual code can be represented with a **vector,** an ordered list of numbers. Consider a population of four neurons, a, b, c, and d. The rate of firing (spikes per second) of each of these four neurons could be represented by a single number (a = 8.2, b = 2.7, c = 9.3, d = 4.0). To represent the state of this population of four neurons, a single vector consisting of four ordered values can be used, {8.2, 2.7, 9.3, 4.0}. The values in this example are scalar, but they could just as well be binary, indicating, for example, whether each neuron is active or inactive (has a firing rate above or below some threshold level). If we pick a threshold of 3.0, the last example becomes {1, 0, 1, 1}. Local coding is simply a special case of vector coding in which the vector has only one element.

Contrary to intuition, the amount of information that can be coded for with an across-fiber coding scheme based on individual neurons that act as tuned filters is most efficient if the filters are tuned broadly rather than narrowly. This is called **coarse coding,** meaning that the individual neurons have tuning curves so broad that they overlap with one another. Consider two hypothetical neurons with broad overlapping tuning curves along some arbitrary stimulus dimension, as illustrated in Figure 6.10.

Neuron A is most sensitive to a particular value of this dimension, P1, and neuron B to a different stimulus value, P2. To make this example more concrete, consider that the values along this dimension are wavelengths of light, and the neurons are coding for color. Perhaps wavelength P1 is seen as blue and wavelength P2 is seen as red. As feature detectors, each of these tuned neurons would be

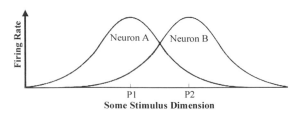

FIGURE 6.10. Illustration of across-fiber, coarse coding based on two neurons that each act as broadly tuned filters.

able to code only for the presence of one of these colors. If neuron A has a high rate of activity, this indicates that P1 is likely to be present and "blue" is signaled. If neuron B is active, then P2 is likely to be present and "red" is signaled. If confronted with a decision about whether P1 or P2 is *more* likely to be present, the relative rates of activity in neurons A and B would have to be compared. However, consider a potential code that could be used based on taking the ratio of the firing rates in neurons A and B. The ratio produced for P1 will be different from the ratio produced for P2, allowing blue to be discriminated from red based on this ratio code. This has a potential advantage. Note that in addition to coding specifically for colors associated with wavelengths P1 and P2, the ratio code is also different for every other wavelength along this dimension, allowing the potential for literally hundreds of wavelengths along this dimension (P1, P2, P3 . . . P*n*) to be given a unique color code. Further discussion of the use of coarse vector coding based on tuned filters to code color percepts is presented in Chapter 7, which considers perceptual processing of wavelength information.

Vector coding provides us with a slightly different perspective for thinking about how neurons might code for a complex feature like grandmother. Suppose a scientist records from a neuron that responds with a high rate of activity to grandmother. If that were the only neuron recorded from, it might be concluded that a feature detector for grandmother had been discovered. However, if the scientist had continued recording from additional neurons, she might have discovered three others that always responded with no activity whenever grandmother was present and a fifth that also had a high rate of activity whenever grandmother was present. In other words, the true code for grandmother might have been a vector, {1, 0, 0, 0, 1}, and the scientist just happened to record from the first element of this vector during the first experiment. The fact that this first element is active does not

really code for grandmother. It does so only when this element is active in the presence of elements 2, 3, and 4 being inactive and element 5 being active.

This example highlights the idea that if vector coding is being used, this fact puts a different interpretation on neurons that are not active during presentation of a particular feature. The values of zero would be treated as "no value" for feature coding. However, for vector coding, a value of 0 defines one of the elements within the vector just as much as a value of 1 does.

It is possible to imagine a large number of ways sensory codes could be designed based on constructing vectors whose individual elements are based on some aspect of the outputs of specific neurons in the population. When coding schemes are based on vectors constructed from the outputs of large numbers of neurons, they are generally referred to as **population codes.** Population codes can easily become so complex that they are hard to analyze and evaluate unless they are modeled formally. The next section focuses on some of the kinds of methods that can be used to carry out formal evaluations of complex sensory codes involving interactions among large numbers of individual neurons.

3 Neural Network Models of Sensory Processing

Chapter 5 introduced the concept of a biological micronetwork, a small number of highly interconnected neurons that perform an elementary processing function. It is sometimes possible to use qualitative intuition to construct hypothetical models of micronetworks. However, intuition fails as the number of neurons or the complexity of their connections increases. At these higher levels of complexity, formal models provide useful tools. One way to test many hypotheses about whether or not a particular form of perceptual processing could be performed, in principle, by a particular biological micronetwork is to construct a formal model of the micronetwork and show that the model performs the desired computation.

Formal models that try to mimic certain essential aspects of brain tissue are often referred to as **neural network models.** Small neural networks can be used to model simple biological micronetworks. Several networks can also be joined together to model processing that takes place in higher-order forms of biological organization, such as hypercolumns, nuclei, and processing streams, or in even

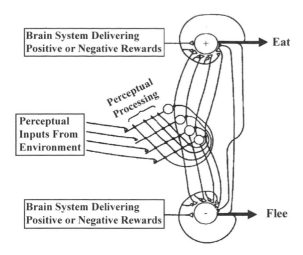

FIGURE 6.11. An example of a higher-order network model that relates behavioral responses to perceptual input. Adapted from A. Gelperin, et al, The logic of *Limax* learning. In *Model Neural Networks and Behavior*, edited by A.E. Selverston, pp. 237–261, © 1986, by permission of Plenum Publishing Corporation.

higher-order perceptual behavior engaged in by whole organisms.

Various levels of details of biological circuitry can be built into formal models to determine the computational consequences of detailed circuit design. However, formal models can be instructive even when not enough is known about the details of the underlying cellular physiology and anatomy to make the model realistic at a detailed level, as in the higher-order model shown in Figure 6.11.

This network model does not attempt to incorporate detailed anatomy or physiology. Nevertheless, models along these lines have been used to help researchers understand how perceptual information is processed and fed into brain subsystems to cause general behavioral response patterns that can be categorized as "eat" or "flee."

Neural networks are typically implemented as simulations on computers. However, a neural network that performs well is also a candidate for implementation in nonbiological hardware as a physical system that carries out this particular form of perceptual processing.

Formal Neurons

The information processing function of a neuron can be described abstractly as a **formal neuron,** as first described by Warren **McCulloch** and Walter **Pitts** in a paper published in 1943. They demon-

strated how neurons could be modeled as logic devices that can perform simple operations of Boolean logic such as AND, OR, and NOT.

Features present in a typical model of a formal neuron are shown in Figure 6.12.

The formal neuron shown here, i, receives inputs from some finite number of other formal neurons (σ_1 through σ_N). The input from each neuron is given a weight W that can be thought of conceptually as the synaptic efficacy of the connection. Values for these weights can be set to 0, in which case the input is ignored, or can vary from maximally inhibitory (−1) to maximally excitatory (+1). The weighted inputs are collected at the cell body of the formal neuron according to some function, typically summation:

$$h_i = \sum_{J=1}^{N} W_{ij}\sigma_j ,$$

where h_i is the effect produced at the cell body of neuron i after the weighted inputs have been collected from neurons σ_1 through σ_N. This function can be thought of conceptually as the generation of a graded membrane potential in a neuron based on its synaptic inputs. When sensory coding is expressed in terms of vectors, this function can be expressed as calculating the inner product of the vectors specifying the inputs and their associated weights.

The simple summation function is adequate for many purposes but leads to a problem, both mathematically and in terms of biological plausibility, in more complex models. The summation function is not bounded, allowing the value of h_i to grow to

Formal Neuron

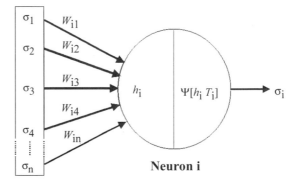

FIGURE 6.12. This schematic shows the primary elements involved in a formal neuron model as described by McCulloch and Pitts. Symbols and other details are described in text.

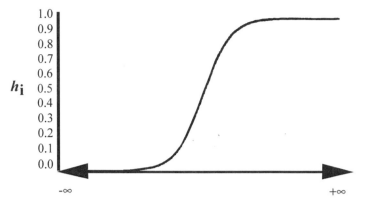

h_i

Value of the summed weighted inputs

FIGURE 6.13. The problem of very large valves being formed in formal neurons can be eliminated by replacing simple summation of weighted inputs with a sigmoid function, as illustrated here.

potentially huge values. Consequently, models based on this function can "run out of control" towards infinity. This cannot happen in biological tissue, because membrane voltages have a limited range. Formal models can be made more biologically realistic, and eliminate this problem at the same time, by using a function that has a smooth sigmoid shape to limit the range of values that are allowed, as illustrated in Figure 6.13.

In formal neurons, once the value of h_i has been determined, an output, σ_i, is generated. The decision about what output to produce given a particular value for h_i is based on some function Ψ. In cases where neurons communicate with one another via graded potentials, as in the retina, Ψ can simply transmit the value of h_i as the output. However, neurons more typically communicate via action potentials; this can be modeled as follows:

$$\sigma_i = \Psi[h_i] = \quad 1 \text{ if } h_i > T_i$$
$$0 \text{ if } h_i \leq T_i,$$

where T_i is the **threshold** of neuron i, conceptualized as the activation level needed by the neuron to generate a spike. The neuron generates an output spike, designated 1, only if its inputs exceed this threshold. Note that h_i is a scalar value that can be expressed as a single real number. The output value, σ_i, is a binary value that takes on only the values of 0 or 1. The output values of σ_i can be treated as scalar only if time is included as a parameter and some measure is made of the rate of output over time. The models that will be dealt with first are all static and consider the output only at a particular instant in time. Dynamic neural network models are discussed below.

This formal model operates in a deterministic fashion, always producing the exact same output

for the identical input. This is probably not totally biologically realistic, because neurons respond slightly differently from trial to trial even when subjected to the same inputs. This variability, or **noise,** can be easily incorporated into the formal neuron model by treating the threshold not as a deterministic value, but as the probability that the neuron will fire. For example, if the probability is set to 50%, then the decision about whether the neuron will fire on a given trial is determined by a random process such as flipping a coin. This specifies that over the long run the neuron will fire half of the time, but on any given trial it is impossible to know whether it will fire or not. One way to accomplish a probabilistic function is to treat the values of h_i associated with the sigmoid function in Figure 6.13 as probabilities that the neuron will fire on the current trial.

Basic Perceptron

One of the first attempts to apply formal networks to perceptual problems was the **perceptron,** illustrated in Figure 6.14.

The perceptron was proposed by Frank **Rosenblatt** in the 1950s, based on a formal neuron, i in Figure 6.14, that receives weighted inputs from several elements. Each of these input elements responds to patterns of light from particular portions of the visual scene analogously to a biological neuron with a receptive field. A perceptron evaluates the visual inputs and makes a binary decision, σ_i in Figure 6.14, for its output.

Rosenblatt argued that a properly constructed perceptron could serve, in principle, as a pattern-analyzing perceptual system. For example, a perceptron might be able to be constructed to detect the

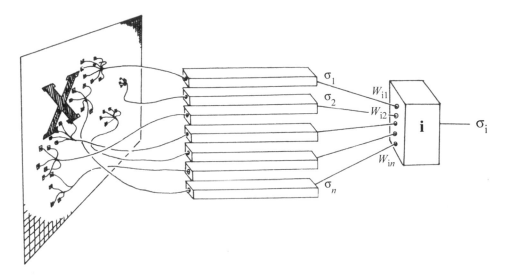

FIGURE 6.14. Rosenblatt's perceptron model. Symbols and other details are described in text. Adapted from M. Minsky and S. Papert, *Perceptrons*, © 1969, by permission of MIT Press.

presence or absence of a circle, as illustrated in Figure 6.15.

A perceptron that performed this function successfully would respond 1 (true) whenever a circle was present and 0 (false) otherwise. More generally, it was argued that, in principle, it should be possible to construct perceptrons to respond as feature detectors for the same kinds of complex perceptual patterns, such as grandmother, that can be perceived by human observers.

Marvin **Minsky** and Seymour **Papert**, in a book published in 1969, did a formal analysis of the capabilities of the perceptron model. They discovered that as a logic device, the perceptron could not solve

a simple Boolean **exclusive OR** (XOR) problem. An example of an XOR perceptual task would be deciding whether a square or a circle, but not both, is present in a scene, as illustrated in Figure 6.16.

The proper response is 1 (true) if a square alone is present, 1 (true) if the circle alone is present, 0 (false) if neither is present, and finally 0 (false) if both are present. The fact that a perceptron could not be constructed to solve this elementary, yet fundamental, problem of Boolean logic was widely considered a fatal flaw.

It turns out that the analysis of Minsky and Papert was technically correct for the original perceptron but was not correct in general for all classes of perceptron models. However, most researchers accepted the Minsky and Papert criticism at face value and abandoned further work with perceptron models. In the mid-1980s, David **Rumelhart** and colleagues proved that more complicated percep-

$$\sigma_i = \begin{cases} 1 \text{ if the figure is a circle} \\ 0 \text{ if the figure is not a circle} \end{cases}$$

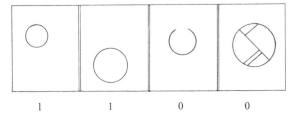

FIGURE 6.15. The top of the figure shows a rule that might be implemented by a perceptron model. The bottom of the figure illustrates the output that would be formed by this perceptron for each of four specific stimuli. Adapted from M. Minsky and S. Papert, *Perceptrons*, © 1969, by permission of MIT Press.

$$\sigma_i = \text{XOR (circle, square)}$$

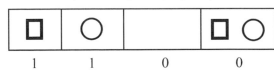

FIGURE 6.16. The top of the figure shows an XOR rule for deciding whether a square or a circle is exclusively present. The bottom of the figure shows the response that should be generated by a perceptron carrying out this rule.

tron models can solve the XOR problem. It has now been proven that neural networks like the perceptron, based on nothing more than connections between formal neurons, can solve any **well-behaved** mathematical function – that is, one whose solution can be characterized formally in terms of an algorithm, a topic considered further below in the section on processing as carrying out an algorithm. Over the past couple of decades there has been a resurgence of interest in applying neural network models to perceptual processing.

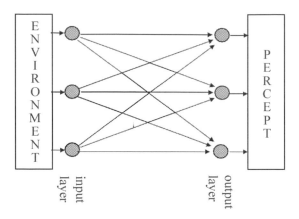

FIGURE 6.18. Architecture of a general two-layered, feedforward, fully connected neural network model.

Architectures of Neural Networks

The capabilities of a neural network depend on the connectivity among the individual processing elements. This is sometimes referred to as the network's **architecture**.

Architecture of the Basic Perceptron

The architecture of the basic perceptron can be represented in the format shown in Figure 6.17.

There are two layers of processing elements. An **input layer** samples the environment, and an **output layer** contributes to some aspect of a percept. The network is characterized as **feedforward,** meaning that elements in each layer project only to an element(s) in the succeeding layer, not to an element(s) in the same or preceding layers. This network architecture is also referred to as being **fully connected,** meaning every element in one layer is connected to every element in the next layer. The output layer for the basic perceptron uses **local coding** to act as a **feature detector** by responding with 1 if the feature is present, 0 if not.

Vector Code

	Output neuron #1	Output neuron #2	Output neuron #3
Feature 1	0	0	0
Feature 2	0	0	1
Feature 3	0	1	0
Feature 4	0	1	1
Feature 5	1	0	0
Feature 6	1	0	1
Feature 7	1	1	0
Feature 8	1	1	1

FIGURE 6.19. A network model with only three output neurons can potentially code eight features using a binary code, as illustrated here.

Two-Layered, Feedforward, Fully-Connected Architecture

The architecture of the basic perceptron only needs to be changed slightly, as illustrated in Figure 6.18, to accommodate perceptual processing models based on vector coding rather than local coding.

This architecture is the same as the basic perceptron in terms of being two-layered, feedforward, and fully connected. The only difference is that more than one neuron is present in the output layers, and therefore that the output signal is a vector code rather than a binary code. The simple network illustrated in Figure 6.18 has three output elements and thus can potentially code eight features, as illustrated in Figure 6.19.

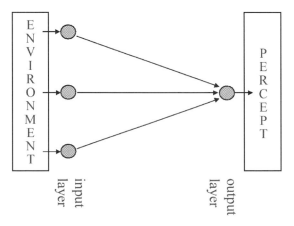

FIGURE 6.17. Architecture of the basic perceptron model.

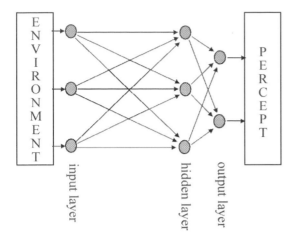

FIGURE 6.20. Architecture of a neural network that incorporates a hidden layer of units that do not communicate directly with either the input or the output.

Architectures with Hidden Layers

Figure 6.20 illustrates an architecture with a **hidden layer** between the input and output layers.

This network has three elements forming an input layer and two elements forming an output layer. However, internal processing elements are also present that do not receive any direct input from outside the network and do not contribute any outputs that leave the network. Neural network architectures with hidden layers are needed to solve certain classes of problems whose solutions depend on nonlinear mathematical functions including the XOR problem described earlier.

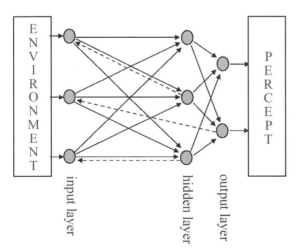

FIGURE 6.21. Architecture of a neural network that includes recurrent as well as feedforward connections.

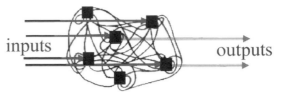

FIGURE 6.22. Architecture of a fully interconnected neural network.

Architectures with Recurrent Connections

Recall for Chapter 5 that the connections in biological brains are not strictly feedforward but include return pathways that connect higher-order processing areas with earlier stages of processing. These are modeled in neural networks with architectures that include some number of recurrent as well as feedforward connections, as illustrated in Figure 6.21.

In the limiting case, a neural network is fully interconnected, as illustrated in Figure 6.22.

Every element within the network is connected to every other element and with itself. At this level of complexity, the designation of layers is no longer meaningful, except to identify the elements that receive from outside the network as the input layer, and those that project outside the network as the output layer.

Adjusting the Weights of a Network

The relationships between the inputs and outputs of a neural network can be described formally by a mathematical function. The particular mathematical function being carried out by a network depends on the network parameters. These include the weights applied to the inputs of each formal neuron, the summation, sigmoid, or other function that determines the value for h_i, and the threshold or other parameters associated with Ψ. However, all of the parameters of a neural network are typically held constant except the weights, so the parameters to be adjusted are often simply referred to as the weights.

It can be shown that, in principle, some adjustment of weights has to exist that will allow a network to be constructed that can compute the solution to any mathematical function. However, the specific parameters that are needed to implement a particular function are usually unknown. Thus, the task of finding the parameters that are needed to solve a given problem is analogous to looking for a needle in a haystack when we can be certain that the needle is located somewhere in the haystack but have no idea where.

The process that is involved in constructing a neural network to model a perceptual function can be illustrated with an example. Suppose a network model is being used to evaluate how accurately a perceptual system that implements a particular type of processing could perform on a two-alternative forced-choice psychophysical task. To make this evaluation, a neural network is designed that models the essentials of the type of processing to be studied in terms of its inputs, outputs, and internal architecture. Then the model must be provided with various inputs to evaluate its performance. Unfortunately, this evaluation will not be very meaningful in terms of addressing the original question, because performance will vary depending on how the weights of the network are set. This model will be useful for addressing this question only once its weights have been set such that performance is optimized.

In general, there is no known analytical solution for specifying what settings of the weights will result in optimized performance. These optimal settings have to be established during an initial **training stage** before the network is actually used to test the question of interest. Discussion of training methods will be deferred until Chapter 11, which considers models of perceptual learning. For now, it can just be assumed that the training stage has already taken place. Thus, when the model is used, its performance will be **optimal** for the **network architecture that is being employed** on

the **class of problems for which it was trained.** Performance of the optimized network model can be used as a standard for comparison of performance of other neural networks with different architectures, performance of other types of models, or performance of human or animal observers on the same task.

Interpretation of Hidden Units

Once a neural network is trained to solve a particular class of perceptual problems, it is of interest to examine the internal structure of the processing elements in the model to see how each contributes to its overall operation. This is analogous to the approach that might be taken by a neuroscientist who examines the activities of neurons in the brain with the goal of understanding how those neurons contribute to perceptual processing. An evaluation of a neural network model carried out by Terrence **Sejnowski** and colleagues provides an example of this approach. They designed a neural network model to solve a particular perceptual problem called **shape-from-shading,** illustrated in Figure 6.23.

A concave shape projecting above the surface is illustrated on the left, and a concave shape dimpling below the surface is shown on the right. Both surfaces are being illuminated by a small light source at a fixed position. The perceptual problem to be solved is to determine characteristics of the surface shape based only on information about the

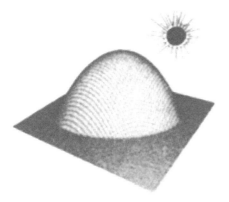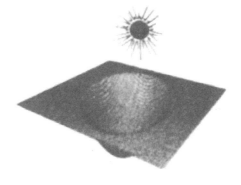

FIGURE 6.23. The perceptual problem of figuring out shape based on shading involves comparing areas of an image that appear bright, and can be assumed to be directly illuminated by light, with areas that are dark, and assumed to fall within shadows. The left panel illustrates the luminance reflected from a light source for a dome rising above the surface and the right panel for a pit below the surface. Adapted from S.R. Lehky and T.J. Sejnowski, Neural network model of visual cortex for determining surface curvature from images of shaded surfaces. *Proc. Roy. Soc. Lond. B* 240:251–278, © 1990, by permission of the Royal Society.

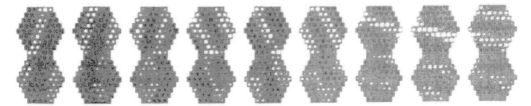

FIGURE 6.24. Outputs of hidden units in the Lehky and Sejnowski shape-from-shading model when probed with spots of light. Additional details are presented in the text. Adapted from S.R. Lehky and T.J. Sejnowski, Neural network model of visual cortex for determining surface curvature from images of shaded surfaces. *Proc. Roy. Soc. Lond. B* 240:251–278, © 1990, by permission of the Royal Society.

distribution of light reflected from the surface. Biological visual systems solve this problem effortlessly.

The network was designed to be somewhat biologically realistic in its architecture. An input layer sampled the brightness of the surface with receptive fields similar to those present in dLGN neurons. This input was fed into a hidden layer of units. The properties of the hidden units were not specified in the model. The hidden layer fed into an output layer that, following training, provided a solution in the form of an accurate specification of the shape of the surface.

During training, the weights of the hidden units were allowed to form automatically in a manner that would provide a solution to the problem. When training was complete, units in the hidden layer were probed by measuring their activity while spots or bars of light were presented in the environment. This procedure is conceptually similar to the way a neurophysiologist would record from single neurons in the brain while presenting stimuli to the retina to map out their receptive fields. Results are shown in Figure 6.24.

Each hexagon shows the "receptive field" of one particular hidden unit with the light spots demonstration the locations of maximum sensitivity to light stimulation. Each of these receptive fields exhibits orientation selectivity, in the sense of responding best to bars of a particular orientation, much like the receptive fields of single neurons in the monkey striate cortex.

When neuroscientists have encountered neurons with these kinds of oriented receptive field properties, they have often drawn the conclusion that the function of these neurons is to act as oriented-line detectors. Based on the same logic, one might infer that the function of the hidden units in the neural network is to detect oriented lines. However, in the case of the neural network, we know the function of the hidden units was to assist the network with solving the perceptual problem of determining shape-from-shading, because it was explicitly trained to solve that problem. Any functions involving detection of oriented lines in the case of the neural network are simply an incidental consequence of solving the problem of shape-from-shading. These results suggest caution when inferring function in biological brains based solely on examination of receptive field properties. As discussed in Chapter 1, explanations for perception need to be sought at several levels of description simultaneously. Any interpretation of results obtained from within a particular level of description, such as the receptive field properties of single neurons, must be evaluated within a larger context, including trying to understand the sets of problems the neurons were trained, over evolutionary time, to solve.

Evaluations of properties of hidden units in neural network models have also provided some novel insights into how one might design feature detectors. When thinking about how to construct a feature detector from the point of view of an individual neuron, the traditional view has been to consider how one might adjust the specific properties of that neuron. However, modeling based on neural networks reveals that it would also be possible to construct neurons that act as feature detectors by designing at the level of the weights of the entire network. This has potential relevance both in terms of how engineers might design perceiving machines and in terms of how we think about the biological units on which evolution might be acting.

Vector-Vector Transformation Through a Network

Neural networks are used to model how perceptual information is represented and processed. **Representation** is conceptualized as **vector coding** and

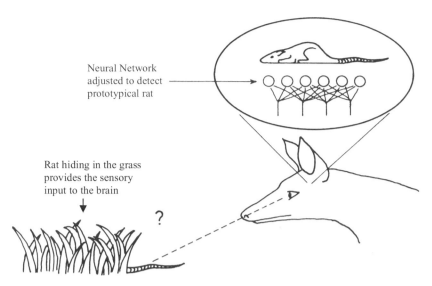

Neural Network
adjusted to detect
prototypical rat

Rat hiding in the grass
provides the sensory
input to the brain

FIGURE 6.25. Neural networks can exploit vector completion to form percepts based on incomplete or noisy information as illustrated here for a network intended to detect a prototypical rat. The network responds to the partial input of the hidden rat and completes this partial input to generate a percept of a rat. Adapted from P.S. Churchland and T. Sejnowski, *Computational Neuroscience*, © 1992, by permission of MIT Press.

processing as **vector-vector transformation.** Thus, computationally, the operation of a neural network can be analyzed strictly in terms of vectors and operations on vectors. We have already noted that both the input and the output of a neural network can be specified as vectors. The adjustment of the weights can also be specified as a vector, or more generally as a matrix specifying the weights associated with each connection between formal neurons. Thus, the operations performed by a neural network can be fully characterized starting with an input vector, continuing with intermediate transformations as the vector passes through the individual elements of the network, and ending with an output vector.

Vector coding and processing in a neural network is **robust.** Once the weights have been set properly, elimination of a few of the individual processing elements may cause some loss in accuracy, but the network does not usually come to a total halt. This is different from solutions based on a computer program running on a digital computer, in which loss of any element is likely to cause the program not to run at all. Thus, neural network models have an intuitive appeal for having the potential to model the effects of conditions such as strokes or brain injuries in biological brains more accurately. These typically lead to impaired or altered perceptual function but not a total loss of function.

Neural networks exhibit another type of robustness with regard to their inputs, called **vector completion,** that also has relevance for perception, as illustrated in Figure 6.25.

If a neural network has weights set to detect a particular pattern, then it will also respond to, or complete, a partially degraded input from the original pattern. In Figure 6.25, the brain of the coyote has a neural network set for detecting a rat. The sensory input does not match that expected for a prototypical rat exactly, because the rat is hiding in the grass. However, the sensory input is similar enough to the prototypical rat to cause the "rat" detector to respond.

Dynamic Neural Network Systems

The neural networks described so far have been static. They exhibit no intrinsic knowledge of order or time. The inputs are instantaneous snapshots of the world, and the outputs are a single state rather than temporally extended patterns of activity. This form of representation and processing ignores dynamics, a prominent feature of neurons in biological brains. This has led many biologically oriented perceptual scientists to feel that neural networks along the lines of the perceptron, regardless of how complex they become, provide an inherently inadequate model of perceptual processing as it occurs in biological organisms.

Neurons involved in biological processing of perceptual information are constantly firing rather than passively sitting and waiting for an input to come in from the sensory receptors. They generate outputs based on their inputs as a function of time. Thus, these outputs do not have to depend on simple binary codes that are produced at a specific moment in time but can also carry dynamic scalar codes, such as variation in numbers of spikes over time.

Another type of formal model, called an **attractor neural network** (ANN), has somewhat more biological plausibility with respect to these dynamic factors. An ANN models a dynamic system that consists of several interacting parts whose state evolves continuously with time. An intuitive understanding of the computations that are performed by an ANN can be attained with reference to a landscape metaphor, as illustrated in Figure 6.26.

The x-y position on the terrain represents the output state of the network at a given moment. Although shown here in two dimensions, the actual output state will be a vector that could be represented only in multidimensional space. The locations of the valleys and hills are determined by the settings of the weights. When the input of a network is set to a particular value, it is as though a ball has been dropped onto this surface at a given point. The outputs will then change dynamically over time, following the same trajectory that the ball would follow as it rolls downhill into a valley. If the weights have been set properly, the valley into which the ball rolls, called an **attractor,** will be the solution to a particular problem.

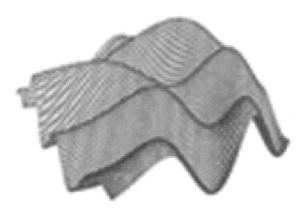

FIGURE 6.26. The outputs of an attractor neural network over time can be conceptualized with a landscape metaphor like this one.

Consider the example of vector completion for detecting a rat in the grass that was described from the point of view of a static network in the previous section (Figure 6.25). In an ANN, all of the input states that are similar to the prototypical rat would form sloped hillsides around the valley that constitutes the attractor for the presence of a rat. Vector completion takes place dynamically in the form of the dropped ball's rolling into the valley.

ANN models are particularly useful for modeling dynamic aspects of How-perception and will be discussed in more detail in Chapter 10.

4 Processing Conceptualized as Carrying Out an Algorithm

When analyzing perceptual processing as in the previous section, from the point of view of a neural network model, the solution to a perceptual problem is conceptualized as finding an appropriate set of connections and their weights. Another way to characterize processing is as an explicit set of rules applied to perceptual information. These two levels of description appear to be quite different from one another, at least when analyzed at a superficial level. However, the mathematician and artificial intelligence researcher **Alan Turing** demonstrated that these two levels of description are equivalent when analyzed at a deeper level. Turing showed that in principle, all solutions for a computable problem can be translated into a single general set of procedures involving manipulations of symbols and data that could be stored on a single tape. The tape might have to be very long as illustrated in Figure 6.27, depending on the problem to be solved. A device that is capable of carrying out these operations is referred to as a **universal Turing machine.**

Thus, the perceptual processing being carried out in any neural network model, or indeed in any biological micronetwork, can be given a formal level of description in terms of operations implemented by a Turing machine. The tape of the Turing machine can be conceptualized as the environment surrounding an observer. Reading from the tape corresponds to sampling the environment and writing to the tape corresponds to a behavioral response that alters the environment. Perceptual processing is simply the **algorithm** governing the operations performed by the Turing machine.

Box 6.4 provides more information about algorithms.

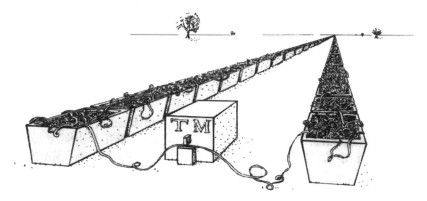

FIGURE 6.27. A universal Turing machine (TM) that performs operations on information stored on a (very long) tape. Reproduced from R. Penrose, *The Emperor's* *New Mind: Concerning Computers, Minds, and the Laws of Physics*, © 1989, by permission of Oxford University Press.

Box 6.4
Algorithms Are Well-Defined and Guaranteed to Solve a Certain Class of Problems

An algorithm is a systematic procedure in which the operations that need to be performed at each step are clearcut, with no ambiguity. The procedure has to be able to be described in finite terms and must be guaranteed to arrive at a solution to the problem within a finite number of steps. Finally, there also has to be a clearcut way to determine when the solution has been found so that the result can be read. Figure 1 shows a simple example of an algorithm that can be used to find the remainder resulting from the division of any two number, x/y.

By following this algorithm, even a person who has no knowledge of formal mathematics can solve difficult division problems such as what is the remainder if 1,768 is divided by 27. The procedures to be followed are clearly defined for any agent that knows how to perform subtraction and evaluate which of two numbers is larger. The procedures are stated in finite terms as symbols, boxes, and arrows. Finally, it can be demonstrated that this procedure is guaranteed to arrive at a solution to a problem of division in finite time, and it will be known when the solution has been found because the answer will be printed out.

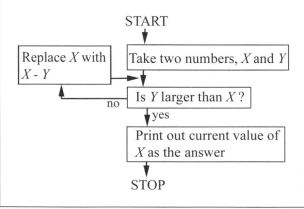

FIGURE 1. This schematic shows an algorithm in the form of a flow-chart that can carry out the operation of finding the remainder following a division.

The processing performed by a Turing machine governed by an algorithm might be incredibly slow, depending on the speed and architecture of the hardware, and may not be a practical solution to a problem that must be solved in real time. However, it is sure that given sufficient time, the Turing machine would find the solution to any problem governed by this algorithm. Modern digital computers are all variations on the general universal Turing machine, and their function can be described abstractly in terms of the algorithm they are carrying out.

Algorithms can be simple, like the one for finding the remainly following division, but they can also be exceedingly complex. The algorithm for perceptual processing in humans, if it exists, is expected to be very complex, perhaps taking billions of pages to express. However, there are also some deep philosophical issues having to do with whether such an algorithm even exists. It has been demonstrated within the domain of mathematics that there are problems whose solutions cannot be computed by any algorithm, as elaborated further in Box 6.5.

There are two theories about the kinds of processing that are carried out by human brains. One is that brains are just carrying out algorithms, very complicated algorithms to be sure, but still just algorithms. This position is held by many researchers in the field of **artificial intelligence** (AI), where it is called the **strong AI position**. This position carries with it a promissory note that eventually machines that carry out algorithms will be able to do everything a human can do. Minds are nothing more than algorithms running on computers made of biological tissue. The strong AI position argues that as soon as we know what those algorithms are, and can construct sufficiently powerful computing machines to carry them out in real time, the machines will exhibit the same properties that we attribute to the mind in humans.

The alternative theory about the processing that takes place in brains is that it involves "mysterious" processes such as insight that no algorithm could ever achieve. A proponent of this position is the mathematician **Roger Penrose.** In a book published in 1989, he constructed an elaborate argument that human minds are able to engage in forms of activities that are not computable by any algorithm.

If the first theory is correct, then it might be possible, in principle, to build a robot that carries out an algorithm and perceives in the same way humans perceive. These machines will be able to process perceptual information to form a knowl-

Box 6.5
Some Problems Do Not Compute

Within the domain of mathematics, it has been demonstrated that certain classes of problems cannot be solved by any algorithm, no matter how complicated. Problems that cannot be solved by a given algorithm are sometimes solvable by other means. For example, a human mathematician can sometimes look at the algorithm and at the problem(s) it cannot solve and use "insight" to figure out how the algorithm could be modified to solve the new class of problems. This can go on indefinitely in an iterative manner, because it has been proven that once the modifications have been made, a new class of problems will exist that defeat the new algorithm. Furthermore, it has been demonstrated that it will never be possible to systematize this iterative procedure into an algorithm.

An example of an important class of problems that cannot be solved by an algorithm is referred to collectively as the **halting problem.** These are problems for which the algorithm enters an infinite loop such that it will never arrive at the solution. However, the algorithm itself has no way to determine that it is in an infinite loop. It just steps through its sequence of well-defined steps one at a time. Only someone from outside the algorithm looking in and using insight is able to figure out that it is a waste of time to wait for the solution because the algorithm will run forever without coming up with the answer. Those of you who have tried to do computer programming have perhaps encountered this problem. A program is written that you think is an algorithm for solving a particular problem. You set the program running, and wait, and wait, and wait. At some point you have to decide: Is the program just implementing an algorithm that is taking a long time, or is it in an infinite loop that will continue forever unless you terminate the program?

edge base to be used for What-perception and How-perception and will also be able to generate **subjective** perceptual experiences. If the second theory is correct, then no matter how complicated the algorithm being carried out, machines will never be able to perceive the way that humans do. These issues will be addressed further in Chapter 12, which deals with the topic of subjective perceptual experiences in more detail.

Irrespective of the broader philosophical issues about whether *all* perceptual processing can be subsumed under an algorithm, it is possible to pursue a more limited enterprise of searching for algorithms to describe specific aspects of perceptual processing. The exemplar of this approach wad **David Marr,** who offered a rigorous analysis of formal operations that might be carried out during certain types of perceptual processing. In Marr's approach, algorithms are applied to information derived from the retinal image. The original information undergoes a number of transformations and intermediate representations during processing and in the end has qualities that reflect our percepts. Marr's rule-based approach to analyzing perceptual processing will be highlighted in the discussion of form perception in Chapter 8.

In biological perceivers, it is important to carry out processing that not only gives the correct answer but does so in real time. Researchers working in the field of artificial intelligence have largely ignored this issue until recently. Given that an algorithm has been demonstrated to provide the correct answer to a given problem, it was considered to be a "mere" engineering problem to design hardware fast enough to carry out that algorithm in real time. Surprisingly, the practical issue of achieving algorithmic solutions to problems in perceptual processing in real time has turned out to be much more intractable than most perceptual scientists originally anticipated. For over 20 years, starting in the 1960s, there was an optimism that the problem of obtaining solutions in real time would eventually be solved by sufficiently fast and powerful hardware. In recent years, that optimism has started to wane. Many problems in perceptual processing have proved to be computationally intensive to an unexpected degree – for example, recognizing a predator in a noisy environment, adjusting gaze in an appropriate and timely manner when confronting a member of the same species, distinguishing food from nonfood and mates from nonmates, and navigating through complex environments. Algorithms running on the most

powerful hardware available today are unable to compute solutions to these kinds of problems in anything close to real time. In fact researchers are now just beginning to be able to simulate perceptual activities of a simple biological organism such as an insect in real time with the most powerful machines available.

These difficulties have forced some humility on most perceptual scientists and engineers. It is no longer considered to be a mere engineering problem to extrapolate algorithms that simulate perceptual processing very slowly on the current generation of digital computers to hardware fast enough to carry out this processing in real time. At best, these are exceedingly difficult problems, and at worst, no solution is possible. The primary source of residual optimism is the success of simulating some aspects of the operations of the peripheral sensory organs in real time with massively parallel analog devices, as described at the beginning of this chapter. Thus, one approach to trying to build a machine that can perceive may be a hybrid approach that combines special-purpose analog and general-purpose digital components, for example, neuromorphic sensors that feed their outputs to digital electronics for subsequent processing. This might allow design of machines that capitalize on the benefits of specialized analog processing, as occurs in biological organisms, as well as the benefits of digital electronics for carrying out general-purpose algorithms.

5 Processing Conceptualized as an Application of Statistical Procedures

Another useful abstraction for characterizing perceptual processing is as an application of statistics to the inputs from the environment. **Statistical models of perception** are based on either descriptive or inferential statistics.

Models Based on Descriptive Statistics

In models based on **descriptive statistics,** the images impinging on the retina are conceptualized as measurements made on a coherent stream of events occurring in the distal environment and carried into our eyes via light rays. Perceptual processing reduces the complex information present in the retinal image to qualities of our percepts. In many cases, this processing involves selective loss of information, or perceptual filtering, as described qualitatively in Chapter 4. From a computational perspective, perceptual filtering is analogous to procedures used in descriptive statistics in which a small number of parameters, such as the mean and standard deviation, are used to characterize a complex mixture of numbers. The reduction of the complex population to a few parameters clearly loses much detailed information that was present in the original distribution. However, for some purposes, use of a small number of statistics proves to be a convenient way to characterize a complex distribution of numbers in a compact manner.

Color vision processing provides a good example of this characterization of perceptual processing as descriptive statistics. Color vision can be conceptualized from a computational perspective as reducing the **complex distribution of wavelengths** reflected from an object in the environment into **color,** a **low-dimensional quality** that can be described with only a few degrees of freedom. These issues are discussed in more detail in Chapter 7.

Models Based on Inferential Statistics

Other aspects of perceptual processing are analogous to **inferential statistics.** Visual processing can be considered an active process whose function is to infer useful descriptions of the world from changing patterns of light falling on the retina.

Inferential statistics do not have to be complex. For example, inferences can be facilitated with very simple filtering operations in which redundant information is eliminated so that only information that is predictive is passed on for subsequent processing. Consider the neurons in the retina with concentric center-surround receptive fields that were described in Chapter 5. Recall that these neurons respond to contrast but not to uniform intensities. The function of these neurons can be conceptualized as applying statistics to the null hypothesis that images are uniform in intensity. Only when the null hypothesis is rejected is information sent on from the retina to central processing areas of the brain.

Inferential statistics can also be helpful under conditions where perceptual information is noisy. In the presence of noise, the task of perceptual processing can be compared to the task of a scientist confronted with noisy data. The best way to decide

whether the data sets from an experimental and a control group are the same or different is to apply a battery of statistical tests to them. Similarly, perceptual processing can be conceptualized as applying statistical tests to make decisions about properties of environmental stimuli. Whenever a person is confronted with deciding between two or more alternatives about "what is out there," perceptual processing reduces the information in the spatiotemporal retinal image, a three-dimensional array, down to a single number, the decision, which is manifested in the observer's performance. For example, the observer responds, "I see a light." Performance under these conditions is limited by strength of the signals produced by the environmental conditions that must be discriminated and also by the amount of inherent variability, or noise, within which these signals are embedded.

Evaluating Signal-to-Noise Ratios and Quantum Efficiency

It is impossible, even in principle, to eliminate variability from visual signals. As discussed in Chapter 3, the fundamental event that initiates vision, absorption of a photon, is not deterministic but instead a stochastic event governed by probabilities. Transmission through the nervous system is also inherently variable, because of factors that are governed by random components such as opening and closing of channels in membranes of photoreceptors and neurons and release of vesicles across synapses. These sources of variability referred to as **noise,** produce uncertainty about the true magnitude of visual signals.

The size of a signal that is extracted during perceptual processing can be related to the magnitude of the noise from all sources (external and internal). This can be quantified in terms of a **signal-to-noise ratio (SNR).** One important way of evaluating the statistical properties of perceptual processing involves comparing the magnitudes of the SNR at various stages of processing.

At the initial stage, when photons in the retinal image are absorbed in photoreceptors, statistical probabilities associated with photon absorptions, called **photon statistics,** impose a fundamental limit to the reliability of information that can be extracted from the retinal image. The SNR at this stage of processing is simply

$$S/N_{photon}$$

where S describes the magnitude of the signal that could be extracted from the receptor absorptions by an ideal observer and N_{photon} describes the inherent variability based on photon statistics.

The SNR can also be evaluated at a later stage of processing that takes into account the amount of intrinsic noise added during neural processing as well as photon noise. This stage is sometimes referred to as the **neural image,** and the noise measured at this stage is called the equivalent input noise, N_{eq}. It can be evaluated based on psychophysical methods that are analogous to those in common practice in electrical engineering, as described in Box 6.6.

When N_{eq} is measured for a human observer, set amounts of noise are added to a visual stimulus while thresholds are measured. An example of a typical psychophysical result for a human observer is shown in Figure 6.28.

Thresholds are not affected by low amounts of noise added to the input, Nin, because these added noise levels are insignificant compared to the amount of photon noise and the intrinsic noise of the observer. At higher levels of added noise, the input noise dominates, so the thresholds rise. The knee of the curve occurs where the effect of the added noise is equal to that of the sum of the photon noise and the observer's intrinsic noise. This measure, N_{eq}, estimates the amount of noise that is present in the neural image. The SNR in the neural image is simply

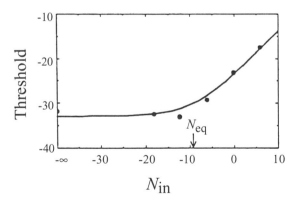

FIGURE 6.28. Illustration of the results of a typical measurement of equivalent noise measured with psychophysical methods in a human observer. This psychophysical measurement is analogous to measurement of the equivalent noise of a physical device by an electrical engineer as described in Box 6.6. Symbols and details are described in text. Reproduced from D.G. Pelli, The quantum efficiency of vision. In *Vision: Coding and Efficiency,* edited by C. Blakemore, pp. 3–24, © 1990, by permission of Cambridge University Press.

$$S/N_{eq}$$

where S describes the signal that could be extracted from the neural image by an ideal observer and N_{eq} is the equivalent input noise of the observer, which includes both photon noise and intrinsic noise added during neural processing.

The ratio of the SNRs at two stages of processing is a statistical measure of the **efficiency** of the transformation relating the two stages. **Transduction efficiency** is defined as $(S/N_{photon})/(S/N_{eq})$ and refers to the statistical efficiency of the process that transforms the photon image into a neural image.

The SNR can also be evaluated at the "output stage" of an observer treated as a diagnostic system using SDT methods as described in Chapter 2. Recall that signal detection theory provides an index d' for quantifying the magnitude of the signal relative to the noise. For technical reasons that need not concern us here, this value must be squared when used to evaluate statistical efficiency. Thus, the value d'^2 provides an estimate of the SNR for an entire human observer. **Calculation efficiency** is defined as $(S/N_{eq})/(d'^2)$, and refers to the statistical efficiency of the transformation of the neural image into a behavioral decision. The ratio of the first to the third SNR, $(S/N_{photo})/(d'^2)$, which measures efficiency across all stages of perceptual processing is commonly referred to within the perception literature as **quantum efficiency.**

The SNR can only stay the same or go down as visual signals move through the system. Thus, how close the statistical efficiency ratios stay near 1.0 is a reflection of how good of a statistician the observer's visual system is, i.e., how efficiently she makes perceptual decisions based on noisy data.

Stochastic Resonance

The traditional analyses of signals and noise derived from signal detection theory assume that noise is always detrimental to perceptual processing and that perceptual processing is linear until the decision stage. However, recent theoretical approaches suggest, somewhat surprisingly, that if nonlinear stages of processing are involved, the presence of small amounts of noise do not have to always be detrimental. In fact, under specialized conditions referred to as **stochastic resonance,** noise can be beneficial for extracting weak signals, as elaborated in Box 6.7.

Box 6.6
Measurement of Equivalent Input Noise

Equivalent input noise, N_{eq}, is a common measurement made by engineers on a device such as an amplifier to specify its intrinsic noise. Figure 1 illustrates how this measurement is made for an amplifier.

A calibrated amount of white noise is added to the input, and total amount of noise is measured at the output. The amount of noise that is present may vary with temporal frequency (see Chapter 3), so filters are used to allow the measurements to be made for a set band of frequencies. The measured noise at the output will contain contributions from both the externally applied noise and the intrinsic noise of the amplifier. The results

of these measurements will be as shown in Figure 2.

Noise as measured at the output (Nout), specified in units of decibels (dB) that need not concern us here, is shown as a function of the amount of noise added at the input (Nin). The curve is flat at low input noise values, because the amount of noise that is added is insignificant compared to the intrinsic noise of the amplifier. At higher levels of input noise, the external noise dominates, so the curve rises. The knee of the curve occurs at the point where the effect of the external noise is equal to that of the intrinsic noise, and this specifies the equivalent input noise, N_{eq}, of the amplifier.

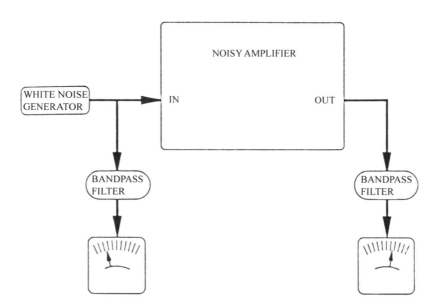

FIGURE 1. Schematic illustrating the methods used by engineers to measure equivalent input noise. See text for symbols and details. Reproduced from D.G. Pelli, The quantum efficiency of vision. In *Vision: Coding and Efficiency,* edited by C. Blakemore, pp. 3–24, © 1990, by permission of Cambridge University Press.

FIGURE 2. Illustration of the results of a typical measurement obtained from the setup in Figure 1. Symbols and details are described in text. Reproduced from D.G. Pelli, The quantum efficiency of vision. In *Vision: Coding and Efficiency*, edited by C. Blakemore, pp. 3–24, © 1990, by permission of Cambridge University Press.

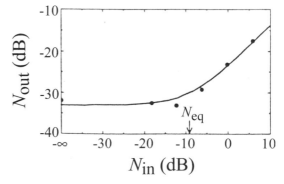

Box 6.7
Special Processing Allows Noise to Be Beneficial in Detecting Weak Signals

Computational models based on a nonlinear effect called **stochastic resonance** demonstrate that, with the right kind of processing, noise can actually be used to help detect weak signals. This finding is counterintuitive but has been demonstrated both computationally and in empirical studies. In order to understand the strategy that is involved, consider a ball in a bowl that has a divider down the middle, as illustrated in Figure 1.

The divider prevents the ball from rolling from one side to the other. Next, suppose the bowl is rocking periodically, but so slightly that the motion cannot be detected. The following procedure will allow us to detect the rate of rocking motion. We slowly add "noise" to the bowl in the form of vibration. The amount of noise is critical. It has to be just enough that the ball changes sides sometimes. Every time the ball changes sides, we mark the time. After we have collected a large number of samples, we use Fourier analysis to determine the amplitude spectrum of the marked times (as described in Chapter 3). This spectrum will reveal power at the frequency at which the bowl is rocking. Every dynamic system with a threshold has an optimum noise level at which stochastic resonance allows it, in principle, to become sensitive to otherwise undetectable signals.

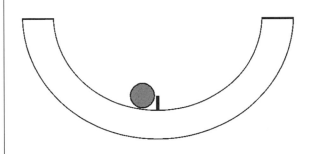

FIGURE 1. The operation of stochastic resonance can be illustrated by the example of using a divider and a ball to detect the rate of rocking motion in a bowl.

Recall that every neuron has a threshold in the form of a certain magnitude of membrane voltage that is necessary to generate an action potential. This threshold is a nonlinearity that can be thought of as being analogous to the energy required to move the ball over the barrier in the example in the Box 6.7. In the absence of noise, small amounts of signal would never generate an action potential and thus could never be detected. However, most neurons have spontaneous activity, meaning that they respond with an ongoing, variable rate of firing even in the absence of a stimulus. This spontaneous activity is analogous to the noise added to the bowl in the example in Box 6.7. Now, spikes sometimes occur whether or not a signal is present, just as the ball sometimes changes sides in the bowl in the presence of added noise. This spontaneous activity sets up the conditions in which stochastic resonance can, in principle, pick up small signals that would be undetectable in the total absence of noise.

6 Solving the Forward and Inverse Problems

Information processing can be described in terms of a forward flow of information. For example, the information that is present in the retinal image is transformed and processed through a number of stages until it reaches the final stage, where it forms a percept that has certain qualities. This can be stated formally as

$$O = F(I),$$

meaning simply that the output (O) stage is some function (F) of input (I). However, as we have seen when describing perception as an inferential process, many aspects of perception are better described by the inverse process. Given a certain output in the form of properties of the percept, what can the observer infer about the input? This can be stated simply as

$$I = F(O).$$

A fact that is sometimes underappreciated is that, computationally, the formal equations that describe the information transfer going forward are not equivalent to the equations that describe the inverse process. Even if one had a formal system that specified O given I, that formal system would not necessarily be able to specify I given O. This is because many inverse problems are **ill-posed,** meaning there is no unique solution. We will provide some explicit examples of ill-posed perceptual problems in Chapter 8 in the context of shape perception and in Chapter 9 in the context of stereoscopic vision.

A fascinating question has to do with why perceptual inferences are seldom wrong despite the fact that they have to be derived based on an inverse process. We do not fully understand how our brains accomplish this. However, a promising lead has to do with the facts that the visible world has exhibited numerous regularities over evolutionary time and that biological brains are adapted to exploit these regularities by engaging in niche-dependent perception. Unlike a general-purpose machine that must try to infer the most likely inverse solution from among all logical possibilities, biological perceivers simply infer the explanation that is most credible based on known regularities.

Summary

Perceptual processing in biological organisms differs in fundamental ways from the kinds of general-purpose processing that are carried out in a digital computer. Neuromorphics is an emerging field that tries to design special-purpose analog devices that process information in a manner that is more similar to what occurs in biological brains. There has been some success in using this approach to design artificial sensor devices, including a silicon retina.

Perceptual information must be coded before it can be processed. Coding schemes can include spatial representations that use distorted topographical mappings. Coding schemes within individual neurons can be based on binary digital coding over labeled lines, scalar coding, tuned filters, or temporal patterns. The amount of coded perceptual information carried by a single neuron can be quantified based on a Shannon model of information transmission. A similar approach is to use the concept of an ideal observer, derived from SDT, to quantify how well two or more stimuli could be distinguished based strictly on the information carried in a single neuron.

In addition to coding within spike trains of single neurons, coding of perceptual information can also be based on patterns of activity across more than one neuron, referred to as "across-fiber coding." Across-fiber coding schemes based on comparison of the results of an entire population of neurons simultaneously are often referred to as "population coding." One convenient way to describe and analyze complex schemes of coding that involve many neurons is with vectors, and thus these schemes are also sometimes referred to as "vector coding."

Coding and processing of perceptual information can be modeled with neural networks that are based on abstractions, called "formal neurons," and their inter-connectivity. Neural networks come in a variety of architectures, ranging from the simple two-layered perception, through inclusion of hidden layers and recurrent connections, to fully interconnected networks. Processing in neural networks can be expressed mathematically as operations on vectors that specify the inputs, outputs, and weights of the network.

Perceptual processing can be conceptualized abstractly as carrying out an algorithm. Many, including Marr, have used algorithms to describe certain aspects of perceptual processing. Some researchers approaching perception from the point of view of artificial intelligence have argued that algorithms can explain the totality of perceptual processing. This point of view is controversial, and others have raised fundamental issues about whether algorithms can apply, in principle, to all aspects of human perception.

Perceptual processing has also been conceptualized abstractly as the application of statistical procedures. There is evidence that some aspects of perceptual processing can be modeled by either descriptive or inferential statistics or combinations of both. Statistical procedures are particularly helpful for reducing uncertainty due to noise and for eliminating redundant information. Quantum efficiency refers to the efficiency with which perceptual processing extracts statistical signals from noise. Somewhat surprisingly, there is evidence that under certain conditions small amounts of noise can be beneficial for detecting weak signals.

Issues of forward flow have to do with characterizing how information passes from the environment through the brain to the behavioral response. Inverse problems involve drawing inferences about what must be present in the environment based on the properties of the percept. Inverse problems are often ill-posed and have no unique solution unless certain assumptions are made. In biological observers, many of these assumptions are built in based on properties of the environment that have been stable over evolutionary time.

Selected Reading List

Adrian, E.D. 1928. *The Basis of Sensation*. New York: Norton.

Attneave, F. 1954. Some informational aspects of visual perception. *Psychol. Rev.* 61:183–193.

Barlow, H.B. 1961. Possible principles underlying the transformations of sensory messages. In *Sensory Communication*, ed. W.A. Rosenblith, pp. 217–234. Cambridge, MA: MIT Press.

Barlow, H.B. 1962. A method of determining the overall quantum efficiency of visual discriminations. *J. Physiol.* 160:155–168.

Collins, J.J. 1995. Stochastic resonance without tuning. *Nature* 376:236–238.

Craik, K.J.W. 1966. *The Nature of Psychology*, ed. S. Sherwood. Cambridge, England: Cambridge University Press.

Crick, F. 1989. The recent excitement about neural networks. *Nature* 337:129–132.

Denning, P.J. 1990. The science of computing. *Am. Sci.* 78:100–102.

Fitzhugh, R. 1957. The statistical detection of threshold signals in the retina. *J. Gen. Physiol.* 40:925–949.

Geisler, W.S., Albrecht, D.G., Salvi, R.J., and Saunders, S.S. 1991. Discrimination performance of single neurons: Rate and temporal-pattern information. *J. Neurophysiol.* 66:334–362.

Gur, M., Beylin, A., and Snodderly, D.M. 1997. Response variability of neurons in primary visual cortex (V1) of alert monkeys. *J. Neurosci.* 17:2914–2920.

Hopfield, J.J., and Tank, D.W. 1986. Computing with neural circuits: A model. *Science* 233:625–633.

Koch, C., and Laurent, G. 1999. Complexity and the nervous system. *Science* 284:96–98.

Lehky, S.R., and Sejnowski, T.J. 1990. Neural network model of visual cortex for determining surface curvature from images of shaded surfaces. *Proc. Roy. Soc. Lond. B* 240:251–278.

Mahowald, M.A., and Mead, C. 1991. The silicon retina. *Sci. Am.* 264:76–82.

McCulloch, W.S., and Pitts, W. 1943. A logical calculus of the ideas immanent in nervous activity. *Bull. Math. Biophysics* 5:115–133.

Minsky, M., and Papert, S. 1969. *Perceptrons*. Cambridge, MA: The MIT Press.

Optican, L.M., and Richmond, B.J. 1987. Temporal encoding of two-dimensional patterns by single units in primate inferior temporal cortex. III. Information theoretic analysis. *J. Neurophysiol.* 57:162–178.

Pelli, D.G. 1990. The quantum efficiency of vision. In *Vision: Coding and Efficiency*, ed. C. Blakemore, pp. 3–24. Cambridge, MA: Cambridge University Press.

Penrose, R. 1989. *The Emperor's New Mind: Concerning Computers, Minds, and the Laws of Physics*. Oxford: Oxford University Press.

Perkel, D.H., and Bullock, T.H. 1969. Neural coding. In *Neurosciences Research Symposium Summaries*, eds. F.O. Schmidt, T. Meinechuk, G.C. Buarton, and G. Adelman, Vol. 3, pp. 405–527. Cambridge, MA: MIT Press.

Poggio, T., and Koch, C. 1985. Ill-posed problems in early vision: From computational theory to analogue networks. *Proc. R. Soc. Lond. B* 226:303–323.

Rosenblatt, F. 1958. The perceptron: A probabilistic model for information storage and organization in the brain. *Psychol. Rev.* 65:386–408.

Rumelhart, D.E., Hinton, G.E., and Williams, R.J. 1986. Learning internal representations by error propagation. In *Parallel Distributed Processing: Explorations in the Microstructures of Cognition*, ed. D.E. Rumelhart and J.L. McClelland, Vol. 1, pp. 318–362. Cambridge, MA: MIT Press.

Shannon, C.E., and Weaver, W. 1949. *The Mathematical Theory of Communication*. Urbana, IL: University of Illinois Press.

Tan, F.C. 1997. Sharpening the senses with neural "noise." *Science* 277:1759.

Tank, D.W., and Hopfield, J.J. 1987. Collective computation in neuronlike circuits. *Sci. Am.* 257:104–114.

7

Color Vision: How Are Objective Wavelengths of Light Transformed into Secondary Qualities of Percepts Called Colors?

Questions

After reading Chapter 7, you should be able to answer the following questions:

1. What does it mean to state that color is a secondary quality of a percept?
2. What are hue, saturation, and brightness?
3. Compare and contrast brightness and lightness.
4. Explain how the simple mechanism of the ratio principle allows the visual system to achieve lightness constancy in ordinary environments.

5. Describe the three-dimensional color solid and explain how it organizes basic facts of color experience.
6. Compare and contrast the trichromatic and opponent process theories of color vision.
7. How well can monochromats, dichromats, and trichromats discriminate between mixtures of wavelengths?
8. Compare scientific and engineering approaches to the constraints imposed by the formal statement of trichromacy.
9. Relate the functional defects present in the common forms of human color blindness to their underlying biological deficiencies.

at the retina and continuing into
astriate cortical areas, describe the
ing streams involved in color
tion.

1. de examples of potential neural explanations of higher-order color phenomena.

12. Why might primates have evolved a color-
processing system in which the complex information about the wavelength composition of
light entering the eye is reduced to just three
numbers (trichromacy)?

The topic of color perception has fascinated scientists for hundreds of years. Because of this early
fascination, scientific studies of color vision have a
long history; a large body of color vision facts has
accumulated; and theories that try to account for
these facts are the most highly developed area of
perception studies.

Experimental studies of color vision provide a
prime example of fruitful interaction among psychology, biology, and the physical sciences. The
psychological, biological, and physical descriptions
of color come from widely different levels of
discourse. However, our understanding of how
color perception operates allows us to link descriptions of color from all of these realms.

1 Color Is a Secondary Quality

Sir Isaac Newton, in a famous quote, about the
relationship between physical properties of light
and color perception asserted: "The rays, to speak
properly, are not colored." This quote underscores
a fundamental fact about color perception. Color is
a property of percepts produced in observers rather
than a property of light energy. Thus, even though
we speak colloquially using terms such as "red
light," we really mean "light that has an appearance
of being red when viewed by a human observer
with normal color vision."

The philosopher John Locke, writing in the 17th
century, made a distinction between **primary
(physical) qualities** and **secondary (psychological)
qualities** of percepts. He used as an example of a
primary quality the three-dimensional shapes of
objects, a property that he asserted exists both in the
world and in our percepts. When we state that an
object appears round, we mean that the quality
roundness applies both to the physical shape and to
the observer's percept of the shape. Secondary
qualities apply to percepts but not to the physical
world. Color was considered to be a secondary

quality because color percepts, although caused by
certain combinations of physical wavelengths, do
not resemble the wavelengths.

As we learn more about how color perception
operates, it will become apparent that the traditional categorization of color as being strictly a
secondary quality is an oversimplification. It is
probably more appropriate to consider the distinction between primary and secondary qualities
as a continuum, with color falling somewhere in
the middle. This can be illustrated by contrasting
three examples of perceptual qualities: perceived
speed, color, and beauty. Percepts having each
of these qualities arise following sensory processing of physical stimulation on the retina (Figure
7.1).

Under conditions in which the eye does not
move, **perceived speed** can be related, at least to
a first-order approximation, to a relatively simple
physical property, **retinal velocity,** as illustrated in
Figure 7.2.

There is a straightforward, and easily understood,
monotonic relationship between retinal velocity
and perceived speed. In general, as the velocity of
an image across the retina increases, the perceived
speed also increases. Most primary qualities of
percepts can be characterized similarly. They have a
relatively simple, often intuitive, relationship with
certain low-level, and easily characterized, physical
properties of retinal stimulation.

Next, consider a percept described by an
observer as having a quality of appearing beautiful.
Perceived beauty is closer to being a pure
secondary quality. It seems unlikely that there exists
any lawful mapping between specific patterns of

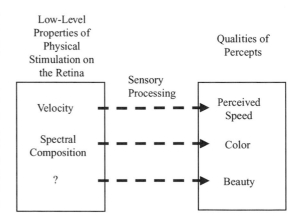

**FIGURE 7.1. Relationship between low-level properties of
physical stimulation on the retina and perceptual qualities.**

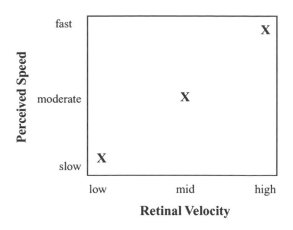

FIGURE 7.2. Illustration of expected hypothetical relationship between the low-level property of physical velocity on the retina and perceived speed.

physical stimulation on the retina and properties of perceived beauty. Furthermore, this mapping, if it exists, is likely to be so complicated as to defy intuitive understanding even if it were to be spelled out.

The perceptual quality of color falls somewhere between these two examples. Some mappings can be made between the wavelength composition of light stimulation on the retina and properties of perceived color. However, these mappings are not always simple or intuitively obvious. Instead, those aspects of the psychophysical mapping that are understood are often complicated and described adequately only in schemes utilizing multiple dimensions.

2 Properties of Color Percepts

Measurements of difference thresholds reveal that humans have the potential to distinguish over a million different colors. Attempts to systematize color experiences in the historical psychological literature used three dimensions, **hue, saturation,** and **brightness.**

Hue

The dimension of hue is most closely related to what is typically meant by the term "color" in ordinary language. Hue can be related in a rough way to a low-level physical property, wavelength of light. When English-speaking human subjects

are asked to look at wavelengths between 490 and 570 nm and name the resulting sensation, most report **green. Yellow** is reported between 570 and 600 nm. **Blue** is reported for visible wavelengths below 490 nm and **red** above 600 nm.

Humans can distinguish about 200 different hues when tested with psychophysical methods in which they simply have to discriminate between two stimuli using Class A observations, even though most report only a small number of unique color names when allowed to report Class B observations (see Chapter 2 for the distinction between Class A and Class B observations). Comparisons across languages and cultures show many similarities among all humans in the application of color names. All languages contain terms for what English-speaking observers call "black" and "white." If a language includes three terms for colors, then a term to designate red is added. Next come green and yellow, and finally blue. These cross-cultural similarities have been cited as evidence that perception of color is biologically based. Something about the way human visual systems are organized causes the percepts labeled black, white, red, green, yellow, and blue to be special.

The relationship between wavelength and hue becomes complicated when human subjects are shown spots of light composed of mixtures of wavelengths. Many mixtures lead to perception of the same hue as that produced by a single wavelength, demonstrating that the mapping from wavelengths to hues is many-to-one. In addition, some hues are experienced when looking at mixtures of wavelengths but never when looking at a single wavelength. An example is the hue labeled **purple,** experienced only when looking at a mixture of blue and red. Another example is the hue experienced when looking at **white light,** a mixture of all wavelengths. The hue experienced when looking at white light is called various names, including white or gray, but is most appropriately labeled **achromatic,** meaning that there is an experience of color but an absence of any distinct hue.

Saturation

When humans look at mixtures of wavelengths that consist of bands of the spectrum wider than a single wavelength but narrower than white light, a hue is perceived. The particular hue reported can be predicted based on the central wavelength of the band. As the bandwidth of wavelengths being viewed is increased, hue stays the same, but observers report that the color experience changes gradually along another dimension. The hue appears to become less distinct and more similar to the achromatic experience produced by white light. This psychological dimension of how distinct from white a hue appears is called "saturation." Highly distinct hues are **saturated,** while less distinct ones are **desaturated.**

Saturation can be related in a rough manner to the physical property of **purity,** which refers to the number of wavelengths combined in a mixture. A single wavelength, called **monochromatic** light, has the highest purity and white light, a mixture of all wavelengths, the least. However, it is easy to demonstrate that the dimension of saturation is based on more factors than simply the degree of purity by having observers make a distinction along the dimension of saturation when comparing across individual wavelengths. For example, a yellow hue elicited by a 570-nm wavelength looks less saturated than a blue hue elicited by a 490-nm wavelength, even though both stimuli are equally pure.

Degree of saturation can be quantified psychophysically by measuring the lumens of white light that must be added to the more saturated member of a pair of stimuli to make them equally saturated. Observers can distinguish about twenty steps of saturation when viewing short or long wavelengths but only about six in the midspectral region.

Brightness and Lightness

The third color dimension used to characterize color experience has traditionally been called "brightness" in the psychological literature. A closely related term is **lightness,** and there is some confusion in the use of these terms. The confusion comes about largely because traditional studies of color vision used simple stimuli that were each composed of a spot of light of uniform color against a dark or dim homogeneous background. The Class B experience reported by observers reviewing these spots of light is called **film color.** The color is located indistinctly somewhere in space rather than appearing to be part of an object. Film color occurs occasionally in natural environments, as in the color of the cloudless sky, but is relatively rare. Most of our color experiences are of **surface colors** perceived as properties of objects, as in *green leaves,* in which the greenness appears to be either permeating the leaves or coating their surface. Technically,

the term "brightness" should be used to describe the intensity dimension of color experience associated with film color and "lightness" the intensity dimension with reference to a surface color. In practice, these two terms are often interchanged or applied indiscriminately.

When asked to report how intense a film color appears, human observers can distinguish about 500 steps, ranging from dimly illuminated, **dark,** through intense, **bright.** The level of brightness can be related, roughly, to a low-level stimulus property, the rate at which photons are being absorbed in the eye. However, brightness is also influenced by other factors, such as the recent history of light stimulation and the intensity of stimulation of surrounding regions.

When asked to rate surface colors along the same dimension of intensity, observers can also distinguish about 500 steps, ranging from black to white. The physical property of a surface that is related most closely to its lightness is not photon absorptions at the eye but rather its **reflectance,** the proportion of incident photons reflected from the surface towards the eye.

With proper training or instructions, human observers can report semiindependent values for the lightness and the brightness of an object. Consider a piece of black coal and a sheet of white paper viewed indoors and outdoors, as illustrated in Figure 7.3.

The dim illumination from the light bulb is depicted in this cartoon as emitting only 100 photons, while outdoors, the sun is emitting 100,000 photons. The numbers of photons reflected to the eye from the paper under indoor and outdoor conditions are vastly different, 90 and 90,000. A similar difference exists for the numbers of photons reflected to the eye from the coal, 10 and 10,000. However, the reflectance values (R) for the white paper (0.9) and for the coal (0.1) remain constant under the two conditions.

Brightness is related to number of photons entering the eye, while lightness is more closely related to reflectance values of surfaces. Indoors, the coal appears to be dimly illuminated (brightness) and black (lightness). When brought outside, it appears intensely illuminated but still black. Similarly, the piece of paper continues to look white whether viewed indoors or outdoors.

This example illustrates **lightness constancy,** meaning that the lightness of surfaces does not change appreciably with changes in background illumination. Theories of perception consider lightness constancy interesting because it demonstrates that a quality of our percept can be more similar to

FIGURE 7.3. This figure illustrates the relationships responsible for lightness constancy when the same objects, in this hypothetical cartoon example a piece of coal and a sheet of paper, are seen under indoor conditions, illuminated by a light bulb, and outdoors, illuminated by the sun. See text for additional details.

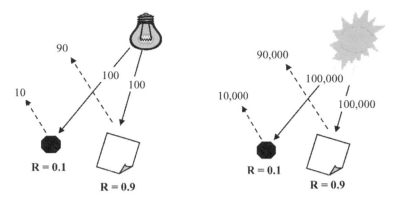

a property of the distal stimulus (reflectance of an object's surface) than one of the proximal stimulus (photons impinging on the retina).

A basic mechanism used by the visual system to maintain lightness constancy is the **ratio principle.** As a first-order approximation, the lightness of a small homogeneous region of a scene is determined by the ratio of its luminance to that of the surrounding area. This simple mechanism usually works to maintain lightness constancy in ordinary environments, because all of the objects are illuminated by the same source. However, the visual system can be easily fooled, as demonstrated in specially designed laboratory environments.

The Germain psychologist Adhémar **Gelb** performed studies in the 1920s in which subjects viewed a black disk in a dimly lit room through a peephole in a wall. The layout of the experiment is illustrated in Figure 7.4.

Objects in the room were illuminated by a dim light in the ceiling. However, unbeknownst to the observer, a small track light that was hidden from view had been focused to provide additional illumination to just the black disk and not the rest of the room. The observers reported that the black disk appeared white under these conditions. The interpretation of this result is that the visual system makes an assumption (an unconscious inference) that the disk is illuminated by the dim light bulb. Since the black disk is reflecting a high number of photons into the eye (presumably from this dim light source), the visual system assumes it must have a high reflectance, and the percept of the surface is given the lightness quality "white."

In a further study, Gelb had an experimenter move a piece of white paper to a position revealing the presence of the light beam from the track light, as illustrated in Figure 7.5.

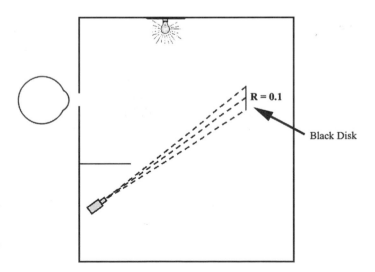

FIGURE 7.4. Layout of the room used by Gelb for studies of lightness constancy in the 1920s. A black disk appears white when viewed under these conditions. See text for details.

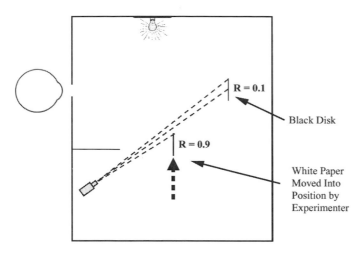

The observer reported that the appearance of the disk suddenly changed to black, revealing the operation of the unconscious inference, which takes into account the new information about the existence of the beam from the track light. However, the unconscious inference process is not subject to higher-order cognitive influences. The black disk appears to return to white as soon as the paper is removed, despite the fact that the observer is now aware of the existence of the hidden light beam.

In complex environments that have surfaces with many different reflectance values, the visual system computes relative reflectance for each surface based on a complex set of comparisons of adjacent surfaces. Simple mechanisms such as the ratio principle are not sufficient to account for lightness of surfaces in these complex environments. Sets of computations that are sufficient for these conditions were initially described and studied by **Edwin Land,** a scientist and inventor who was also responsible for inventing the Polaroid Land Camera. Land referred to these computations as **retinex theory.** The term "retinex" was derived by combining RETINa with cortEX to emphasize that these computations were presumed to reflect multiple stages of biological color processing. As will be elaborated below, our brains calculate a separate lightness value semiindependently for each local region of a scene, perhaps within each hypercolumn during early biological stages of processing. At later stages, the visual system is able to compare lightness values of surfaces throughout larger regions of the visual scene simultaneously to carry out computations of the kind described by retinex theory.

The more complex global computations described by retinex theory do not come into play for simple visual stimuli of the type that have been typically employed in classic studies of color vision. This has led to an emphasis in the historical perception literature on more simple color phenomena that can be explained by more simple mechanisms, such as the ratio principle. An example is **simultaneous lightness contrast.** This can be demonstrated as shown in Figure 7.6.

The two small gray circles within the larger circles shown on the left and right sides of the figure appear to differ in lightness, with the small circle on the left appearing darker. In fact, both small circles reflect the same number of photons to the eye. The two small circles appear to have differing degrees of lightness because the lightness of each small square is being computed based primarily on the simple mechanism of the ratio principle.

FIGURE 7.6. Demonstration of simultaneous lightness contrast. See text for details.

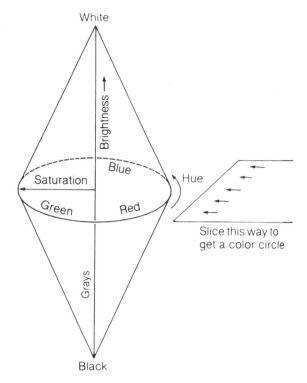

FIGURE 7.7. The three dimension of color experience, hue, saturation, and brightness, can be represented in a color solid. A slice through the color solid produces a color circle, as shown in Figure 7.8. Reproduced from S. Coren, C. Porec, and L.M. Ward, Sensation and Perception, Second Edition, © 1984, by permission of Academic Press.

The larger surrounding squares cause the luminance ratio between the small center square and its surrounding region to be different for the stimuli on the left and right sides of the figure. Consequently, the small squares appear to have different degrees of lightness.

The Color Solid

One scheme that has been used for representing the three dimensions of psychological color experience, hue, saturation, and brightness, is a **color solid**, as illustrated in Figure 7.7.

Brightness is represented along the vertical axis, saturation as distance from the vertical axis, and hue as direction around the axis. The brightness dimension extends from black at the bottom to white at the top. Saturation is zero at the vertical axis and maximum at the circumference. A slice through the color solid at its widest point produces a two-dimensional **color circle,** as illustrated in Figure 7.8.

The color wheel wraps around in such a way that long wavelengths run back into short wavelengths. The hues produced in this region are called **extraspectral.** They can be produced by mixtures of long and short wavelengths but not by any single wavelengths. Purple is an example of a extraspectral color. Surprisingly, so is pure red. If a small amount of short-wavelength light is added to a reddish-appearing stimulus composed of long wavelengths, the mixture appears redder than the long wavelengths alone. A physiological explanation for this phenomenon will be given in the section "Color Processing in the Retina."

Each hue is represented on the color circle by a straight line that passes from the circumference of the circle to its center. The saturation of each hue decreases gradually from a maximum at the circumference to an achromatic white at the center. The representation of saturation in the color circle is relative rather than absolute, with the circumference of the circle representing the maximum saturation possible for the corresponding hue. If saturation were represented on an absolute scale, the circumference would form a distorted shape rather than a circle due to the fact that some wavelengths give rise to more saturated percepts than others.

The color solid organizes some of the facts of Class B color experiences based on an assumption

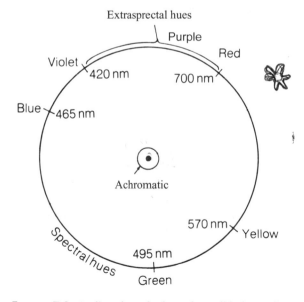

FIGURE 7.8. A slice through the color solid shown in Figure 7.7 at its widest point produces the two-dimensional color circle shown here. Reproduced from S. Coren, C. Porec, and L.M. Ward, Sensation and Perception, Second Edition, © 1984, by permission of Academic Press.

that the three dimensions, hue, saturation, and brightness, are independent of one another and do not interact. That assumption breaks down under extreme conditions. For example, a light of moderate luminance that appears red will appear green when made very intense, a perceptual phenomenon called the **Bezold-Brucke effect.** Nevertheless, the three-dimensional color solid provides a reasonable metric of many aspects of color experience that can be reported with Class B Observations over a moderate range of conditions.

The color circle is convenient for making predictions about the effects of color mixing when only two wavelengths of light are involved. One simply draws a straight line between the two points on the circumference corresponding to the wavelengths that are involved in the mixture. Then one moves

along this line by an amount that is proportional to the intensities of the two wavelengths. For example, if the two wavelengths are mixed in equal parts, then the hue of the percept will fall halfway between the two wavelengths. If the two wavelengths are mixed in a proportion of 10 to 1, then the point will fall at the same proportionate distance between the two wavelengths, and so on. The perceived hue of the mixture can be predicted by simply drawing a line from the center of the circle through the same point and determining the hue where this line intersects the circumference. Similarly, relative saturation is predicted by simply noting the relative distances from this point to the center and to the circumference. Mixtures of any two wavelengths can be used to mix any hue that falls between them on the circle, but with

Box 7.1
Colors Are Based on Additive Mixtures of Lights and Subtractive Mixtures of Pigments

Mixing of pigments is different from mixing wavelengths of light, as illustrated in Figure 1.

Mixing pigments that look yellow and blue produces a pigment that appears green. This result would not be predicted based on what is shown in the color wheel. The color wheel shows what happens under conditions of **additive color mixture,** when wavelengths of light add together to form a mixture. The appearance of pigments is based on **subtractive color mixture,** in which the wave-

lengths the pigments absorb are subtracted from a mixture. For example, the reason a blue pigment looks blue is that it absorbs wavelengths *except* those around the blue portion of the spectrum. Pigment mixtures can be related to the color wheel by figuring out which wavelengths make it through both pigments without being absorbed. Then these wavelengths can be combined using the rules of the color wheel.

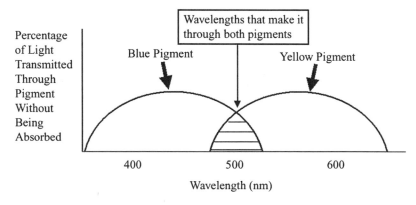

FIGURE 1. Demonstration of the difference between colors formed by mixtures of wavelengths and by mixtures of pigments.

reduced saturation. Wavelengths that lie opposite one another on the color circle are said to be **complementary.** When complementary wavelengths are mixed, the percept will include only the two hues of the components, or an achromatic gray if the complementaries are mixed in equal proportions.

Readers with experience mixing pigments to produce colors may be confused by this color circle and feel that it appears to be in error. That confusion is due to a difference between mixing pigments and mixing lights, as explained in Box 7.1.

The color wheel is limited to making predictions about color experiences that result from simple mixtures of wavelengths. Other forms of representation, described in the section on colorimetry, are more general and allow quantitative predictions to be made about complex mixtures of wavelengths.

Trichromatic and Opponent Theories

Two major theories of color vision were developed in the 1800s to try to explain causal relationships involved in perception of color. One, called **trichromatic,** is associated historically with **Thomas Young** (1773–1829) and Helmholtz (1821–1894), and the other, called **opponent,** is associated with Hering (1834–1918). These are psychological theories formulated based on results obtained from human observers using psychophysical methods based on Class A and Class B observations as defined in Chapter 2.

The trichromatic theory tries to account for the fact that specific combinations of wavelengths, when mixed, give rise to particular colors. It proposes that three processing mechanisms are involved in color perception of mixtures.

The opponent theory was developed to account for the fact that certain color experiences appear to be opposites of one another. To account for these facts, the opponent process theory proposes three pairs of processing mechanisms in which the two components within each pair work in opposition to one another.

In both theories, the properties of the processing mechanisms, and their interactions, are supposed to account for the facts of color perception. For a long period, the trichromatic and opponent theories were considered to be in competition with one another, and the goal of evaluating the two theories was to determine which one was correct. However, modern neuroscience findings described at the

level of neural mechanisms that underlie color perception have now allowed these two theories to be synthesized. The initial absorption of light occurs in three types of cone photoreceptors. This fact gives rise to the observations that are accounted for by trichromacy theory. The trichromatic information that is present at the level of the receptors is then reorganized by the neurons that transmit color information through the visual processing areas of the brain. The receptive fields of these neurons operate in an antagonistic, or opponent, manner with respect to wavelengths of light. Properties of these neurons can explain many opponent properties of color vision.

3 Properties of Receptors Explain Some Aspects of Color Vision

Metamers, Channels, and the Principle of Univariance

The basic facts in support of trichromatic theory were collected using primarily Class A observations in which human observers were asked to report whether two lights looked the same or different. A fundamental finding is that many pairs of lights look identical, although they are composed of different wavelengths. These pairs of lights are called **metamers,** defined as stimuli that are physically different but perceptually identical. Metamers play a major role in understanding color vision. Any given experience of color can be caused by any of a large number of metamers. In other words, there is a many-to-one mapping of physical wavelengths onto color experiences. A fundamental problem for any color vision theory is to explain why metamers map to the same psychological color experience.

A **channel** is a method of transmitting information from one stage of visual processing to the next, characterized by the fact that only values of a single continuous variable can be transmitted. Consequently, information concerning two or more variables, such as intensity and wavelength, is always confounded when these values must be transmitted at the same time within a single channel. Described abstractly at the level of information processing, trichromatic color theory proposes the existence of three channels to account for color metamers. Described at the level of neuroscience, the three channels proposed by trichromatic theory correspond to the three types of cone photoreceptors.

FIGURE 7.9. The principle of univariance.
The receptor on the right follows the principle of univariance, but the receptor on
the left does not.

"I just absorbed 1 photon
of 450 nm light and
1 photon of 600 nm light."

"I just absorbed 2 photons
of light. I have no idea what
their wavelengths are."

The fact that photoreceptors act as channels is sometimes referred to as the **principle of univariance,** first articulated by **K. Naka** and **W.A.H. Rushton** in 1966. The principle of univariance is illustrated schematically in Figure 7.9.

Two hypothetical kinds of photoreceptors and the information transmitted to the retina by each are illustrated. Each of these photoreceptors has just absorbed two photons, one having a wavelength of 450 nm and the other having a wavelength of 600 nm. The hypothetical photoreceptor on the left signals information to the retina (for further processing and transmission to the brain) that specifies both how many photons were absorbed and their wavelengths. The hypothetical photoreceptor on the right signals information to the brain only about how many photons have been absorbed, not about their wavelengths.

The photoreceptor on the right follows the principle of univariance and illustrates the limited kind of information transmitted by biological photoreceptors. Biological photoreceptors follow this principle because of the mechanisms of transduction that were discussed in Chapter 4. The wavelength of a photon simply determines its probability of being absorbed by a given photopigment molecule. However, once it is absorbed, the transduction events triggered within a photoreceptor are identical, whatever the wavelength of the photon.

Despite univariance, the visual systems of humans and monkeys are not destined to remain totally ignorant about the wavelength composition of the photons that are being absorbed in the retina. The brain utilizes a sensory coding scheme for wavelength information that can be characterized as **across-fiber coding using broadly tuned filters,** as described in Chapter 6. How much information can be potentially recovered with this type of coding scheme depends on factors such as the number of receptor types and the shapes of the underlying spectral absorption curves. The shapes of the spectral absorption curves in human and monkey receptors are determined by their photopigments. Most humans and many species of monkeys are **trichromats,** which means they have three types of cone photoreceptors, each of which has a unique photopigment. The cone types that are present in humans with normal trichromatic color vision are typically labeled **S cones, M cones,** and **L cones** to designate that they are most sensitive to short (430 nm), middle (530 nm), and long (560 nm) wavelengths. The relative absorption curves for the photopigments in these three types of cones are illustrated in Figure 7.10.

Some "color blind" humans and some species of monkeys, called **dichromats,** have only two types of cone receptor. Rare individuals with only one cone type are called **monochromats.**

The sections that follow analyze the types of color information that can be recovered, in principle, in monochromats, dichromats, and trichromats.

Monochromacy

A spectral absorption curve for the single type of photoreceptor found in a hypothetical monochromat is illustrated on the left side of Figure 7.11.

FIGURE 7.10. Spectral absorption curves of the photopigments present in three kinds of human cone photoreceptors, which are most sensitive to short (S), middle (M), or long (L) wavelengths.

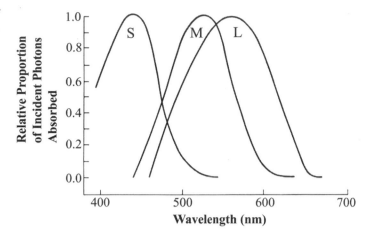

The dashed lines illustrate the proportion of photons passing through the outer segments of the photoreceptor that will be absorbed for two specific wavelengths of light, 500 nm and 575 nm. When 500-nm photons pass through the photoreceptors, about 10% of them will be absorbed, compared to only about 1.5% of 575-nm photons. Given this spectral absorption curve, we can calculate what will happen in the photoreceptors when the monochromatic observer looks at the two spots of light shown in the inset on the right side of Figure 7.11.

These spots of light may be flashed briefly such that each emits 1,000 photons that impinge on the outer segments. Equivalently, the observer may be looking at steady spots of light and the luminance of each is adjusted so 1,000 photons are incident per unit time (e.g., 1,000 photons/msec). Either way, the luminance of each spot will be specified as 1,000 photons. The numbers in the first row of the diagram on the right side of the figure indicate how many photons will be absorbed by the photoreceptors for each spot of light. When the 500-nm light arrives at the receptors of this monochromat, 100 photons (10% of 1,000) are absorbed, and when the 575-nm light arrives 15 photons (1.5% of 1,000) are absorbed.

Could this monochromat observer, in principle, discriminate between these two particular spots of

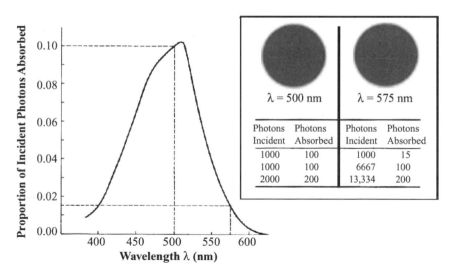

FIGURE 7.11. Left panel: Spectral absorption curve for the photopigment present in a hypothetical monochromat. The dashed lines show the proportions of photons that would be absorbed for 500 nm and 575 nm light. Right panel: Demonstration of how many photons will by absorbed by the photopigment while the monochromat observer views stimuli composed of 500-nm and 575-nm light. Adapted from T.N. Cornsweet, *Visual Perception,* © 1970, by permission of Academic Press.

light? The answer is "yes," if the observer follows the following strategy: Photoreceptors are known to generate a signal that is related to the number of photons that are absorbed. Thus, an observer could adopt the strategy of monitoring that signal, and responding "500-nm spot" every time the retina reports a signal strength of 100 and "575-nm spot" for every signal with a strength of 15. This observer could even be taught to respond with color names, "green" for a signal strength of 100 and "red" for a signal strength of 15. A monochromatic observer following this strategy would apply color names appropriately, i.e., with the same names that would be reported by a human with normal color vision, whenever confronted with these two particular stimuli.

However, let's ask a slightly different question. Could this observer, in principle, discriminate between any spots of light having physical wavelengths of 500 nm and 575 nm? The answer is no! All we have to do to fool this observer is change the luminance of one or both of the light spots. The second row of the diagram in Figure 7.11 shows an example in which the 575-nm light spot is adjusted to have a luminance of 6,667. When viewing this spot of light, the signal that is generated will be 100 (1.5% of 6,667), and our trained observer will respond (incorrectly) "500 nm spot" or "green." Stated another way, we have just succeeded in making two physically different lights (500-nm light with a luminance of 1,000 and 575-nm light with a luminance of 6,667) into metamers for this monochromat.

This example illustrates the general rule that luminance and wavelength are always confounded for a monochromat. Suppose we double the luminance of the 500-nm light to 2,000, as illustrated in the third row of the diagram. This would result in absorption of 200 photons. If the observer were asked to give a Class B observation of how the appearance of the light spot changed, he would probably state that the light spot now looks "brighter." But the 575-nm light spot can be made to look similarly bright by increasing its luminance to 1,334, which will result in an identical 200 absorptions. Thus, there is no way for this observer to disentangle whether any given signal strength (such as 15 or 100 or 200) was produced by the 500-nm or the 575-nm wavelength of light. It could have been either one depending on the intensity.

These conclusions, illustrated for this particular hypothetical monochromat in this example, are true in general. Under many conditions that occur in the natural environment, monochromats report color names correctly, e.g., that the sky appears blue and grass green. However, this is just learned behavior, and if the observer is confronted with conditions in which color names must be applied strictly on the basis of wavelength with no other information available, then he will fail.

This hypothetical example demonstrates that a monochromat cannot discriminate a single wavelength (λ_1) from a single other wavelength (λ_2). It should be intuitively obvious that a monochromat will also fail when trying to discriminate between a single wavelength and a mixture of two other wavelengths, a mixture of three other wavelengths, or any higher-order mixture. The limiting case will be discrimination from white light, formed from a mixture of all wavelengths. Regardless of the number of wavelengths that are mixed together, the photoreceptor will simply report a single number that reflects the total number of absorptions. If the luminance is adjusted appropriately, the number of absorptions from the mixture can be made identical to the number of absorptions from the single wavelength to which the mixture is being compared, and the two stimuli will become metamers.

Dichromacy

Let's now conduct a similar assessment of what kinds of wavelength discriminations can be accomplished, in principle, by a hypothetical dichromatic observer with two retinal receptor types. The spectral absorption curves for a dichromat with two receptor types, called "A" and "B," are illustrated in Figure 7.12.

The proportions of photons that will be absorbed for two representative wavelengths of light, λ_1 and λ_2, by each photoreceptor type are illustrated by the dashed lines. Figure 7.13 illustrates the signals that will be generated in the retina by each photoreceptor type while viewing two spots of light, one composed of 1,000 photons of λ_1 and the other of 1,000 photons of λ_2.

Note that two signals will be generated in the retina for each light spot, one by the A receptor type and the other by the B receptor type. The signal strength generated for the λ_1 spot when it has a luminance of 1,000 will be 460 for the A receptor and 150 for the B receptor. The strengths of the two signals generated by the λ_2 spot when it has an luminance of 1,000 will be 260 and 410 for the A and B receptors, respectively.

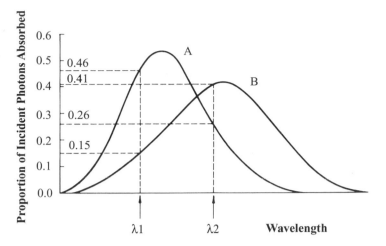

FIGURE 7.12. Spectral absorption curves for two photopigments, A and B, found in the photoreceptors of a hypothetical dichromat. The dashed lines show the proportions of photons that would be absorbed for two specific wavelengths, λ_1 and λ_2. Adapted from T.N. Cornsweet, *Visual Perception,* © 1970, by permission of Academic Press.

This dichromatic observer can use simple strategies to discriminate between these two particular stimuli. It is obvious that they are different based on a comparison of the strengths of the signals generated by the A receptors (460 vs 260) or of the signals generated by the B receptors (150 vs 410). However, a more important question is whether this dichromatic observer can discriminate between these two wavelengths in general, irrespective of their luminance. In other words, is it possible to adjust the intensities and fool the dichromat in the same way monochromat was fooled in the previous example?

By adjusting the luminance of the λ_2 spot of light to 1,770, the signals generated by the A receptors can be matched. However, the dichromat will still be able to discriminate between the two spots by comparing the signals generated in the B receptors (150 vs 725). Similarly, by adjusting the luminance of λ_2 to 366, the signals for the B receptors can be matched, but now the signals generated by the A photoreceptors are different.

The finding for this specific example holds in general for a dichromat. Sufficient information is available, based on monitoring of the signals from both receptor types, to discriminate any single wavelength from any other single wavelength, regardless of their relative intensities.

The strategy of simply comparing the outputs of the two receptor types to see if they are the same or different is sufficient to allow a dichromat to avoid being fooled when discriminating between any two separate wavelengths. However, another potential strategy that a dichromat can use is more powerful. The ratio of the signals from the A and B receptors is unique for every wavelength and does not vary with luminance. For example, a ratio of 3:1 might designate 500 nm and a ratio of 3:4

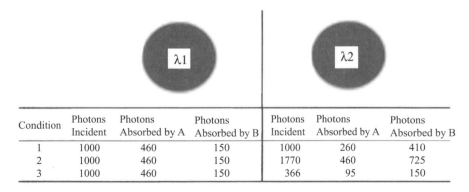

Condition	Photons Incident	Photons Absorbed by A	Photons Absorbed by B	Photons Incident	Photons Absorbed by A	Photons Absorbed by B
1	1000	460	150	1000	260	410
2	1000	460	150	1770	460	725
3	1000	460	150	366	95	150

FIGURE 7.13. Numbers of photons absorbed by the photopigments referred to in Figure 7.12 while the hypothetical dichromat views stimuli composed of light of wavelengths λ_1 and λ_2. Adapted from T.N. Cornsweet, *Visual Perception,* © 1970, by permission of Academic Press.

might designate 575 nm. If a dichromatic observer can implement an algorithm that allows this ratio to be computed, the observer has the capability of responding with unique color names, such as "green" whenever the ratio is 3:1 and "red" whenever it is 3:4. A dichromat following this strategy would be able to consistently apply color names to single wavelengths and would not be fooled by changes in luminance. One possible way an algorithm that computes signal ratios might be implemented by a biological neuron is described in Box 7.2.

Consider next how well a dichromat can do, in principle, when it comes to discriminating mixtures of light. Let's start out by analyzing the limiting case, in which the dichromat tries to discriminate one particular wavelength from white light, a mixture of all wavelengths. White light will produce a signal in receptor type A that is proportional to the area under, that is, the integral of, its spectral absorption curve. Similarly, the effects of white light on receptor type B will produce a signal proportional to the integral of its spectral absorption curve. If the two types of receptor have spectral absorption curves that are about the same width and height, their integrals will be of about the same magnitude. Consequently, the ratio of the signals produced by white light will be near 1:1. One particular single wavelength will also produce a signal ratio of 1:1, as illustrated by the vertical line in Figure 7.14.

This wavelength corresponds to the point where the spectral absorption curves cross. In general, even when the integrals of the absorption curves are not identical and produce ratios other than 1:1, there will still be some single wavelength that produces a ratio identical to that of white light (see Box 7.3). This single wavelength will not be able to be discriminated from white light and is called the dichromat's **neutral point.** However, every other single wavelength except the neutral point will produce a signal ratio that is different from that of white light.

Finally, how well can a dichromat be expected to do when trying to discriminate between a single wavelength and a mixture of two wavelengths? Figure 7.15 shows the relative absorptions of three specific wavelengths, λ_1, λ_2, and λ_3, in receptor types A and B.

Figure 7.16 shows a specific example of what would happen if such a dichromat were to be presented with two spots of light, one of single wavelength λ_1 and the second a mixture of λ_2 and λ_3.

Box 7.2
Calculating a Ratio with a Neuron

Consider the hypothetical neuron shown in Figure 1, which receives, directly or indirectly, excitatory synaptic inputs from receptor type A and inhibitory inputs from receptor type B.

The neuron simply sums its inputs, and consequently its output is proportional to the synaptic inputs from A minus the synaptic inputs from B. However, suppose the relationship between the number of photons absorbed in a photoreceptor and the signal produced at the synapse is not linear but approximates a logarithmic relationship. Then the difference signal will be roughly equivalent to the log of the photons absorbed in A minus the log of the photons absorbed in B. Consequently, the output corresponds to the signal ratio, since $\log(A) - \log(B) = A/B$.

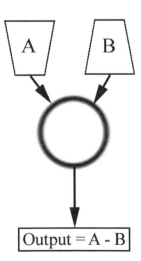

FIGURE 1. Illustration of a hypothetical neural circuit that could potentially compute a ratio. See text for details.

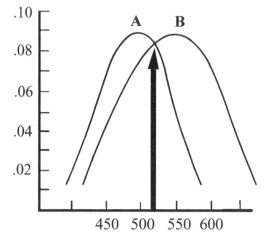

Box 7.3
Why a Neutral Point Must Exist for a Dichromat

At each wavelength, a ratio can be determined between the heights of the two spectral sensitivity curves. The ratio of the integrals of the two functions will be equal to the mean, over all wavelengths, of these ratios produced at each wavelength. Since the mean always falls between the extremes, this ratio between the integrals cannot be larger than the largest ratio for an individual wavelength or smaller than the smallest. The wavelength function is continuous, and thus there must exist a wavelength that produces the same ratio as the integrals.

FIGURE 7.14. Light having a wavelength corresponding to the point where the spectral absorption curves of two photopigments cross results in a ratio of absorptions of 1:1.

Condition 1 demonstrates that when all three wavelengths are at a magnitude of 1,000, the two spots can be discriminated, as for single wavelengths, by simply comparing the outputs of A and B. However, condition 2 demonstrates that when the intensities of λ_2 and λ_3 are adjusted to magnitudes of 1,250 and 300, their combined photon absorptions match that of single wavelength λ_1. The fact that a match can be found in this particular example is also true in general for all wavelengths for a dichromatic and can be expressed in a **formal statement of dichromacy:**

Given two spots of light, one composed of wavelength λ_1 and the other of wavelengths λ_2 and λ_3, it is always

possible, by varying the intensities of λ_1, λ_2, and λ_3, to turn the two spots into metamers.

Trichromacy

By the same logic used in discussing dichromacy in the previous section, it should be obvious that trichromats can, in principle, discriminate any single wavelength from any other wavelength. Furthermore, a trichromat with receptor types A, B, and C can compute ratios of A/B, A/C, or B/C, any one of which will be sufficient to allow unique color names to be applied to individual wavelengths.

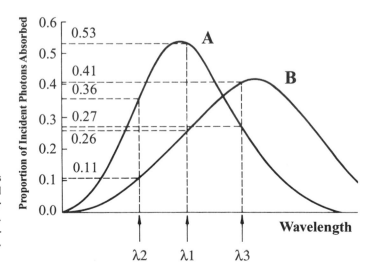

FIGURE 7.15. Proportion of absorptions that would occur in the same hypothetical photopigments shown in Figure 7.12 for three specific wavelengths, λ_1, λ_2, and λ_3. Adapted from T.N. Cornsweet, *Visual Perception*, © 1970, by permission of Academic Press.

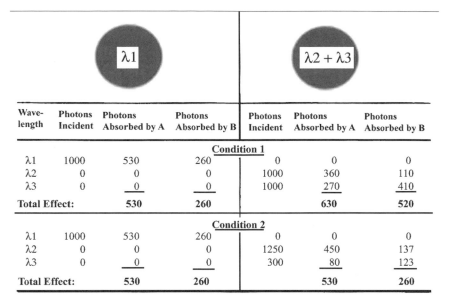

Wave-length	Photons Incident	Photons Absorbed by A	Photons Absorbed by B	Photons Incident	Photons Absorbed by A	Photons Absorbed by B
			Condition 1			
$\lambda 1$	1000	530	260	0	0	0
$\lambda 2$	0	0	0	1000	360	110
$\lambda 3$	0	0	0	1000	270	410
Total Effect:		530	260		630	520
			Condition 2			
$\lambda 1$	1000	530	260	0	0	0
$\lambda 2$	0	0	0	1250	450	137
$\lambda 3$	0	0	0	300	80	123
Total Effect:		530	260		530	260

FIGURE 7.16. Numbers of photons absorbed by the photoreceptors illustrated in Figure 7.15 while the hypothetical dichromat views two stimuli, one composed of wavelength λ_1 and the other of wavelengths λ_2 and λ_3. Adapted from T.N. Cornsweet, *Visual Perception*, © 1970, by permission of Academic Press.

White light, since it has approximately equal effects on all three receptors, results in signal ratios of approximately 1:1:1. The likelihood of any single wavelength's producing this ratio is so remote that it can safely be ignored (see Box 7.4).

Thus, the expectation is that a trichromat will not have a neutral point. When stated the opposite way, the assertion is a logical certainty: If no neutral point is present, the observe cannot be a monochromat or a dichromat and thus must be (at least) a trichromat. This is sometimes used as a diagnostic test to establish that an observer is not a monochromat or a dichromat.

It can be demonstrated that, in principle, trichromats *can* discriminate any single wavelength from a mixture of only two other wavelengths but *not* from a mixture of three or more other wavelengths. The rationale for this conclusion is based on the same line of reasoning discussed above for dichromacy. Stated as a **formal statement of trichromacy:**

"Given two spots of light, one composed of wavelength λ_1, the other of wavelengths λ_2, λ_3, and λ_4, it is possible, by varying the luminances of λ_1, λ_2, λ_3, and λ_4, to turn the two spots into metamers."

Table 7.1 summarizes how well trichromats can discriminate wavelengths of light from one another in comparison to monochromats and dichromats.

The first column designates whether or not a given type of observer can use appropriate color names. When viewing complex mixtures composed of many wavelengths, neither monochromats nor dichromats can apply appropriate color names consistently, although both may report correctly under conditions in which extraneous information is available. Trichromats include most humans. By definition, trichromatic humans apply "appropriate" color names, if by appropriate we mean the same names applied by other normal humans who speak the same language. However, if we define "appropriate" as the ability to apply unique labels that accurately characterize the wavelength composition of complex mixtures, then trichromats fail just as miserably as monochromats and dichromats.

The second column reports whether a given type of observer can discriminate, in general, between any two wavelengths. In this regard, the dichromat and trichromat are each superior to the monochromat. Dichromat and trichromat observers can discriminate between any two wavelengths and can learn to apply color names consistently to single wavelengths.

The third column reports whether an observer can discriminate single wavelengths from white light. In this regard monochromats, dichromats, and trichromats can be ordered by rank. A trichromat can discriminate all wavelengths from white

Box 7.4
Signal Ratios of 1:1:1 Are Highly Unlikely

The only way a single wavelength can produce a signal ratio of 1:1:1 is if the three spectral sensitivity curves are the same height at some wavelength.

In Figure 1, the wavelength at which the ratio between A and B is 1:1 is illustrated by the thick solid arrow. The corresponding wavelength for A and C is shown by the thick dashed arrow and that for B and C by the thin dotted arrow. The only way these three arrows could fall at the same wavelength would be if the three curves happened to cross at a single point, a highly unlikely event.

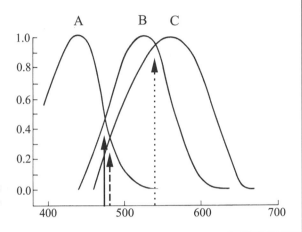

Figure 1. Illustration that no single wavelength is likely to produce a ratio of absorptions in three pigments of 1:1:1.

light; a dichromat all except one, the neutral point; and a monochromat none.

For discriminations of a single wavelength from a mixture of two wavelengths, the fourth column shows that the trichromat has an advantage over dichromats and monochromats. However, the fifth column reveals that the wavelength discrimination capabilities of trichromats are not all that much better. A trichromat can be completely fooled into thinking a mixture of as few as three wavelengths is some totally different single wavelength.

Much of the technology in our society depends on exploiting the deficiencies of trichromacy. If human observers' abilities to perceive wavelengths accurately were more advanced, it would not be possible to design and produce "color" televisions, movies, and magazines based on cheap technologies. As it is, manufacturers of these devices do not have to accurately reproduce the wavelengths that were present in the original scene. All they have to do in order to satisfy even their most discerning customers is produce three wavelengths in various mixtures. The applied science that deals with spec-

TABLE 7.1. Summary of the kinds of color discriminations that can be performed by monochromats, dichromats, and trichromats.

	"Proper" Color Names	Discriminate λ_1 from λ_2	Discriminate λ_1 from White	Discriminate λ_1 from $\lambda_2 + \lambda_3$	Discriminate λ_1 from $\lambda_2 + \lambda_3 + \lambda_4$
Monochromats	No	No	No	No	No
Dichromats	No	Yes	All except neutral point	No	No
Trichromats	?	Yes	Yes	Yes	No

ification and measurement of mixtures of wavelengths with respect to human color perception is **colorimetry.**

Colorimetry

The formal statement of trichromacy presented in words in the previous section can also be stated in terms of mathematical symbols:

$$w_s(\lambda_1) = x_s(\lambda_2) + y_s(\lambda_3) + z_s(\lambda_4)$$
$$w_m(\lambda_1) = x_m(\lambda_2) + y_m(\lambda_3) + z_m(\lambda_4)$$
$$w_l(\lambda_1) = x_l(\lambda_2) + y_l(\lambda_3) + z_l(\lambda_4),$$

Where x, y, and z are the luminances the three wavelengths λ_2, λ_3, and λ_4 required to match wavelength λ_1 when its luminance is at w. The three equations are for the short (s), middle (m), and long (l) photopigments. The mathematical symbol "=" means "matches." The mathematical symbol "+" means the wavelengths are mixed together. The values for each term are in units of photons absorbed. When trichromacy is stated in terms of these mathematical [symbols], it refers to a solution, in the form of values for x, y, and z, that simultaneously satisfies all three equations. An algebraic solution must exist, because the problem is in the form of three equations in three unknowns. **Grassman's laws** of color theory, formulated by H. Grassman in the mid-1800s, assert that this set of mathematical equations provides an accurate description of the facts of human color matching.

There is a subtle difference between the way the formal statement of trichromacy was stated in words in the previous section and the form of these equations. The formal statement in words does not actually state that a given wavelength λ_1 at a given luminance w can always be matched. The formal statement asserts only that a match can be made if the intensities of *all four* wavelengths are allowed to vary. The above equations imply that a solution needs to be found in which only the intensities of the three wavelengths on the right side of the equations are allowed to vary.

The difference has to do with a distinction between mathematical and physical solutions. When a mathematical solution is found for the above equations, it may involve negative values for x, y, or z. Negative values are acceptable mathematical solutions but are not realizable in terms of physical light. Grassman's laws deal with this situation operationally by stating that if one of the mathematical quantities on the right side of an equation is negative, it means that light having an

intensity of this absolute amount must be added to λ_1 on the left side of the equation.

That operational definition works for scientific studies of color vision, and experimental studies using these methods have demonstrated that, as a first approximation, Grassman's laws hold. However, this operational definition is not very practical as an engineering solution. Suppose your goal is to build a color television that appears (to trichromatic human observers) to display all colors but in fact displays only mixtures of various amounts of three wavelengths. It is impossible to produce negative amounts of light so mathematical solution requiring negative value are not helpful in achieving this goal.

Engineering solutions need particular wavelengths, called **primaries,** having a property that makes it possible to produce matches for most colors without resorting to negative values. Specifications of color based on three primaries, X, Y, and Z, are commonly represented graphically in the form of a chromaticity diagram, as illustrated in Figure 7.17.

Chromaticity diagrams have been devised based on a number of different sets of primaries. However, in order to promote uniformity in the way colors are specified by scientists and engineers, an **International Commission on Illumination** (CIE) adopted a set of standards for a chromaticity diagram in 1931 based on primaries X, Y, and Z, corresponding to physical wavelengths at 460, 530, and 650 nm. The primaries specified by the CIE are spread apart sufficiently to allow their mixtures to reproduce a large range of colors using only positive values. The CIE standards are now commonly used by scientists when specifying color stimuli and by engineers when building devices such as color televisions.

Color Blindness

As stated above, the retinas of most humans contain four photopigments: rhodopsin, which is present in rods, and three types of **cone opsin,** each of which is present in a particular type of cone. Under conditions of dim illumination, when only the rods are operating, humans have no color vision and operate according to the rules described above for monochromats. However, during daytime conditions, when the cones receptor types are operating, most humans are trichromats. The exceptions are humans with various forms of color blindness. The presence of individuals with color blindness has been documented for hundreds of years. For

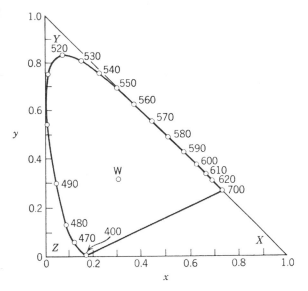

FIGURE 7.17. A chromaticity diagram can be used to specify all colors based on their location within a three-dimensional space, defined by the hypothetical primaries X, Y, and z. The relative amount of primary X that is needed is specified by the location (X) along the horizontal axis and the relative amount of primary y by the location (Y) along the vertical axis. The amount of primary Z is conceptualized as a magnitude of depth (Z) coming out of the page. Its value can be calculated algebraicly once the values for X and Y are known, since the values are constrained by the formula: $X + Y + Z = 1$. The locations of the colors produced by all wavelengths in the visible spectrum from 400 to 700 nm fall along the smooth curve. The straight line between 400 nm and 700 nm defines the extraspectral region. Colors associated with all mixtures of wavelengths, including the achromatic color, W, correspond to a single point in the chromaticity diagram. Adapted from C.H. Graham (ed.), *Vision and Visual Perception*, 1965, by permission of Wiley, New York.

example, in 1770, the **Reverend J. Priestly** reported the case of a subject who

"...[stated] that he had reason to believe other people saw something in objects which he could not see; that their language [involving color names] seems to mark qualities with confidence and precision which he could only guess at with hesitation, and frequently with error."

Most color blindness is inherited. About 8% of males and 0.5% of females have inherited color vision deficits. In recent years great strides have been made in understanding the molecular genetics of both normal color vision and color blindness. The molecular geneticist **Jeremy Nathans** and colleagues have documented many of the genetic details of cone pigment expression.

Color deficiencies occur mostly in men because of recessive mutations on the X chromosome. The genes for rhodopsin and for the three cone photopigments are all similar, and shifts in the spectral absorbance of the various pigments are due to amino acid substitutions in the opsin. Genes that reside on the **chromosome** 3 code for **rhodopsin** and the S-cone photopigments. The M-cone and L-cone pigment genes are on the X chromosome and show variation in males. Each male has a single copy of the gene for the L-cone pigment and one, two, or three copies of the gene for the M-cone pigment, located next to it. However, not all of these genes are expressed. In males with normal trichromatic color vision, two X-linked pigments, with peaks around 530 and 560 nm, are expressed, the same as in females.

Cone monochromats, in whom two out of three cone pigments are missing, are extremely rare. The majority of color vision defects involve a defect in only one of the three cone pigments. When one of the three pigments is missing, leaving only two, the colorblind observer is a **dichromat.** When three pigments are present but the spectral absorbance of one is abnormal, the observer is referred to as an **anomalous trichromat.**

In the case of either dichromacy or anomalous trichromacy, the defect is named according to which photopigment is missing or altered. The first type, **protanopia,** involves loss of the L-cone pigments. **Protanomaly** is a corresponding abnormality in the L-cone pigments. The spectral absorption curves for the remaining two photopigments of a **protanope** are shown in Figure 7.18.

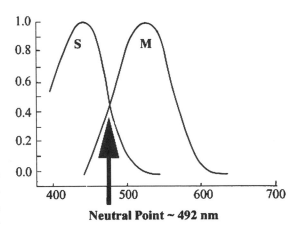

Neutral Point ~ 492 nm

FIGURE 7.18. A human with color blindness in the form of protanopia is missing the L cones and has only S and M cones present, resulting in a neutral point where the spectral absorption curves of the remaining pigments cross.

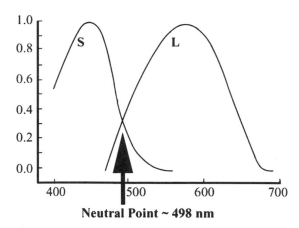

Neutral Point ~ 498 nm

FIGURE 7.19. A human with color blindness in the form of deuteranopia is missing the M cones and has only S and L cones present, resulting in a neutral point where the spectral absorption curves of the remaining pigments cross.

The neutral point is near 492 nm, where the spectral absorption curves for the two remaining photopigments cross. This makes intuitive sense, because white light will produce a signal ratio in these two photopigments of approximately 1:1, and the only single wavelength producing that ratio has to be one near where the absorption curves cross. Observers with a protanomaly have L-cone pigment, but its spectral sensitivity has shifted to be more similar to that found in the M cones.

In both protanopia and protonomalous trichromacy, there are both a weakness in responding to long wavelengths and a deficit in the ability to discriminate between medium and long wavelengths. Since monochromatic middle wavelengths look green and long wavelengths red when viewed by normal trichromatic observers, these defects are often referred to as being one form of **red-green color blindness.**

Another form of color deficiency that falls within the general category of red-green color blindness is **deuteranopia,** which involves loss of the medium-wavelength pigment. **Deuteranomaly** is the name for an abnormality of the medium-wavelength pigment. More than half of men with genetic color disorders fall into this category. The spectral absorptions of the remaining pigments in deuteranopia are illustrated in Figure 7.19.

Deuteranopic and deuteranomalous observers have a weakness in detecting medium wavelengths and have trouble making discriminations between medium and long wavelengths.

In the case of deuteranomaly, the M-cone pigment is missing but two other pigments are present, with peak sensitivities near that of the normal L-cone pigment. Somewhat puzzling is the fact that the gene coding for M-cone pigments is present in these observers, so the reason the proper M-cone photopigment is not expressed remains a mystery.

Finally, the third type of deficit, **tritanopia,** involves loss of the short-wavelength receptors, leaving two receptors with spectral absorption curves as illustrated in Figure 7.20.

Tritanopia is an autosomal rather than an X-linked defect. It affects both males and females but is rare. Pigment abnormality in the same receptors is named **tritanomaly.** Tritanopic and tritanomalous observers have the most problems perceiving and discriminating wavelengths in the short end of the spectrum.

4 Neural Processing Explains Some Aspects of Color Vision

The trichromatic theory and an understanding of its implementation in terms of three types of cone photopigments can explain most of the perceptual phenomena of color mixing and can also help in understanding the various forms of genetic color blindness. However, other color phenomena are not as easily accounted for by the trichromatic theory. For example, Class B observations of human observers suggest that certain pairs of colors, specifi-

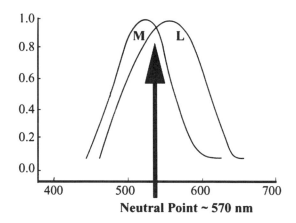

Neutral Point ~ 570 nm

FIGURE 7.20. A human with color blindness in the form of tritanopia is missing the S cones and has only M and L cones present, resulting in a neutral point where the spectral absorption curves of the remaining pigments cross.

**Organization of Three Color
Opponency Channels**

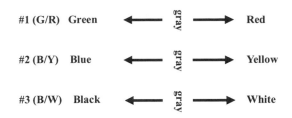

#1 (G/R)   Green   ◄——— gray ———►   Red

#2 (B/Y)   Blue   ◄——— gray ———►   Yellow

#3 (B/W)   Black   ◄——— gray ———►   White

FIGURE 7.21. The opponent theory of color vision proposes that during processing our visual systems organize color information into three opponent channels: green-red (G/R), blue-yellow (B/Y), and black-white (B/W).

cally **red and green, blue and yellow,** and **black and white,** have special relationships.

When asked to use combinations of color names, observers almost never report reddish-green, bluish-yellow, or blackish-white. This implies that these pairs of colors are not experienced simultaneously. These pairs of colors also lead to special aftereffects. If an eye is stimulated by an intense light for a short period, as with a flashbulb, a visual sensation called an **afterimage** persists, in which a spot of light floating in space is perceived. If the flash looks red, the afterimage may initially appear red but after a few seconds often turns green. A blue flash produces an afterimage that turns yellow. A similar phenomenon occurs for **color adaptation aftereffects.** Following adaptation to a moderately intense light that looks red, a neutral white light appears green for a period that can last many second; following adaptation to yellow, a white light may look blue; and following adaptation to white, a medium gray may seem to shift to a darker gray.

These kinds of Class B observations are what originally led to the formulation of the opponent theory of color vision. This theory proposes three processing channels for color information, the same number proposed by the trichromatic theory. However, in the opponent theory, the three channels are organized into three antagonistic pairs along the lines illustrated in Figure 7.21.

Each of the three channels responds in one direction to code for one color experience and in the opposite direction to code for the other. The first channel responds in one direction to green and in the opposite direction to red. A second channel responds antagonistically to blue and to yellow. A

third channel is organized for opposite responses to black and to white. Each of these channels, when properly balanced, will produce no output, corresponding to an experience that can be described as either gray or achromatic.

At the level of description of biology, the color opponent channels can be conceptualized, roughly speaking, in terms of neurons with receptive fields that respond in an antagonistic manner to different wavelengths of light. Consider a neuron with a spontaneous firing rate. When the cell's output is at its spontaneous rate, it signals a neutral gray. When the neuron's firing rate increases above the spontaneous rate, it might signal red and when it decreases below that rate, green. Note that neurons acting in this manner would be *unable* to transmit red and green simultaneously. This provides a potential explanation for the Class B observations that certain pairs are not perceived simultaneously.

These neuronal receptive field properties also have the potential to explain color adaptation aftereffects. Consider a neuron that signals red with a high rate of firing and green with a low rate of firing. After responding for a period to an intense stimulus that signals red, this neuron may readjust its adaptation level causing it to fire at a slower rate in response to the ongoing stimulus. Speaking colloquially, it might be stated that the neuron becomes **fatigued.** Subsequently, when confronted with no stimulus or with a neutral, achromatic light, this neuron will respond with a firing rate that is below its ordinary level of spontaneous activity. This lower rate will be interpreted by the brain as green, and this aftereffect will continue until the neuron recovers to its ordinary adaptation level.

Color Processing in the Retina

Beginning in the retina, information from the three receptor types is organized into a sensory coding scheme that has opponent properties. Recent studies by the visual neuroscientist **Barry Lee** and colleagues have demonstrated that retinal ganglion cells in the monkey are organized into three classes that have wavelength opponent properties. Roughly speaking, these three neural subsystems allow three kinds of color experiences to be differentiated: black from white (B/W), blue from yellow (B/Y), and red from green (R/G).

The B/W neural subsystem of ganglion cells receives input from L- and M-cone receptors in both

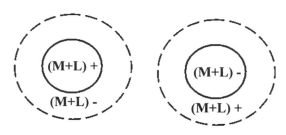

FIGURE 7.22. This diagram illustrates the receptive fields of two variants of retinal ganglion cells that together form the concentric broadband channel that carries information about black and white (B/W) contrast. The receptive field shown on the left is On-center and that on the right Off-center. Inputs to the center and surround portions of the receptive field come from M and L cones.

the center and surround areas of the reception field. This class of ganglion cell comes in two variants, as indicated in Figure 7.22.

These variants correspond to the on-center and off-center ganglion cells that were discussed in Chapter 5. Both the center and the surround portion of each receptive field sum inputs from the M and L cones, and there is a spatial antagonism between the center and surround regions. Somewhat surprisingly, the S-cones do not appear to contribute to this channel, so the short wavelengths are underrepresented. Nevertheless, the combined wavelengths absorbed by the M- and L-cone photopigments have broad enough bandwidth that the term "black-white" seems an appropriate characterization of the color information carried in this channel.

The output of the B/W neurons, even at this early stage of color processing, largely disregards absolute levels of stimulation and signals the brightness of the center region relative to its immediate surround region. However, these comparisons involve only those small regions of the visual field making up the receptive field of a single ganglion cell. Computations of brightness or lightness of a more complex nature as described by the retinex theory do not occur until subsequent stages of processing.

The B/Y retinal ganglion cells are probably the oldest class in terms of phylogenetic origin, and many, perhaps all, do not have a concentric center-surround spatial organization. They receive inputs from all three receptor types, but their reception fields are organized in a specific fashion, in two variants, as illustrated in Figure 7.23.

The first type of B/Y retinal ganglion cell increases its firing rate for inputs from S cones and decreases

its firing rate based on the sum of the inputs from M and L cones. The second variant responds in the opposite manner. Note that white light has little effect on either variant of these B/Y retinal ganglion cells, because it simultaneously stimulates S, M, and L cones, whose antagonistic effects tend to cancel each other out. Thus, they respond to color contrast but not to luminance contrast.

The third class of ganglion cell, R/G, receives input from either the M or the L cones in its center area and from the other cone type in its surround area. There are at least four variants of this type of R/G ganglion cells, as illustrated in Figure 7.24.

In all variants of R/G retinal ganglion cells, there is an antagonism between receptor types as well as

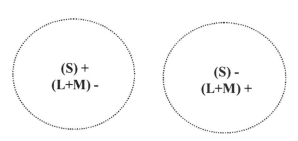

FIGURE 7.23. Receptive fields of two variants of retinal ganglion cells that together form a channel that carries information about blue and yellow (B/Y) color contrast by comparing the rate of absorption in the S cones with that of the M and L cones.

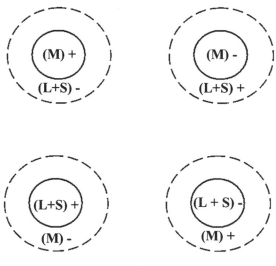

FIGURE 7.24. Receptive fields of four variants of retinal ganglion cells that together form a channel that carries information about green and red (G/R) color contrast by comparing absorption rates in the M cores compared to the L and S cores.

between spatial regions of the receptive fields. The center region receives inputs either from the M cones or from a combination of the L and S cones. The response to the center region can be either excitatory or inhibitory. There is some uncertainty about the inputs to this type. Some evidence suggests that they receive only the inputs shown in Figure 7.24. Other evidence suggests that the surround areas sometimes receive at least small amounts of input from combinations of M, L, and S cones. Regardless of the specific types of input contributing to the surround region, there is an antagonistic relationship between the responses elicited by the center and by the surround areas.

When illuminated by white light, all variants of R/G retinal ganglion cells respond in a manner similar to the B/W subsystem of ganglion cells and signal when there is more or less overall light falling in the center than in the surround region. However, R/G neurons also respond differentially to stimuli that differ only in wavelength even when the overall light levels are the same in the center and surround areas. These **strictly chromatic differences** would not be picked up by the B/W neurons, which receive broadband input.

The input from the M cones is signaled to the brain as **greenish.** The combined input from the L⁻ and S cones is signaled to the brain as **reddish.** The fact that the S cones contribute to the percept of redness is counterintuitive, since short wavelengths, when presented alone, are seen as blue. However, this receptive field organization provides a potential explanation for the fact of color mixture, described in the section "The Color Solid," that adding a small amount of short-wavelength light makes a red stimulus look redder.

Color Processing in the M-, P-, and K-Streams

Recall from Chapter 5 that the ganglion cells leaving the retina are organized into M-, P-, and K-streams. These streams have to transmit all of the information from the retina that is used to form qualities of percepts such as form and motion in addition to color. Thus, the color channels have to be distributed within these more general parallel streams. Current evidence indicates that the B/W neuron axons are located mostly within the M-stream, R/G within the P-stream, and B/Y within the K-stream.

At the level of the dLGN, neurons participating in color processing have receptive field properties very similar to the properties of the corresponding retinal ganglion cells providing their inputs. Magnocellular (M-stream) layers have many B/W neurons. Parvocellular (P-stream) layers have many R/G concentric opponent-type neurons. There has been some confusion about where the B/Y neurons are located in the dLGN. However, accumulating evidence now indicates that the dLGN neurons with B/Y receptive field properties are part of the koniocellular (K-stream) population of neurons. This population of neurons is located primarily in the interlaminar zones between the parvocellular layers, but some of them may also fall within the parvocellular layers proper, making it difficult to separate out the P-stream and K-stream populations of neurons.

Following lesions to the P-stream layers of the dLGN (which inevitably destroy an indeterminate number of neurons associated with the K-stream as well), monkeys have a reduced ability to discriminate between wavelengths. This is illustrated in Figure 7.25.

The lesion was made to a small portion of tissue within the parvocellular layers, affecting only the small portion of the visual field indicated by the shaded region in part A of the figure. Within the area of the visual field affected by the lesion, the animals made many errors when asked to discriminate between stimuli based on wavelength, as illustrated in part B of the figure. The animals performed normally on color discriminations in the remainder of the visual field.

Following lesions restricted to the M-stream layers of the dLGN, no deficits were found in the ability to discriminate between wavelengths. These results are consistent with the hypothesis that only the P- and K-streams are essential for making color discriminations based on wavelengths. Technically speaking, the M-stream also carries information about color, but only for the B/W channel, and thus its absence does not impair the ability to discriminate between wavelengths.

Recall from Chapter 5 that the P-, K-, and M-streams continue from the dLGN to cortical area V1, where they terminate in separate layers, as summarized in Figure 7.26.

Color processing seems to be funneled into the **blobs and interblob regions in layers 2 and 3** of V1. At least some of the B/Y neurons in the K-stream appear to project directly to the blobs. The R/G neurons in the P-stream project first to 4Cβ and 4A, and their signals are carried to layers 2 and 3 via interneurons. The B/W color channel carried

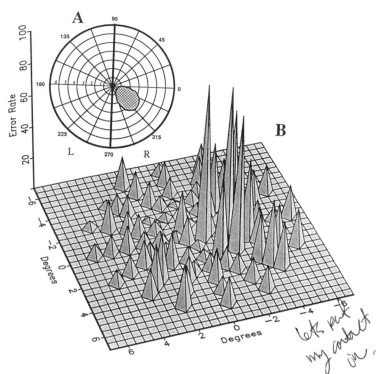

FIGURE 7.25. Monkeys with a small lesion to the parvocellular layers of the dLGN have difficulty discriminating wavelengths of light for stimuli presented within the part of the visual field affected by the lesion. The top panel, A, shows the location of the lesion in the visual field. The bottom panel, B, shows the number of errors made when discriminating wavelengths at various locations in the visual field, including the portion affected by the lesion. Reproduced from P.H. Schiller, et al, Role of the color-opponent and broadband channels in vision. *Visual Neurosci.* 5:321–346, © 1990, by permission of Cambridge University Press.

in the M-steam terminates in layer 4Cα, and interneurons allow this color information to reach layers 2 and 3.

Individual neurons within the blob and interblob regions have some properties that are similar to those of neurons feeding into them from the M-, P-, and K-streams. However, there is also some reorganization of color information at this stage of neural processing.

The receptive fields of most neurons in the blobs are **monocular** and **unoriented**. Some neurons in the blobs exhibit **color opponency but no center-surround organization,** in a manner similar to that seen in some retinal ganglion cells and dLGN neurons. These neurons do not respond well to spots of white light of any size but respond to selected wavelengths regardless of spot size. An example, recorded from a neuron in the blob region of a monkey, is illustrated in Figure 7.27.

This neuron has a circular receptive field, as illustrated on the left. It has no center-surround organization, but it does exhibit an antagonism between wavelengths that appears to be organized to signal **B/Y color channel information.** Specific examples

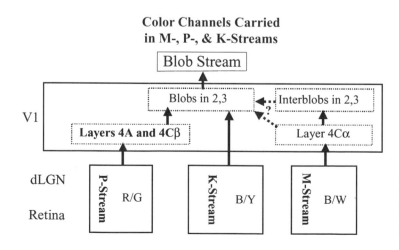

**Color Channels Carried
in M-, P-, & K-Streams**

FIGURE 7.26. Summary diagram illustrating the flow of information from M-, P-, and K-streams into cortical area V1 and the color information carried by each stream.

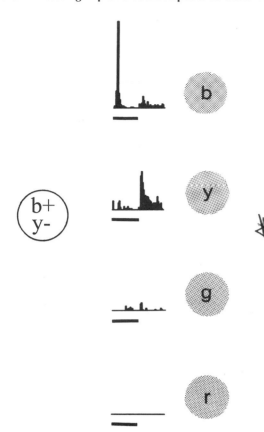

FIGURE 7.27. This figure shows the responses to colored stimuli of a neuron located in the blob region of monkey cortical area V1. The receptive field of the neuron is shown on the left, and the responses to lights of various colors are shown on the right. Adapted from D.Y. Ts'o and C.D. Gilbert, The organization of chromatic and spatial interactions in the primate striate cortex. *J. Neurosci.* 8:1712–1727, © 1988, by permission of Society for Neuroscience.

of responses to particular colored stimuli are illustrated in the panels on the right. For each row, the firing rate of the neuron is plotted for a period during which a particular stimulus is turned on, illustrated by the horizontal thick bar, and then for a period after the stimulus is turned off. This particular neuron exhibits a burst of activity whenever a blue light is turned on or a yellow light is turned off. No major changes in activity are elicited by other stimuli.

Other cortical cells in the blobs show a color and spatial opponent receptive field organization, similar in many respects to that seen in some retinal ganglion cells and dLGN neurons. However, there is also a transformation of color opponent information, the details of which go beyond the scope of this book. Briefly, this transformation allows the

crude versions of wavelength opponency coded for by the B/W, R/G, and B/Y neural subsystems to take on more complex color opponency properties that are reflected in psychophysical measurements of color percepts.

One example of a more complex opponent processing type of receptive field is called **double opponency.** Double opponent cells have spatial opponency organized into center and surround regions and color opponency in both the center and the surround areas. For example, the center region might be organized to signal R/G and the surround region to signal G/R in the opposite manner, as illustrated in Figure 7.28.

This particular double opponent cell would respond best to a small spot of red positioned on a green background. In general, **double opponent color cells** respond strongly only to **chromatic contrast** (contrast based on wavelength differences) and not to contrast based simply on differences in luminance. This is because inputs that vary in luminance but not in wavelength produce effects that cancel each other out at all locations in the receptive field.

Some of the kinds of more complex computations required by the retinex theory may begin to be carried out at this stage. For example, some of the intrinsic connections between blobs are long range, spanning more than one hypercolumn. This may reflect a reorganization of color processing in central visual pathways that begins to put more emphasis on comparisons across larger regions of the scene, as distinct from the processing up until this stage, which has been involved primarily with analyzing wavelength information from small regions that fall within a localized receptive field. Also, merging of information about color and spatial properties is beginning to take place in the

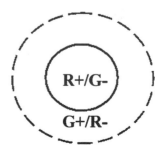

FIGURE 7.28. One variant of a receptive field of a cortical neuron that exhibits double opponency, characterized by a center-surround spatial opponency and a color opponency (R/G) in the center and the surround that are the opposite of one another.

interblob regions, where some neurons have receptive fields tuned to both orientation and color. Color and spatial information is funneled into extrastriate areas primarily via two new streams of processing, called the **blob** and **interblob streams.**

Color Processing in the Blob and Interblob Streams

The term "**blob stream**" is used to describe the continuation of color processing beyond the blobs of the striate cortex to extrastriate areas. Neurons in the blobs of V1 project to neurons in the thin stripes in V2, as described in Chapter 5. The receptive fields of these neurons appear to be specialized for color processing and do not exhibit orientation tuning. The term "**interblob stream**" is used to describe continuation of processing of information about low-level properties of shape, such as orientation, from the interblob regions of V1 to neurons in the interstripe regions of V2. The continuation of the blob and interblob streams into extrastriate cortical areas is summarized in Figure 7.29.

The blob and interblob streams join in extrastriate area V4, which some have proposed to be a major processing area for combining information about properties of surfaces such as color and shape. There also appears to be a minor projection from color processing neurons in the dLGN (probably K-stream neurons) that passes directly to V4, bypassing V1 and V2.

Little is known about color processing at or beyond V4. Monkeys with lesions to V4 have minor deficits in discriminating stimuli based on color.

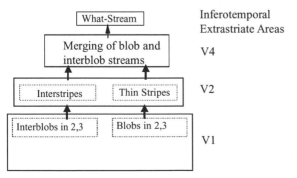

FIGURE 7.29. Summary of information flow from blob and interblob portions of cortical area V1 to extrastriate cortical areas, and then into the what stream.

Humans with brain damage that may include V4 sometimes exhibit **cortical color blindness.** In addition to mild deficits on low-level tasks such as discriminating between spots of light composed of different wavelengths, these patients often exhibit more severe deficits in higher-order aspects of color perception.

Potential Explanations of Higher-Order Color Phenomena

Color constancy refers to our ability to see an object as the same color under conditions in which the distributions of wavelengths that reach the eye are different. A specific subtype of color constancy, lightness constancy, was described earlier in this chapter. Recall that a piece of coal appears to have the same lightness (black) regardless of whether it is viewed outdoors or indoors. Lightness constancy involves only the B/W color channel. Color constancy is simply a more general phenomenon that involves all of the color channels and applies to hue as well as lightness.

An example of color constancy is that objects appear to have the same color whether viewed outdoors in sunlight or indoors with an incandescent light bulb. This is unexpected based on properties of the proximal stimulus, because the wavelengths that reach the eye under these two conditions differ.

Compared to sunlight, the light bulb produces more long-wavelength photons relative to short-wavelength photons. Thus, relatively more long-wavelength photons reach the eye indoors. In order for color constancy to operate, the visual system has to discount these differences and compute colors based on the spectral reflectance of the surface. The rules involved in doing this are complex and specified in the retinex theory. The details go beyond the scope of this book.

Briefly, before the rules of the retinex theory can operate, the scene must first be analyzed to figure out what surfaces are present. Comparisons are then made across surfaces rather than across locations defined by receptive fields of neurons of the type found in the retina, the dLGN, or at stages of processing up to and including layers 2 and 3 of V1. The receptive fields of double opponent cells that operate in the blob stream are also inadequate for this task, but may provide building blocks out of which more general color constancy could be constructed. As noted above, a shift in ambient light levels towards longer or shorter wavelengths has little effect on a double opponent cell, since the increase is the same for both the center and the surround areas of the receptive field. Presumably, higher-order neurons in the blob and interblob streams have double opponent properties but operate on higher-order inputs that correspond to surfaces in a scene rather than on inputs that simply correspond to small areas of retinal stimulation.

Even the simplest kind of double opponent cells may be sufficient to explain a simple perceptual phenomenon called **simultaneous color contrast** that involves interactions between simple patterns involving only two spatially segregated areas. For example, a gray object seen on a background of red will appear to have a green tinge. A double opponent cell of an appropriate size centered on the gray object would be expected to signal green, as in the percept, rather than gray, as in the stimulus. The phenomenon of simultaneous color contrast can have practical impact, as illustrated in Box 7.5.

Box 7.5
Practical Implications of Simultaneous Color Contrast

Suppose you want to buy curtains for your living room and you want them to appear gray. Your living room has yellow walls. You go to the department store at the mall to purchase the curtains. You find curtains that look perfect in the store. The display in the store is on a wall painted white. You order the curtains and wait for them to be delivered. The delivery truck arrives, and you immediately hang the curtains in your living room. They appear blue! You hate blue curtains! Do you have grounds to bring a lawsuit against the store for delivering blue curtains when you ordered and paid for gray curtains? Before hiring a lawyer, consider an explanation based on simultaneous color contrast.

Our visual systems can differentiate between an **illumination edge,** as in a shadow, and a **reflectance edge,** in which the reflectance of a surface changes, as at a corner of a room. When a shadow is cast across an object, the brightness changes but the lightness of the surface does not. Color-processing neurons in the interblob stream

that combine information about shape and color are thought to be involved in making these kinds of distinctions.

Color-processing neurons with orientation tuning, as found in the interblob stream, may provide an explanation for a perceptual phenomenon called the **McCollough effect.** A pattern of black vertical lines is presented on a yellow background, followed by a pattern of black horizontal lines on a blue background. The two patterns are shown alternately for several seconds each, for a total duration of several minutes. At the end of this adaptation period, a subject views black vertical and horizontal lines on a white background. The vertical lines appear bluish and the horizontal lines yellowish. When the McCollough effect has been induced in humans undergoing brain imaging, increased activity has been reported in cortical area V4.

Neurons that respond to **motion as well as color** might be involved in the perception of **subjective colors induced by moving or flickering black and white stimuli.** A well-known example is **Benham's top,** often constructed as a toy for children. Black line segments are painted on the white background of the top. When the top is stationary, no color is seen. However, when the top is spun, various colors are seen depending on the speed of rotation.

Ultimately, color information that passes through extrastriate processing areas of the brain merges with other types of information to form the general **What-perception stream** of processing described in Chapter 5.

5 Computational and Ecological Significance of Color

Color vision processing can be conceptualized from a computational perspective as reducing the **complex distribution of wavelengths** reflected from an object in the environment into **color,** a **low-dimensional quality** that can be described with only a few degrees of freedom. This reduction depends on regularities in the spectral environment.

Regularities in the Spectral Environment

The amount of information that is lost when reducing the raw data in a complex distribution of wavelengths to a low-dimensional statistic depends on how much the magnitudes of photon absorptions associated with different wavelengths are correlated with one another. If the absorptions that occur at one wavelength are independent of those occurring at other wavelengths, then much information will be lost in this process. However, if the absorptions at various wavelengths exhibit **covariance,** then less information will be lost.

In natural environments, as opposed to artificial laboratory situations in which studies of color perception are typically carried out, the distributions of wavelengths absorbed in the eye have a high amount of covariance. For example, if the amount of 530-nm wavelength light reflected by an object increases, there is a very high likelihood that the amount of 531-nm reflected also increases.

Thus, in ecological environments the visual system does not have to monitor all wavelengths independently in order to gain a great deal of information about the distributions of wavelengths reflected by objects.

Statistical procedures, used in combination with quantitative analyses of wavelength distributions reflected by objects in the ecological environment in which primates evolved, have been used to evaluate the extent to which the visual system has adapted to the predictable regularities in the spectral environment. Results of one such study are illustrated in Figure 7.30.

The top panel shows empirical results from human observers who used psychophysical

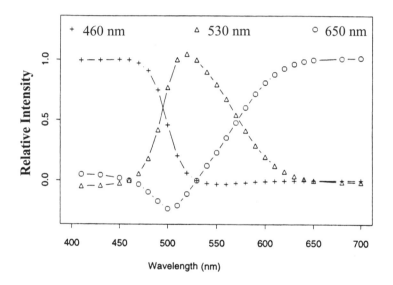

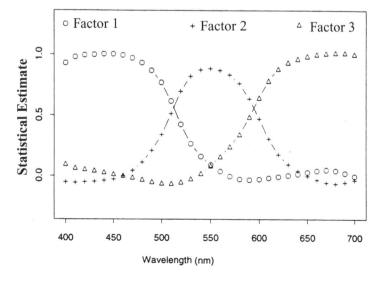

FIGURE 7.30. This figure compares psychophysical (top panel) and statistical (bottom panel) estimates of the relative intensities of three primaries needed to match wavelengths across the visible spectrum. See text for details. Adapted from S.M. Boker, A measurement of the adaptation of color vision to the spectral environment. *Psychol. Sci.* 8:130–134, © 1997, by permission of American Psychological Society.

methods to adjust three primaries to match wavelengths from across the visible spectrum. The bottom panel shows the results of a statistical procedure, which provided an answer to the following question: Given that wavelength information must be reduced to only three factors, what proportions of those factors would be most efficient based on wavelengths reflected from natural objects? The similarity of the statistical and the perceptual results supports the hypothesis that evolution of the human visual system has attained a high degree of efficiency in exploiting the covariance of spectral distributions in the natural environment.

Nevertheless, a question remains as to why the human visual system evolved to use three channels, (trichromacy). It could have exploited covariance somewhat with only two factors (dichromacy) and could have exploited it still more had it evolved to use four or more channels. Some keys to understanding why humans are trichromats come from an evaluation of the evolutionary history of primates.

Evolution of Color Perception in Primates

Genes for two cone photopigments emerged in the course of evolution of animals perhaps 500 to 800 million years ago. This division substantially predates the appearance of the earliest mammals, and thus dichromacy is the baseline arrangement for mammals. The primates are unique among mammals in having species that possess trichromatic color vision. However, trichromacy is not present in all primates.

The primordial primate color system appears to be dichromacy in the form of a B/Y channel that combines an S-pigment and a second pigment somewhere in the middle-to-long-wavelength range. The emergence of trichromacy, in which two separate pigments, L cones and M cones, operate in the middle-to-long-wavelength range is much more recent, perhaps 30 to 50 million years ago.

The basic organization of the primate visual system into separate M-, P-, and K- streams of processing formed prior to trichromatic color vision. The K-stream is most ancient and appears to be closely associated with the primordial B/Y color channel. The M-stream and P-stream probably originally carried only B/W channel information. The P-stream adapted to handle the R/G channel after the L- and M-cone photopigments diverged.

A major split between Old World (Asia and Africa) and New World (South and Central America) monkeys occurred about 50 million years ago. All Old World monkeys are trichromats, while New World monkeys exhibit a variety of forms of color vision, including trichromacy. The preponderance of the evidence indicates that Old and New World monkeys were both dichromats at the time they split. Thus, the separation into modern L-cone and M-cone types in all Old World and some New World monkeys has occurred more than once and is an example of convergent evolution. This fact raises a fascinating question regarding what kinds of selection pressures would facilitate a change from dichromacy to trichromacy in primates.

One hypothesis about why trichromacy arose in several primate species is that it is a consequence of visual adaptation to foraging for yellow and orange fruits among green foliage. In fact, there is some evidence that trichromacy in primates and the spectral reflectance properties of certain fruits coevolved. Some tropical trees depend almost exclusively on primates for dispersal of their seeds, and when ripe, these fruits typically offer a yellow or orange color signal to trichromatic observers.

The task facing primates searching for these fruits can be conceptualized as a signal detection task. Primates must detect a signal (fruit) embedded in noise (background foliage). Computationally, it is possible to ask what pair of photopigments, acting as inputs to an opponent channel of color processing, maximizes the signal to noise ratio (SNR) for detecting these fruits. A computational study conducted with howler monkeys found that the highest SNRs would be obtained with an R/G color system based on having one pigment in the range 554–576 nm and the other in the range 516–532 nm. The actual photopigments of the R/G color system in this species have peaks at 562 and 530 nm, offering an SNR that is near optimal.

Color vision in many species of primates is polymorphic, with trichromacy present in females but dichromacy existing in some or all males. Species living in social groups that cooperate in finding food and detecting predators might enjoy an advantage in having some individual members (some males) with a different type of color vision than others (other males and all females). This could imply that differences in color vision in different species are related to the methods used for sharing or signaling found food. Species whose individual members need to find their own food would be most likely to stabilize on trichromacy. Species with more cooperative searching or food sharing would be more likely to remain polymorphic.

secondary quality that exists in our percepts but does not resemble low-level stimulus properties of light. The color experiences of humans can be systematized along three dimensions, hue, saturation, and brightness, and these dimensions can be represented in the form of a color solid.

Two historical theories of color vision, called trichromatic and opponent have been developed in the attempt to explain color perception. Trichromatic theory is convenient for understanding how monochromatic, dichromatic, and trichromatic observers discriminate between mixtures of wavelengths. Its precepts are summarized in a formal statement of trichromacy that has been derived from scientific studies of color vision and implemented in applied engineering disciplines concerned with building devices such as color televisions. The biological basis for trichromatic color vision rests on the presence of three kinds of photopigments in the receptors. Certain genetic defects result in losses or abnormalities of one or more photopigments and in corresponding types of color blindness.

Opponent theories of color vision are more convenient for understanding many aspects of postreceptor processing of color information. Neurons in the retina and dLGN have receptive fields that allow them to serve as B/Y, R/G, and B/W channels for transmitting color information along the geniculostriate pathway. These three color channels are incorporated into the K-, P-, and M-streams, which are involved in transmitting various kinds of information, including color, to the striate cortex.

In the striate cortex, color information appears to be funneled to the blob and interblob regions of layers 2 and 3, which are perhaps responsible for some higher-order color phenomena. A blob stream of color processing proceeds from V1 to the thin stripes of V2 and then to V4. Neurons in the interblob regions of V1 also carry some color information but are more specialized for low-level spatial properties, such as orientation. The blob and interblob streams are combined in extrastriate cortical area V4, leading to speculation that this area is involved in qualities of percepts that pertain to surface properties including both shape and color. The color processing streams merge into the general dorsal processing stream, which is heavily involved in all aspects of What-perception.

Computational and ecological approaches to color vision have been taken to try to determine why color processing in primates reduces the complex information about the wavelength composition of light entering the eye to a specification based on only three numbers. One suggested reason is that trichromacy is an optimized system for detecting food sources consisting of yellow and orange fruits against a background of green foliage, needed in the natural environments of many primate species.

Selected Reading List

Derrington, A.M., Krauskopf, J., and Lennie, P. 1984. Chromatic mechanisms in lateral geniculate nucleus of macaque. *J. Physiol.* 357:241–265.

Gegenfurtner, K.R., and Sharpe, L.T. (eds.) 1999. *Color Vision: From Genes to Perception.* New York: Cambridge University Press.

Helmholtz, H. von. 1828. On the theory of compound colors. *Philos. Mag.* 4:519–534.

Hunt, D.M., Dulai, K.S., Cowing, J.A., Julliot, C., Mollon, J.D., Bowmaker, J.K., Li, W.-H., and Hewett-Emmett, D. 1998. Molecular evolution of trichromacy in primates. *Vision Res.* 38:3299–3306.

Jacobs, G.H., and Harwerth, R.S. 1989. Color vision variations in Old and New World primates. *Am. J. Primatol.* 18:35–44.

Kainz, P.M., Neitz, J., and Neitz, M. 1998. Recent evolution of uniform trichromacy in a New World monkey. *Vision Res.* 38:3315–3320.

Krauskopf, J., Williams, D.R., and Heeley, D.W. 1982. The cardinal directions of color space. *Vision Res.* 22:1123–1131.

Kremers, J., and Lee, B.B. 1998. Comparative retinal physiology in anthropoids. *Vision Res.* 38:3339–3344.

Land, E.H., and McCann, J.J. 1971. Lightness and retinex theory. *J. Opt. Soc. Am.* 61:1–11.

Lee, B.B. 1996. Receptive field structure in the primate retina. *Vision Res.* 36:631–644.

Mollon, J.D. 1989. Tho' she kneel'd in that Place where they grew. *J. Exp. Biol.* 146:21–38.

Naka, K.I., and Rushton, W.A. 1966. An attempt to analyse color reception by electrophysiology. *J. Physiol.* 185:556–586.

Nathans, J. 1987. Molecular biology of visual pigments. *Annu. Rev. Neurosci.* 10:163–194.

Neitz, J., Neitz, M., and Kainz, P.M. 1996. Visual pigment gene structure and the severity of color vision defects. *Science* 274:801–804.

Regan, B.C., Julliot, C., Simmen, B., Vienot, F., Charles-Dominique, P., and Mollon, J.D. 1998. Frugivory and colour vision in *Alouatta seniculus*, a trichromatic platyrrhine monkey. *Vision Res.* 38:3321–3327.

Shyue, S.-K., Hewett-Emmett, D., Sperling, H.G., Hunt, D.M., Bowmaker, J.K., Mollon, J.D., and Li, W.-H. 1995.

Adaptive evolution of color vision genes in higher primates. *Science* 269:1265–1267.

Silveira, L.C.L., Lee, B.B., Yamada, E.S., Kremers, J., and Hunt, D.M. 1998. Post-receptoral mechanisms of colour vision in new world primates. *Vision Res.* 38:3329–3337.

Stockman, A., and Sharpe, L.T. 1998. Human cone spectral sensitivities: A progress report. *Vision Res.* 38:3193–3206.

Wald, G. 1964. The receptors of human color vision. *Science* 145:1007–1016.

Wiesel, T.N., and Hubel, D.H. 1966. Spatial and chromatic interactions in the lateral geniculate body of the rhesus monkey. *J. Neurophysiol.* 29:1115–1156.

Young, T. 1801. On the theory of light and colours. *Philos. Trans. R. Soc. Lond.* 92:12–48.

8

Form Vision: How Is Information About Shapes of Objects Transferred from the Environment to Our Percepts?

Questions

After reading Chapter 8, you should be able to answer the following questions.

1. Compare and contrast forward flow of information with solving the inverse problem in form vision.
2. Describe how recurrent information flow allows a combination of bottom-up and top-down influences in form vision processing.

3. Give some examples of how form perception in biological organisms differs from what might be expected in a general-purpose form-processing system.
4. Summarize some of the evidence that visible surfaces have primacy over other forms of representation during form vision processing.
5. Speculate about why nature did not give our eyes better-quality optics.

6. What implications does retinal image sampling by the receptors have for snellen acuity and for hyperacuity?
7. Describe how information about the retinal image is band-pass filtered by neurons with DOG receptive fields, and discuss some of the behavioral correlates of this filtering.
8. How do the M- and P-streams participate in transmitting spatial information along the geniculostriate pathway?
9. Describe some neural mechanisms in V1 that might be able to extract the primitives that make up Marr's primal sketch.
10. What areas of the brain are most heavily involved in higher-order processing of form vision?
11. What kinds of strategies can be employed to try to infer properties of surfaces from pieces of contour?
12. Summarize what is known about spatial properties at the level of processing Marr called the "2½D sketch."
13. Describe how nonaccidental properties can be used to help match two-dimensional properties of an image to three-dimensional geons.
14. What is special about face perception?

1 Characterizations of Form Vision

In contradistinction to the perception of color, discussed in chapter 7, that has traditionally been considered a secondary quality, perceived shape is usually considered a primary quality. The term **shape** refers to the geometry of an object's physical surface. The topic of **form vision** is concerned with **what kinds of information** observers can extract from the environment about **shape** and with **how** this information is processed (Figure 8.1).

Studies of form vision are carried out within several different traditions. **Psychophysical studies** specify the kinds of information human observers can extract from the environment about shape. **Neuroscience studies** look for the biological mechanisms involved in sensory coding of shape. **Computational approaches** try to understand the information-processing problems that must be solved to achieve form perception and remain neutral regarding psychological or biological details. **Machine vision** tries to design machines that can recognize shapes of objects.

As we look around our immediate surrounding environment, it seems as though we are able, without any great effort, to automatically perceive shapes of objects. However, the ease with which form vision is accomplished in biological observers is in stark contrast to the seemingly intractable problems that have been encountered in machine vision. No machine comes even close to accomplishing form vision in real time in the types of complex visual environments where biological organisms routinely live and perceive (Figure 8.2).

Forward Flow and Inverse Problems in Form Perception

One way of conceptualizing form vision processing is from the point of view of **forward flow along a causal chain** (Figure 8.3), as discussed in Chapter 6.

Research conducted from this theoretical approach starts from the perspective of the environmental input and addresses questions about how information about shape travels in a forward direction, first to the retina and then through the brain. Most neuroscience research has been conducted from this perspective.

Machine vision approaches have been more concerned with trying to solve the **computational**

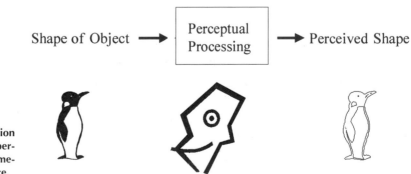

FIGURE 8.1. Form perception generates a primary quality of perceived shape based on the geometry of an object's physical surface.

FIGURE 8.2. Biological organisms must perform form perception in real time in complex environments like the one shown here.

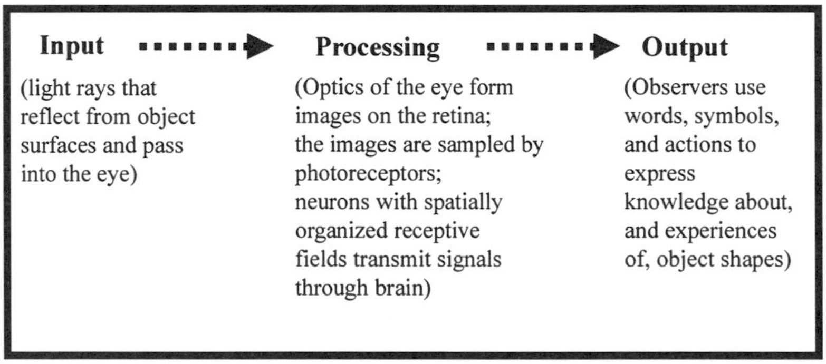

Input ••••••▶	Processing ••••••▶	Output
(light rays that reflect from object surfaces and pass into the eye)	(Optics of the eye form images on the retina; the images are sampled by photoreceptors; neurons with spatially organized receptive fields transmit signals through brain)	(Observers use words, symbols, and actions to express knowledge about, and experiences of, object shapes)

FIGURE 8.3. One way of conceptualizing form perception is in terms of forward flow along a causal chain.

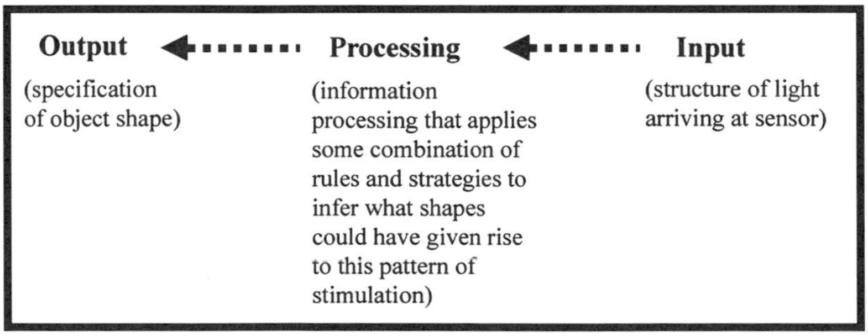

Output ◀••••••	Processing ◀••••••	Input
(specification of object shape)	(information processing that applies some combination of rules and strategies to infer what shapes could have given rise to this pattern of stimulation)	(structure of light arriving at sensor)

FIGURE 8.4. Information processing associated with form perception can be conceptualized as solving an inverse problem.

FIGURE 8.5. No unique solution exists for the inverse optics problem of specifying shapes of objects based on the shapes of the retinal images.

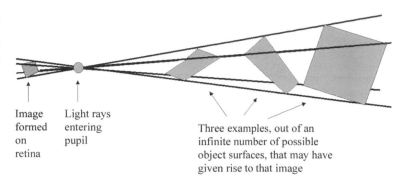

Image formed on retina

Light rays entering pupil

Three examples, out of an infinite number of possible object surfaces, that may have given rise to that image

inverse problem that was described in Chapter 6. Given the information available at the sensor, how can one infer what shape is present in the environment (Figure 8.4)?

There is no unique geometrical solution to the inverse problem for shape, as illustrated in Figure 8.5.

For any given shape in the environment, the formal equations of geometrical optics, described in Chapter 3, specify the image that will be formed on the retina. However, the inverse problem, specifying the object in the environment given the retinal image shape, is ill posed, because many possible configurations of three-dimensional shapes in the world are consistent with any given retinal image shape.

In order to solve the inverse problem, perceptual systems must apply **constraints** to the inferences that can be made. In other words, a perceptual system does not consider every possibility consistent with geometry but only possibilities that fall within certain boundaries. Operation of these constraints by the human visual system is easy to demonstrate, as illustrated in Box 8.1.

Many examples of the constraints that are imposed on form perception were noted and studied extensively during the first half of the 20th century by the **Gestalt psychologists.** They noted a strong agreement among observers concerning the spatial organization of two-dimensional patterns and described a number of **Gestalt principles** that were supposed to be responsible, three of which are illustrated in Figure 8.6.

The elements seen in each row are perceived to be organized into pairs. The Gestalt principles responsible for this grouping are **spatial contiguity** in the top row, **similar size** in second row, and **similar orientation** in the third. Although extensively studied, the organization principles, or constraints, that operate during form perception remain somewhat mysterious. One hypothesis is that they reflect deep knowledge about spatial structure of our three-dimensional environment, presumably acquired over evolutionary time. The

general idea is that the organizing principles facilitate, in some incompletely understood manner, correct inferences to be drawn, more often than not, regarding three-dimensional shapes from two-dimensional patterns of retinal stimulation.

Bottom-Up and Top-Down Influences During Recurrent Processing

The topic of form processing will be approached in this chapter primarily with a forward-flow approach, starting with the environment, then the retinal image, then neural processing, and so on. Approaching the topic in this manner is convenient as a heuristic but fails to capture the complexities of information flow that occur during biological processing. Information flow does not follow sequential stages. Instead, it involves numerous **parallel pathways** that have a **recurrent** organization along the lines illustrated in Figure 8.7.

The sections that follow will describe some of the activities that take place in the various boxes shown in Figure 8.7. However, our description will not be able to capture the overall complexity of the processing that results from the fact that activities at various stages are interacting with one another simultaneously. This allows a simultaneous appli-

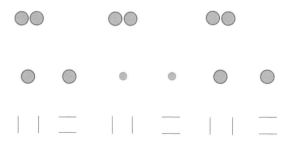

FIGURE 8.6. Illustration of Gestalt principles. Top row: spatial contiguity. Middle row: similar size. Bottom row: similar orientation.

Box 8.1
Subjective Contours Demonstrate Operation of Built-In Constraints

Look at the pattern inside the box of Figure 1 and consider two interpretations of environmental conditions that could have given rise to the image formed on your retina by this pattern. The first is that five black ovals are present but are partially obstructed from view by an irregularly shaped white surface that lies in front of them. A second interpretation is that there are five irregularly shaped black figures, all set on a single white surface inside the box. The shapes of two of these black figures are illustrated where the solid arrows point outside of the box. With the second interpretation, there should be no difference between the portions of the white surface where the two dashed arrows point. However, the first interpretation requires that the white area in the center region of the box be seen as a separate surface distinct from the rest of the background white area within the box.

Most normal human observers, when they view this pattern, report the first interpretation rather than the second. A white surface is seen in the center of the box. It is separated from the remainder of the background white area by **subjective contours,** i.e., contours present in the percept but not in the physical stimulus. These subjective contours are referred to as being **modally completed,** meaning they appear as part of the front surface. The black shapes are perceived as partially occluded oval-shaped discs, the hidden subjective contours forming the oval shape being **amodally completed** behind the white surface.

This demonstration illustrates the general principle that given two or more possible interpretations based on geometry, our visual system usually applies a set of constraints that allows us to settle on one particular interpretation immediately and without conscious effort. Note how this differs from what would be expected from a general-purpose perceptual processing system, which would have to systematically examine all geometric possibilities.

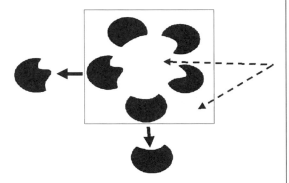

FIGURE 1. The perceptual phenomenon of illusory contours. See text for details.

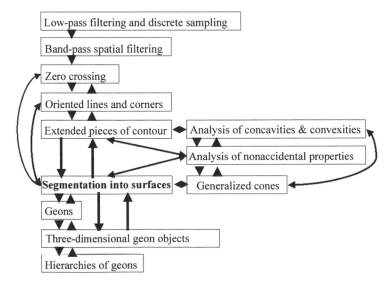

FIGURE 8.7. Some of the major stages of processing involved in form perception and their interconnections.

FIGURE 8.8. Stimulus that can be perceived as a vase or as two faces.

cation of stimulus driven **bottom-up** and higher-order cognitive driven **top-down** mechanisms to influence form vision processing.

An example of the complexities involved in information flow during form vision processing can be illustrated by considering what happens when Figure 8.8 is viewed.

If you stare at this figure for awhile, you will discover that it leads to an unstable percept. At times a vase is perceived and at other times two faces are seen. While the vase is seen, the portion of the pattern that would form two faces dissolves into the background. While the faces are seen, the vase dissolves into the background.

This example highlights a fundamental problem of perception called **figure-ground segregation,** deciding which parts of the environment compose the shape of interest and which the background. When initially confronted with a stimulus, before any processing has been done to determine details about its shape, how does the visual system decide which parts of the stimulus constitute the figure and which are simply the background?

A bootstrapping issue is immediately confronted when trying to decide when in the sequence of form processing figure-ground segregation should be performed. On the one hand, it would seem that figure-ground segregation is a prerequisite before any other processing of information about shape can even begin. On the other hand, it is hard to see how the figure-ground segregation can be performed until some processing of form information has already taken place. The answer to this paradox is most likely that figure-ground segregation does not take place at any particular stage but takes place throughout, via recurrent processing that allows simultaneous bottom-up and top-down influences to operate. In examples such as Figure 8.8, processes consistent with two potential solutions continue competing with one another and the system cannot converge on a stable solution.

Neuroscience studies provide some clues about the neural substrates for these kinds of interactions. Consider the results from a study of neurons in monkey striate cortex reported by **Lamme** and colleagues. They presented spatial patterns of texture to the receptive fields of single neurons. This presentation was done within a larger context in which some texture formed a spatial target and the remainder formed a background around the target. During the initial 40–60 ms after texture was placed in the receptive field of a neuron, its neural response was identical whether the texture formed part of the figure or part of the background. However, longer-latency responses (80–350 ms) depended on contextual information and were stronger for texture when it was part of the figure than the background. The interpretation of these results is that the initial response reflects a response to bottom-up information flow regarding the local pattern of texture within the receptive field. The longer-latency response reflects modulation by a top-down process, perhaps coming from extrastriate cortical areas V2 and V4, as discussed further below.

A second example that illustrates influences of top-down processes on form perception is shown in Figure 8.9.

FIGURE 8.9. The figure shown on the left is referred to as a Necker cube. This is an example of an ambiguous figure that can be perceived in more than one configuration. Two possible interpretations are diagrammed in the middle and on the right.

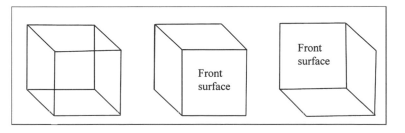

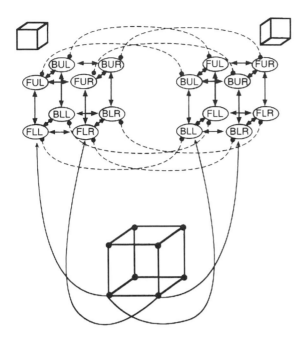

FIGURE 8.10. Diagrammatic view of a neural network model of perceptual processing that tries to account for the alternating percepts that arise while viewing the Necker cube. This model involves excitatory (solid lines) and inhibitory (dashed lines) connections that provide built-in knowledge about possible relationships between vertices of three-dimensional cubes (FLL: front lower left, BUR: back upper right, and so on). Reproduced from J.L. McClelland, D.E. Rumelhart, and the PDP Research Group. *Parallel Distributed Processing: Explorations in the Microstructure of Cognition, vol. 2: Psychological and Biological Models,* © 1988, by permission of MIT Press.

The pattern shown on the left is called the **Necker cube,** after **I. A. Necker,** the Swiss naturalist who described it in the early 1800s. An observer looking at this pattern for a period of several seconds will notice that the percept alternates spontaneously from time to time between the three-dimensional shapes depicted in the middle and right panels. The explanation for this alternating percept is that the geometry of the two-dimensional representational pattern of the Necker cube is equally consistent with two possible three-dimensional objects. An interaction between bottom-up and top-down influences settles on a segregation of the pattern into one three-dimensional shape that persists for a while. However, some subtle changes in the incoming bottom-up information, perhaps triggered by the idiosyncrasics of the scanning patterns of the eyes over the figure, causes the visual system to switch to the alternate interpretation.

The bi-stable appearance of the Necker cube has been modeled with a neural network approach, as illustrated in Figure 8.10.

The model receives inputs from selected parts of the Necker cube pattern, as illustrated at the bottom of the figure. The strengths of these inputs can change over time to reflect factors such as scanning eye movements to different locations in the stimulus. Highly interconnected hidden nodes carry implicit built-in knowledge about possible and impossible three-dimensional relationships. Given any particular input, interactions between these hidden nodes cause the model to settle into one of two output states, shown on the upper left and right, representing two different percepts. Changes in the inputs are sufficient to cause the network to change to the alternative output state. This neural network provides a good illustration of how form vision processing can be modeled based on interactions among a large number of highly interconnected elements.

The Necker cube also underscores the general principle that only certain logical possibilities are considered by biological form vision processing. Note that there is an equally likely (based on geometry) two-dimensional alternative to the two three-dimensional percepts that are perceived when looking at the Necker cube stimulus. This alternative is illustrated in exploded form in Figure 8.11.

If the visual system were systematically switching among logical geometrical alternatives, the pattern should sometimes be seen as consisting of these seven simple geometric shapes positioned directly adjacent to one another. The fact that we do not perceive this two-dimensional pattern of shapes provides additional evidence that the visual system considers only certain classes of potential geomet-

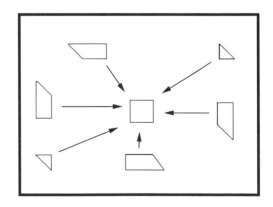

FIGURE 8.11. Based simply on geometry, the Necker cube stimulus is composed of these seven simple two-dimensional shapes, positioned adjacent to one another.

rical solutions when processing spatial qualities of percepts.

Primacy of Visible Surfaces

Several lines of evidence are consistent with the idea that although there are interactions among all levels of form processing, stages of processing where a surface is represented carry the most weight. The visual system seems more concerned with surface properties than with properties at lower levels, such as contours, and properties at higher levels, such as objects. Form processing is often discovered to extract information about visual surfaces automatically, without particular regard to how they are defined (e.g., luminances, textures, color) or what they belong to (e.g., person, penguin, automobile).

Psychophysical studies have demonstrated the importance of surface information even while observers perform low-level psychophysical tasks that have traditionally been considered to be related to early stages of visual processing. For example, **He** and **Nakayama** discovered that observers could not ignore information regarding surface layout while carrying out tasks that involved detecting low-level spatial properties, such as oriented contours. The interpretation of these studies is that information about surface properties is intrinsically intermingled in some way with other types of spatial information at many levels of form vision processing.

The **primacy of surfaces** makes sense from an evolutionary perspective, because it is surfaces that **photons** are **reflected from** and **carry information about.** Thus, vision has evolved deep knowledge about the ways surfaces structure light. Gibson, in his theory of direct perception that was introduced in Chapters 1 and 3, took this idea to an extreme by arguing that understanding perception involves, fundamentally, nothing more than **understanding how surfaces structure light.** Even though Gibson's theory about direct perception is not widely accepted, he had a tremendous influence on the field of form perception by directing attention to the primacy of visible surfaces.

2 Preneural Imaging and Sampling

As described in Chapter 4, the optical components at the front of the eye, the cornea and lens, form an inverted image on the retina of the light rays that pass into the eye (Figure 8.12).

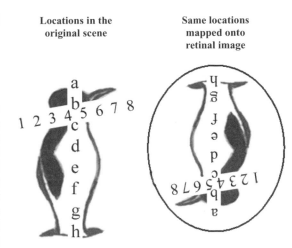

Locations in the original scene   Same locations mapped onto retinal image

FIGURE 8.12. Locations in the visual field, shown on the left, are mapped onto the retinal image, shown on the right.

This retinal image forms a **continuous map** that preserves certain geometrical relationships, as described in Chapter 5. However, in the process of forming the image, some spatial information is lost because the lens acts as a **low-pass filter.**

Low-Pass Filtering in Optics

Low-pass filtering is best understood by describing its properties in the spatial frequency domain. In the sections that follow, it will be assumed that the reader is familiar with the concepts described in Chapter 3 regarding representations in the **spatial and spatial frequency domains,** and the use of **Fourier's theorem** to translate between them (Figure 8.13).

The amount of low-pass filtering performed by a lens can be characterized in a formal manner by its **optical modulation transfer function,** or, more simply, its **transfer function.** An intuitive understanding of a transfer function can be attained based on how it is measured, as described in Box 8.2.

A lens's transfer function characterizes how well contrast is transferred from the object to the image within the frequency domain. However, it is computationally straightforward to use a transfer function to make predictions about how lenses operate on object intensities in the spatial domain as well, as illustrated in Figure 8.14.

One simply begins with a **description of an object in the spatial domain.** Fourier analysis converts this to a **description of the object in the spatial frequency domain.** Then the transfer

Box 8.2
Measuring the Transfer Function of a Lens

The transfer function of a lens can be measured with the following procedures. The first step is to assemble a series of sine wave grating slides that vary in spatial frequency. For this example, assume that the slides are all oriented vertically and have the same phase. Then one forms an image of these slides with the lens one at a time and measures the luminance profile along the horizontal axis in both the original slide and the image. The results of these measurements, as they might be entered into a lab notebook, are illustrated in Figure 1.

In the top of the diagram, the luminance profile for each of four slides is noted on the left and that for the corresponding images formed by those slides on the right. Contrast has been measured for each luminance profile. All of the slides have 50% contrast and all of the images have less than 50%.

At the bottom of the diagram, a plot has been generated that shows the percentage of the contrast in each of the four slides that was transferred to its image. The plot demonstrates that 98% (49/50) of the contrast of the low-spatial-frequency sine wave grating shown in the top row (#1) was transferred to the image. For the grating slides of somewhat higher-spatial-frequency diagrammed in the second and third rows, only 80% and 50%, respectively, of the contrast was transferred. No measurable contrast was transferred for the slide with the highest spatial frequency (#4).

Examination of the plot reveals that the results fall along a smooth curve that can be estimated with reasonable accuracy following measurement of some limited number of sine wave gratings. This smooth curve is the **transfer function** of this lens. In this example, the transfer of contrast is measured only as a function of spatial frequency. Recall from Chapter 3 that in characterizing transfer functions in general, how contrast is transferred would have to be measured for slides having different orientations and phases as well.

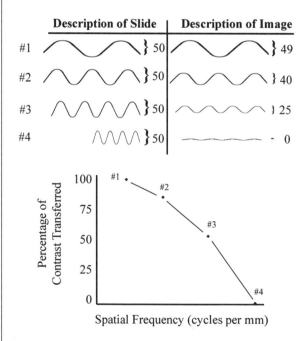

FIGURE 1. Hypothetical measurements that could be used to characterize the transfer function of a lens. See text for details.

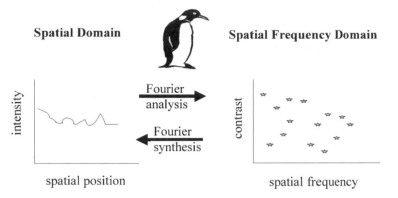

FIGURE 8.13. Shapes of objects, such as a penguin, can be described in two equivalent ways as described in detail in Chapter 3. A hypothetical description of the penguin shape is shown for the spatial domain on the left. This is a description of the intensities at each spatial location. A corresponding description of the same penguin in the spatial frequency domain would be in the form shown the right. This is a description of the spatial frequencies that would have to be superimposed to produce the penguin shape. Fourier analysis and Fourier Synthesis can be used to translate between descriptions in the spatial and spatial frequency domains.

function is applied to the spatial frequencies that make up the object. This involves simple multiplication of the contrast in the original spatial frequencies by the percentage transferred, specified at the corresponding spatial frequency by the transfer function. The result is a **description of the image in the frequency domain.** The last step is to use Fourier synthesis to translate the result back to a **description of the image in the spatial domain.**

It is possible to make these same predictions entirely within the spatial domain by carrying out mathematical procedures based on **convolution** that will not be described here. The reason these operations based on transfer functions in the frequency domain are described here is that this approach has certain advantages, as described in Box 8.3.

The transfer function of any lens has a prototypical shape such that a high percentage of the contrast of an object is transferred to the image for low spatial frequencies, but the contrast transfer is progressively attenuated at higher frequencies. This feature gives rise to the term **low-pass filter.** All lenses have an upper limit beyond which spatial frequencies are not transferred at all, called the **high-frequency cutoff.** In a cheap lens, that cutoff occurs at relatively low frequencies, it moves to progressively higher frequencies in lenses of better optical quality. An appreciation of the qualitative effects of low-pass filtering on an image can be attained from Figure 8.15.

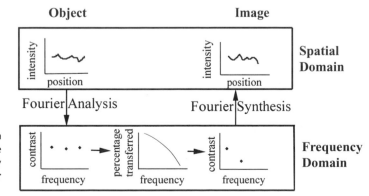

FIGURE 8.14. Series of steps that can be followed to characterize the image that will be formed by a lens for any given object based on the transfer function of the lens.

Box 8.3
Linear Systems Approach Versus Catalog Approach

The procedure of using a transfer function for predicting the image a lens will form is called a **linear systems approach.** The advantage of this approach can be illustrated by contrasting it with what might be called a **catalog approach.**

Image that you are a manufacturer of lenses. You could try to form an image of every shape that a customer might conceivably want to look at with a lens. These images could then be put in a catalog. When a customer came in and asked what lens to purchase, the salesperson would ask, "What things do you intend to image with the lens?" If the customer replied "Penguins," then the catalogs for various lenses would be opened to the "penguin page" and the customer could compare the images. If the customer replied, "My grandmother," then the catalogs would be opened to the "grandmother" page, and so on and so forth. Clearly, one problem with this approach is that the catalogs would have to be huge.

An advantage of the linear systems approach is that a manufacturer of lenses does not have to test lenses on every object that might conceivably be imaged. Instead, a manufacturer

simply applies a stereotypical set of measurements involving sine wave gratings in order to determine the transfer function for the lens. Then the manufacturer of the lens can specify its properties by supplying the transfer function. Any potential customer can use this transfer function along with Fourier analysis and Fourier synthesis to determine ahead of time exactly what any images produced by that lens will look like.

Some vision scientists have argued that a similar advantage holds for application of the linear systems approach to perception. A transfer function that describes how well a human perceives sine wave gratings can be measured using methods that are described later in this chapter in the section Behavioral Correlates of Band-pass filtering. If human perception is approximately linear, it should be possible to use the same procedures as for lenses to make predictions about perception of any spatial pattern. This approach has had some success in predicting performance of human observers on tasks in which perception is near the threshold.

The contributions of the entire range of spatial frequencies transferred to an image by an expensive lens are illustrated in the image of Groucho Marx shown in the middle panel. The left panel illustrates how the image formed by a cheap lens with a lower cutoff frequency would look. Also illustrated, in the right panel, is the image that would be formed if one superimposed only the high spatial frequencies eliminated by the low-quality lens. This figure illustrates that **coarse features** in an object can still be seen in an image that has been low-pass filtered by a cheap lens. However, some of the **finer spatial details** that are imaged by an expensive lens are lost by low-pass filtering.

Low-pass filtering affects form perception because the optics of the **human eyeball** eliminate all spatial frequencies higher than about **60 cy/deg** from the retinal image. Because of this filtering, human observers can have no perceptual knowledge about spatial properties of the world that are carried by spatial frequencies higher

than 60 cy/deg. The vast amounts of information that reach the front surface of the eye about properties of the environment at the scales of molecules and atoms are immediately eliminated from consideration.

The high-spatial-frequency cutoff of the eyeball places an upper limit on **visual acuity,** typically measured in humans by determining the smallest letters that can be read on an eye chart. Acuity is traditionally specified in units based on a **Snellen acuity chart** where a value worse than **20/20** is used as a diagnostic measure to identify individuals with abnormally poor acuity, i.e., acuity poorer than the normal range found in the population. The spatial frequencies that carry the information that enables humans to discriminate the shapes of the letters in the 20/20 line of the chart are near **30 cy/deg.** Most individuals can actually see somewhat better than this and have an acuity that corresponds to about **50 cy/deg,** very near the upper limit allowed by the eye's optics. The next section, which discusses the

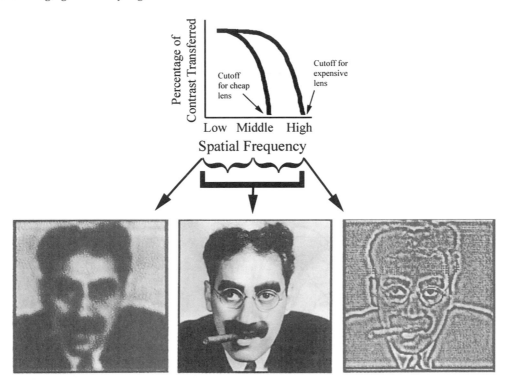

FIGURE 8.15. Top panel: Transfer functions for a cheap and an expensive lens. Bottom panel: Contributions of low and high spatial frequencies to an image. Adapted from J.P. Frisby, 1979. *Seeing. Illusion, Brain and Mind.* Oxford: Oxford University Press, © 1979, Roxby & Linsey Ltd.

implications of receptor sampling, says more about this.

Sampling of the Image by Photoreceptors

The **continuous, low-pass filtered, topographic mapping of the environment** that is present in the **retinal image** is transformed into **discrete samples** by the **photoreceptors.** The form of the information present at this stage can be characterized formally in the spatial domain as an **array,** $I(x,y)$, where I is the intensity value at each x and y position occupied by a photoreceptor.

The fact that sensory coding of the image takes the form of discrete samples has two potential implications for form vision. First, biological organisms have had to solve a potential problem called **aliasing** that arises from discrete sampling. Second, the spacing between the individual samples, corresponding to the spacing between photoreceptors, puts some limits on the spatial resolution capabilities of form vision.

Aliasing

Whenever engineers build a physical device that takes discrete samples of a continuous pattern, they have to guard against aliasing, an artifact in which small details that are present in the input pattern are reproduced as illusory larger patterns. The principle of aliasing is illustrated in Figure 8.16.

The thin smooth line depicts a luminance profile of a continuous image of a high-spatial-frequency grating on the retina. The vertical tick marks at the bottom of the figure designate the locations of the photoreceptors that sample this image. The small filled circle data points show the intensities of the image at the points where it is being sampled. Given the form of the information being sampled by the photoreceptors and sent to central brain areas for further processing, based on this input, the brain is likely to infer that the pattern on the retina is something along the lines depicted by the thick smooth line. This illusory impression of a low spatial frequency when in actuality a much higher spatial frequency was present is an example of aliasing.

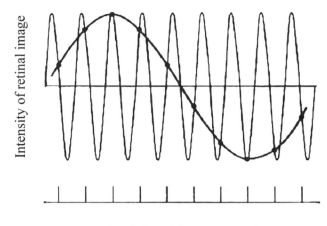

Spatial position across retina

FIGURE 8.16. How discrete sampling of an image can lead to aliasing. See text for details.

Aliasing can occur whenever a continuous spatial pattern is changing over a finer spatial grain than the distance between the samples that are obtained. For any given spacing of the samples, determined in the case of the retinal image by the spacing between the photoreceptors, there is an upper limit, called the **Nyquist limit,** named after the American physicist and engineer Harry Nyquist who first described it, on the spatial frequencies that can be resolved veridically. Any spatial frequencies in the continuous image that exceed the Nyquist limit have the potential to lead to aliasing.

The Nyquist limit set by the spacing of the photoreceptors in the human eye is about 60 cy/deg, the same as the high-spatial-frequency cutoff of the optics of the eye. This is highly unlikely to be a coincidence. More likely, evolution resulted in a solution to eliminate aliasing by adjusting the quality of the optics to be **just** *bad* **enough** to filter out all spatial frequencies that might lead to aliasing. The solution to the problem of aliasing discovered by nature over the course of evolution is very similar to the solution used by modern engineers, which is referred to as using a lens as an antialiasing filter.

There is limited support from comparative studies for the idea that primates may have reached some biological limit on the packing density of their photoreceptors. The only species that have achieved acuities better than those of primates are certain birds, such as the bald eagle, which has acuity better than 120 cy/deg.

The photoreceptors in these species are not packed appreciably closer together than in primates. Evolution has perhaps discovered a "trick" to achieve a higher resolution in these species, as described in Box 8.4.

The presence of aliasing can be demonstrated in human observers by using a laser to bypass the eye's optics and generate high spatial frequencies in the form of interference patterns directly on the retina. Under these conditions, observers perceive illusory low spatial frequencies when the luminance profile on the retina has a high spatial frequency that exceeds Nyquist's limit.

Hyperacuity

The precision with which observers can align the positions of two or more small elements such as dots, bars, or edges is referred to as **vernier acuity.** Some examples of vernier tasks are shown in Figure 8.17.

Humans are able to perform on these vernier tasks with an unexpectedly high level of precision. Relative displacements in location as small as a few seconds of arc can be detected. This precision in localization is much finer than the spacing between human photoreceptors, which are separated by 30 seconds of arc or more. This ability to localize with more precision than the sampling distance between photoreceptors seems, on first analysis, impossible; this has given rise to the label **hyperacuity.** An understanding of why hyperacuity is not

Box 8.4
Nature's "Trick" to Defeat Nyquist's Limit

The photograph on the right of Figure 1 shows the foveal region from the retina of a bald eagle. This fovea is exceedingly deep. One possible function of this specialized anatomical adaptation is to defeat the limitations of Nyquist's limit on acuity. The thick lines on the left represent the positions of three hypothetical receptors and the thin arrows three hypothetical light rays. If those light rays are coming from successive peaks and troughs of a fine sinusoidal grating, the hypothetical photoreceptors will not be able to resolve the grating, because it is above their Nyquist limit. Suppose there is a compelling reason to achieve veridical perception of this grating but also a biological limitation that will not allow the receptors to be packed any closer together. One trick that will accomplish this goal is to tilt the surface on which the receptors are placed, as indicated by the lines superimposed on the retina on the right side of the figure. Nature may have played just such a trick with the eagle retina by tilting the receptor surface into a deep pit.

 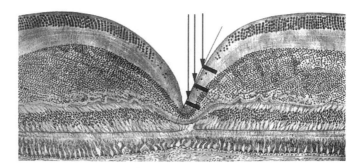

FIGURE 1. Left: A sampling rate below Nyquist's limit that would be expected to lead to aliasing. Right: The same sampling rate placed at an angle to the incoming light rays, as illustrated in this superimposed image from an eagle retina, allows a higher Nyquist limit. Photograph of eagle retina adapted from S. Polyak, The Vertebrate Visual System, © 1957, by permission of University of Chicago Press.

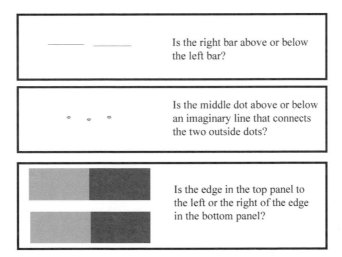

FIGURE 8.17. Illustration of three vernier tasks on which humans exhibit levels of performance referred to as hyperacuity.

Image on Retina

Luminance Profile

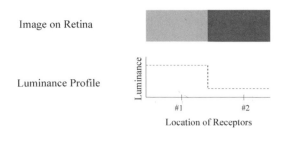

Location of Receptors

Photoreceptor code

Receptor #1	Receptor #2
ON	OFF

FIGURE 8.18. One coding scheme that could be used by the visual system to localize the position of the luminance edge shown in the top panel.

impossible given the receptor spacing requires a detailed analysis of the task. Consider how an image of an edge might be coded in photoreceptors (Figure 8.18).

The top panel shows an image of an edge on a retina. The left portion of the image is a brighter gray than the portion on the right, thus forming a luminance edge that separates the two portions of the image. The luminance profile of the stimulus is shown in the middle panel. The two tick marks on the horizontal axis of the luminance profile plot indicate where the receptors that sample this image are located on the retina. Finally, the bottom panel indicates one way this image could be coded by these two photoreceptors. Receptor #1 codes an ON signal because it is sampling a bright portion of the image, while receptor #2 signals OFF because it is sampling a dark region of the image, yielding a receptor code of ON, OFF.

Next, consider how the receptor code will vary as the position of the image is moved across the retina. Suppose that the image is moved to the left, as illustrated in Figure 8.19.

In this position, both photoreceptors are sampling dark regions, and the receptor code is OFF, OFF. It should be obvious that if the edge is moved to the far right, the code will be ON, ON.

Suppose a hypothetical observer with a visual system containing only these two photoreceptors were taught this strategy: Respond "left" when the receptors signal OFF, OFF, "right" for ON, ON, and "middle" for ON, OFF. How much precision could be achieved by following this strategy? Consider the two edges shown in Figure 8.20.

The signals coded by the two photoreceptors for these two edges are identical, because both edges fall within a **dead zone** between the two receptors. The rationale just described makes it appear that resolution should be impossible within the dead zone and is the basis for using the term "hyperacuity" for tasks in which humans discriminate between edges that differ in position by less than the dead zone width.

A number of explanations have been proposed over the years to try to account for hyperacuity. One is based on the rationale that an edge stimulates other receptors in addition to the two that have been shown and that perhaps a strategy can be based on the outputs of a larger population of receptors. Another possibility is that an observer might be able to use a strategy based on temporal properties of receptor stimulation as small eye movements sweep the image across the dead zone.

These types of strategies can account, in principle, for a small degree of hyperacuity. However, measured levels of hyperacuity are much too fine to be accounted for based on these strategies. Furthermore, hyperacuity levels are relatively little impaired when small points of light rather than lines are used and when strobe light presentations eliminate the effects of eye movements. A more general explanation for hyperacuity comes from the same source as the explanation for lack of aliasing: low-pass filtering of the image by the optics of the eye. Consider the effects of low-pass filtering on a point source of light, as illustrated in Figure 8.21.

Image on Retina

Luminance Profile

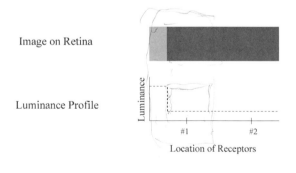

Location of Receptors

Photoreceptor code

Receptor #1	Receptor #2
OFF	OFF

FIGURE 8.19. The same coding scheme illustrated in Figure 8.18 will respond as shown here if the luminance edge is moved to the left.

FIGURE 8.20. The same coding scheme illustrated in Figures 8.18 and 8.19 will respond identically to the luminance edges shown in the top and bottom panels.

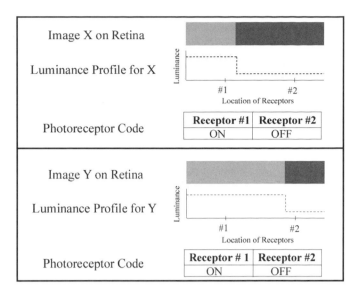

The top panels illustrate a point source of light in the spatial and spatial frequency domains. The bottom panels illustrate the effects of low-pass filtering on the point source. Characterized in the spatial frequency domain, low-pass filtering by the eye eliminates spatial frequencies above 60 cy/deg. Characterized in the spatial domain, a point source is turned into a spot of light of a given width. Note that the width of the spot following low-pass filtering of frequencies above 60 cy/deg is about the same as the width of the space between the photoreceptors, indicated by the two tick marks on the horizontal axis. This has the effect of eliminating the dead zone between receptors. Now even a small change in position of the dot will result in a potential change in the signals sent to the brain from two adjacent photoreceptors, as illustrated in Figure 8.22.

Based on this analysis, it should be obvious that the limits on vernier acuity have to do simply with the ability of the brain to discern small differences in magnitudes of intensity signals from adjacent receptors. There is no inherent limit due to a dead zone.

The properties of the spatial information that is transmitted from photoreceptors to neurons in the retina may be summarized as follows: Globally, there is a topographic mapping of locations in the environment onto an image array of discrete pixels. Information carried by spatial frequencies higher than 60 cy/deg has been eliminated. This places a limit on resolution of fine spatial detail as measured

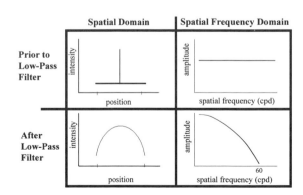

FIGURE 8.21. How descriptions of a point source in the spatial and frequency domains are altered by low-pass filtering. See text for details.

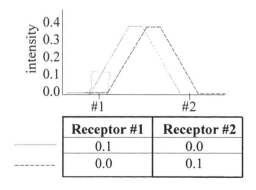

FIGURE 8.22. There is no dead zone between receptors when appropriately low-pass-filtered stimuli are displaced by small amounts.

in tasks such as those involving Snellen acuity. This limit corresponds to the Nyquist limit based on receptor spacing, but it is imposed by the eye's optics. Somewhat surprisingly, vernier acuity and other forms of hyperacuity tasks are not affected by this limit. Localization of features can be carried out with a precision that is a small fraction of the distance between receptors, as long as sufficient information about these features is carried by spatial frequencies lower than 60 cy/deg.

3 Geniculostriate Neurons as Band-Pass Spatial Frequency Filters

The spatial information that is coded for in the receptors is passed on to neurons in the retina for further processing. As discussed in Chapter 5, there are a number of different types of neurons in the retina. However, in this section we will concentrate on the properties of spatial processing that are generic to neurons with concentric center-surround receptive field organization. Retinal ganglion cells, which form the output layer of retinal processing, can be considered prototypical of this receptive field organization.

Retinal Ganglion Cells

Recall that a concentric center-surround receptive field shape appears, when viewed from the surface of the retina, as a circle surrounded by an annulus. This is illustrated in the upper right-hand panel of Figure 8.23.

The example shown is an on-center neuron, which is excited when light comes on in the center and inhibited when light comes on in the surround. Recall that there is also a corresponding type with the opposite polarity of response, called an off-center neuron. This discussion will be confined to the on-center type. The upper left panel of the diagram illustrates the amount of excitation and inhibition that would be created by a small spot of light as it is moved across the receptive field passing through the center. This profile of the receptive field approximates a mathematical function called a **difference of Gaussians (DOG).**

Both the two-dimensional (right) and one-dimensional (left) descriptions of receptive fields shown in the top row are in the spatial domain, because they describe responses to intensity as a function of retinal position. Receptive fields can also be characterized in the spatial frequency

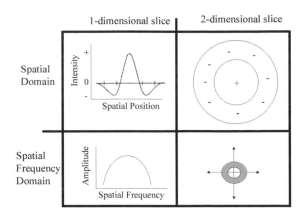

FIGURE 8.23. How center-surround receptive fields can be characterized in the spatial and spatial frequency domains. The left panels show a one-dimensional description of a slice through the receptive field, and the right panels show a two-dimensional description.

domain. The Fourier transform for the one-dimensional cross-section is illustrated in the form of an amplitude spectrum in the lower left panel. The two-dimensional case is illustrated in the form of a polar plot in the lower right panel (see Chapter 3). Examination of these receptive field properties in the spatial frequency domain reveals that only a band of spatial frequencies is represented; all spatial frequencies higher or lower than this band are eliminated. This reveals a fundamental property of neurons with concentric center-surround receptive fields: They act as **spatial frequency band-pass filters.**

An intuitive impression about why the concentric center-surround receptive field structure confers this property can be attained by considering how receptive fields respond to sine wave gratings in the spatial domain. The strongest response to a sinusoidal grating occurs when the stripes are just the right size so that one bright bar stimulates the entire center of the receptive field and the two flanking dark bars both fall on the surround. The cell responds less well to sinusoidal gratings with stripes that are either larger or smaller than this optimum size.

Ganglion cells have receptive fields of different sizes, as illustrated in Figure 8.24. A ganglion cell with a large receptive field will have its pass band centered on low spatial frequencies. Ganglion cells with progressively smaller receptive fields will have pass bands centered on increasingly higher spatial frequencies. The **upper envelope** that encompasses the responses of ganglion cells of all sizes can be used to characterize the totality of the

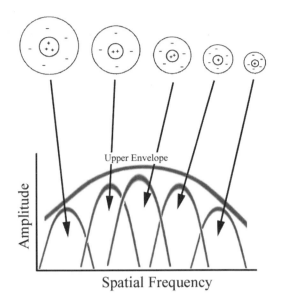

FIGURE 8.24. Concentric center-surround receptive fields of different sizes correspond to band-pass filters for different ranges of spatial frequencies when described in the frequency domain.

spatial frequencies transmitted from the retina to higher visual processing areas of the brain.

Ganglion cells vary in size as a function of eccentricity. However, within any small region of the retina corresponding to a functional hypercolumn, there are perhaps four to seven sizes of ganglion cells operating in parallel, similarly to what is illustrated in Figure 8.24. This has led some investigators to propose that early stages of form processing use a sensory coding scheme in which spatial information is represented in the spatial frequency domain.

Consider two alternative models about how spatial information might be encoded and transmitted through the visual system. The first would be that the intensity of every point sampled by a photoreceptor is encoded and transmitted to the brain. This would involve coding in the spatial domain. Alternatively, the information sampled by the photoreceptors might be immediately coded in terms of the spatial frequencies that are present. In this case, the information that is transmitted from the retina to the central visual processing centers of the brain is encoded in the frequency domain. Since any receptive field can be described equivalently in terms of its spatial extent (spatial domain) or its filtering characteristics (spatial frequency domain), these two alternatives are best thought of as different (but compatible)

levels of description rather than as mutually exclusive alternatives.

A number of models of spatial processing have been proposed based on the general idea that information is coded and transmitted at early stages of processing in band-pass filters that operate in parallel. These filters can be thought of as **spatial frequency channels** analogous to the channels responsible for transmitting information about color that were discussed in Chapter 7. This organization of spatial processing into a small number of band-pass filters, or channels, has implications for spatial perception.

Behavioral Correlates of Band-Pass Filtering

A **contrast sensitivity function (CSF)** is a form of transfer function, roughly analogous to the transfer functions that characterize formation of images by lenses, described earlier in this chapter. Recall that the transfer function for a lens specifies the amount of contrast transferred from the input (object) to the output (image). The behavioral transfer function in the form of a CSF specifies the amount of contrast that is transferred from a stimulus to the observer's percept. Thus, the rationale for how one measures a CSF is fundamentally the same as that for measuring the transfer function for a lens.

Recall that the transfer function for a lens is measured by imaging sine wave gratings. If the amount of contrast in the input grating is labeled X and the amount in the image Y, then the percentage transferred for a lens is calculated as Y/X. For a CSF, the input, X, is specified as the amount of contrast in the sine wave grating that is being viewed. The output, Y, is how visible the grating is to the observer. During measurement of the CSF, visibility for each viewed grating is adjusted to be at the threshold. Thus, visibility of all spatial frequencies is made **perceptually identical** in the sense of being just visible. An exact calculation of Y/X cannot be made because the actual value of Y is unknown. All that is known is the Y remains (perceptually) constant across all measurements. Thus, instead of calculating Y/X, we calculate just X, called **threshold contrast,** a value that will be proportional to Y/X. A CSF is simply a plot of threshold contrast or its reciprocal, **sensitivity,** as a function of spatial frequency.

The basic form of an observer's CSF can be demonstrated by viewing the bottom panel of Figure 8.25 at a distance of about arm's length.

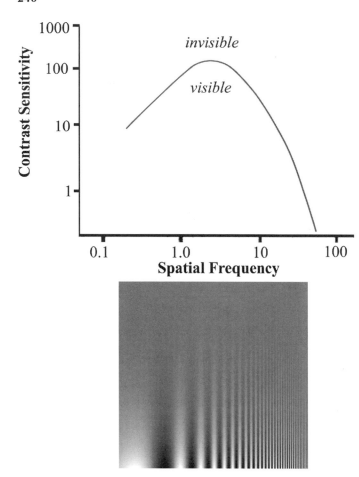

FIGURE 8.25. Viewing the stimulus shown
in the bottom panel allows a perceptual
demonstration of one's own CSF, as described in the text. Adapted from T.N.
Cornsweet, *Visual Perception*, © 1970, by
permission of Academic Press.

The pattern of stripes varies gradually from a low spatial frequency on the left to a high spatial frequency on the right. Contrast varies gradually along the vertical axis, being high at the bottom and low at the top. Thus, for any given spatial frequency (horizontal position), the observer should see a stripe in the bottom portion of the figure that becomes invisible above some height where its contrast drops below the threshold. The shape of the dividing line between visible and invisible is the observer's personal CSF, and it should take the general form of the CSF shown in the top panel.

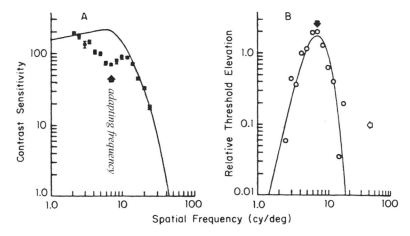

FIGURE 8.26. Results of a behavioral study supporting the idea
that the overall shape of the CSF
is derived from the operation of
narrower band-pass channels. See
text for further details. Reproduced from C. Blakemore and F.W.
Campbell, On the existence of
neurons in the human visual
system selectively responsive to
the orientation and size of retinal
images. *J. Physiol.* 203:237–260,
© 1969, by permission of the
Physiological Society and C.
Blakemore.

Sensitivity is highest for mid-spatial frequencies, meaning that only a small amount of contrast is needed for visibility. Stated another way, there is an efficient transfer of contrast to perception at these frequencies. The human eye is less sensitive to both higher and lower spatial frequencies, demonstrating an attenuated transfer of their contrast to perception. Thus, the overall shape of the behavioral transfer function operates as a **band-pass filter**. Very low spatial frequencies are attenuated as well as very high ones. Behaviorally, humans are able to see only a band of spatial frequencies that fall between about 0.1 and 50 cy/deg. Some implication of this fact are discussed further in Box 8.5.

The overall shape of the behavioral CSF reflects the upper envelope of the four to seven narrower neural band-pass filters that transmit information from the retina along the geniculostriate pathway, as illustrated in Figure 8.24. The operation of the individual band-pass filters that make up the overall CSF can be demonstrated psychophysically as illustrated in Figure 8.26.

This figure shows the results of a behavioral study performed by Colin **Blakemore** and Fergus **Campbell** in 1968. The CSF of human subjects was measured before and immediately after their eyes adapted to a high-contrast spatial frequency. The CSF measured prior to adaptation is indicated by the solid line and that after adaptation by the filled symbols in the left panel. The right panel shows the amount of threshold elevation caused by the adaptation as a function of spatial frequency. The rationale for this study is that the underlying mechanisms responsible for seeing the adapting frequency should become "fatigued" during the adaptation and thus become less sensitive to contrast for a short period, until they recover. If a single mechanism is responsible for detection of contrast at all spatial frequencies, then fatiguing that mechanism should cause a uniform elevation of all spatial frequencies. On the other hand, if contrast at various spatial frequencies is detected by independent channels, then only those channels near the adapting frequency should be fatigued. This would be expected to produce threshold elevation only near the adapting frequency. The results of the study shown in the right panel are consistent with the second expectation.

As discussed above, acuity in humans typically falls between 30 and 50 cy/deg. The acuity value corresponds to a single point along the horizontal axis of the CSF near its **high-frequency cutoff.** However, measurement of the entire CSF provides a large amount of additional diagnostic information about the spatial vision of an observer

> **Box 8.5**
> **Are There Large Things That Humans Cannot See Lurking in the Environment?**
>
> The fact that we cannot see very high spatial frequencies even if they have a large amount of contrast seems intuitively obvious. After all, we are aware that we cannot see extremely fine features in our environment, such as molecules, so it is not surprising that a sine wave grating in which the individual stripes are extremely narrow is not visible. An intuitive appreciation that there might be objects present in the environment that have features too large to see is harder to attain. An example that may help is to consider what the ocean looks like when details such as waves on the surface can not be seen. This might occur when viewing the ocean from an airplane when there is mist or fog in the air, or near dawn or dusk. Pilots have to be careful when flying over the ocean without using instruments so that they do not fly into the water. If it is a windy day and there are waves, or if there is debris floating in the water, then these items can be seen, because information about them is carried by mid-spatial frequencies. But the ocean itself is so big that information about its presence is carried only in the very low spatial frequencies that are attenuated by the CSF.
>
> So the next time you are out walking, think about what might be "out there in the environment" that you cannot see. It is obvious that there are all kinds of small things present, such as molecules, of which we are unaware. However, there might also be large things present of which we are also unaware! We are aware only of things whose shapes are carried by mid-spatial frequencies— things within a range only somewhat larger and smaller than a breadbox.

that would not be captured by simple specification of acuity.

Performance on many perceptual or visual-motor tasks depends on the spatial frequencies that carry the predominant information required for that task. An analysis of spatial frequencies required for the task in combination with the observer's sensitivities

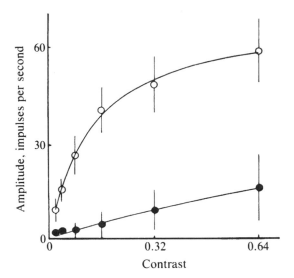

FIGURE 8.27. Responses of P-stream (filled symbols) and M-stream (open symbols) neurons in monkey dLGN to contrast. Reproduced from E. Kaplan and R.M. Shapley, The primate retina contains two types of ganglion cells, with high and low contrast sensitivity. *Proc. Natl. Acad. Sci.* 83:2755–2757, © 1986, by permission of the National Academy of Science and E Kaplan.

at those frequencies offers the potential of more accurate predictions about expected performance than would be obtained by a simple measure of acuity. An example involving the task of flying a plane is discussed in Box 8.6.

P- and M-Streams

Information about spatial patterns is transmitted along the geniculostriate neural pathways along the P- and M-streams of processing delineated in previous chapters. High-spatial-frequency information is carried exclusively by the P-stream, while low- and mid-spatial-frequency data are carried by both the M- and the P-stream neurons. Over the range of spatial frequencies that are carried by both systems, the M-system is more sensitive to contrast. This is illustrated in Figure 8.27, which shows the changes in responses of P-stream (filled symbol) and M-stream (open symbol) neurons in monkey dLGN to contrast.

Behavioral studies in monkeys following lesions confined to the P-layers of the dLGN reveal severe deficits in shape discrimination based on fine spatial detail. Such discrimination is unimpaired by M-stream lesions. These results demonstrate that information carried in the M-stream can be used

Box 8.6
Which Pilot Would You Prefer to Have Flying Your Plane?

Consider the CSFs of two hypothetical pilots (Figure 1).

Pilot B has a higher value for the high-frequency cutoff. Consequently, this pilot will have a higher acuity score on a traditional test of trying to read the smallest print on an eye chart. However, Pilot A has better sensitivity to mid-spatial frequencies.

Consider a situation in which the pilot is on a collision course with a blue airplane seen against a blue sky. While still a reasonable distance away, the other plane is likely to be visible as a low-contrast blob. Information about this blob will be carried primarily by mid-spatial frequencies. These spatial frequencies are visible to Pilot A, even at low contrast, but they are invisible to Pilot B unless they are at very high contrast. Pilot A is more likely to see the airplane at this time than Pilot B.

Next, consider a situation in which the collision course was not detected and the two planes are 1 cm away from making a head-on collision. At that instant, suppose there is an ant crawling across the inside of the cockpit window of the other plane. The details of the shape of the ant's body are carried primarily by high spatial frequencies. Pilot B will be able to see these details better than Pilot A.

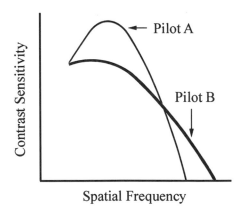

FIGURE 1. Contrast sensitivity functions for two hypothetical pilots.

only for spatial tasks that involve discrimination of coarse shapes. Spatial information used for the analysis of fine spatial detail is carried almost exclusively by the P-stream. However, the M-stream must come into play whenever shape discrimination must be made based on information carried in mid- or low spatial frequencies with very low contrast.

4 Cortical Processing of Zero Crossings and Oriented Edges

Information traveling along the P- and M-streams of the geniculostriate pathway as described in the previous section is next passed to visual cortical areas for further processing. Consider the inverse optics problem that the visual system has to solve based on processing this information. Ultimately, this information is going to be used to infer the **shapes** of **surfaces** of **objects.**

As described at the beginning of this chapter, a fundamental aspect of form perception is that the visible world is regarded as being composed of smooth surfaces. However, smooth surfaces do not really exist in the physical world. A **visible smooth surface** is defined only with respect to a particular spatial scale. The coat of a cat may be perceived as a surface when the cat is viewed from a distance but in fact, as seen under a magnifying glass, consists of individual strands of hair. Similarly, from a far enough distance the trees on the side of the mountain form a surface, but this surface vanishes on approach. Consequently, the visual system needs to analyze a scene at an appropriate spatial scale in order to define visible surfaces.

Finding a surface based on the intensity pattern in the retinal image is a nontrivial task. The reflectance patterns coming from a smooth surface do not always clearly define its borders, as illustrated in Figure 8.28.

Notice that when the image at the top of the figure is viewed, it is immediately perceived as depicting distinct surfaces of two separate leaves that partially overlap. The portion of the image enclosed within the box is shown at the bottom of the figure as an intensity array. Perusal of the numbers in the array underscores the fact that high-level perceptual structures such as surfaces are not always obvious based simply on the pattern of light intensities imaged onto the retina. The intensity distribution is determined by many factors, including the spatial structure of the reflectance pattern of the surface, the light source, and the point of observa-tion of the viewer. The task of form vision at this stage of processing is to ignore the reflectance changes caused by all of these other changes and try to find a pattern that defines a surface. **Marr** described the operations that need to be performed at this stage of processing as extracting **primitives** called **zero crossings.**

Zero Crossings

Recall that the information passing up the geniculostriate pathways is carried in neurons with DOG receptive field properties that act as band-pass filters. Now we take note of another characteristic of DOG receptive fields. When a spatial pattern is swept over a DOG receptive field, the signal generated by the neuron approximates the second derivative of the pattern. An intuitive understanding of this relationship is presented in Box 8.7.

Since the neuron is acting as a band-pass filter, the second derivative that it computes is based only on the spatial frequencies within its pass band. The locations where the second derivative crosses the baseline firing level, called "zero crossings," correspond to abrupt luminance changes within the spatial scale of the band-pass filter. For example, a band-pass filter at the spatial scale of the individual hairs of a cat will detect an abrupt change at every hair, but a filter at the spatial scale of a cat will see the coat of hair as a smooth surface.

Clearly, one strategy of form vision is to discount abrupt intensity changes at spatial scales finer than the one that will be used to define a visible surface. A computational strategy proposed by Marr to try to accomplish this goal is to examine zero crossings of second derivatives at several spatial scales simultaneously. Figure 8.29 shows the zero crossings of a scene that are based on mathematical spatial filtering at three different spatial scales.

The scene is a sculpture by Henry Moore (A). Zero crossings are shown at three spatial scales: fine (B), medium (C), and coarse (D). If a zero crossing is present in one filter, the outputs of filters of other sizes can be examined for zero crossings at the same location. In principle, this allows a strategy by which visible surfaces could be built up from only information that is consistent across spatial scales of interest for the task at hand.

How might this kind of strategy be implemented in biology? Recall that DOG receptive fields transmit information along the geniculostriate pathway about zero crossings at four to seven different spatial scales. This information is delivered to cortical area V1. However, there are some important

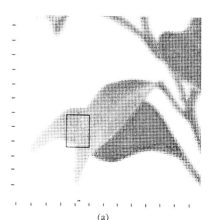

FIGURE 8.28. Visibly distinct surfaces cannot be easily discriminated based simply on the pattern of light intensities reflected into the eye. Adapted from D. Marr, *Vision*, © 1982, by permission of W.H. Freeman and Company.

(a)

X =	34	35	36	37	38	39	40	41	42	43	44	45	46	47	48	49
Y
58	171	169	167	167	166	165	166	164	167	171	171	174	174	175	173	171
57	168	168	168	167	166	167	167	165	169	168	174	176	175	175	175	172
56	168	167	167	165	166	166	167	167	168	170	178	177	176	174	174	173
55	168	168	165	169	167	168	167	165	168	175	177	177	175	175	172	171
54	169	170	167	169	169	168	163	166	172	169	174	173	175	178	173	173
53	171	169	170	168	169	168	169	168	168	170	175	173	175	177	178	176
52	172	171	170	168	169	169	167	168	173	172	173	177	174	175	178	176
51	172	174	171	170	166	168	167	168	172	172	172	177	179	172	175	175
50	171	167	176	169	170	169	168	169	171	172	174	174	173	173	174	178
49	174	172	173	173	173	174	171	171	172	174	172	172	172	169	173	173
48	173	173	173	176	178	172	171	174	174	173	175	175	175	173	173	171
47	173	175	178	173	173	171	171	175	175	177	178	175	174	173	175	178
46	178	175	174	169	173	175	177	175	177	177	174	175	176	177	177	174
45	173	175	173	174	172	173	174	175	174	171	173	174	175	174	172	171
44	177	174	175	175	172	171	172	176	172	173	172	172	173	170	170	175
43	173	171	174	168	176	172	173	173	173	174	171	174	175	173	174	174
42	175	173	171	172	170	171	176	175	178	172	174	175	175	175	175	172
41	181	179	177	172	170	170	169	179	175	174	175	174	172	175	174	175
40	188	184	179	178	176	176	176	174	172	178	172	174	173	172	174	173
39	195	191	188	186	185	183	180	177	178	175	174	176	175	174.	176	176
38	200	199	197	193	190	187	185	180	176	175	180	177	175	175	176	177
37	202	202	199	202	199	194	187	180	175	179	177	176.	174	175	176	173

(b)

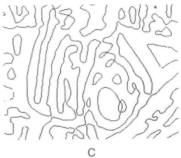

A

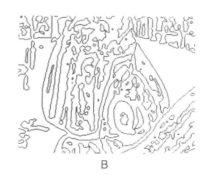

B

C

D

FIGURE 8.29. Descriptions of a complex scene based on zero crossings at different spatial scales. See text for details. Adapted from D. Marr and E. Hildreth, Theory of edge detection. *Proc. Roy. Soc. Lond. D* 290:199–218, © 1980, by permission of the Royal Society.

Box 8.7
DOG Receptive Fields Compute the Second Derivative of Input Patterns

The top panel (Figure 1) shows a luminance edge with a brighter region to the right and a darker region to the left, and the relationship of this edge to the DOG receptive field of a neuron (as described in Figure 8.24). The middle panel shows the luminance profile of the edge, and the bottom panel depicts the firing rate of the neuron as the edge moves across the receptive field. When the edge is at position #1, the neuron fires at its baseline level, because the bright region stimulates the center and surround portions of the visual field equally. As the edge moves towards position #2, the firing rate increases, because the dark region is moving into the surround area, causing inhibition to be decreased, but the center area remains fully excited by the bright region. As the edge moves between #2 and #3, the firing rate decreases to its baseline level again, since the center and surround regions are totally balanced between excitatory and inhibitory inputs. The firing rate moves maximally below its baseline level at position #4 and back to baseline at position #5. Examination of the luminance profile and the firing rate pattern of the neuron reveals that the output of the neuron approximates the second derivative of

the luminance edge. The zero crossing of the second derivative occurs when the edge is located at the center of the receptive field.

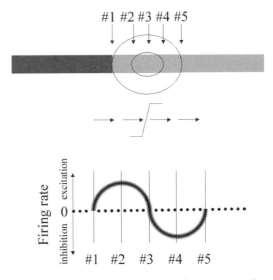

FIGURE 1. Responses of neurons with DOG receptive fields perform the equivalent of taking the second derivative of a spatial pattern. See text for details.

---

differences between the sensory coding of the information delivered to V1 and what we see when we look at Figure 8.29. Biological information about zero crossings in V1 is not in the form of extended contours but strictly local, occurring separately within each hypercolumn. Two additional stages of neural processing will be needed to get from the information provided by DOG receptive fields to that illustrated in Figure 8.29. First the orientations of local edges will need to be extracted within each hypercolumn. Then it will be necessary to start piecing together the local oriented zero crossings to obtain representations of extended contours.

Local Oriented Edges, Line Segments, and Corners

Recall from Chapter 5 that neurons in area V1 called "simple cells" have receptive fields like those illustrated in Figure 8.30.

The receptive field on the left would respond best to a vertical edge passing down the center of the receptive field with the brighter region to the left and the darker region to the right. Receptive fields that are the mirror image of this type would respond best to edges of the opposite polarity. The receptive field on the right would respond best to a vertically oriented dark bar that was just the right width to fill the center region of the receptive field. This field is that of an off-center neuron; it's opposite, on-center companion would respond best to a similar-sized bright bar. Each hypercolumn contains simple cells with these kinds of receptive fields. The simple cells represent all orientations and come in four to seven sizes corresponding to the scales of the band-pass filters feeding into them from the geniculostrate pathway.

Simple cells seem well suited to analyze the local region of a scene for oriented edges and line segments. A second type of neuron in V1 has similar

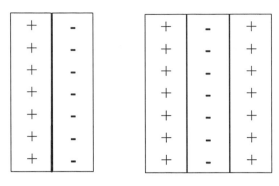

FIGURE 8.30. Diagram of receptive field shapes of representative simple cells in cortical area V1.

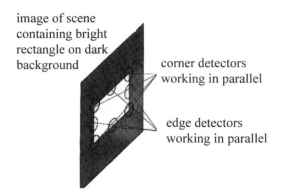

image of scene containing bright rectangle on dark background

corner detectors working in parallel

edge detectors working in parallel

FIGURE 8.32. Individual hypercolumns analyze the entire scene in parallel, looking for features such as edges and corners. Adapted from J.P. Frisby, *Seeing: Illusion, Brain and Mind.* Oxford: Oxford University Press, 1979, by permission of J.P. Frisby.

characteristics but in addition exhibits a property called **end-stopping,** illustrated in Figure 8.31.

The neuron whose response is illustrated does not respond when a luminance edge is passed over its receptive field. However, it does respond when one end of the edge is terminated so that it does not pass outside of the receptive field. Terminated edges are often seen in natural scenes at a corner. Based on this property, neurons that exhibit end-stopping are sometimes called **corner detectors.** Neurons with simple receptive fields and those with end-stopping properties are potential candidates for those that begin analyzing the scene for **edges** and **corners,** as illustrated in Figure 8.32.

The next stage of form vision processing needs to start putting primitives detected within individual

hypercolumns together to form what Marr called the **primal sketch.**

Extended Contours Make Up the Primal Sketch

The primitives that have been discovered up to this stage of processing are still strictly local: a piece of oriented contour here, a potential corner here, and so on. A computational strategy is needed to begin putting these primitives together to form extended contours, as illustrated in Figure 8.33.

An example of a computational strategy that can potentially build up an extended contour based on strictly local relationships is illustrated in Box 8.8.

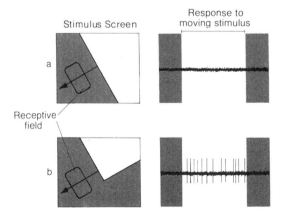

Stimulus Screen

Response to moving stimulus

Receptive field

FIGURE 8.31. Expected response to two visual stimuli of a cortical neuron with a receptive field property called end-stopping. Top: Response to a simple edge. Bottom: Response to a corner. Adapted from J.P. Frisby, *Seeing: Illusion, Brain and Mind.* Oxford: Oxford University Press, 1979, by permission of J.P. Frisby.

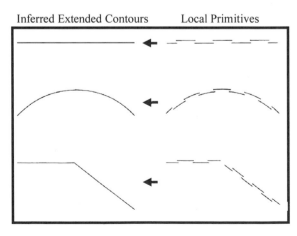

Inferred Extended Contours　　　Local Primitives

FIGURE 8.33. Local primitives as illustrated in the right panel must be put together by the visual system to form extended contours as illustrated in the left panel.

Box 8.8
A Local Algorithm Can Potentially Discover Global Contours

Figure 1 illustrates the operation of an algorithm designed to build up extended contours from simple texture patterns. The bottom panel illustrates on the left side two texture patterns that were provided as input and on the right side the output of the algorithm. The operation of the algorithm is illustrated in the top panel. The first step is to construct a virtual line from a given dot to each neighboring dot within some local neighborhood centered on the dot. This is shown for two dots, A and D, in the figure. A weighting function is applied that gives lines of certain lengths more weight than others. Then a histogram is made for the virtual lines contributed by all of the neighbors of each particular dot. The figure illustrates the histogram for dot A. For example, neighbor D contributes AD to the histogram. Then a smoothing algorithm is applied to the histogram to find the orientation at which it peaks. The virtual line closest to that orientation, AB in the example, is taken as the solution.

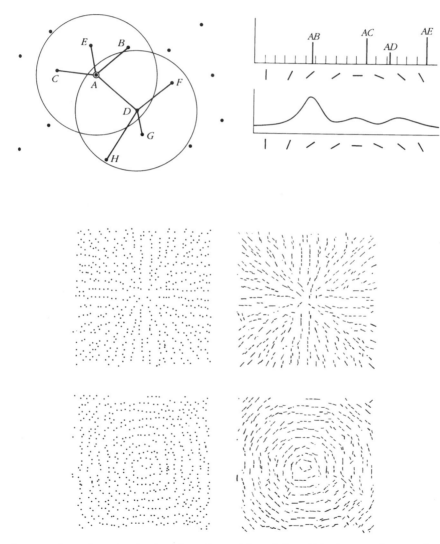

FIGURE 1. Operation of a local algorithm that can potentially build up extended contours based on local information, as described in the text.

Adapted from K.A. Stevens, Computation of locally parallel structure. *Biol. Cybern.* 29:19–28, 1978.

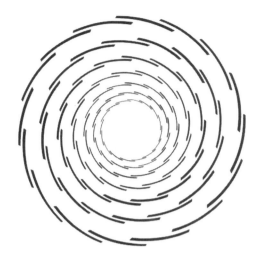

FIGURE 8.34. Demonstration of an illusion called Fraser's spiral. See text for description.

Demonstrations such as the perceptual illusion referred to as **Fraser's spiral,** shown in Figure 8.34, provide evidence that the human visual system can produce percepts of extended contours based on strictly local information.

An observer's percept when viewing Fraser's spiral is that of an extended contour that follows a global spiral pattern. However, close scrutiny of this stimulus will reveal that there is no extended contour defining the spiral. This illusory stimulus is simply constructed in such a manner that it plants a suggestion in the brain in the form of local oriented line segments presented to nearby hypercolumns.

Biological processes that could potentially begin to carry out this stage of processing are probably present at stages of processing as early as V1 in the form of medium- and long-range intrinsic connections between hypercolumns. For example, it is known that selective connections pass between orientation columns processing similar orientations that are located in nearby hypercolumns.

5 Higher-Order Processing

The stages of processing to this point have presented form vision with nothing more than a piece of contour, as in Figure 8.35.

When approaching form perception from a feedforward causal chain viewpoint, the complexity and intricacies of the biological mechanisms that have been followed to reach this point are

impressive. However, from the point of view of the inverse problem that form vision needs to solve, the more important question remains, what can be inferred from this piece of contour? What is now known about the shape(s) of whatever object(s) happen to be present in the environment that gave rise to that piece of contour? The answer: Not much.

Obviously, a great deal of higher-order form processing remains to be done to get from an inference about a piece of contour to a percept of an object shape. However, relatively few details are known about how this higher-order form processing is performed in biological tissue. The next section summarizes a few general facts that the known about form vision processing as information fans out from area V1 into extrastriate cortical areas. Then the chapter returns to a discussion of the hard computational issues that must still be solved in order to get from a stage of processing that provides a simple piece of contour to a stage that can specify shapes of objects.

Biological Processing in the What-Perception Stream

The interblob stream that originates from the interblob regions of layers 2 and 3 of V1 passes to the interstripe regions of V2 and then on to other extrastriate areas and appears to be heavily involved in processing information about shape (Figure 8.36).

As discussed in Chapter 7, the interblob stream merges with the blob stream, which is primarily concerned with processing color, within extrastriate area V4. Processing in V4 is located in an ideal location to combine information about surface shape derived from the interblob stream with surface characteristics based on color derived from the blob stream.

Lesions to V4 in the monkey cause only slight disruptions of basic spatial vision capacities such as acuity and mild effects on discrimination of simple geometrical shapes. However, these lesions result in profound deficits on more complex form discrimi-

FIGURE 8.35. A piece of contour.

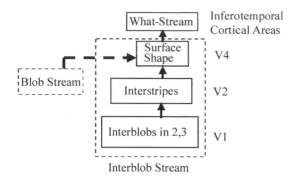

FIGURE 8.36. Information flow from the interblob region of V1 to extrastriate visual processing areas, and eventually into the What-perception processing stream.

nation tasks, such as detecting leftward-oriented texture elements within a background of rightward-oriented texture elements.

Studies of receptive field properties of V4 neurons have discovered that their receptive fields are considerably larger than those of V1 neurons. Furthermore, these receptive fields often combine more than one orientation such that they respond to shapes that are curved or contain orthogonal orientations such as a "+" or an "×" pattern. This is accomplished by pooling information from several V1 hypercolumns in V2 and then further selectively pooling V2 neurons to form the receptive field of a single V4 neuron.

After leaving V4, the combined information from the blob and interblob streams continues as part of the general What-perception stream of visual processing that extends into inferotemporal cortex.

Occluding Contours

Pieces of extended contours discovered by the stages of processing that have been described in previous sections might have come from any number of environmental sources, such as a portion of an edge of an object surface. Biologically important object shapes are usually enclosed by smoothly curving contours called **occluding contours**, as illustrated in Figure 8.37.

A three-dimensional object seen from an unobstructed view will project a two-dimensional image onto the retina. An occluding contour is a piece of **enclosing contour** that surrounds the silhouette of that image. "Enclosing" means that the contour **closes back onto itself** so that tracing around it will eventually lead back to the starting point.

However, even when an enclosing contour is detected in the retinal image, a difficult inverse problem is encountered in making the inference that it is an occluding contour. Although all three-dimensional objects seen from an unobstructed view will produce an occluding contour in the retinal image in the form of an enclosing contour, not all pieces of enclosing contour in the retinal image are occluding contours. Figure 8.38 shows a piece of enclosing contour, but it is produced by the combined silhouettes from two objects, one behind the other, as illustrated in Figure 8.39.

It would be fruitless to spend processing time trying to infer what three-dimensional object shape could have given rise to this enclosing contour, because there is no such single object present. Marr described a set of constraints on possible viewing conditions under which enclosing contours can be interpreted as occluding contours, but those constraints do not allow a general solution to the problem under ordinary viewing conditions. An iterative strategy that works under a moderate range of viewing conditions is summarized in the next few sections. The strategy starts with the observation, seen in Figure 8.39, that the place where the two surfaces overlap produces concavities in the enclosing contour.

FIGURE 8.37. Marr's concept of an occluding contour. See text. Adapted from D. Marr, Analysis of occluding contour. *Proc. R. Soc. Lond. B.* 197:441–475, © 1977, by permission of the Royal Society.

Three-dimensional object shape

Surface occupied on two-dimensional retinal image

Occluding contour in retinal image

FIGURE 8.38. This enclosing contour does not necessarily correspond to an occluding contour. One example where it does not is illustrated in Figure 8.39.

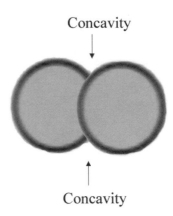

FIGURE 8.39. The silhouette shown here gives rise to the enclosing contour shown in Figure 8.38. However, the silhouette is formed by two separate objects, one of which partially obstructs the view of the other. Each of the two objects has its own occluding contour; the enclosing contour shown in Figure 8.38 represents portions of each. The locations where the enclosing contour makes a transition from representing one occluding contour to representing the other are marked by concavities.

Concavities and Convexities

The relationships between the shape of an object surface and the portions of its enclosing contour where the curvature is concave or convex are complex, and a discussion of the generalized rules that can be used to make inferences about object surface shapes based on shapes of enclosing contours goes beyond the scope of our interests here. However, one simple example involving **matched concavities** can be used to illustrate the general approach.

Whenever one object partially occludes the view of a second, the enclosing contour is highly likely to be marked by matched concavities at locations corresponding to discontinuities from one surface to the other as was observed in the example shown in Figure 8.39. The existence of paired concavities for an enclosing contour of a single object surface would be a relatively rare occurrence. Thus, locations where matched concavities are present establish probable boundaries in the environment where there are discontinuities in the occluding contour as it jumps from one surface to another.

Generalized Cones and the 2½D Sketch

A number of relationships between properties of an occluding contour and the surfaces of the three-dimensional object giving rise to the contour can be specified if it can be assumed that the object shape approximates a **generalized cone**, as illustrated in Figure 8.40.

A generalized cone is a surface swept out along an axis by a moving cross section of constant shape. The contour defined by a generalized cone can vary in size along its axis. By applying the geometry of generalized cones, certain inferences can be drawn about properties of an object surface based on analysis of the occluding contour. For example, it is possible to infer a surface's **orientation in space relative to the viewer.**

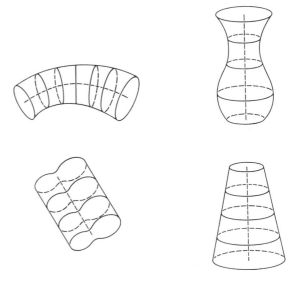

FIGURE 8.40. Examples of generalized cones. See text. Adapted from D. Marr, *Vision*, © 1982, by permission of W.H. Freeman and Company.

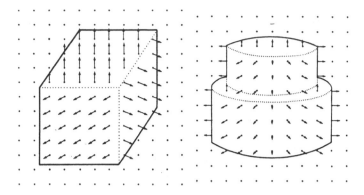

Marr referred to the stage of form vision processing where inferences have been drawn about orientations of visible surfaces in the scene relative to the point of view of the observer as a **2½D sketch.** Two examples of descriptions of a three-dimensional (i.e., 3D) shape at the level of the 2½D sketch are illustrated in Figure 8.41.

Descriptions of shape at the level of the 2½D sketch provide information about the orientation of each small patch of the surface. Each arrow represents a unit vector that can be thought of as a needle of fixed length sticking straight out of the surface at that point. The direction and length of the arrows depict the orientation of the surface at that point relative to the observer. The longer the arrow, the farther the surface dips out of the image plane.

At the level of description of the 2½D sketch, inferences about boundaries between surfaces can be based on properties such as abrupt changes in orientation, in addition to the analyses of paired concavities in the occluding contour, as described in the last section. Further segmentation can be done based on an analysis of nonaccidental properties, as described in the next section.

Nonaccidental Properties

The psychologist **Irving Biederman** has recently formulated a set of rules based on a **principle of nonaccidental properties** that allows one to infer the presence of particular forms of generalized cones from properties of pieces of contour in the retinal image. When certain primitive features appear in the two-dimensional retinal image, they are highly unlikely to be **accidental** and much more likely to signify that a specific three-dimensional feature is present in the distal stimulus. Some examples of nonaccidental properties are shown in Figure 8.42.

For example, if collinear points or straight lines are present in the image, form processing can infer that the edge producing that line in the three-dimensional world is also straight. Smooth curved (curvilinear) elements in the image are similarly inferred to arise from smoothly curved features in the three-dimensional world. If the image is symmetrical, form processing can assume that the object projecting that image is also symmetrical. When edges in the image are parallel or coterminate, it can assume that the real-world edges are also parallel or coterminate, respectively. These and the other properties shown in Figure 8.42 are called "nonaccidental" because they would only rarely be produced by accidental alignments of viewpoint and object features.

In many instances, the presence of primitive three-dimensional object structure based on generalized cones can be inferred from an analysis of **nonaccidental** features. An example illustrating nonaccidental differences that differentiate between a brick and a cylinder is illustrated in Figure 8.43.

The description of surfaces at the level of the 2½D sketch as described in the previous section is still **viewer-centered** and depends on the vantage point of the viewer. However, once a description of the surfaces at the level of a 2½D sketch is matched to a primitive three-dimensional object structure, form perception can be restated in an **object-centered** coordinate system. This makes it possible to describe a "brick" or a "cylinder" in a coordinate system that is independent of the viewer. Biederman has proposed a scheme that allows descriptions at the level of the 2½D sketch to be converted into object-centered descriptions of primitive three-dimensional structures called **geons.**

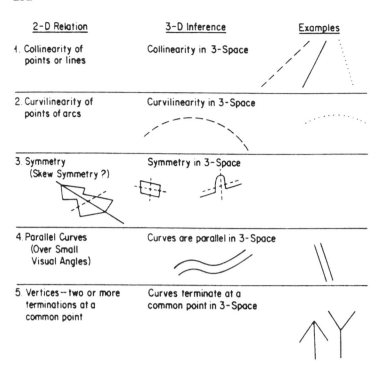

FIGURE 8.42. Examples of nonaccidental properties of retinal images. Adapted from I. Biederman, Human image understanding: Recent research and a theory. *Comput. Vision Graphics and Image Process.* 32:29–73, © 1985, by permission of Academic Press.

Geons and Three-Dimensional Objects

The last stage of form processing considered here is a three-dimensional model representation that describes objects and their shapes in an object-centered coordinate system. Primitives at this last stage are called geons. They provide **primitive volumetric descriptions** (dealing with the volume of space occupied by a shape) of objects that are based on surface characteristics specified by generalized cones. These descriptions may begin to emerge during biological form processing as early

as V4, where some receptive field properties appear to be tuned to respond to geons.

Biederman describes 36 geons that are capable of being combined in various ways to model millions of ordinary objects. Figure 8.44 illustrates five geons along the top row.

The second row illustrates some three-dimensional objects that can be modeled by combinations of only two geons and the third row one that can be modeled by combinations of only three geons. Even relatively complex objects can be constructed from only a small number of geons. For

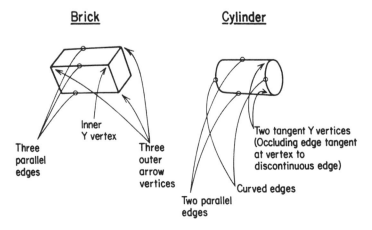

FIGURE 8.43. A brick and a cylinder can be differentiated based on these nonaccidental features in their retinal images. Reproduced from I. Biederman, Recognition-by-components: A theory of human image understanding. *Psychol. Rev.* 94: 115–147, © 1987, by permission of the American Psychological Association.

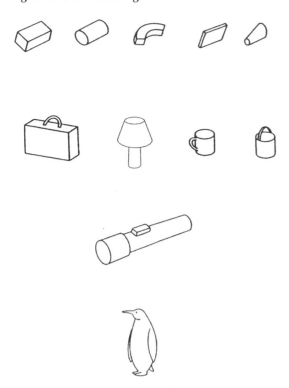

example, the penguin in the bottom row requires only nine geons.

More complex three-dimensional models can also be built up in the form of nested hierarchies, as illustrated in Figure 8.45.

Approaching object perception as the process of building a three-dimensional descriptive model requires a final step in which the model that is built is matched to some **template stored in a library.** For example, presumably, once the model shown in Figure 8.45 was finished, the observer would compare that model to templates of shapes stored in a library and recognize a match between the model shape and a stored "human object shape." Issues of visual recognition memory go beyond the scope of this book. However, the very fact that 36 geons can be combined to form models for millions of three-dimensional object shapes makes this last step a nontrivial issue. No machine vision system yet built can match a model shape against millions of templates in real time. The types of machine vision systems that have had some success have limited the domain of objects to be recognized to some small set. An example of a domain where this approach has had some success is illustrated in Figure 8.46.

This machine vision system tries to recognize only a limited range of objects, automobiles on the freeway, but can do so with reasonable success in real time. A captured image of an object on the freeway is compared to stored templates of an automobile. Each stored template is displaced along some number of degrees of freedom and evaluated

FIGURE 8.44. Examples of geons (top row) and objects whose descriptions can be built up from combinations of only two geons (second row) or three geons (third row). As illustrated in the bottom row, even complex objects can be built up from as few as nine primitive geons. Adapted from I. Biederman, Human image understanding: Recent research and a theory. *Comput. Vision Graphics and Image Process.* 32:29–73, © 1985, by permission of Academic Press.

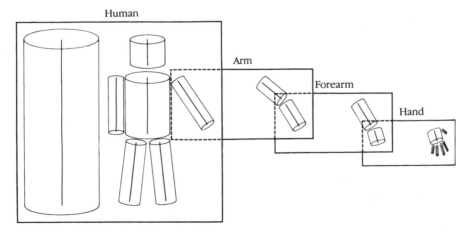

FIGURE 8.45. This figure illustrates how a three-dimensional model description of a human can be described at different levels of detail using a nested hierarchy of generalized cones. Reproduced from D. Marr and H.K. Nishihara, Representation and recognition of the spatial organization of three-dimensional shapes. *Proc. R. Soc. Lond. B.* 200:269–294, © 1978, by permission of the Royal Society.

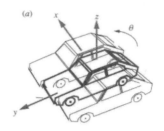
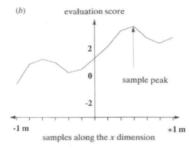

Initial Fit   Best Fit

FIGURE 8.46. An example of machine form vision applied to the limited domain of recognizing automobiles based on their images. Adapted from G.D. Sullivan, Visual interpre-tation of known objects in constrained scenes. *Phil. Trans. R. Soc. Lond. B.* 337:361–370, © 1992, by permission of the Royal Society.

for "goodness of fit" at each point along each dimension. The top left panel illustrates three dimensions for one template, and the right panel provides an evaluation of goodness of fit scores along a single dimension. The matching process in real time is illustrated in the bottom panels.

The approach to form perception emphasized in this chapter is based on an initial detection of luminance edges and the use of properties of those edges to construct contours until finally a percept is formed about three-dimensional object shapes. However, this discussion has ignored other strategies for accomplishing form perception. There are strategies that ignore the initial step of looking for luminance edges. For example, such a scheme may infer surface shape from motion of texture elements in the absence of any luminance contours that define surfaces. This is referred to as percep-tion of **structure-from-motion** and will be dis-cussed in Chapter 10. Another example is perception of surface shape defined based only on gradual changes in luminance rather than on lumi-nance edges. This is referred to as perception of **shape-from-shading;** Chapter 6 described a neural network algorithm that can recover shape-from-shading without using a strategy based on an

explicit analysis of edge information as described in this chapter. Information that bears on three-dimensional properties is also derived from binocular information that will be described in Chapter 9.

It is simply not possible in a single chapter to cover all of the theoretical approaches that have been and are currently applied to the topic of form perception. The approach emphasized here was chosen because of the ease with which its early processing stages can be mapped onto known receptive field properties in the geniculostriate pathway. However, it is likely that many complex coding and processing schemes are implemented in parallel in biological form vision processing. The approach described here should be taken as illus-trative of those strategies.

What and How

In this chapter form perception processing has been treated as though it were strictly part of What-perception. Obviously, that is too simplistic. Consider the perceptual processing that forms a percept of "coffee cup." Some of the perceptual knowledge in that percept is in a form suitable for describing explicit facts. For example, an observer might be able to report that the cup has a round shape. However, other knowledge about the coffee cup is in an implicit form appropriate for guiding actions. For example, an observer might reach out and pick up the cup, thus demonstrating, implicitly, knowledge about where the handle is located. Issues of How-perception will be deferred until Chapter 10.

6 Perception of Faces

Biological form perception is general purpose in the sense that it is able to recover shapes of arbitrary objects. However, certain classes of biologically relevant shapes are processed in a special way. A prime example is perception of faces. Human faces can be described generically with the same kinds of terms used for arbitrary object shapes. For example, a face is simply an ovoid shape containing internal contours defining the eyes, nose, and mouth. However, there is abundant evidence that information about faces is processed in special ways by the human brain.

A simple manipulation such as turning stimuli upside down or displaying them with contrast reversal produces a much more dramatic impairment of recognition of faces than a similar manipulation does for arbitrary shapes. In addition, subtle changes in pictures of faces, such as in the amount of bilateral symmetry, alter not only the low-level stimulus feature that was altered but also complex qualities such as perceived beauty. Humans are exquisitely sensitive to subtle changes in the shape of a face, and this enables them to engage in highly complex social behavior. The state of mind of other individuals is inferred in real time based on slight changes in curvature of the mouth and direction of gaze. It does not seem unreasonable to extrapolate from what this chapter says about form processing to imagine, in principle, designing a machine that might be able to discriminate among some number of arbitrary shapes, such as a brick or a cylinder. However, it is hard even to fathom the complexities of form processing that would be required to build a machine that could carry out the kinds of form vision that human beings use effortlessly to engage in complex activities such as visual social communication.

Form vision processing of biologically relevant stimuli such as faces appears to take place along the What-perception stream in portions of the inferotemporal cortex. For example, the neurophysiologist Charles **Gross** and colleagues described neurons in the inferotemporal cortex that respond strongly to complex shapes, including some that respond preferentially to faces.

Perhaps the neuron generating the percept of grandmother discussed in previous chapters lurks somewhere within the What-perception stream. But we now know that it will not be sufficient for this grandmother cell to generate a percept of general grandmother shape. The percept must convey many complex facts to the observer in real time, such as that it is *my* grandmother, and that she is glad to see me but is also irritated with me, but not too much, and also that her attention is not fully on me but simultaneously directed at the television set, and so forth.

Summary

The topic of form vision is concerned with what kinds of information observers can extract from the environment about shapes of objects and with how this information is processed. Psychophysical studies document how well human observers can accomplish form vision. Neuroscience studies conducted with animals have been concerned primarily with causal mechanisms involved in forward flow of information through the brain. Machine vision has had to be more concerned with the inverse problem of inferring shapes of objects in the world based on information available at various stages of processing. All of these approaches are confronted with solving similar kinds of computational issues.

Processing involves a combination of bottom-up and top-down influences that interact with one another due to recurrent information flow. Information is represented in various forms during processing, but the representation of visible surfaces appears to have special importance.

Initial processing in biological form perception involves low-pass filtering and sampling of the image. These early stages of processing place constraints on some aspects of form vision, such as Snellen acuity, and solve other potential problems, such as aliasing. Somewhat surprisingly, low-pass filtering facilitates rather than hinders hyperacuity tasks.

A second stage of processing involves band-pass filtering into four to seven parallel processing channels that cover a range of spatial scales. This information is carried along the neural P- and M-streams to cortical area V1. Processing within individual hypercolumns in V1 uses zero crossings to extract the locations of local oriented contours. Interactions between hypercolumns allow extraction of information in the form of basic primitives such as the locations of extended pieces of contour that make up what Marr called the "primal sketch."

Higher-order processing, beginning in V1 and extending into the extrastriate interblob stream, is involved in extracting information about surfaces of the type Marr described as the "$2\frac{1}{2}$D sketch." From this sketch, nonaccidental properties of images can be detected and matched to three-dimensional representations that can be related to generalized cones called "geons." Ultimately, form perception processing continues primarily along the What-perception stream of neural processing, although information about shape is also provided to the How-perception stream to allow observers to interact with objects having various three-dimensional shapes.

Certain biologically relevant shapes, such as faces, are processed in a special manner by biological observers, enabling them to engage in elaborate social interactions based on detecting complex, subtle changes in face shape in real time.

Selected Reading List

Biederman, I. 1987. Recognition-by-components: A theory of human image understanding. *Psychol. Rev.* 94:115–147.

Blakemore, C., and Campbell, F.W. 1969. On the existence of neurons in the human visual system selectively responsive to the orientation and size of retinal images. *J. Physiol.* 203:237–260.

Campbell, F.W., and Green, D.G. 1965. Optical and retinal factors affecting visual resolution. *J. Physiol.* 181:576–593.

Campbell, F.W., and Robson, J.G. 1968. Application of Fourier analysis to the visibility of gratings. *J. Physiol.* 197:551–566.

Gallant, J.L., Conner, C.E., Rakshit, S., Lewis, J.W., and Van Essen, D.C. 1996. Neural responses to polar, hyperbolic, and Cartesian gratings in area V4 of the macaque monkey. *J. Neurophysiol.* 76:2718–2739.

Graham, N. 1979. Does the brain perform a Fourier analysis of the visual scene? *Trends Neurosci.* 2:207–208.

Gross, C.G., Rodman, H.R., Gochin, P.M., and Colombo, M.W. 1993. Inferior temporal cortex as a pattern recognition device. In *Computational Learning and Cognition: Proceedings of the Third NEG Research Symposium*, ed. E. Baum, pp. 44–73. Philadelpia: SIAM Press.

Grossberg, S. 1994. 3-D vision and figure-ground separation by visual cortex. *Percep. Psychophys.* 55:48–121.

He, Z.J., and Nakayama, K. 1994. Perceiving textures: Beyond filtering. *Vision Res.* 34:151–162.

Kanizsa, G. 1979. *Organization in Vision: Essays on Gestalt Perception.* New York: Praeger.

Kobatake, E., and Tanaka, K. 1994. Neuronal selectivities to complex object features in the ventral visual pathway of the macaque cerebral cortex. *J. Neurophys.* 71:856–867.

Koenderink, J.J. 1984. What does the occluding contour tell us about solid shape? *Perception* 13:321–330.

Lamme, V.A.F. 1995. The neurophysiology of figure-ground segregation in primary visual cortex. *J. Neurosci.* 15:1605–1615.

Marr, D. 1982. *Vision: A Computational Investigation into the Human Representation and Processing of Visual Information.* San Francisco: W.H. Freeman.

Merigan, W.H. 1996. Basic visual capacities and shape discrimination after lesions of the extrastriate area V4 in macaques. *Visual Neurosci.* 13:51–60.

Schiller, P.H. 1993. The effects of V4 and middle temporal (MT) area lesions on visual performance in the rhesus monkey. *Visual Neurosci.* 10:717–746.

Wilson, H.R., and Wilkinson, F. 1998. Detection of global structure in Glass patterns: implications for form vision. *Vision Res.* 38:2933–2947.

9

Perception of Three-Dimensional Space: How Do We Use Information Derived from One or Both Eyes to Perceive the Spatial Layout of Our Surroundings?

Questions

After reading Chapter 9, you should be able to answer the following questions:

1. Elaborate on the two fundamental problems associated with three-dimensional space perception that have concerned philosophers and scientists for hundreds of years.

2. What is single binocular vision and how is it achieved?
3. Which portions of the visual field are seen by each eye and by each side of the brain?
4. Which brain areas and streams are involved in processing information from the two eyes to achieve stereovision?
5. Describe how binocular fixation on any point in three-dimensional space can be accomplished by combining two types of eye movement.
6. Explain the concept of a local sign for visual direction.
7. Compare and contrast the binocular perceptual qualities of diplopia, confusion, suppression, and rivalry.
8. Compare and contrast motor fusion, sensory fusion on the horopter, and sensory fusion within Panum's area.
9. How can projection be used to specify depth based on disparity?
10. What is the correspondence problem regarding stereopsis?
11. Compare and contrast cooperative models and coarse-to-fine models of stereopsis.
12. Compare and contrast position-shift and phase-shift models of how monocular receptive fields can be combined to produce disparity-sensitive binocular vision.
13. What role, if any, does vertical disparity play in depth perception?
14. Describe what it means to assert that stereovision cannot be based exclusively on either space or time but must be based on space-time.
15. Describe some of the environmental information about three-dimensional space that can be derived based on input from only one eye.
16. Speculate about the functional significance of stereopsis that might have led to its evolution in primates.

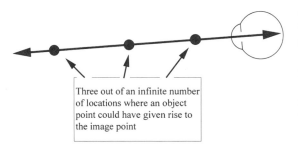

Three out of an infinite number of locations where an object point could have given rise to the image point

FIGURE 9.1. The location of an image point on the retina can specify the direction of the object point relative to the eye but not its distance.

1 Classic Problems Regarding Stereovision

Stereovision is the capacity of an organism to perceive a three-dimensional visual world. There are two classic problems regarding stereovision that have been discussed by philosophers and scientists for hundreds of years: perception of three dimensions based on a flat image, and single binocular vision.

How Is Three-Dimensional Space Perceived Based on a Flat Retinal Image?

The task of inferring the location of an object point in three-dimensional space based the location of its image point is an example of an ill-posed problem, as illustrated in Figure 9.1.

Two kinds of solutions have been proposed to explain how visual processing recovers three-dimensional information from a flat image. The first is that the retinal image consists not of isolated points but of **structured light,** whose structure encodes many three-dimensional aspects of the environment. A good example of this was discussed in the section on geons in Chapter 8.

The second solution is that our visual systems carry out complicated analyses of differences between information derived from the left and right eyes. This includes a process called **stereopsis** that provides information about depth. However, these binocular processes, which help solve the first classic problem, give rise to other issues that form the second classic problem.

How Can Single Binocular Vision Be Accounted For?

Single binocular vision refers to the perception of only one world although the information about the world is conveyed to the brain from two separate retinal images. The term **binocular fusion,** or simply **fusion,** is used to refer to the processes involved in achieving single binocular vision.

Fusion is exceedingly complex in terms of both its computational requirements and the required neural circuitry. Furthermore, its operation depends upon an exquisite coordination of binocular motor control and binocular sensory processing. These two processes are so intertwined that it is impossible to understand either without reference to the other (Figure 9.2).

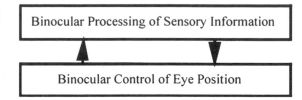

FIGURE 9.2. Processing associated with binocular sensory information coming from the eyes and processing associated with control of binocular movements of the eyes are highly interrelated.

Humans are usually unaware of the complex processes that are involved in fusion. The world just "looks single" without any conscious effort on the observer's part. In fact, it takes special effort or unusual conditions before one even becomes aware of the fact that one's percept is actually a fusion of two separate monocular inputs. Similarly, the coordination of the movements and positions of the two eyes seems so effortless as one looks around the world that humans seldom give any thought to the complexities that are involved in achieving this coordinated binocular control.

These complexities become apparent under two conditions. One is a breakdown of binocular coordination. This happens permanently in the clinical condition of **strabismus,** in which the patient is cross-eyed or wall-eyed, and can happen to anyone temporarily when under the influence of drugs such as alcohol. The second condition occurs when roboticists try to build machines that can achieve binocular fusion. The most complicated robots built to date are unable to achieve anything close to binocular fusion and stereopsis in real time in complex environments. For these reasons, ophthalmologists, roboticists, and, increasingly in recent years, computational neuroscientists have developed a deep appreciation of the complexities of the binocular processes that participate in stereovision.

This chapter first describes the binocular organization of the visual system. Second, it deals with issues of how the brain uses a combination of motor and sensory processes to fuse the information provided by the left and right eyes to achieve single binocular vision. Third, it deals with how the brain takes advantage of disparities between the images from the two eyes to achieve stereoscopic depth perception. Fourth, it describes some of the rich information about the three-dimensional structure of the environment that can be extracted equally well with one or two eyes. Finally, the chapter addresses the issue of why complex binocular processing might have evolved in primates.

2 Basic Concepts of Binocular Vision

Visual Fields

In many species, the eyes are positioned laterally in the skull and there is little or no overlap in the **visual fields** that can be seen by the two separate eyes. However, in primates there is a large area of overlap between the visual fields seen by the two eyes. Information about various portions of the visual field is processed by specific portions of the left or right eye, as illustrated in Figure 9.3.

The hypothetical disembodied observer's gaze is fixated at the girl's nose, designated as location 5. The left half of the visual field is designated by the numbers 1–5 and the right half by 5–9. Both eyes see portions of both halves of the visual field. The left eye sees positions 1–8 and the right eye positions 2–9. Locations 2–8 constitute the **binocular portion of the visual field,** which can be seen simultaneously by both eyes. The portion of the visual field between positions 1 and 2 constitutes the left **monocular crescent,** which can be seen only by the left eye; similarly, the portion between 8 and 9 delineates the right monocular crescent.

The **left visual field** maps onto the **right half of each retina,** the nasal retina of the left eye and the temporal retina of the right eye, and similarly the **right visual field** maps onto the **left half of each retina.** At the optic chiasm, the ganglion cell axons from each nasal retina cross to the opposite side of the brain, while those from the temporal retina remain ipsilateral. Consequently, the **left brain** processes information about the **right visual field** as seen by both eyes and the **right brain** information about the **left visual field,** again as seen by both eyes. These relationships are shown for the left eye in Figure 9.4.

The **visual field for a single eye** can be divided into **three segments.** A **binocular nasal segment** of the field projects to the temporal retina. Information about this portion of the visual field travels up uncrossed neural pathways from the eye to the central visual processing regions of the brain. A **binocular temporal segment** of the visual field projects to the nasal retina and travels up crossed neural pathways to central processing regions. Finally, the **monocular temporal segment** of the visual field also projects to the nasal retina and along crossed neural pathways. The primary concern of this chapter is with the binocular segments from each eye.

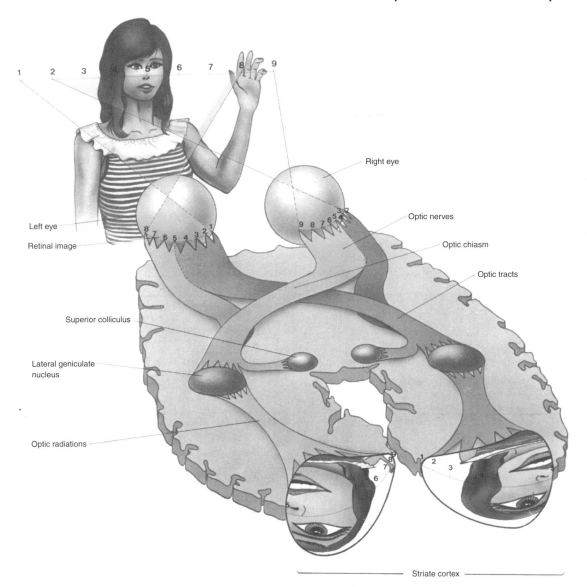

FIGURE 9.3. As described in more detail in Chapter 5, there is a mapping of positions in the visual field onto the two eyes and subsequently onto brain structures. Adapted from J.P. Frisby, *Seeing: Illusion, Brain and Mind.* Oxford: Oxford University Press, 1979, by permission of J.P. Frisby.

The total visual field as seen by one eye extends both directions from straight ahead; from about 95° on the temporal side to about 65° on the nasal. The nasal limitation corresponds roughly to where the field is blocked by the nose (see Box 9.1).

The visual field is typically measured in humans with a procedure called **perimetry.** The observer is instructed to maintain fixation on a small target at a central location on a large screen. Then small stimuli, typically spots of light, are presented at various positions on the screen. The observer is asked to report, without moving the eye from the central fixation point, whether each eccentric stimulus was seen. Similar procedures have been used to measure the visual fields of monkeys, as illustrated in Figure 9.5.

A monkey is sitting in a plexiglass chair facing a circular white screen. In the top panel (A) a piece of food, a raisin, is introduced in front of the screen at a central location. The monkey is trained to fixate on the food stimulus for 1–2 seconds and is then given the food to eat as a reward. On some trials,

Box 9.1
Do Our Noses Spite Our Visual Fields?

The nasal extent of the visual field is ordinarily blocked by the nose. However, the limitation of the visual field is not due to shape of the nose. Nature, in another example of its wisdom obtained over evolutionary time, declined to provide us with neural machinery for processing portions of the visual fields that would be used mostly for looking at the side of our noses.

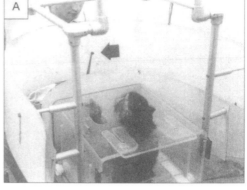

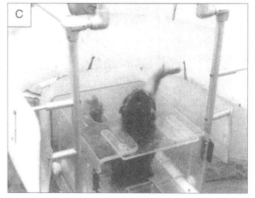

as shown in the second panel (B) a second food stimulus is introduced at an eccentric location immediately after the monkey takes up fixation of the central stimulus. If the monkey looks toward the eccentric location, it is immediately given the food stimulus to eat, as illustrated in the third panel (C) without having to wait the 1–2 seconds for the central reward. Blank trials are also included, during which only the central stimulus is presented. If the monkey moves its eyes during a blank trial, no food stimulus is provided and the trial is scored as a false alarm.

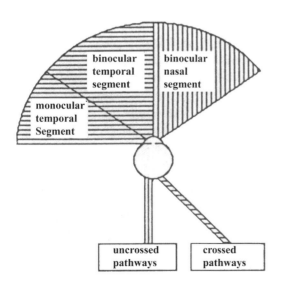

FIGURE 9.4. Monocular and binocular segments of the visual field project to contralateral brain via crossed pathways and to ipsilateral brain via uncrossed pathways, as shown here for a left eye.

FIGURE 9.5. The procedure used to measure the visual fields of a monkey. See text for details. Reproduced from J.R. Wilson, et al, Visual fields of monocularly deprived macaque monkeys. *Behav. Brain Res.* **33:13–22, © 1989, by permission of Elsevier Science.**

Binocular Neural Pathways

Two major neural systems are of concern in trying to understand binocular function (Figure 9.6).

Sensory processing of binocular information that is involved in stereovision takes place along the geniculostriate pathway that passes from the eyes

Binocular Sensory Processing System

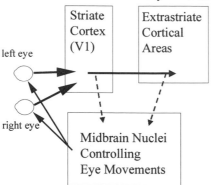

Binocular Motor Processing System

FIGURE 9.6. Central neural pathways responsible for binocular vision include geniculostriate and extrastriate cortical pathways involved in sensory processing of binocular information and also midbrain nuclei involved in binocular oculomotor control.

through striate cortex and extrastriate cortical pathways. Motor processing responsible for binocular control of eye movements takes place in nuclei in the midbrain. The midbrain nuclei are influenced by signals generated by sensory binocular processing in cortical areas. In turn, binocular motor control of the eyes influences the kinds of information that travel up the sensory processing pathways.

Sensory Processing of Binocular Information

Information traveling up the geniculostriate pathways from the eyes terminates in the dLGN, as described in Chapter 5. The relationships among visual fields, eyes, and layers in the dLGN are summarized in Figure 9.7.

The figure shows the inputs to the six layers of the right dLGN. The inputs to the left dLGN are a mirror image of this. The topographic organization of the dLGN layers is such that information from corresponding regions of the visual field as seen by each eye is kept in register. However, no binocular processing takes place in the dLGN, because neurons carrying input from the left and right eyes are segregated into different layers.

The projection from the dLGN to layer 4 of primary visual cortex is illustrated in Figure 9.8.

The segregation of left and right eye inputs is maintained at this first stage of processing in V1.

Inputs from the left and right eyes are distributed into separate anatomical columns where the dLGN fibers terminate in layer 4. Binocular neurons that process inputs from both the left and the right eyes are first encountered in the superficial and deep layers that lie above and below layer 4.

Some binocular neurons respond to input from the left eye, the right eye, or both eyes. Others are exclusively binocular, responding only to simultaneous stimulation of both eyes, as illustrated in Figure 9.9.

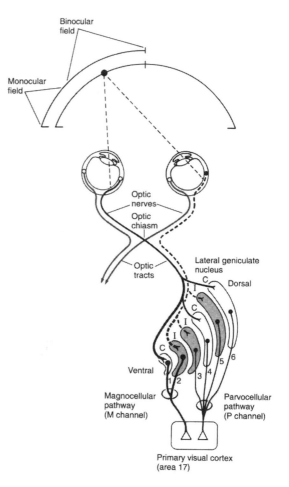

FIGURE 9.7. Information derived from the left and right eyes is processed in different layers of the dLGN. Pathways used to transmit information from the left binocular field to the right brain are illustrated. Crossed (c) pathways from the left eye terminate in dLGN layers 1, 4, and 6. Uncrossed pathways from the ipsilateral eye (I) terminate in layers 2, 3, and 5. Adapted from E.R. Kandel, et al, *Principles of Neural Science*, Third Edition. Norwalk, CT: Appleton & Lange, © 1991, by permission of McGraw-Hill Companies.

what is the dLGN

[handwritten: P and M streams??]

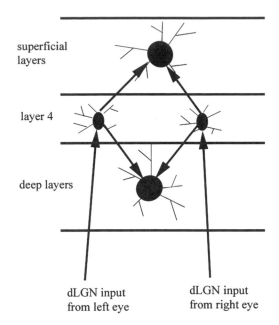

FIGURE 9.8. Information derived from the left and right eyes continues to be segregated when it is transmitted from the dLGN to layer 4 of cortical area V1. Only when information is projected from layer 4 to the superficial and deep layers of V1 does processing begin to take place in binocular cells that receive inputs from both eyes.

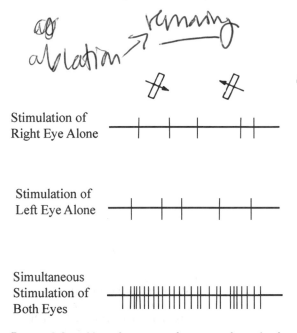

[handwritten: all ablation → remaining]

FIGURE 9.9. A binocular neuron that responds to simultaneous inputs from the left eye and the right eye but not to input from either eye alone. Adapted from results originally published by D.H. Hubel and T.N. Wiesel, Stereoscopic vision in macaque monkey. *Nature* 225:41–42, © 1970, by permission of Macmillan Magazines Ltd.

The top two panels illustrate that only a slow firing rate of spikes is produced for this neuron when a stimulus is swept in both directions across the receptive field of only one eye. However, the bottom panel shows that there is an increase in firing rate when the stimuli are swept across the receptive fields of the left and right eyes simultaneously.

The flow of stereovision information along the extrastriate streams of processing is summarized in Figure 9.10.

Both the P- and the M-streams participate in specialized stereovision processing. Information derived from the M-stream is transmitted from layer 4B of V1 to the thick stripes of V2 and then on to V3, V5, and other portions of the dorsal neural stream involved in the How-perception. Information derived from the P-stream travels from the interblob regions of V1 to the pale interstripes of V2 and then on to V4 and other inferotemporal areas in the ventral neural stream involved in What-perception. There is some intermixing of stereovision information within the extrastriate pathways. However, some stereovision functions are localized within specific cortical areas. For example, ablation of V5 in monkeys has no effect on stereovision based on contours but impairs performance when special stimuli called random dot stereograms, described below in the section "The Correspondence Problem," are used as stimuli.

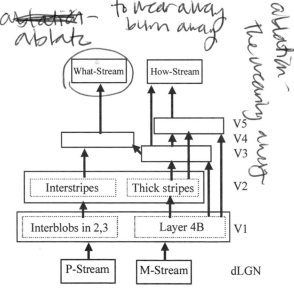

[handwritten: ablation — ablate; to wear away, burn away; ablation - the wearing away]

FIGURE 9.10. Flow of stereovision information through extrastriate processing streams, and eventually into the neural streams associated with What-perception and How-perception.

[handwritten: what stream + how stream run parallel.]

Motor Control of Binocular Eye Movements

There are two major types of binocular eye movements. Some examples of the first type, called **versions,** are illustrated in Figure 9.11.

Versions are **conjunctive** binocular movements, in which the eyes move the same distance in the same direction. These movements can be under the direction of the smooth pursuit or the fast saccadic eye movement neural control systems discussed in Chapter 4. Binocular coordination is accomplished simply by having the same neural signal directed to muscles with opposing actions in the two eyes. For example, to direct a binocular eye movement from left to right, the brain sends the identical signal to the lateral rectus muscle of the right eye and to the medial rectus muscle of the left eye, as illustrated in Figure 9.12.

In this example, a signal originates in the **abducens nucleus** and travels along the **abducens nerve** to the lateral rectus muscle of the right eye. The same signal is sent along the **medial longitudinal fasciculus (MLF)** fiber bundle to the oculomotor nucleus on the opposite side of the brain. The oculomotor nucleus sends the signal along the oculomotor nerve to the medial rectus muscle of the left eye. An understanding of this circuitry allows neurologists and ophthalmologists to diagnose lesions to these regions of the brain based on abnormalities in binocular eye movements, as illustrated in Figure 9.13.

A lesion to the abducens nerve, at the location labeled #1 in Figure 9.12, allows the rightward signal that is generated in the abducens nucleus to travel normally to the left eye but not to the right eye. Damage to the abducens nucleus (#2) disrupts signals to both eyes. Damage to the MLF (#3) disrupts the signal to the left eye only.

Disjunctive binocular movements, called **vergences,** cause the eyes to move in opposite directions. This text will concern itself only with horizontal forms of disjunctive movements, called **convergence** and **divergence,** illustrated in Figure 9.14.

The **vergence control system** coordinates binocular movements of the eyes when the gaze is changed between far and near locations. It is one of the four eye movement control systems described in Chapter 4. It operates by directing signals of the same size to the same muscles in each eye. For example, a convergence movement is produced by sending equal signals to the medial rectus muscles in each eye Figure 9.15 summarizes the differences between the signals generated by the brain for version and vergence eye movements.

Combinations of vergence and version movements cause the location where the lines of sight from the two eyes cross to move to different locations in three-dimensional space, as illustrated in Figure 9.16.

A conceptual geometrical circle, called a **Vieth-Mueller (VM) circle,** can be specified by three points: the center of rotation of each eye and the location where the lines of sight from the two eyes cross. A pure version movement moves binocular fixation from one point to another along a VM circle. For example, a version could be used to move the eyes from a to b to c, m to n to o, or x to y to z. A pure vergence movement moves along a straight line radiating from a midpoint between the centers of rotation of the two eyes, called the **cyclopian center.** For example, a vergence movement could be used to move binocular fixation from a to m to x, b to n to y, or c to o to z. The location where the lines of sight cross following each vergence movement defines a new VM circle.

A combination of a pure version and a pure vergence movement can be used to position the lines of sight at any location in three-dimensional space. This basic fact was emphasized by **Hering** over 100 years ago when he noted that the binocular system appears to achieve its objectives with a small number of control mechanisms "compared to the inexhaustible variety of possible eye movements." The principles governing these simplifying

Rightward

Leftward

Upward

Downward

how the brain gets the eye pointing to where it needs to.

FIGURE 9.11. Binocular eye movements called "versions," in which both eyes move in the same direction. Adapted from G.K. Von Noorden, *Atlas of Strabismus,* Fourth Edition, © 1983, by permission of C.V. Mosby Company.

FIGURE 9.12. The top panel shows the muscles and their cranial nerve innervations that are responsible for controlling eye movements. These are described in more detail in Chapter 4. The eyes are aimed to the left. The bottom panel shows neural circuitry that would be used to initiate a binocular movement to the right. The positions of the eyes following activation of this neural circuitry in a normal individual are illustrated in the top panel of Figure 9.13. Adapted from J.H. Martin, *Neuroanatomy*, Second Edition. Stamford, CT: Appleton & Lange, © 1996, by permission of McGraw-Hill Companies.

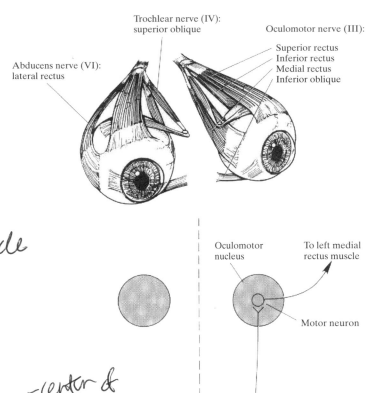

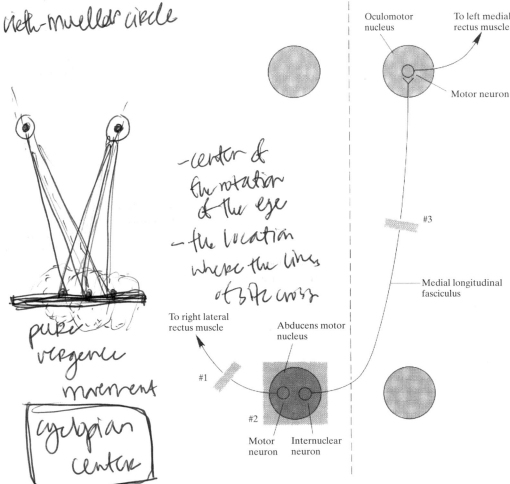

vieth–mueller circle

–center of the rotation of the eye

–the location where the lines of sight cross

pupil

vergence

movement

cyclopian center

binocular control mechanisms are often referred to in modern times as **Hering's laws.**

A diagram from a book published by Hering is shown in Figure 9.17.

The eyes are originally pointing straight ahead, and the lines of sight from the left eye (L) and the right eye (R) are parallel to one another. The figure illustrates how the motor system might direct two simple eye movements that result in crossing of the lines of sight at location F. A vergence movement is programmed that, by itself, would move the line of sight for the left eye to l' and that of the right eye

Attempt to gaze right

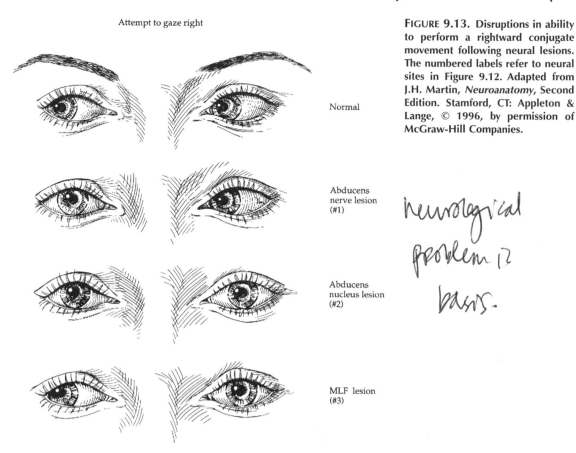

Normal

Abducens
nerve lesion
(#1)

Abducens
nucleus lesion
(#2)

MLF lesion
(#3)

*neurological
problem i?
basis.*

FIGURE 9.13. Disruptions in ability to perform a rightward conjugate movement following neural lesions. The numbered labels refer to neural sites in Figure 9.12. Adapted from J.H. Martin, *Neuroanatomy*, Second Edition. Stamford, CT: Appleton & Lange, © 1996, by permission of McGraw-Hill Companies.

to r', causing the lines of sight to cross at F'. However, simultaneously, a version movement is programmed that, by itself, would have the effect of rotating the line of sight for the left eye to l" and that of the right eye to r". The combination of these two movements has the effect of causing the lines of sight to cross at F.

Local Sign for Visual Direction

A basic concept derived from the classic perception and clinical ophthalmology literatures is that a percept always carries with it certain primitive attributes called **local signs.** One of the local signs is **visual direction,** a psychological impression that a percept is localized as existing in some direction with respect to oneself. This is the direction in which one would point if asked where the object one is seeing is located.

Each visual direction is associated with a **unique anatomical position on the retina.** To understand how this works, imagine that you are sitting in a completely dark room with only one eye open. A local anesthetic has been given so that you cannot feel where your head or eyes are pointing. Now imagine that a mad scientist has grasped your eye with a forceps and can aim it in various directions.

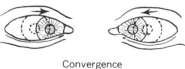

Convergence   Divergence

FIGURE 9.14. Binocular eye movements called "vergences," in which the eyes move in opposite directions. Adapted from G.K. Von Noorden, *Atlas of Strabismus*, Fourth Edition, © 1983, by permission of C. V. Mosby Company.

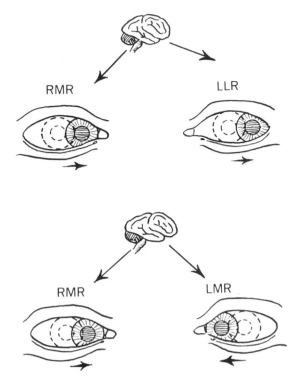

FIGURE 9.15. How the neural control of binocular eye movements differs for version and vergence movements. In versions, illustrated in the top panel, identical signals are sent to muscles that act in the same direction: In this example of a left version, signals are sent to the medial rectus in the right eye (RMR) and the lateral rectus in the left eye (LLR). Vergence movements, illustrated in the bottom panel, are accomplished by directing the same neural signal to muscles that have opposing directions of action in the two eyes: In this example of a convergence movement, signals are sent to the medial rectus of the right eye (RMR) and the medial rectus of the left eye (LMR). Adapted from G.K. Von Noorden, *Atlas of Strabismus*, Fourth Edition, © 1983, by permission of C.V. Mosby Company.

Suddenly a flashbulb goes off somewhere in the room, and you are asked to point in the direction where you saw the flash. Can you guess what the relationship will be between where the flash was located in the room and the direction in which you point? It turns out that the factor determining the direction in which you would point under these conditions is the anatomical location on your retina where the image of the flash falls as illustrated in Figure 9.18.

If the flash of light falls on the **fovea,** you will perceive the light as coming from **straight ahead.** Thus, even if the flashbulb was actually located off to the side, above, or below your head, you will never-

theless point straight ahead whenever the image of the flash falls on the fovea. Similarly, if the spot of light falls **above the fovea** it will be seen as **down,** if **below the fovea** as **up,** if to the **left** as **right,** and if to the **right** as **left.**

The mapping between visual directions and locations on the retina of a *single* eye is **one to one.** In other words, there is a unique anatomical location on the retina for each psychological visual direction and a unique visual direction for each anatomical location. However, humans have two eyes.

Corresponding and Disparate Pairs of Points

For the **binocular visual field,** that is, the portion of the visual field that can be seen simultaneously with both eyes, there is a **two-to-one** relationship between anatomical locations and visual directions. For each visual direction, one anatomical location gives rise to that direction when viewing through the left eye and a second anatomical location gives rise to that same visual direction when viewing through the right eye. These pairs of anatomical positions, one on the left retina and the other on the right retina, that share the same visual direction are called **corresponding points.**

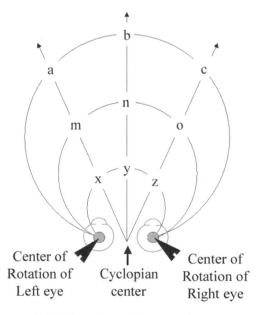

FIGURE 9.16. The effects of pure versions, pure vergences, and combinations of the two kinds of eye movements on the location in space where binocular fixation occurs. See text for description.

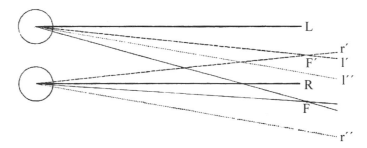

FIGURE 9.17. Drawing originally published by Hering in 1868 illustrating how a change of binocular fixation from one point in three-dimensional space to another can be accomplished with a combination of a pure version and a pure vergence movement. Reproduced from English translation, E. Hering, *The Theory of Binocular Vision*, Translated and edited by B. Bridgeman and L. Stark, © 1977, by permission of Plenum Press.

Every visual direction in the binocular visual field shares a pair of anatomical corresponding points. Figure 9.19 illustrates two pairs of corresponding points.

The fovea in the left eye and the fovea in the right eye are corresponding points because they share the visual direction "straight ahead." In other words, if an observer is viewing an object that is imaged on the fovea of the left eye, she will point straight ahead when asked to indicate its visual direction. If she views the same object imaged on the fovea of the right eye, she will point in the same direction. Similarly, if an object falls onto a specific location on the lower retina of the left eye, the observer will point upward in a certain direction. If an object falls at the corresponding point on the lower retina of the right eye, she will point upward in the same direction.

Keep in mind that these psychological visual directions are not necessarily being caused by light coming from the same direction in three-dimensional space. The mad scientist described in our earlier example might have been using forceps to aim your left eye upward while you viewed the first light flash and to aim your right eye downward while you viewed the second flash. Your arm pointed in the same direction for each flash because their images fell on corresponding points, but the two flashes actually came from different locations in the room.

This example demonstrates that visual directions are relative to the eye. In order for the brain to be able to interpret visual directions relative to the head, it must have information about rotational positions of the eyes within their orbits. In order to interpret direction relative to the trunk, it must

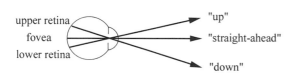

FIGURE 9.18. Each visual direction is associated with a unique anatomical location on the retina. Top panel illustrates an eye viewed from above in which retinal locations to the left of the fovea are associated with the visual direction "right," and to the right of the fovea with the visual direction "left." Bottom panel shows an eye viewed from the side in which retinal locations above fovea signal visual direction "down," and locations below fovea with visual direction "up."

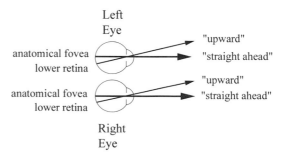

FIGURE 9.19. Corresponding points in the two eyes are pairs of anatomical locations that share the same visual direction. Two examples are illustrated. The anatomical fovea in each eye signals the visual direction "straight ahead." Corresponding points in the lower retina signal the visual direction "upward."

have information about the orientation of the head at the neck, and so on. This underscores the exquisite interplay that is required between motor and sensory processing to interpret information about three-dimensional space. In these examples, the term "straight ahead" really means "straight ahead with respect to where the brain thinks the eyes are pointed." In the mad scientist examples, the brain has no knowledge about where the eyes are pointing. Under these conditions, the brain typically resorts to a default assumption that the eyes are pointing straight ahead with respect to the body.

A variety of environmental conditions will give rise to stimulation of noncorresponding points. Whenever this happens, the binocular percept that is experienced takes on one of the following special qualities: **diplopia, confusion, suppression** or **rivalry,** or some form of **fusion.**

Diplopia

Diplopia is commonly referred to as **double vision.** Consider another example involving the mad scientist, illustrated in Figure 9.20.

A single light is viewed while the mad scientist aims the right eye so the image falls onto its fovea and the left eye downward so the flash falls onto its lower retina. In terms of the information about visual direction coming from the two eyes, your brain is confronted with the following information: The right eye is reporting that there is a flash located straight ahead, and the left eye is simulta-

neously reporting a flash located above. One possible interpretation is that two flashes of light are present. When the brain adopts this interpretation, one experiences diplopia. This is a failure of binocularity, since the condition that existed in the real world was that only one flash was present but the percept indicates two.

A condition that leads to diplopia more frequently than being under the influence of a mad scientist is being under the influence of alcohol. Under these conditions, the brain loses its capacity to maintain the images of the intended object of regard on each fovea simultaneously. The images end up falling onto noncorresponding points, and the interpretation reflected in the percept is that there must be two objects present.

Another condition that can lead to diplopia is strabismus. The view of a strabismic patient with diplopia is illustrated in Figure 9.21.

The bottom panel illustrates the views the patient sees when looking at a scene with only one or the other eye. The top panel illustrates the percept of the strabismic patient when viewing the same scene with both eyes open.

Confusion

Going hand-in-hand with diplopia when viewing in a complex environment where many objects are present is **confusion**, a condition that occurs whenever the fovea of the left eye and the fovea of the right eye are pointing at different objects. The brain becomes confused about what is located straight ahead. Consider the penguin scene in Figure 9.21. The two small open circles indicate the locations in the image that are projected onto each fovea. Each of these two separate locations in space is being signaled to the brain as being straight ahead. Thus, the brain is being subjected to confusion about what portion of the environment is straight ahead as well as to diplopia. Confusion comes about whenever the brain thinks two objects have the same visual direction, while diplopia comes about whenever the brain thinks a single object has two visual directions.

Both diplopia and confusion are very disruptive to visual function, and individuals afflicted with conditions such as strabismus that lead to these abnormal percepts report that the symptoms are extremely onerous. Fortunately, when these conditions are present for an extended period, in most individuals the sensory processing parts of the brain adopt adaptations that serve to eliminate the adverse effects of diplopia and confusion.

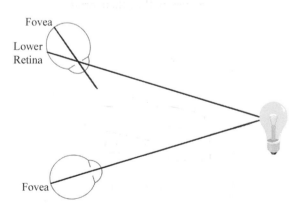

FIGURE 9.20. When the same object is imaged onto disparate points, as in this example where the lightbulb falls on the fovea of the right eye and the lower retina of the left eye, one interpretation that can be made by the visual system is that two objects are present. This is called "diplopia."

FIGURE 9.21. Top panel illustrates the binocular percept formed by a subject with strabismus. Bottom panel shows views of the two individual eyes. Adapted from M.V. Joosse, *et al*, Quantitative visual fields under binocular viewing conditions in primary and consecutive divergent strabismus. Graefe's Arch Clin Exp Ophthalmol., 237:535–545, 1998.

Suppression and Rivalry

Suppression is a sensory adaptation to abnormal binocular stimulation in which the input coming from one eye is simply blocked so that it is not able to contribute to the percept.

In some individuals, one particular eye is always suppressed and the image formed in the dominant eye is the one that is always seen. In Figure 9.22, the percept reflects an "A" viewed by the dominant eye, and the "B" viewed by the suppressed eye is not seen. In other individuals, the suppression alternates over time, eliminating the input from one eye for a while and later the input from the other eye. An individual with this form of **alternating suppression** would report seeing an "A" for some period of time, alternating with a "B" the rest of the time.

Permanent forms of suppression ordinarily develop only in individuals with chronic abnormal binocular stimulation, such as occurs in strabismus. However, temporary forms of suppression can be revealed in individuals with normal visual systems by using special laboratory methods to present different images to the two eyes. This leads to a perceptual phenomenon referred to as **binocular**

rivalry. The classic stimulus for demonstrating rivalry in normal adult humans is presentation of two orthogonal gratings to the two eyes. This perceptual phenomenon was first reported by **Panum** near the turn of the 20th century. The

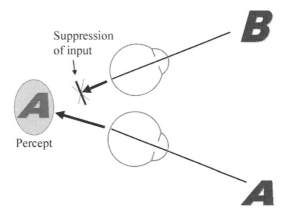

FIGURE 9.22. A viewing condition in which a different object is imaged onto the fovea of each eye and leads to confusion about what is located straight ahead. The binocular system of this hypothetical observer eliminates the confusion by suppressing the input coming from one eye.

FIGURE 9.23. Three types of binocular experiences reported by observers while viewing orthoganal gratings under viewing conditions as demonstrated in text box in Chapter 1. These conditions produce binocular rivalry. See text for additional details. Reproduced from J.M. Wolfe, Resolving perceptual ambiguity. *Nature* 380:587–588, © 1996, by permission of Macmillan Magazines Ltd.

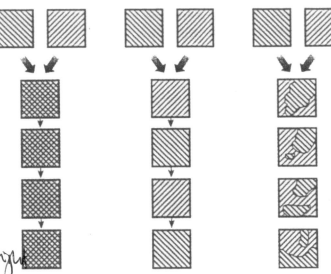

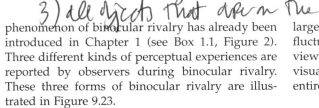

phenomenon of binocular rivalry has already been introduced in Chapter 1 (see Box 1.1, Figure 2). Three different kinds of perceptual experiences are reported by observers during binocular rivalry. These three forms of binocular rivalry are illustrated in Figure 9.23.

When the gratings are flashed on such that they can be seen for only a brief period, no suppression takes place of the images presented to either eye, and the percept is of a fused plaid as shown in the left column. This is presumably because the sensory mechanisms that are used to eliminate diplopia and confusion take some time to operate. When observers view the orthogonal gratings for longer than a momentary flash, suppression of the image in one or both eyes begins to operate. When the gratings subtend a relatively small visual angle, observers usually report a percept that fluctuates over time, as illustrated in the middle column. A grating in one orientation as viewed by one eye is seen for several seconds, and then at unpredictable times the percept fluctuates so that the orthogonal orientation viewed by the opposite eye is seen. When larger orthogonal gratings are viewed, dominance is often patchy, with bits and pieces from the left- and the right-eye views visible at the same time, giving rise to a dynamically changing pattern, as illustrated in the right column. This is most likely because the mechanisms responsible for rivalry can operate semiindependently within each hypercolumn. You can experience these two types of percepts while viewing the gratings in Box 1.1 by changing viewing distance. At near distances the gratings subtend a

large visual angle and most observers report patchy fluctuations of the binocular percept. At greater viewing distances, the gratings subtend a smaller visual angle and the perceived orientation of the entire grating fluctuates.

3 Using Binocular Fusion to Achieve Single Binocular Vision

Motor Fusion

In individuals with normal binocular visual systems, an intended object of regard that forms disparate images leads to eye movements that direct the line of sight from each eye so that the two intersect at the object. An example is illustrated in Figure 9.24.

Initially, the image of the intended object of regard, in this example a light bulb, falls onto disparate points on the two retinas, as shown in the left panel. However, a binocular eye movement rotates the eyes by an amount that moves the image onto the fovea of the left eye and simultaneously onto the fovea of the right eye. The light bulb is now seen as "single," because it is seen by both eyes as being located in the same visual direction in space, straight ahead. This binocular process is called **motor fusion,** and it allows single binocular vision of the object of regard. In other words, the object of regard is perceived as single even though it is being seen separately by the two eyes.

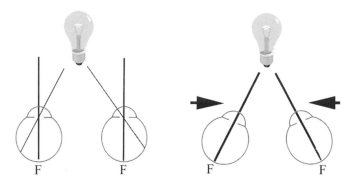

FIGURE 9.24. How motor fusion is achieved. The motor response of converging the eyes causes the lightbulb to be imaged onto the fovea of each eye simultaneously.

Sensory Fusion Based on the Horopter

Once motor fusion has allowed single binocular vision to be achieved for the object of regard, fusion comes "for free" for certain other objects in the field of view. An example is shown in Figure 9.25.

Motor fusion has placed the cross-hair target onto the fovea of each eye, where it is seen as single in the straight-ahead direction. The straight line passing out of the left eye has a visual direction of 10° upward and to the left. Any object in space that happens to fall anywhere along that line will have the visual direction of "10° upward and to the left" for the left eye. Any object in space that happens to fall anywhere along the corresponding line that passes out of the right eye will have the visual direction of "10° upward and to the left" for the right eye. Note that those two lines cross at a particular location in space. If an object happens to be located at that point, it will be seen as "10° upward and to the left" simultaneously by the left and right eyes. Thus, the images for this object are fused in such a way that only one single object is seen.

Similarly, for every location in space where lines projecting from corresponding points in the two eyes cross fusion is achieved automatically if an object happens to be present at that location. The locus in three-dimensional space where all of the pairs of lines having the same visual direction intersect is called the **horopter**. This was first described by **Aguilonius** in the 1600s, although he mistakenly thought its shape was a flat surface in the frontal plane (Figure 9.26).

The theoretical shape of the horopter can be determined from geometry. It is a VM circle, identical for all practical purposes to the ones described for specifying binocular eye movements in Figure 9.16.

However, technically, the VM circle that defines the horopter passes through the anterior nodal point of each eye (Figure 9.27) rather than through

their centers of rotation as defined in Figure 9.16. All objects that fall on the horopter produce images that fall on corresponding points on the two retinas, as illustrated in Figure 9.28.

The observer is looking at point F. This causes the image of point F to fall on the fovea of the left eye (F_l) and that of the right eye (F_r). Thus, point F is fused by motor fusion and seen as straight ahead by both eyes. Points A and B fall on the horopter. They are fused automatically by sensory fusion of corresponding points. Point A falls on corresponding points A_l and A_r, point B on corresponding points B_l and B_r. Similar relationships hold for all points on the horopter.

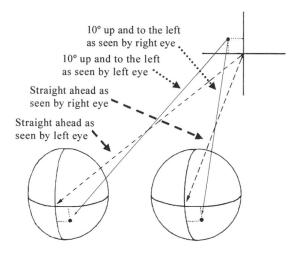

FIGURE 9.25. How sensory fusion works. Motor fusion has caused the center of the cross to be imaged onto the fovea of each eye. A point in space is shown that is seen as 10° up and to the left in both eyes simultaneously. Adapted from C.W. Tyler, Sensory processing of binocular disparity. In *Vergence Eye Movements: Basic and Clinical Aspects*, eds. C.M. Schor and K.J. Ciuffreda, pp. 199–295. Boston: Butterworths, 1983, by permission of C.M. Schor.

FIGURE 9.26. Photograph of a print by Rubens that was included in the book Opticorum, published by Franciscus Aguilonius in 1613, and currently located at Museum Plantin-Moretus. Reproduced from M.V. Joosse, Visual Fields in Strabismic Suppression and Amblyopia, doctoral dissertation from Erasmus University, © 1999, by permission of M.V. Joosse and Museum Plantin-Moretus, Antwerp Belgium.

In some human patients with strabismus, an unusual sensory adaptation takes place to compensate for the fact that anatomical corresponding points are continually out of register. This is described in Box 9.2.

The horopter is not fixed in space but is defined relative to where the lines of sight from the foveas of the eyes cross. Thus, the horopter is defined at each moment based on the distance of binocular fixation (Figure 9.29).

When motor fusion places binocular fixation anywhere on the innermost VM circle, as illustrated at #1, all objects that happen to fall on the horopter defined by that circle are fused automatically. If motor fusion moves binocular fixation to anywhere on the middle VM circle, as shown at #2, only objects located along the horopter defined by that circle will be fused automatically. Similar relationships hold for the outermost VM circle, as shown at #3, and for others.

Sensory Fusion Within Panum's Area

Objects that fall in front or behind the horopter always form **disparate retinal images.** An example is illustrated in Figure 9.30.

The tail of the arrow falls on the current horopter and is fused. However, the head of the arrow falls in front of the current horopter, and its image falls on disparate points that are different distances from the fovea in the two eyes.

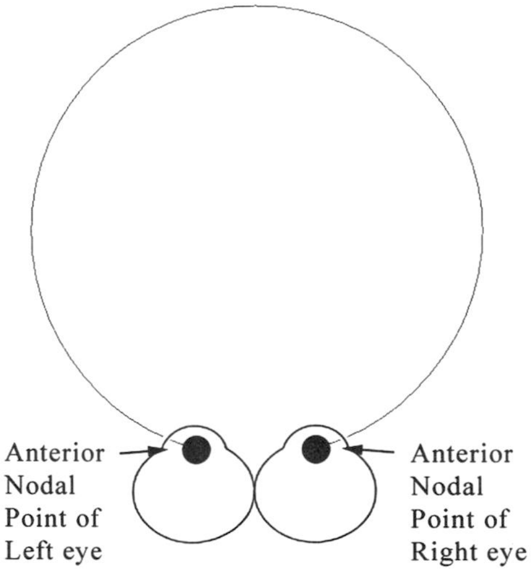

Anterior Nodal Point of Left eye

Anterior Nodal Point of Right eye

FIGURE 9.27. The theoretical shape of the horopter is a VM circle that passes through the anterior nodal point of each eye.

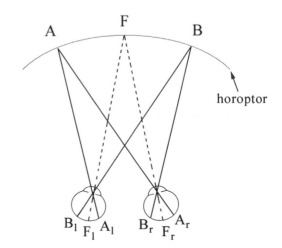

FIGURE 9.28. Objects on the horopter produce images on corresponding points in the two eyes. See text.

Box 9.2
Anomalous Correspondence

In individuals with a constant strabismus, the anatomical corresponding points on the retina no longer serve as functional corresponding points. This can give rise to a sensory adaptation called **anomalous correspondence,** as illustrated in Figure 1.

The hypothetical individual has a strabismus in which the left eye is turned chronically inward. The individual uses the right eye to look at objects, and this causes the object of regard to form an image on the fovea of the right eye (F) and simultaneously at an eccentric location on the left eye (AC). In an individual with a normal binocular visual system, the pair of locations F in the right and left eyes would form corresponding points that designated "straight ahead". However, in anomalous correspondence, location AC has taken on a new (anomalous) correspondence with location F in the right eye. Its visual direction during binocular viewing becomes straight ahead. Combinations of sensory adaptations also occur. For example, in many strabismic patients, anomalous correspondence develops in peripheral vision, while suppression of the input from one eye eliminates diplopia and confusion from the central field.

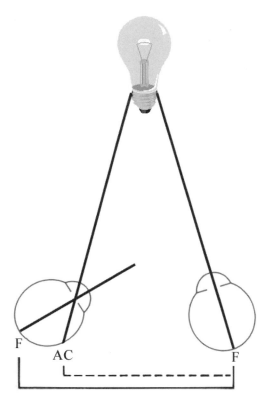

FIGURE 1. Anomalous correspondence. See text. Adapted from J. Lang, *Strabismus.* Thorofare, NJ: Slack Incorporated, 1984, by permission of J. Lang.

Objects that fall in front of the horopter are said to have **crossed disparity,** meaning that the left eye's view is to the right of the right eye's view and vice versa. Objects that fall behind the horopter are said to have **uncrossed disparity,** because the left eye's view is to the left of the right eye's view and vice versa. The magnitude of the disparity varies with distance from the horopter (Figure 9.31).

Objects that have only a small amount of disparity can be fused with the sensory process of **stereopsis.** The region of three-dimensional space from slightly in front of to behind the horopter where stereopsis can operate to fuse stimuli is called **Panum's area.** The relationships among the horopter, Panum's area, and qualities of binocular percepts are summarized in Figure 9.32.

A simple exercise can demonstrate the operation of fusion within Panum's area. Hold your finger a

few inches in front of your nose while fixating on a distant object straight across the room. This places your horopter at the distant location, causing your finger to be so far in front of the horopter that it falls in front of Panum's area. You will experience some combination of diplopia and/or suppression as you hold your finger in this close position (Figure 9.33).

By alternately closing your left and right eyes, you should be able to demonstrate that the images of the finger are creating a crossed disparity. You will see your finger to the left of the fixation point when viewed by the right eye and to the right of the point when viewed by the left eye. If your brain is allowing the percept to have the quality of diplopia, you will see two images of the finger simultaneously. If your brain is suppressing one of the two images, then only one image will be seen, but you can bring the opposite image into view by

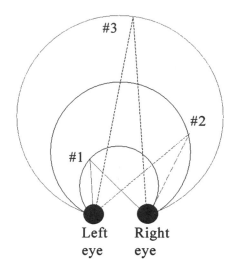

FIGURE 9.29. The horopter is not fixed in space but defined at each moment by the convergence angle of the eyes. See text.

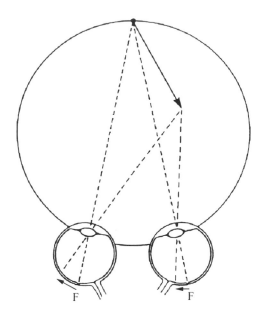

FIGURE 9.30. The tail of the arrow falls on the horopter and is imaged onto corresponding points. The head of the arrow falls in front of the horopter and produces disparate retinal images. Adapted from C.W. Tyler, Sensory processing of binocular disparity. In *Vergence Eye Movements: Basic and Clinical Aspects*, eds. C.M. Schor and K.J. Ciuffreda, pp. 199–295. London: Butterworths, 1983, by permission of C.M. Schor.

alternately closing each eye. When the eye that is being suppressed is closed, no change takes place in the percept. However, when the eye that is not being suppressed is closed, the formerly suppressed image suddenly comes into view.

Next, change the viewing condition so that your finger falls within Panum's area. You can do this with some combination of changing your fixation (and thus the horopter) from across the room to a closer location and moving your finger farther away from your nose (Figure 9.34).

When you have changed your viewing conditions sufficiently that your finger falls within

Panum's area, it no longer looks double. You now see only a single finger, despite that fact that its image is still falling on disparate retinal regions. We can rephrase what happens during stereopsis in

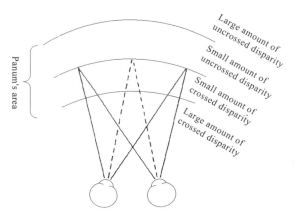

FIGURE 9.31. The greater the distance from the horopter, the greater the amount of retinal disparity. Only stimuli that fall within Panum's area produce amounts of disparity that are small enough to be fused.

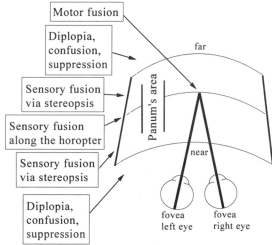

FIGURE 9.32. Summary of relationships between location of stimuli with reference to horopter and qualities of the binocular percept.

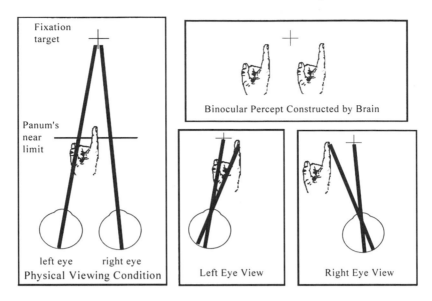

FIGURE 9.33. Viewing conditions that can be used to demonstrate failure of fusion.

terms of a **change in visual direction.** When you look at your finger with your left eye, it appears to be located slightly to the right. When you look at your finger with your right eye, it appears to be located slightly to the left. When you look at your finger with both eyes at the same time, you see a single finger that is not in the position seen by either eye alone. Instead, it is seen in a position that is midway between the visual directions seen by the two individual eyes.

When stimuli are fused via stereopsis, an additional quality is added to the percept. The object is seen as having a specific depth in three-dimensional space relative to the horopter. The processes responsible for this added depth quality of the percept are covered in the next section.

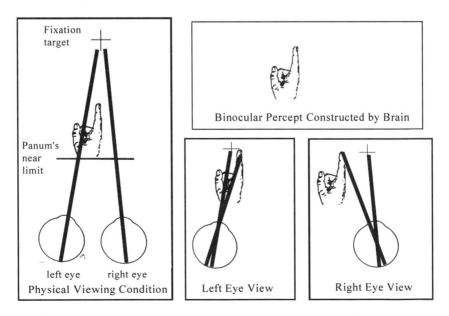

FIGURE 9.34. Viewing conditions shown in Figure 9.33 can be altered as shown here to demonstrate fusion.

FIGURE 9.35. Information that is available as input to the sensory process that must accomplish fusion under a specific condition of binocular viewing.

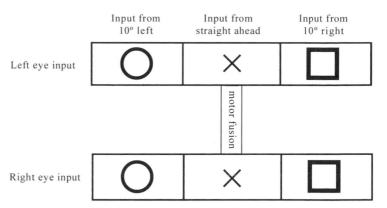

4 Stereoscopic Depth Perception

The perceptual process of obtaining sensory fusion of disparate images by stereopsis gives the binocular percept that is formed a more intense quality of depth than is apparent when it is viewed with only one eye.

Determining Depth from Disparity

Consider how the brain might draw inferences about depth based on sensory fusion of disparate images. Assume that the inputs from each eye each come with a labeled line that specifies a visual direction based on the location of the image in the retina. To make things simple, this section's examples assume inputs from a simple retina that conveys only three visual directions: 10° left, straight ahead, and 10° right. These examples also assume that motor fusion has already taken place, so whatever is imaged on the foveas of the left and right eyes is already fused. The information that is available as input to the sensory process that must

accomplish stereopsis can be conceptualized as shown in Figure 9.35.

A fixation target that is located straight ahead for both eyes has already been fused via motor fusion. Additional information received from the eyes specifies that a circle is imaged at 10° left in the left eye and a circle is imaged at 10° left in the right eye. There are also a square imaged at 10° to the right in the left eye and a square imaged at 10° to the right in the right eye.

Given these inputs, the task that needs to be accomplished to achieve sensory fusion appears straightforward, as illustrated in Figure 9.36.

The circle images in the two eyes can be fused to form a single binocular percept of a circle located 10° to the left. Similarly, the square images in the two eyes can be fused to form a single binocular percept of a square located 10° to the right. The process of sensory fusion is illustrated in Figure 9.36 by a connecting fusion line between each pair of images, one coming from the left eye and one from the right, that will be fused into a single percept. The pairs of monocular images are represented in the figure lying one above the other, and thus the fusion lines are oriented vertically.

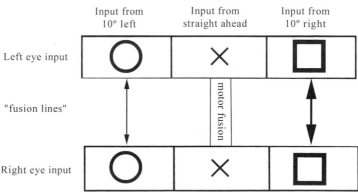

FIGURE 9.36. One way sensory fusion could be accomplished, given the binocular input shown in Figure 9.35.

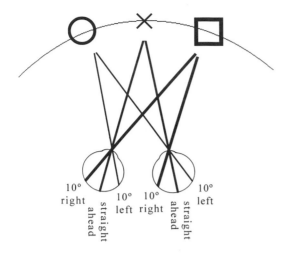

FIGURE 9.37. Projection of the fused stimuli shown in Figure 9.36 into three-dimensional space.

Once fusion of a single binocular percept has taken place, a computational procedure can be used to infer the location of the circle and the square in three-dimensional space. The circle that was imaged in the left eye could have been, in principle, located at any distance that lies along the line projecting out of the left eye at a visual angle of 10°. Similarly, the circle that was imaged in the right eye could have been located at any distance that lies along the line projecting out of the right eye at a visual angle of 10°. However, following fusion, there is only one location where the object could be located. It has to be at the point that falls where the 10° lines from the two eyes cross. This inverse optic geometrical method of determining where the object must be located in three-dimensional space is called **projection** and is illustrated in Figure 9.37.

Once the brain has accomplished sensory fusion, it has enough information, in principle, to use pro-

jection to calculate the distance of each object relative to the distance of the horopter. In this example the circle and square are each projected to fall on the horopter, or equivalently, have zero depth with respect to the horopter.

Consider another example. Suppose the sensory process that must carry out stereopsis is provided with the inputs shown in Figure 9.38.

In this example, the images formed in the left eye are identical to those used in the previous example. However, the images in the right eye for the square and the circle are now reversed. Can these monocular inputs be fused? They can if the binocular processing system is allowed to fuse images from noncorresponding points. The circle image formed at the 10° left location in the left eye can be fused with the 10° right location in the right eye, as can the square images, as demonstrated in Figure 9.39.

Note that in this example, the fusion lines for the 10° left and right images have to be represented in the figure with oblique instead of vertical lines, because they do not connect corresponding points. The image of the square in the left eye appears to the right of the image of the square in the right eye. This relationship designates a crossed disparity. Thus, the square has to appear in front of the horopter. The fusion line connecting the circles is angled leftward, combining uncrossed disparities, and thus the circle has to be located behind the horopter. By using projection, it is possible to determine where the circle and the square have to be located in space relative to the horopter, as demonstrated in Figure 9.40.

These examples demonstrate that it is possible, in principle, to use information based on disparity in the images formed in the two eyes to fuse the images in the two eyes into a single binocular percept and in the process gain information about depth.

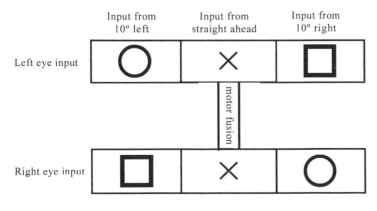

FIGURE 9.38. Information that is available as input to the sensory process that must accomplish fusion under a specific condition of binocular viewing.

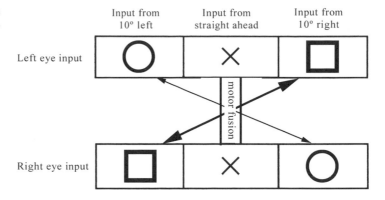

FIGURE 9.39. One way sensory fusion could be accomplished, given the binocular input shown in Figure 9.38.

Empirical evidence that the brain does in fact make use of disparity information to give a fused binocular percept a quality of depth was first demonstrated by **Wheatstone** in the 1800s. He constructed pairs of two-dimensional images called **stereograms** that can be shown to observers under **dichoptic** viewing conditions in which each eye sees only one image.

A stereogram is produced using a procedure that is the reverse of projection. With projection, one starts with the images formed in the two eyes and projects out of the eye to infer what must be present in three-dimensional space. When constructing stereograms, one starts with the three-dimensional world being viewed and determines the two-dimensional image that will be seen by each eye, as illustrated in Figure 9.41.

Wheatstone used these methods to create stimuli that were replicas of the monocular view as seen by each eye. Then he presented these two monocular views separately to each eye of an observer using mirrors, as illustrated in Figure 9.42.

The subject's head is positioned so that the left eye views only the monocular stimulus positioned at E' as reflected in mirror A'. Similarly, the right eye views stimulus E as reflected in mirror A. Wheatstone's rather amazing discovery was that this apparatus yielded a binocular percept of the original three-dimensional scene in depth, as illustrated in Figure 9.43.

Wheatstone's trick of making the brain see depth by presenting only disparity to the eyes is the basis for creating a three-dimensional impression of depth in 3D movies and in modern virtual reality displays. Instead of mirrors, special glasses are used to achieve dichoptic presentation of the two stimuli separately to the two eyes. For example, in 3D movies the stimulus for one eye is presented with long wavelengths and that for the other eye with short wavelengths. Then,

through "red-green glasses", only one stimulus is seen by each eye. Virtual reality computer displays operate by using glasses that operate in synch with a video monitor to alternate rapidly (e.g., 30 times per second) between frames that are seen by the left and by the right eye. Methods for constructing autostereograms that can be seen without using mirrors or special glasses are described in Box 9.3.

The computational operations that are involved in creating depth from disparity seem straightforward based on the discussion to this point. Then why is it asserted at the beginning of this chapter that binocular fusion based on stereopsis is an exceedingly complex process? The complexities arise from a technical difficulty called the **correspondence problem**, which is taken up next.

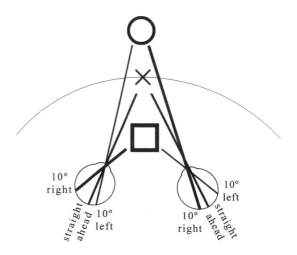

FIGURE 9.40. Projection of the fused stimuli shown in Figure 9.39 into three-dimensional space.

Box 9.3
Magic Eye Stereograms

Methods for constructing autostereograms that can be seen without special glasses were originally described by the visual Scientist Chris Tyler and his colleagues. Similar methods are now used to make posters such as those widely marketed under the name Magic Eye ® stereograms.

These stereograms are constructed by using patterns that have repetitious horizontal structure. This can be illustrated based simply on two elements, as shown in Figure 1.

The distance between any two adjacent solid lines is identical. Consequently, any two solid lines can be fused into a single (illusory) percept by converging or diverging the eyes so that one of their respective images falls onto the fovea of each eye. Consider the two solid lines labeled A and B. In the example shown, the observer has diverged the two eyes so that these two elements each fall on the fovea of one eye and have become perceptually fused. Following this illusory fusion of the solid lines, the nearest dashed line falls to the left in the left eye and to the right in the right eye. We can predict what its (illu-

sory) fused percept will be based on projection, as illustrated in Figure 2.

The illusory percept that will be produced is that a single dashed line is located at a depth behind a single solid line. Experiencing this illusory percept depends on the ability to converge or diverge the eyes by an amount that is unnatural but appropriate for seeing the illusion. Some observers find this task relatively easy, while others need a great deal of practice and effort to achieve the illusory depth percept.

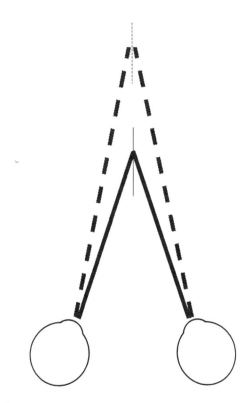

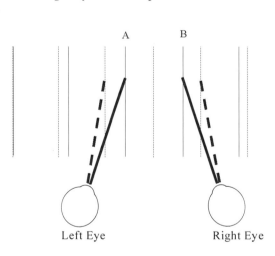

FIGURE 1. Principle used to create an autostereogram.

FIGURE 2. Projection can be used to predict what the percept will look like when viewing an autostereogram as shown in Figure 1.

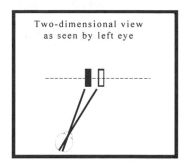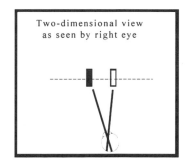

FIGURE 9.41. How stereograms can be created based on reverse projection. The individual stimuli shown in the bottom row form a stereogram pain. When viewed simultaneously, one by each eye, the binocular percept will correspond to the viewing situation from which these stimuli were constructed as illustrated in the top panel.

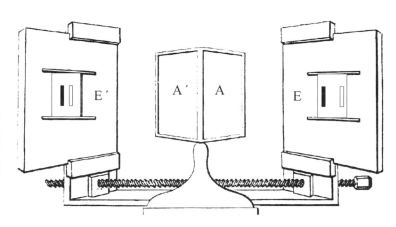

FIGURE 9.42. Method used by Wheatstone in the 1800s to allow viewers to view stereoscopic stimuli. Adapted from C. Wheatstone, Contributions to the physiology of vision. – Part the first. On some remarkable, and hitherto unobserved, phenomena of binocular vision. *Philos. Trans. R. Soc. Lond.* 128:371–394, 1838, by permission of the Royal Society.

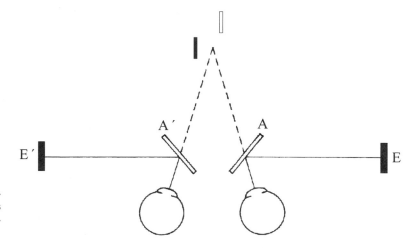

FIGURE 9.43. The binocular percept experienced by subjects viewing the stereograms in Wheatstone's apparatus.

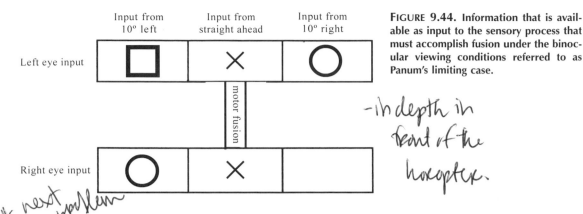

FIGURE 9.44. Information that is available as input to the sensory process that must accomplish fusion under the binocular viewing conditions referred to as Panum's limiting case.

— in depth in front of the horopter.

✱ next problem

The Correspondence Problem

One of the first perceptual scientists to articulate the correspondence problem was the gestalt psychologist Kurt **Koffka**, who posed a number of fundamental questions about perception. **Koffka's fundamental question** about the correspondence problem can be paraphrased with reference to the two simple examples of sensory fusion presented above in Figures 9.36 and 9.39. In the first example, the brain achieved fusion by combining the image at 10° left in the left eye with the one at 10° left in the right eye. In the second example, fusion was achieved by fusing 10° left in the left eye with 10° right in the right eye. Koffka essentially asked, "How did the brain know that it should fuse images from these two retinal locations one way in the first example and the opposite way in the second example?"

Koffka then posed an answer to his own question. He asserted that the brain could fuse only similar images. In the first example, the 10° left point in the left eye was fused with the 10° left point in the right eye because these two locations both had images of a circle. Similarly, in the second example the two locations that were fused both had images of circles.

Koffka's proposed answer led to an assumption accepted by most historical perceptual scientists that there first had to be a stage of monocular processing of the images from each eye. This initial stage of monocular processing identifies the images at each location according to factors such as shape and color. Then the binocular processing stage attempts to fuse only elements that contain similar monocular images.

It has been known for a long time that there are problems with this assumption. One of the classic problems is revealed by a demonstration described by Panum in the mid-1800s, now called **Panum's limiting case,** which can be produced with a stereogram, as illustrated in Figure 9.44.

Panum's contemporaries found it to be a challenging issue to predict what fused binocular percept should be produced given these inputs? The empirical answer that is discovered when this stereogram is viewed by a human observer is illustrated in Figure 9.45.

The observer viewing the stereogram of Panum's limiting case sees a circle in front of the horopter and a square that lies on the horopter in a position such that its view is obscured for the right eye. The fact that the depth for the square is localized tells us something interesting about how sensory fusion operates. The square is localized by combining the element at 10° left in the right eye (containing an image of a circle) with the element at 10° left in the left eye (containing an image of a square). So much for the assumption that only elements containing similar images can be fused! The fusion lines that would have to be produced to arrive at the empirical binocular percept are illustrated in Figure 9.46.

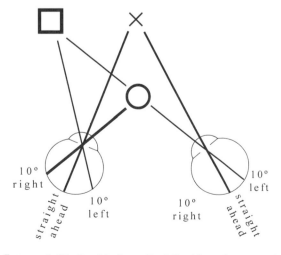

FIGURE 9.45. Empirical result of the binocular percept reported by human observers viewing stimuli shown in Figure 9.44, along with projection lines for this result.

FIGURE 9.46. How sensory fusion is accomplished by the visual system, given the input corresponding to Panum's limiting case as shown in Figure 9.44.

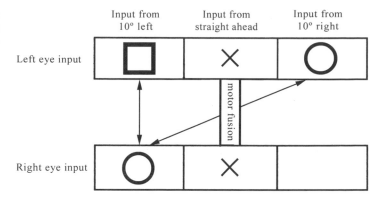

Demonstrations such as Panum's limiting case showed that Koffka's answer to his own question was inadequate. Nevertheless, the assumption that only similar elements can fuse remained common-place among visual scientists as late as 1959. One year later, in 1960, the visual Scientist Bela **Julesz** published a paper in which he described **random-dot stereograms,** a methodology that proved to be the death knell of the assumption that monocular processing must precede fusion.

In a random-dot stereogram, each element is either a single black dot of texture or a single white spot. The stereogram image shown to each eye can contain hundreds or even thousands of dots. The only difference between the left and the right eye images is that some of the dots in one of the images have been shifted a small distance from the positions of the dots shown to the other.

The top panel of Figure 9.47 shows the two images that are presented to the observer. The

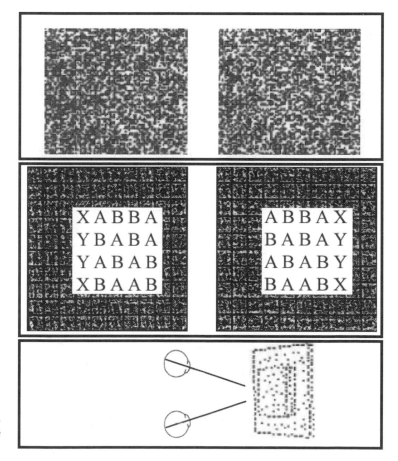

FIGURE 9.47. Method used to construct random-dot stereograms. See text.

Left eye
input

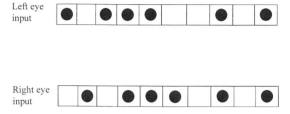

Right eye
input

FIGURE 9.48. Monocular inputs for a sample of a single horizontal row of dots for the right and one for the left eye in a random-dot stereogram.

stimuli contain random-dot texture generated using a two-dimensional matrix. Every location within the matrix is either filled with a black dot or left blank according to a specified probability. Most elements in the matrix are filled in identically in the images shown to the left and right eyes. However, some of the dots in the left image have been shifted relative to the corresponding dots in the right image.

The square boxes in the middle panel illustrate how this shift is done. A 4-by-4 matrix designated by A and B symbols in the left image is shifted to the left by one column in the right image. This shift covers one column of dots in the right image or, stated equivalently, uncovers a similar group in the left image. The regions where the corresponding dots have been covered or uncovered are designated by X and Y symbols. The horizontal shift is some integral multiple of the size of the individual cells of the matrix. Thus, no cell is partly covered or uncovered by the shifted area. Every cell is either completely covered with a new cell or left as is. This procedure guarantees there are no monocular cues that can be used to determine where the region of the shift is located. The only way to determine that a shift has been performed is to perform a dot-by-dot comparison of the two images. In the example shown here, only a 4-by-4 matrix of dots has been shifted, but in the real random-dot stereogram, the matrix might involve hundreds of dots shifted by various amounts in different portions of the stimulus.

When the random-dot stereogram is viewed by a human observer, the regions of the image where a shift has occurred appear to be floating in space at a different depth plane from the rest of the dots. An example where the central region of the stimulus appears to be floating in front of the surrounding stimulus is illustrated in the bottom panel.

Consider the computational requirements that are put on the brain in order to fuse these images and discover depth based on disparity. Figure 9.48

illustrates the monocular inputs for a single horizontal line of dots.

Given these inputs, there is no possible way to know which dots should be fused based on prior monocular processing. All of the dots are identical. Consider the leftmost dot in the left eye input. That dot could be fused with the dot in the second position from the left, with the dot in the 4th position from the left, or with the dot in the 5th position from the left, and so on. Each of these possible matches will give rise to a different inference about depth based on projection. This is illustrated for the case in which only four dots are imaged in each eye in Figure 9.49.

Consider the dot labeled R1 in the right eye. This dot projects to four different depth planes depending on whether it fuses with L1, L2, L3, or L4. All four of these are equally valid based on the geometry of projection, but only one corresponds to the physical three-dimensional location of the dot that was used as the basis for constructing the stereogram image. Similar confusion exists for each of the other dots in the right eye. Yet when normal humans look at random-dot stereograms constructed based on relationships that exist in actual three dimensional scenes, they usually experience the dots as being at the appropriate depth plane. The exact methods that biological brains use to accomplish this are currently a subject of intense investigation by neuroscientists but are still not fully understood. Coming up with an algorithm or form of processing that is efficient at making inferences about which matches are correct and which should be ignored is also a fundamental problem that has not yet been solved for machine vision by engineers.

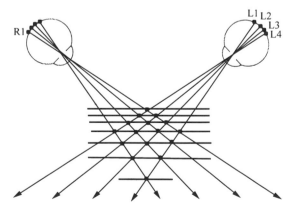

FIGURE 9.49. Projection illustrates that with four dots presented to each eye, each element in one eye can potentially be fused at four different locations in three-dimensional space.

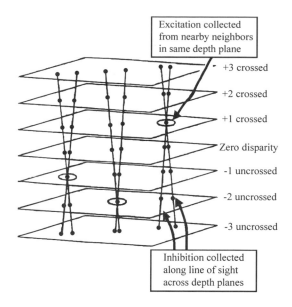

Excitation collected from nearby neighbors in same depth plane

+3 crossed

+2 crossed

+1 crossed

Zero disparity

-1 uncrossed

-2 uncrossed

-3 uncrossed

Inhibition collected along line of sight across depth planes

FIGURE 9.50. Cooperative model of stereopsis proposed by Marr and Poggio. See text.

The correspondence problem is another example of an ill-posed inverse problem that has no unique solution. In order to make correct inferences about what three-dimensional structure is present in the environment given a particular pattern of binocular stimulation, it is necessary to apply constraints. Computational models of stereopsis have tried to build in a number of constraints that facilitate finding a solution.

Computational Models of Stereopsis

Most computational models of stereopsis have been based on a strategy of using excitatory and inhibitory interactions among binocular neural processing elements. These are typically called **cooperative models.** The rationale behind these models is that when the units are stimulated by disparate images, the neurons all interact until the network settles into a steady state in which only units responding to the proper disparity remain active.

Marr and Tomaso Poggio developed one of the earliest models of this type. They attempted to build in two constraints called "uniqueness" and "continuity." The **uniqueness constraint** is an assumption that every object occupies a unique physical position in three-dimensional space. The **continuity constraint** is an assumption that the surfaces of objects are generally smooth, with only occasional abrupt changes in depth.

Their model was a three-dimensional network of processing nodes, called "neurons," organized into layers, with each layer representing a different depth, as illustrated in Figure 9.50.

In order to implement the uniqueness constraint, only one neuron was allowed to be active along each line of sight passing through the depth layers. In order to implement the continuity constraint, neurons within a single depth layer excite one another. The cooperative procedure goes point by point through the nodes in an iterative manner. Each neuron adds up the possible excitatory connections and subtracts from them the possible inhibitory connections. If the result exceeds a given threshold, the neuron is set to one; otherwise, it is zero. This procedure is repeated in an iterative manner until the network settles into a solution.

Cooperative models find the appropriate solutions for some classes of stereograms but fail to arrive at a correct solution for many complex stimuli for which humans infer the correct three-dimensional pattern. Psychophysical studies with humans demonstrate that under some conditions humans can detect multiple depth planes in a given direction. For example, a texture pattern of dots can be painted onto two pieces of glass positioned one behind the other. The dots on the front transparent surface form one depth plane and the dots on the second surface form another depth plane. A random-dot stereogram can be constructed based on the combined dots in these two depth planes. When humans view these stereograms, they perceive two transparent depth surfaces. Cooperative models fail on these stimuli, because the continuity constraint is violated due to the wide range of disparities for nearby dots.

Cooperative models also have trouble with random-dot stereograms in which the image presented to one eye is changed in size relative to the other. However, human subjects can tolerate a 15% expansion of one image. Another problem has to do with maintaining the stability of the binocular depth percept in the face of vergence eye movements. Every vergence movement causes a shift of the entire range of disparities that are present in the image. Biological observers deal with these changes effortlessly, but cooperative models are disrupted, because they take time to settle into a solution and the entire process must start over following each vergence movement. Modern implementations of cooperative models have tried to include a number of additional constraints to improve the speed and accuracy of the inferred solutions, but no model to date can come anywhere close to the performare of biological stereovision.

Cooperative models tend to give many false matches, and attempts to alleviate this problem have led to an approach that applies processing in stages rather than having the entire solution depend on a single cooperative process that gradually settles into a preferred solution. The prototypical model along these lines is the **coarse-to-fine model,** also proposed by Marr and Tomaso Poggio. Images are passed through spatial frequency filters, as discussed in Chapter 8, before stereopsis is attempted. The algorithm involves first looking for matches in a low-pass filtered image. Then successively finer filtering is performed, looking only for matches that are consistent with those found with the coarse filtering. The rationale is that many false matches that show up when fine spatial detail is viewed will be invisible when a low-pass filter is used on the image. This is because the zero crossings for the low-pass spatial filtered images are never very close together, so there is less opportunity for false matches. However, a problem with all stereopsis models that operate on selected spatial frequencies is that disparities larger than a quarter of a cycle are indeterminate, as illustrated in Box 9.4.

Box 9.4
Disparity Processing Based on Spatial Frequencies Has a Quarter Cycle Limit

The spatial frequency shown in the middle row of Figure 1 is imaged onto the left eye. This spatial frequency has a zero crossing at the location indicated by the vertical line. This zero crossing must be matched to a zero crossing from a disparate stimulus of the same spatial frequency in the right eye. The examples directly above and below the center row demonstrate uncrossed and crossed disparity shifts that correspond to a quarter of a cycle. There are four possible matches of equal disparity. One, indicated by the solid arrow, is the proper match in this example. The other three are false matches. The upper and lower rows show shifts of more than a quarter of a cycle, and in this case, the closest match in each case is a false match. Psychophysical studies demonstrate that humans can fuse disparities many times larger than the quarter-cycle limit. The exact mechanisms that allow this are not fully understood except for the general idea that there must be simultaneous comparisons across a number of spatial frequencies.

— wallpaper illusion

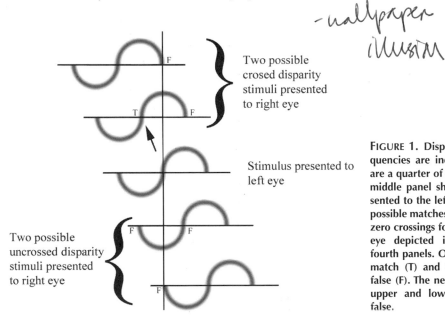

Two possible crosed disparity stimuli presented to right eye

Stimulus presented to left eye

Two possible uncrossed disparity stimuli presented to right eye

FIGURE 1. **Disparities of spatial frequencies are indeterminate if shifts are a quarter of a cycle or more. The middle panel shows a stimulus presented to the left eye. There are four possible matches of nearest-neighbor zero crossings for stimuli in the right eye depicted in the second and fourth panels. One of these is a true match (T) and the other three are false (F). The nearest matches in the upper and lower panels are both false.**

Neural-Based Models of Stereopsis

Disparity-Sensitive Binocular Neurons

The section on Sensory Processing of Binocular Information in this chapter described the existence of binocular cells in the striate and extrastriate cortex that respond better when both eyes are stimulated than when only one eye is stimulated. Most of the cells in the striate cortex have receptive fields in the left and right eyes that fall on corresponding points. Thus, they are most appropriate for detecting objects at the distance of the horopter. Other binocular cells, called **disparity-sensitive neurons**, respond best to binocular stimulation that falls on disparate points on the two retinas. These neurons could form the basis for disparity processing.

Disparity-sensitive neurons in monkeys were described by **Hubel and Wiesel.** An illustration of the pattern of results they found is shown in Figure 9.51.

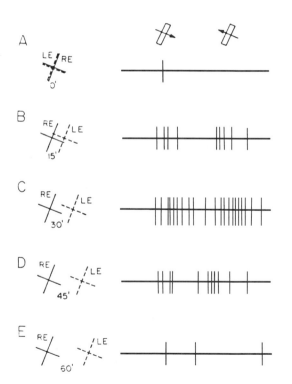

A

B

C

D

E

FIGURE 9.51. Firing rates of a disparity-sensitive neuron in the visual cortex of monkey to stimuli having various amounts of binocular disparity. See text. Adapted from data originally published by D.H. Hubel and T.N. Wiesel, Stereoscopic vision in macaque monkey. *Nature* 225:41–42, © 1970, by permission of Macmillan Magazines Ltd.

Stimuli were swept across the receptive fields of a binocular neuron in each eye simultaneously. The relative locations of the Stimuli in the left (LE) and right (RE) eyes were varied from 0 to 60 minutes of arc as illustrated in the panels on the left side of the figure. The responses of the neuron as the stimuli sweep both directions across the receptive field are shown in the right panels. The firing rate of this binocular neuron rises from some low value when zero-disparity stimuli are presented (A) to a maximum at some moderate disparity (C) and then falls off for large disparities (E).

Gian F. **Poggio and colleagues** discovered that disparity-sensitive neurons tuned to horizontal binocular disparity in V1 fall into three basic types, as illustrated in Figure 9.52.

One class, shown in the center column, is tuned to zero disparity. These come in two subtypes. **Tuned zero neurons,** shown in the top now, respond maximally to objects on the horopter having zero disparity and are often inhibited by objects in front of or behind the horopter. **Tuned inhibitory neurons,** shown in the bottom now, are inhibited by zero disparity. The second basic type, called **Tuned near cells,** illustrated in the left column, respond with excitation to targets in front of the horopter and with inhibition to targets behind the horopter. The top row shows two examples of responses from near cells with narrow tuning of preferred disparities, and the bottom row shows examples of responses from near neurons with broader tuning. Finally, **Tuned far cells,** shown in the right column, are mirror images of the near neurons.

Position-Shift and Phase-Shift Neuron Models

Position-shift models of binocular disparity assume that lateral shifts in position of the right and left eye's receptive fields constitute the mechanism for encoding binocular disparity – in other words, that a disparity-sensitive neuron combines inputs from similarly shaped monocular receptive fields coming from slightly different retinal positions in the left and right eyes.

Phase-shift models of binocular disparity assume that the monocular receptive fields in the two eyes differ in phase instead of in position. The profiles of the receptive fields in the left and right eyes are different in shape, but the center of each receptive field is at a corresponding retinal location. The difference between the two models is illustrated in Figure 9.53.

Computational evaluations reveal that both position-shift and phase-shift models give many

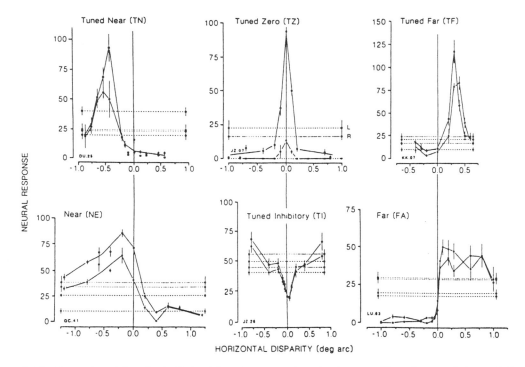

FIGURE 9.52. Illustration of three basic kinds of disparity-sensitive neurons found in monkey visual cortex. Reproduced from G.F. Poggio, et al, Stereoscopic mechanisms in monkey visual cortex: Binocular correlation and disparity selectivity. *J. Neurosci.* 8:4531–4550, © 1988, by permission of the Society for Neuroscience.

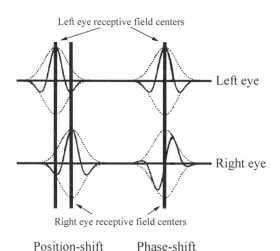

FIGURE 9.53. Position-shift and phase-shift models of disparity-sensitive receptive fields. In the position-shift model, the locations of the receptive field centers in the left and right eyes have been shifted, but the receptive fields have the same shape. In the phase-shift model, the receptive field centers are at the same location, but the receptive fields have different shapes.

false matches. Both models will encode disparity for simple stimuli but break down when confronted with rich stimuli, such as random-dot stereograms or textured surfaces. Somehow, biological perceptual systems eliminate these false matches, but the exact mechanisms by which they accomplish this remain elusive.

Binocular Depth Information That Is Not Based on Horizontal Disparity

Vertical Disparity

The term "stereopsis" is often used as a synonym for depth perception based solely on the ability to detect horizontal disparities. However, this is really a misnomer, because depth perception is influenced by vertical as well as horizontal disparities, a fact that is generally unappreciated.

The horopter that was described earlier in this chapter in the section Sensory Fusion Based on the Horopter applies only to a single plane defined by a VM circle that passes through the eyes and the point of binocular fixation. However, there is another locus in space where disparities are zero.

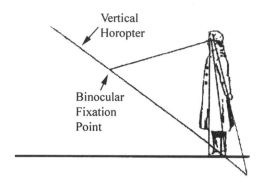

FIGURE 9.54. The vertical horopter falls along a line that passes through the point of binocular fixation and a point near the observer's feet. Adapted from C.W. Tyler, Sensory processing of binocular disparity. In *Vergence Eye Movements: Basic and Clinical Aspects*, edited by C.M. Schor and K.J. Ciuffreda, pp. 199–295. London: Butterworths, © 1983, by permission of C.M. Schor.

That locus is a line that passes through a point near the observer's feet and the point of binocular fixation, as illustrated in Figure 9.54.

Stimuli at any locations in three-dimensional space that do not fall along either the VM circle or this tilted line at the midline will create either horizontal disparity, vertical disparity, or both.

Elegant psychophysical studies by Kenneth Ogle in the mid-1900's and more recently by Barbara Gillam and colleagues have demonstrated examples in which humans take account of this vertical disparity information when inferring properties of three-dimensional depth.

Da Vinci Stereopsis

Leonardo da Vinci had an insight about the role of **binocular parallax,** the fact that the two eyes see the world from different points of view, in creating an impression of depth. With two eyes we can sometimes see around and behind an object that would be hidden if viewed under monocular conditions, as illustrated in Figure 9.55.

The portion of the scene highlighted by the thick black line is hidden from view for the left eye, and the portion highlighted by the thick dashed line is hidden from view for the right eye. However, with the use of both eyes, the entire scene can be viewed. In other words, some parts of a scene that are opaque to monocular vision become transparent to binocular vision. There are a number of demonstrations in the modern perception literature in which forms of dichoptic stimulation based on binocular parallax produce an impression of depth even

though there are no disparities present to be fused based on stereopsis. These perceptual phenomena are based on stereopsis referred to as **da Vinci stereopsis.** It is an interesting historical observation that although the role of binocular parallax in creating depth was described in the Renaissance, the specific role of disparity was not appreciated until the 1800s, when it was described by Wheatstone.

Binocular Motion

The discussion to this point has involved only static disparity, but potential binocular information is also made available by comparing motion in the two eyes, as illustrated in Figure 9.56.

As an object moves from location A to location B, it simultaneously produces rightward motion on the retina of the left eye and leftward motion on the retina of the right eye. This provides information about change in depth that could be extracted, in principle, by comparing relative motion in the two eyes, even if there were no ability to achieve stereopsis based on positional disparity.

Depth Based on Space-Time

Psychophysical studies with humans demonstrate that an impression of depth can be created by stimuli that do not contain strict disparity or strict binocular motion but do create information that is available in a space-time representation. This capability was demonstrated in psychophysical studies carried out by **Shimojo** and colleagues in which humans viewed moving objects through a small aperture, as illustrated in Figure 9.57.

Consider an object that is moving along a trajectory from position A to position B. At time t_1 this

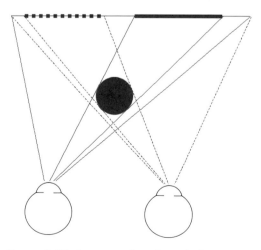

FIGURE 9.55. Demonstration of Da Vinci stereopsis.

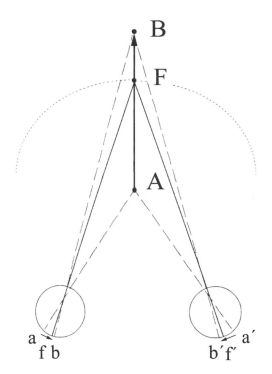

FIGURE 9.56. Demonstration that disparity can result from motion as well as from static differences in position. The observer is fixating benocularly at F causing images to be formed at f and f′ in the left and right eyes respectively. Object location A is imaged at a and a′, and location B is imaged at b and b′. Motion of an object from A to B will result in motion from a to b in the left eye and simultaneously from a′ to b′ in the right eye.

object is seen at position A by the right eye. However, it is invisible to the left eye at this time. At time t_2 this same object is seen at position B by the left eye. However, it is now invisible to the right eye. Classical stereopsis cannot provide any information about depth for this object, because there is no point in time at which the two eyes receive simultaneous images from the object that can be analyzed for binocular disparity. Similarly, there is no point in space where simultaneous motion can be compared in the two eyes. Thus, depth information could not be extracted by any form of binocular processing that analyzed stimulus information over time while holding position constant (local motion processing) or over space while holding time constant (static disparity). Instead, binocular processing that extracts depth information in this viewing situation needs to interpolate over "space-time." Remarkably, the humans tested by Shimojo reported a depth percept while receiving through the aperture. The complexities in process-

ing required to infer depth under these conditions are not well understood and orders of magnitude beyond the capabilities of computational models of stereovision that have been developed to date. This fact will be discussed further below in the section "Evolution of Stereovision in Primates."

5 Monocular Information About Three-Dimensional Space

Monocular Cues to Depth Used by Painters

Painters have known since the Middle Ages that relative depths of objects in a scene could be represented by a number of **monocular cues** – those that can be appreciated just as well with one as with two eyes. One such cue is **masking** or **occlusion** of far objects by near objects. Another is **height above the horizon**, reflecting the difference between looking at the top of a tall tree from a long distance away and when standing near the tree. Specialized light reflections such as **shadows and shading** can also enhance the perception of depth. **Linear perspective** refers to the fact that parallel lines appear to come together in the distance and **aerial perspective** to the fact that distant objects appear hazy and bluish. **Texture gradients** also

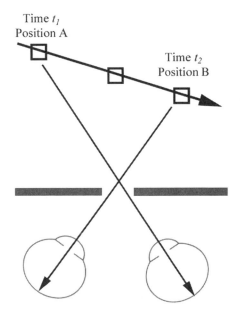

FIGURE 9.57. Viewing situation in experiment carried out by Shimojo and colleagues. See text.

FIGURE 9.58. Photograph illustrating a natural texture gradient.

provide a monocular cue, as elaborated on in the next section.

Texture Gradients

Gibson emphasized the fact that visual space perception consists not of seeing the third dimension but of seeing the laid-out environment composed of surfaces that have texture. These textures provide information about depth in the form of **texture gradients.** Consider the photograph of a landscape in Figure 9.58.

The fact that this landscape is a ground surface receding in depth towards the horizon is apparent in the photograph. The impression is carried not by any specific feature in the photograph but by a texture gradient. Any receding surface imaged onto a two-dimensional surface like the retina gives rise to a regular change in the density and spacing of the elements of texture that form the image. The term "texture gradient" refers to the regularity of the change in the texture elements rather than to the specific elements that are involved. In the photograph, the texture happens to be formed by the ground surface of an open field, but it might just as well have been formed by waves on water or tiles in a hallway. A texture gradient can be captured abstractly, as shown in Figure 9.59.

Abstract textures demonstrate that it is the regularity of the change in size and spacing of the elements rather than any property of the individual elements that constitutes the gradient. Abstract information based on texture can be used to specify

locations and sizes of objects in the environment, as illustrated in Figure 9.60.

The abstract information provided by texture is shown in the form of a grid. This grid can be used to determine that even though the two buildings form images on the retina that have different sizes, they are actually the same size because they cover the same amount of texture (3-by-2 squares in the grid). Information about their locations relative to one another is also provided by texture. Note, for example, that the two buildings are separated by a distance of two grid squares.

Motion Parallax and Optic Flow

In addition to static texture gradients, there are dynamic gradients, formed whenever one moves in the environment. An example is the "center

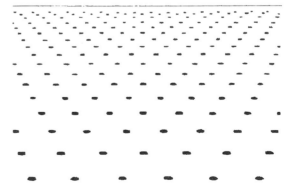

FIGURE 9.59. Abstract texture gradient.

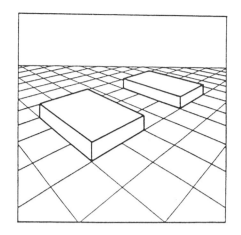

FIGURE 9.60. Information based simply on texture can be used to specify relative locations and sizes of objects at different distances. Reproduced from H.A. Sedgwick, Environment-centered representation of spatial layout: Available visual information from texture and perspective. Adapted from *Human and Machine Vision*, eds. J. Beck, B. Hope, and A. Rosenfeld, pp. 425–458, New York: Academic Press, © 1983, by permission of Academic Press.

of expansion" in the optic flow that specifies an observer's heading, as illustrated in Figure 9.61.

Note that the bird has a basic orientation to the three-dimensional organization of the environment defined by the sky above and the ground below. Static texture gradients along the ground give an impression of distance. In addition, the dynamic optical flow of texture caused by the bird's own motion provides information about other aspects of its three-dimensional environment such as the direction of heading. This topic will be discussed further in Chapter 10 in the section on optic flow.

An example of the role of motion parallax in depth perception is illustrated in Figure 9.62.

The effects of motion parallax can be experienced when riding in an automobile or on a train. Objects beyond the fixation point appear to move in the same direction you are moving in, while objects nearer than the fixation point seem to move in the opposite direction.

6 Evolution of Stereovision in Primates

Previous sections of this chapter have provided many examples demonstrating that stereovision processing in primates is exceedingly complex. Furthermore, many of the specific adaptations that have been imposed on the binocular motor system in the service of stereovision would seem to be disadvantageous. For example, someone whose wrists

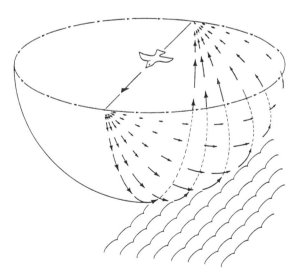

FIGURE 9.61. Dynamic texture gradients are formed whenever an observer moves with respect to the environment. Reproduced from J.J. Gibson, *The Ecological Approach to Visual Perception.* Hillsdale, NJ: Lawrence Erlbaum Associates, © 1986, by permission of Lawrence Erlbaum.

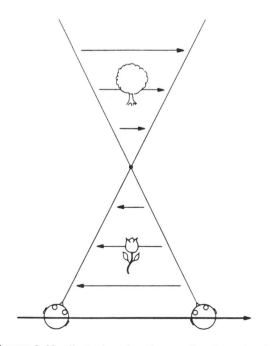

FIGURE 9.62. Illustration of motion parallax. Reproduced from J.-P. Roy, et al, Disparity sensitivity of neurons in monkey extrastriate area MST. *J. Neurosci.* 12:2478–2492, © 1992, by permission of Society for Neuroscience.

were handcuffed so that the positions of the left and right hands were constrained to move only in coordination with one another would be unlikely to consider this a functional advantage for using his arms. Yet evolution has forced these kinds of constraints onto the binocular motor control system, presumably because the functional advantages of having binocular stereovision outweigh the impediments that arise from yoking the movements of the eyes. It is interesting to speculate about what conditions in the environment in which primates evolved might have led to the development of such a complex stereovision system.

One common hypothesis about why stereovision might have evolved involves the fact that primates often **jump or swing between tree branches** and need good depth perception to avoid falling to the ground. This fact probably played a role, but the functional advantage of static disparity processing over and above the information provided about depth that can be obtained from one eye is not overwhelming, especially under these kinds of dynamic conditions.

A more promising hypothesis is that stereopsis allows an organism to **break the camouflage** of a potential predator or of prey. An animal that achieves camouflage based on color and texture will be detectable with stereopsis if it stands out as a distinct surface having a different depth plane from its surroundings.

The breaking-camouflage hypothesis can potentially account for the remarkable finding, illustrated in Figure 9.57, that humans can appreciate depth when looking through an aperture so that the available information must be interpolated across spacetime. Over evolutionary time, primates were confronted with avoiding approaching predators while spending considerable time eating fruits in a leafy forest environment. The view through the foliage surrounding the head in such an environment produces the optical equivalent of viewing the surrounding space through a number of small apertures formed by holes between the leaves. It does not seem unreasonable to speculate that enhancements in the ability to pick up information about approaching predators under these conditions would provide a strong functional advantage to primates, an advantage that perhaps outweighed the complexities and impediments involved in achieving stereovision.

Summary

Stereovision is the capacity of an organism to perceive a three-dimensional visual world. Two classic problems of stereovision are how perception of a three-dimensional environment can be based on two-dimensional retinal images and how a percept of a single world can result from information coming from two separate eyes.

Information about different parts of the visual field is processed by specific portions of each eye. The left brain processes information about the right visual field and vice versa.

Each psychological visual direction is associated with a unique anatomical location on the retina. Anatomical locations in the left and right eyes that share a common visual direction are called "corresponding points." Noncorresponding points process information about disparities in visual direction between the two eyes. When objects in the environment fall on noncorresponding points, this can lead to diplopia, confusion, suppression, rivalry, or fusion.

Motor fusion can be used to achieve single binocular vision for a single intended object of regard. Sensory fusion along the horopter fuses objects with zero disparity. Sensory fusion of objects away from the horopter but within Panum's area depends on binocular processing that changes the object's perceived visual direction. This process, called "stereopsis," not only produces a single binocular percept but also creates a strong impression of depth.

Dichoptic viewing of stereograms can be used to demonstrate that disparity alone is sufficient to produce a perception of depth. Random-dot stereograms demonstrate further that binocular processing of disparity can take place before monocular processing has isolated unique properties of the elements to be fused. The term "correspondence problem" is used to refer to the fact that it is not known exactly how this is accomplished.

Computational models have tried to account for the correspondence problem based on cooperative interactions between binocular processing elements and by using stages of processing involving different spatial frequencies. These models cannot duplicate human stereopsis in real time. It is known that some cortical neurons in V1 and in extrastriate processing areas respond to binocular disparity. Two types of neural models, called position-shift and phase-shift models, try to account for how monocular receptive fields are combined to produce disparity sensitivity.

Stereovision processing utilizes a number of other sources of binocular and monocular information in addition to horizontal disparity. Some of this information, and the specialized processing mechanisms needed to extract the information, is exceedingly complex, leading to a question about why such a complex stereovision system evolved in primates. Speculation has centered on the fact that primates must judge depth accurately when jumping to tree branches. Another idea is that stereovision evolved because it facilitated detection of approaching predators by breaking camouflage.

Selected Reading List

Belsunce, S. de, and Sireteanu, R. 1991. The time course of interocular suppression in normal and amblyopic subjects. *Invest. Ophthalmol. Vis. Sci.* 32:2645–2652.

Blake, R., and Wilson, H.R. 1991. Neural models of stereoscopic vision. *Trends Neurosci.* 14:445–452.

Boothe, R.G., and Brown, R.J. 1996. What happens to binocularity in primate strabismus? *Eye* 10:199–208.

Fleet, D.J., Wagner, H., and Heeger, D.J. 1996. Neural encoding of binocular disparity: Energy models, position shifts and phase shifts. *Vision Res.* 36:1839–1857.

Gillam, B., and Lawergren, B. 1983. The induced effect, vertical disparity, and stereoscopic theory. *Percept. Psychophys.* 34:121–130.

Harwerth, R.S., Smith E.L. III, and Siderov, J. 1995. Behavioral studies of local stereopsis and disparity vergence in monkeys. *Vision Res.* 35:1755–1770.

Hering, E. 1977. *The Theory of Binocular Vision*, eds. B. Bridgeman and L. Stark. New York: Plenum Press.

Hubel, D.H., and Wiesel, T.N. 1970. Stereoscopic vision in macaque monkey. *Nature* 225:41–42.

Liu, L., Stevenson, S.B., and Schor, C.B. 1994. Quantitative stereoscopic depth without binocular correspondence. *Nature* 367:66–69.

Marr, D., and Poggio, T. 1976. Cooperative computation of stereo disparity. *Science* 194:283–287.

Mayhew, J., and Frisby, J. 1981. Psychophysical and computational studies towards a theory of human stereopsis. *Artif. Intelligence* 17:349–385.

Ogle, K.N. 1950. Researches in Binocular Vision. Philadelphia: Saunders.

Ohzawa, I., DeAngelis, G.C., and Freeman, R.D. 1997. The neural coding of stereoscopic depth. *Neuroreport* 8: iii–xii.

Poggio, G.F., Gonzalez, F., and Krause, F. 1988. Stereoscopic mechanisms in monkey visual cortex: Binocular correlation and disparity selectivity. *J. Neurosci.* 8:4531–4550.

Poggio, G.F., and Poggio, T. 1984. The analysis of stereopsis. *Annu. Rev. Neurosci.* 7:379–412.

Qian, N. 1997. Binocular disparity and the perception of depth. *Neuron* 18:359–368.

Regan, D. 1991. *Binocular Vision*, Vol. 9, Vision and Visual Dysfunction. Boca Raton, FL: CRC Press.

Shimojo, S., and Nakayama, K. 1990. Real world occlusion constraints and binocular rivalry. *Vision Res.* 30:69–80.

Shipley, T., and Rawlings, S.C. 1970. The nonius horopter: I. History and theory. *Vision Res.* 10:1225–1262.

Sireteanu, R., and Fronius, M. 1989. Different patterns of retinal correspondence in the central and peripheral visual field of strabismics. *Invest. Ophthalmol. Vis. Sci.* 30:2023–2033.

Tyler, C.W. 1983. Sensory processing of binocular disparity. In *Vergence Eye Movements: Basic and Clinical Aspects*, eds. C.M. Schor and K.J. Ciuffreda, pp. 199–295. London: Butterworths.

Tyler, C.W., and Chang, J.-J. 1977. Visual echoes: The perception of repetition in quasi-random patterns. *Vision Res.* 17:109–116.

Wheatstone, C. 1838. Contributions to the physiology of vision: Part the first. On some remarkable, and hitherto unobserved, phenomena of binocular vision. *Philos. Trans. R. Soc. Lond.* 128:371–394.

Wheatstone, C. 1852. Contributions to the physiology of vision: Part the second. On some remarkable, and hitherto unobserved, phenomena of binocular vision. *Philos. Trans. R. Soc. Lond.* 142:1–17.

10
Dynamic How-Perception: How Do We Perceive and React to Change and Motion?

Questions

After reading Chapter 10, you should be able to answer the following questions.

1. Summarize the neural streams of processing that are particularly concerned with temporal change and motion.

2. Characterize the temporal contrast sensitivity function.

3. What kinds of temporal information are extracted in the M- and P-streams?

4. Describe the computational approaches that can be used to infer local motion from changes in intensity over space and time.

Box 10.1
Perceptual Fading

Equipment that can completely stabilize an image on the retina is expensive and available only in research laboratories. However, even if you do not have access to this equipment, you can still experience some aspects of perceptual fading. One method that requires no special equipment at all is illustrated in Figure 1.

Fixate on the center dot in the top panel while holding the book steady and trying not to blink. If you hold fixation steady enough for an extended period, the larger circle defined by the fuzzy contour will fade away. Eye movements to the X and back to the center dot will cause the circle to reappear and then slowly fade again.

It is impossible to produce the same effect while looking at the stimulus in the bottom panel. The reason for the difference in the two stimuli has to do with dynamic changes in intensity for individual receptors located near the contour that defines the circle, as illustrated in Figure 2.

The intensities signaled by receptors just inside and outside the contour that separates the circle from its surround are illustrated by the arrows. For the stimulus in the bottom panel, which has a sharp contour, tiny movements of the eye that occur even during steady fixation are sufficient to cause large changes in intensity to be signaled. The same small eye movements in the presence of the fuzzy contour represented in the top panel cause only small changes in intensity to be signaled. If the eyes are held sufficiently still, these small changes fall below the threshold for seeing, and the circle fades because the contour defining it is no longer visible.

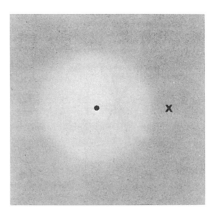

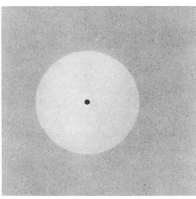

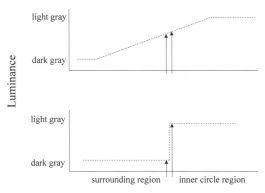

Horizontal Position Across Stimulus

FIGURE 1. Stimuli that can be used to illustrate perceptual fading. Reproduced from T.N. Cornsweet, *Visual Perception*, New York: Academic Press, Inc., © 1970, by permission of Academic Press.

FIGURE 2. The explanation for perceptual fading in the top stimulus but not the bottom stimulus of Figure 1 is related to the changes in luminance caused by small eye movements, as illustrated here. See text.

5. Describe how neurons with specific kinds of receptive field properties might be able to carry out local motion processing.

6. Describe what is meant by the aperture problem and discuss some computational methods that can be used to help alleviate this problem.

7. Describe what is meant by the correspondence problem and discuss some of the constraints biological motion processing systems utilize when searching for solutions to this problem.

8. Describe the motion processing stream and explain how intermediate levels of motion processing within this stream contribute to solving the aperture and correspondence problems.

9. What is the evidence that detection of coherent motion at the threshold can be caused by the activity of a single neuron in V5?

10. Summarize the higher-order types of motion processing that are involved in extracting information about self-motion such as direction of heading.

11. Summarize the higher-order types of motion processing that are involved in extracting structure-from-motion.

12. What kinds of problems must be solved in the How-stream?

13. Summarize the ways neural network models can implement dynamic aspects of perception.

1 Perception That Changes over Time

The ancient Greek philosopher **Heraclitus** proclaimed that one of the fundamental properties of the universe is that everything is constantly changing. Perception has a similar inherent dynamic quality. Indeed, if all change is eliminated from the retinal image, perception fades away and nothing is seen. This is ordinarily impossible to experience, because even when we attempt to hold our eyes perfectly still within a static environment, microsaccades and small drifts cause the retinal image to move by small amounts across the receptors. However, special equipment can be used to produce a static image that compensates for eye movements, thus stabilizing the image on the retina. When such an image is viewed, after a few seconds perception fades and nothing is seen. The phenomenon of **perceptual fading** can also be demonstrated with less sophisticated equipment, as illustrated in Box 10.1.

Observers use information carried by perceptual change for three general purposes. The first is making decisions about what portions of the environment need to be sampled, as described in Chapter 4. Biological perceptual systems are predisposed to take note whenever something in the environment moves or changes.

Second, is to help delineate the two- and three-dimensional structure of the surrounding environment. Information derived from perceptual change makes a significant contribution to higher-order aspects of What-perception, a topic commonly referred to as structure-from-motion. Third, information about change over time is used in the service of How-perception. A prototypical example is an outfielder catching a baseball. The outfielder must monitor the retinal information to detect the baseball, compute its trajectory, and control muscles so ball and hand arrive at the same location at the same time. To be successful, this has to all happen in real time prior to the ball's hitting the ground.

Several stages of processing of the input to perception intervene between the stage where photoreceptors sample changes in intensity over time and the stage where the higher-order properties of temporal change and motion that relate to What-perception and How-perception operate. These stages can be separated conceptually as illustrated in Figure 10.1.

The left panel shows input to perceptual processing in the form of retinal images that change over time. Each successive image is slightly different from the previous one due to the changing geometric relationships between the eye and the environment. These are registered initially as temporal changes in the intensities falling onto individual photoreceptors. An intermediate stage of processing tries to use these temporal changes to infer a **motion flow field,** as shown in the middle panel. The arrows in this diagram depict the inferred motions of individual points in the image. Finally, higher-order processing uses the motion flow field to make inferences about more complex properties such as three-dimensional layout of the surroundings, object motion, observer motion, and shapes of objects.

2 Neural Streams and Changing Information

The discussion in this chapter is organized based on the neural streams that are thought to be primarily involved in processing dynamic informa-

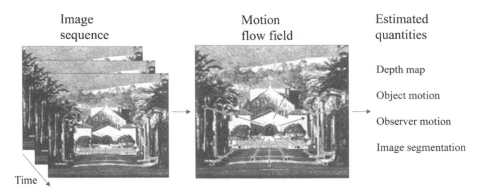

tion. A summary of the primary brain regions that are involved in these streams is presented in Figure 10.2.

This section presents an overview of the organization of these neural streams. Later sections cover the kinds of processing that take place within some of these specific structures in somewhat more detail.

The M-Stream

Information about changes in the retinal image that take place over time is transmitted, along with all other visual information, in the M-, P-, and K-streams that have been described in previous chapters. The K-stream will be ignored in this chapter because relatively little is known about its contribution to temporal properties of percepts. However, it is known that some parts of the K-stream play a significant role in dynamic control of eye movements and, in addition, feed into the extrastriate motion-stream via the pulvinar. Thus, the omission of any discussion of this system is much more a reflection of ignorance than a judgment about its lack of participation in processing temporal perceptual information.

There are two reasons for thinking the M-stream plays a more significant role in processing temporal information than the P-stream. First, the M-stream responds to higher temporal rates of change. Second, the M-stream feeds directly into the cortical motion-stream.

The Cortical Motion-Stream

The cortical motion-stream forms in layer 4B of V1 and passes through a number of extrastriate areas located in the occipital-temporal-parietal regions of the brain. This stream has been studied more carefully in monkeys than in humans. Although some results from humans will be included, the following description is taken primarily from studies of monkeys.

The heaviest input to the motion-stream comes from the geniculostriate M-stream, which terminates in sublayer 4Cα of the striate cortex. Axons of neurons in 4Cα have a dense projection to layer 4B, and that layer contains a high percentage of neurons that respond to motion with **directionally selective** receptive fields.

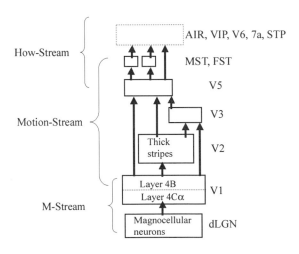

FIGURE 10.2. Diagrammatic summary of neural streams that are particularly involved in processing dynamic information.

Neurons in layer 4b project to the thick stripes of V2 and directly to **V5**. Neurons in the thick stripes of V2 have receptive field properties similar to those of neurons in layer 4B of the striate cortex, including orientation and motion sensitivity. Many of them also show strong responses to binocular disparity and appear to be well suited for processing temporal information about change in depth.

Extrastriate cortical area V3 receives input from layer 4B of the striate cortex and from the thick stripes of V2. Many of its neurons are sensitive to orientation as well as motion, and it has been speculated that this area is particularly concerned with processing information about changes in spatial patterns.

Neural subsystems in cortical areas V1, V2, and V3 that are thought to be involved in processing motion and depth all converge onto V5, also commonly referred to as the middle temporal (MT) area. The functional connectivity of V5 suggests that it is positioned as a bottleneck where motion and stereovision information can be combined. A number of lines of evidence are consistent with the idea that V5 plays a special role in motion processing.

V5 projects heavily to nearby extrastriate areas referred to as the medial superior temporal (MST) area and the fundus of the superior temporal (FST) area, as well as to the posterior parietal cortex. As is elaborated on further below, the receptive fields of neurons in these areas respond to large stimuli that are undergoing complex motions such as rotating, expanding, or contracting. They are probably involved in large field analysis of motion flow patterns that result from movement of the eyes and body, called **optic flow fields,** and with extracting complex structure-from-motion.

In addition to their role in sensory processing of motion information for generating motion percepts, the motion-stream is also involved in regulation of eye movements, as discussed in Chapter 4.

The How-Stream

Information that passes through the cortical motion-stream continues through a number of posterior parietal areas and areas in the frontal lobe that make up the diffusely organized neural stream involved in How-perception. These include the anterior intraparietal (AIP) and ventral intraparietal (VIP) areas, which are thought to be involved in transforming visual information into a form appropriate for guiding reaching and grasping movements. Neurons in the AIP are sensitive both

to three-dimensional properties of objects and to grasping movements towards these objects. The VIP area receives a direct projection from V5 and contains some neurons that are purely visual and others that respond to visual and tactile information.

Another area is V6, a strictly visual area most of whose neurons have receptive fields that are selective for either orientation or direction of motion. However, the visual responses of these neurons are often modulated by the position of the eye. This area is thought to be heavily involved in guiding reaching movements, especially when performed without looking directly at the target. While AIP and VIP are involved in transforming visual information into coordinates appropriate for action, V6 appears to be more specialized for providing information used in the actual guidance of the action.

Posterior parietal area 7a and the superior temporal polysensory area (STP) both respond to complex motion. Many cells in 7a respond to the rotary movement of simple stimuli such as a slit. Neurons in STP respond to size changes or rotation of more complex shapes. Some neurons in these areas show little response to motions of simple stimuli; but respond to motion of biological objects such as a hand or face in a directionally selective manner.

3 Temporal Vision and the M-Stream

Temporal vision refers to our ability to extract information about changes in light intensity over time. We are not equally sensitive to changes in intensity that take place over different time scales. When changes over time are specified in the frequency domain as described in Chapter 3, temporal processing can be characterized in terms of a temporal contrast sensitivity function. The temporal contrast sensitivity function is analogous to the contrast sensitivity function described for form vision in Chapter 8. The exact shape of the temporal contrast sensitivity function varies with mean luminance, but except under very dim lighting, this function acts as a band-pass filter.

Sensitivity to a homogeneous field of light modulated in luminance around its mean level is shown in Figure 10.3.

This figure illustrates that humans are most sensitive to changes at mid-temporal frequencies in the range from about 2 to 20 Hz. Events that occur either more slowly or more rapidly are not as visible.

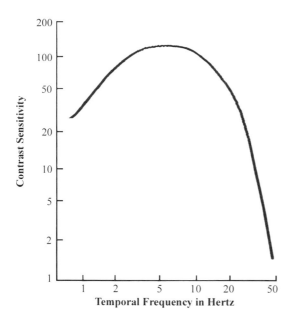

FIGURE 10.3. The temporal contrast sensitivity function for a human observer is a band-pass function.

As expected based on the M-stream's spatiotemporal properties, neural damage to the M-stream causes losses in sensitivity primarily at low spatial and high temporal frequencies, as illustrated in Figure 10.4.

Results are shown for two individual monkeys that had lesions made to the magnocellular layers of the dLGN. In addition to the losses in sensitivity shown here, the lesions also reduced the abilities of the monkeys to discriminate between motion in opposite directions or at different speeds. However, these deficits in motion perception reflect primarily the visibility of the stimuli used to test motion perception rather than direct alterations in motion processing. Lesioned animals can perform normally on motion tasks when stimuli are used that have sufficient spatial and temporal contrast that they can be seen.

4 Local Motion Processing in V1

The upper limit of our ability to see rapid change, called the **critical fusion frequency (CFF)**, is analogous to acuity in the spatial domain. The CFF in humans is typically about 55 Hz. If our temporal resolution were better, we would notice that light emitted from devices such as light bulbs and video display terminals is not really steady but simply changing at a rate more rapid than 55 Hz.

The M- and P-streams are specialized for transmitting different ranges of spatiotemporal information. The M-stream is most sensitive to high temporal and low spatial frequencies. The P-stream is most sensitive to low temporal and high spatial frequencies.

In the natural environment in which evolution of biological organisms took place, fast temporal changes were usually caused by motion of objects rather than by changes in the intensity of light eminating from a stationary object. Consequently, biological visual systems have evolved a predisposition to interpret changes in light intensity as motion. If a particular change in spatiotemporal pattern of stimulation across the retina is consistent with motion, we immediately perceive it as such. In the limiting case, called the **phi phenomenon,** apparent motion can be produced by turning two stationary lights located side by side on and off in a proper sequence. The phi phenomenon was studied extensively by Gestalt psychologists following its description by **Max Wertheimer** in the early 1900s. Modern engineers have been able to

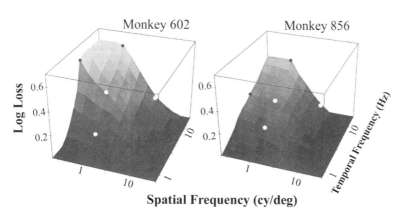

FIGURE 10.4. Deficits in sensitivity to spatial and temporal frequencies following lesions to the magnocellular layers of the dLGN. Measured values are shown by the circles, and the surface was interpolated from the data points. Results are shown for two monkeys, 602 and 856. Reproduced from W.H. Merigan, et al, Does primate motion perception depend on the magnocellular pathway? *J. Neurosci.* 11:3422–3429, © 1991, by permission of Society for Neuroscience.

exploit this biological predisposition to interpret charges in intensity as motion to build devices that simply display rapidly changing spatiotemporal patterns of light intensity but produce a percept of **apparent motion** as discussed in Box 10.2.

Box 10.2
Real Versus Apparent Motion

Much of what humans perceive as motion in a technological society has no physical basis. It is simply apparent motion produced by stationary displays that change in intensity over time. Consider "motion pictures," which actually contain no motion whatsoever. A static frame of intensity stimulation is flashed onto a screen for a brief period. This is followed by another static frame having a slightly different spatial pattern of intensities, and then another, and so on. The same is true for television sets; there is no actual motion on the screen. Individual pixels on the screen simply change their intensities over time but do not move. Nevertheless, when people look at the screen in the movie theatre or on a television set, they do not report a percept of complex spatial patterns of intensity changing over time. They perceive motion, and it is motion of the kind that would most likely have been present if the same spatiotemporal pattern of stimulation had fallen onto their retinas in the environment in which primates evolved.

It does not even take a complex technology to achieve apparent motion. Most people in industrialized societies have encountered storefront signs made of marquee lights that use apparent motion to draw attention to their place of business. There is, of course, no motion present at all in these displays, simply light bulbs that turn on and off in the proper sequence to produce apparent motion.

Biological visual systems have evolved neural mechanisms that convert spatiotemporal patterns of stimulation detected by photoreceptors into a motion signal. Somewhat surprisingly, the primate visual system, in contrast to that of species in many other families of animals, does not make a conversion to a motion signal at the earliest stages of visual processing. Directionally selective neurons are not present (at least not in significant numbers) in the retina, the dLGN, or the input layers of the striate cortex. Thus, the M-stream is concerned with temporal processing but not with motion until information reaches beyond the input layers of the striate cortex.

Directionally selective neurons in the superficial and deep layers of the hypercolumns in striate cortex operate to detect local motion in each small region of the retina. This information, considered globally, forms the basis for a motion flow field of the kind illustrated in Figure 10.1.

The operations that must be carried out to extract motion from spatiotemporal changes in intensity have been described formally in computational models of motion processing.

The Reichardt Correlation Model

Studies of responses to motion in insects carried out by **Werner Reichardt** and colleagues in the 1950s and 1960s demonstrated that two lights falling on adjacent photoreceptors in the proper sequence is the elementary event evoking motion perception. These studies led to development of a model of motion detection now commonly referred to as the **Reichardt correlation model.** The computations performed by this model are equivalent to mathematical autocorrelation of the inputs.

The correlation model is appropriate for insect vision, in which motion is computed based on stimulation of adjacent receptors, but needs to be modified to account for motion processing in mammals, in which signals from several receptors are pooled before motion analysis is performed. Correlation models that have been modified along these lines are sometimes referred to as **elaborated Reichardt correlation models.**

A number of other investigators have developed similar neural models to try to explain how directional selectivity might be attained at the single-unit level in mammals. One of the earliest was proposed by **Horace Barlow** and **William Levick** to explain directional selectivity in the rabbit retina. Similar models have been developed to explain directional selective neurons in the primate striate cortex.

A common element of most neural models of directional selectivity is that a **delay** is introduced into signals generated at some retinal locations and not into others. An intuitive understanding of how a delay can be used to detect motion can be attained

Motion in Space-Time

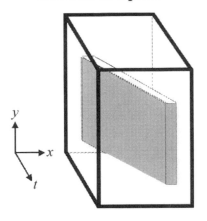

FIGURE 10.5. Motion can be represented in a space-time plot. Adapted from B.A. Wandell, *Foundations of Vision*, Sunderland, MA: Sinauer Associates, Inc., © 1995, by permission of Sinauer Associates, Inc.

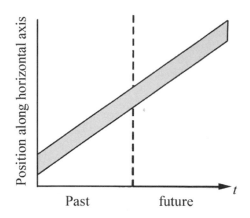

FIGURE 10.6. A simplified space-time plot can be used to represent retinal motion along a single dimension.

by using a space-time representation to specify both motion-sensitive and directionally selective receptive fields.

Specification of Retinal Motion in Space-Time

One convenient way to characterize the direction and magnitude of motion that occurs within the retinal image is a **space-time representation**. This is illustrated in Figure 10.5, which represents an image of a vertical line moving across the retina at a constant velocity.

Each point in this volume represents an individual element of light intensity $I(x,y,t)$ at a particular location x,y on the retina at a particular point in time t. Since the vertical line in this example is moving horizontally across the retina, there is no change in the vertical direction. Thus, a simpler

notation that eliminates the y dimension can be used with no loss of information. This is illustrated in Figure 10.6.

At any moment in time, the spatial pattern of interest is present at some horizontal location on the retina. This location is specified on the vertical axis. Time is plotted on the horizontal axis and runs from the past on the left to the future on the right. The present is designated on this plot by the dashed vertical line. A constant velocity is designated on a plot of this type by a line of a given slope. Some examples are illustrated in Figure 10.7.

An element that does not change position with time shows up as a horizontal line on the space-time plot. If the object is moving at a constant velocity, it will form a straight line of a given slope. Slow velocities map onto shallow slopes and fast velocities onto steep slopes. All of the examples shown here have positive slopes, because they all have the same direction of motion. Motion in the opposite direction would produce similar negative slopes.

Note that a neuron with selectivity for the orientation of the slope in a space-time representation

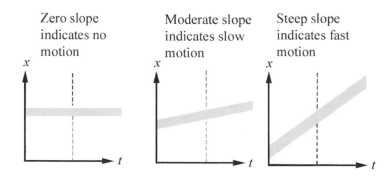

FIGURE 10.7. Motion of constant velocity can be represented by a line of a given slope in a space-time plot.

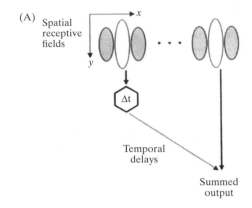

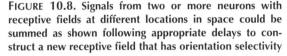

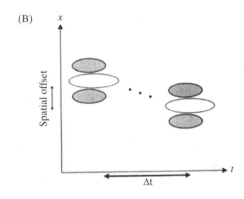

FIGURE 10.8. Signals from two or more neurons with receptive fields at different locations in space could be summed as shown following appropriate delays to construct a new receptive field that has orientation selectivity in space-time. Reproduced from B.A. Wandell, *Foundations of Vision*, Sunderland, MA: Sinauer Associates, Inc., © 1995, by permission of Sinauer Associates, Inc.

could code, in principle, for both direction and velocity of motion.

Specification of Receptive Fields in Space-Time

One method that could be used to construct a receptive field with selectivity for the orientation of the slope in a space-time representation is illustrated in Figure 10.8.

Consider two or more neurons with receptive fields located at different positions in space, as illustrated in the left panel (A). The signals generated by these neurons can be fed as input to a single output neuron, where they are simply summed. However, various amounts of delay are introduced for the signals generated by successive input neurons.

The receptive field of the output neuron can be illustrated in a space-time plot, as illustrated in the right panel (B). The spatial positions of the input to the neuron are separated along the vertical axis (x), and the delay introduces a shift along the time axis (T). Thus, the receptive field of the output neuron will respond best to a stimulus that produces a line of a particular **orientation in this space-time plot** corresponding to a particular direction and velocity. The neuron in this example will be selective for motion represented by the dotted line in the right panel.

Delays can be implemented in biological tissue with two kinds of mechanisms. The first is use of a longer axon to make one of the two connections. The second is alteration of the time constants of the postsynaptic membranes that sum the inputs. A longer time constant acts as a low-pass filter that automatically introduces a delay. The rationale for understanding the effects of a low pass filter in the time domain is analogous to understanding the effects of low pass filters in the spatial domain as discussed in Chapter 8.

Figure 10.9 illustrates the build-up of voltage across the postsynaptic membrane over time when input is received at the time denoted by the tick mark on the horizontal axis. Suppose the postsynaptic cell has a threshold, at the level indicated by the dashed line, for detecting that the voltage across its membrane has changed. The neuron in the top panel will register the fact that input is received quickly. The arrival of the input at the membrane

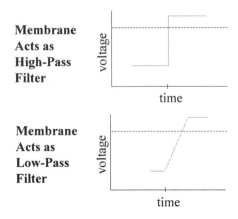

FIGURE 10.9. Illustration demonstrating how membranes with long time constants can be used to introduce delays in registering signals. See text.

shown in the bottom panel will be registered only after a delay.

Computational Approaches Based on Gradient Schemes

Another set of computational approaches to motion processing that were developed independently from correlation models are generally referred to as **gradient schemes.** These are based on carrying out spatial and temporal derivatives of image intensities. In recent years there have been some attempts to compare these two computation approaches to determine the extent to which they make the same or different predictions. It has now been demonstrated that the correlation and gradient approaches are basically equivalent for stimuli having low contrast and make identical predictions for the outcomes of most experiments. The two types of models do make different predictions for certain specially constructed stimuli, but the current evidence from primates that would decide between these two approaches does not yet decisively favor one or the other.

A fundamental assumption of gradient schemes is that all of the intensity changes in an image are due to motion rather than to changes in intensity that occur for some other reason. This gives rise to a potential problem for all gradient schemes, because the intensity values at each point in an image formed of an object change for many reasons not directly related to motion of the object, such as a shadow from a cloud moving overhead. In addition, the neural signals regarding intensities at each point may change over time due simply to neural noise rather than to motion in the image. Because of this problem, capturing motion by taking derivatives of the changes in intensity values signalled by individual neurons, while possible in principle, may not be possible as a practical matter.

Marr suggested that one strategy to alleviate the effects of noise would be to take as the inputs for motion processing the zero crossings of band-pass filters rather than the raw intensity values. This strategy is analogous to the use of zero crossings when looking for edges during form vision processing, as described in Chapter 8.

Neurons with receptive field properties appropriate for taking the spatial and temporal derivatives of band-pass-filtered images are illustrated in a space-time representation in Figure 10.10.

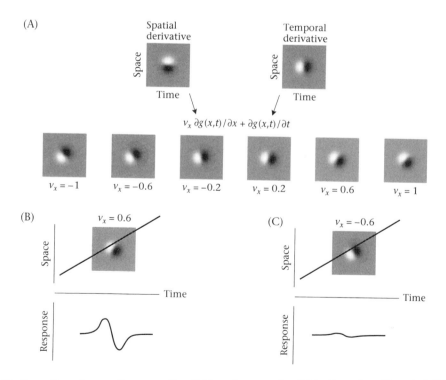

FIGURE 10.10. Formation of space-time receptive fields that take the spatial and temporal derivatives of the retinal image and their combination to form directionally selective receptive fields. Reproduced from B.A. Wandell, *Foundations of Vision,* Sunderland, MA: Sinauer Associates, Inc., © 1995, by permission of Sinauer Associates, Inc.

The outputs of these neurons compute the spatial and temporal derivatives as shown in the top row of panel A. When the derivatives are summed in various combinations, as illustrated in the second row, oriented receptive fields are formed in the space-time plots along the lines discussed in the previous section. For example, panels B and C illustrate a pair of receptive fields, one of which would respond optimally and the other not at all to motion of a particular direction and velocity. These two fields could be combined to produce the receptive field of a neuron that is attuned to motion of this particular direction and velocity.

5 Problems of Local Motion Processing Mechanisms

In principle, neurons with directionally selective receptive fields that are located in the hypercolumns of the striate cortex are sufficient to extract local motion signals. However, all local motion processing mechanisms have certain inherent limitations.

The Aperture Problem

Consider that an observer is looking through a round window and sees a vertical bar move across from left to right as illustrated in the top panel of Figure 10.11.

Based on this observation alone, the observer will not be able to specify either the direction the bar was moving or its velocity. This is because

motion of a bar viewed through an aperture is always ambiguous. The motion that is seen is caused only by motion of the bar orthogonal to its orientation; motion parallel to the bar causes no change in the stimulus. Thus, the bar that produced the perceived motion might have been moving from left to right in front of the window at its perceived speed, as illustrated on the left side of the bottom panel. However, the bar also might have been moving at a faster rate of speed in a direction upward and to the right or downward and to the right, as illustrated in the middle and the right side of the bottom panel. In fact, an infinite number of combinations of directions and velocities for the physical bar could have given rise to the apparent speed of the bar's motion across the window. This fact, called the **aperture problem,** was characterized in an extensive set of studies performed in the 1930s by the psychologist **Hans Wallach.**

Motion Constraint Lines

Seeing the motion of a bar through an aperture does provide some constraints on the possibilities for the true physical motion. All of the potential physical motions that could have given rise to the observed motion are constrained to fall along a **motion constraint line** when plotted in the form of a velocity-space plot as illustrated in Figure 10.12.

This figure represents the physical constraints on motion of a vertical line that appears to be moving through an aperture from left to right as in the top of Figure 10.11. The thick arrows are vectors that represent possible physical motions of the line that might have given rise to this perceived motion. The direction of physical motion is given by the angle of each thick arrow and the velocity by its length. The ends of the arrows fall along the motion constraint line designated by the dashed line. It represents all possible combinations of physical velocities and directions that would be consistent with the motion seen through the aperture.

Motion seen by a local processing mechanism in the form of a neuron with a receptive field always confronts it with this same ambiguity. At best, the local mechanism can specify the velocity of an edge that passes through its receptive field in terms of the velocity in the direction perpendicular to the edge. It can be inferred that the direction of the physical motion that gave rise to it falls along the constraint line, but there are still an infinite number of possibilities that fall along that constraint line.

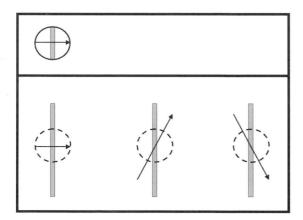

FIGURE 10.11. Demonstration of the aperture problem. See text.

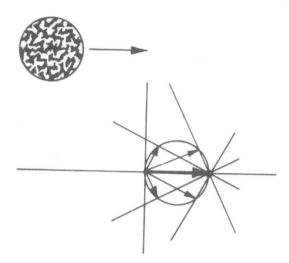

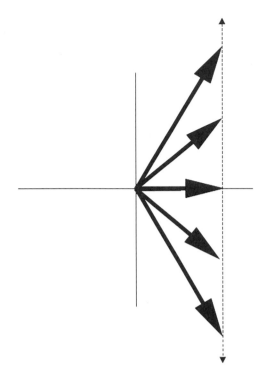

FIGURE 10.13. Oriented texture elements passing through an aperture at the same velocity and direction will have different constraint lines, but all the constraint lines will have to intersect at a point. Adapted from E.H. Adelson and J.A. Movshon, Phenomenal coherence of moving visual patterns. *Nature* **300:**523–525, © 1982, by permission of Macmillan Magazines Ltd.

FIGURE 10.12. All potential physical motions consistent with what is perceived through an aperture fall along a motion constraint line.

Intersection of Constraints

A potential computational solution to the aperture problem, proposed by **Edward Adelson** and **J. Anthony Movshon,** is called the **intersection of constraints.** Figure 10.13 shows a random texture field moving to the right through an aperture.

The edges of the texture elements come in all orientations. Each will produce its own constraint line, as illustrated for four specific contours in the bottom panel. However, if these elements are all part of the same physical object, then they will all be moving at the same direction and velocity. Thus, their constraint lines will have to intersect at a point corresponding to that direction and velocity. This fact offers a strategy that can, in principle, be used by a motion processing system to infer physical motion based on what is seen through an aperture. Confronted with the texture elements shown in Figure 10.13, any individual local motion processing mechanism would be able to specify only that the motion has to fall somewhere along its constraint line. However, a higher-order process that examines the constraint lines that are output by a number of local motion processing mechanisms

could recover the true motion by determining the direction and velocity corresponding to the point where the constraint lines intersect.

The intersection of constraints rule breaks down under conditions in which motion in two or more directions is superimposed at the same location in the retinal image. One way this can happen is under stimulus conditions that produce **transparent motion,** as illustrated in Figure 10.14.

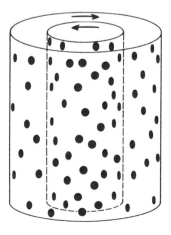

FIGURE 10.14. Humans can see simultaneous motion in two directions when viewing transparencies.

Texture elements have been painted on the surfaces of glass cylinders that rotate in opposite directions. Observers looking at these cylinders perceive transparent motion in which two surfaces moving in opposite directions are seen simultaneously.

This example illustrates that human motion perception does not rely on a simple algorithm based on the intersection of constraints rule. The constraint lines of the motions of the individual texture elements in this display do not intersect at a single point. In order to model human perception of transparent motion, more complicated algorithms would have to be developed in which the constraint lines are examined for evidence of two or more points where intersections cluster.

The fact that humans can detect transparent motion raises a more general issue. In the example of transparent motion perception, some texture elements had to be combined to infer physical motion in one direction and other elements simultaneously combined to infer physical motion in a different direction. The general issue is how higher-order motion processing decides which inputs to combine. In order to try to understand how higher-order motion processing makes these decisions, a large body of research has been carried out with special motion stimuli called **plaids.**

Plaids

Plaid stimuli have been widely studied to try to discover the conditions under which individual texture elements are combined to form a percept of coherent motion of a single physical surface. The rationale for these studies is illustrated in Figure 10.15.

A stimulus displayed on a video monitor is composed of two superimposed gratings that are drifting in different directions and at different speeds. The motion constraint lines show the motion that will be detected when either grating is presented alone. The intersection of constraints shows the motion for a combined plaid pattern. In the example shown in Figure 10.15, the direction and motion of the plaid are radically different from those of either of the component gratings. The question of interest is whether a human observer looking at this display will perceive transparent motion of two separate gratings sliding past one another or coherent motion of a single moving plaid surface.

If the two gratings are of similar spatial frequency and contrast, the empirical answer is that usually

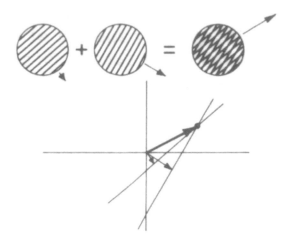

FIGURE 10.15. Illustration of how plaid stimuli can be used to study the operation of the intersection of constraints rule. Adapted from E.H. Adelson and J.A. Movshon, Phenomenal coherence of moving visual patterns. *Nature* 300:523–525, © 1982, by permission of Macmillan Magazines Ltd.

the components are combined and the perceived motion is of a coherent plaid pattern. When the two gratings differ in contrast or spatial frequency, the result is less consistent. In some cases the two component's patterns are combined to produce a percept of coherent plaid motion, and in other cases transparent motion of two gratings slipping past one another is perceived.

Perception of plaids has important implications for perception of motion of more complex patterns, like one's grandmother, as described in Box 10.3.

Questions about what specific stimulus features can be combined to form coherent motion continue to provide the motivation for an active area of psychophysical and computational research. Researchers currently have only a rudimentary understanding about how biological motion processing sorts out the various sources of local motion information to arrive, more often than not, at a veridical percept of moving objects. Machine vision is still struggling with how to deal with these issues.

The Correspondence Problem

The correspondence problem has already been discussed in Chapter 9 in conjunction with stereopsis. In that context, the problem is in deciding what elements in the image from the left eye correspond to elements in the image from the right eye. The same problem exists when processing motion, except that

Box 10.3
Did Grandmother Move?

Recall from Chapter 8 that there is evidence that information about the spatial pattern of an object is transmitted from the eye by four or more parallel channels, each of which carries information about a band of spatial frequencies. Consider information being transmitted from the eye about grandmother. Information about the spatial frequencies that make up grandmother is being processed in parallel by these band-pass filters. What happens when grandmother moves?

One model of motion processing is that the information that grandmother moved must be processed in parallel. The visual system determines that low spatial frequencies moved and, independently, that medium spatial frequencies moved, and so on. The brain infers "Grandmother must have moved because her low spatial frequencies moved, her medium frequencies also moved, etc." This model will work only if the sinusoidal gratings that make up grandmother can be combined to produce coherent motion. One would never know grandmother moved if the different spatial frequencies that make up grandmother couldn't be combined into a single "grandmother plaid" during motion processing.

The second model of motion processing is that the brain has to wait until the spatial frequencies that make up grandmother have been put together. Then motion analysis is performed on the "feature" of grandmother instead of on the component spatial frequencies. This model allows motion processing to figure out that grandmother moved only after performing form vision to know one is looking at grandmother.

have been shifted to the right. Depending on a number of specific parameters, such as the spacing between the dots and the timing used in showing the frames, one of two percepts is reported. The middle column shows a percept called **group motion,** in which the observer reports that all three dots move to the right as a group. The arrows in the figure illustrate which pairs of dots in the first and second frames must have been fused to produce this percept. The right column shows a percept called **element motion,** in which the observer reports that the leftmost dot in the first frame changes position but the other two dots remain stationary. The arrows illustrate the pairs of dots that must have been fused to produce this percept. This example underscores the **motion correspondence problem:** Motion percepts depend on which elements in a scene that is changing over time are fused.

The correspondence problem has been extensively studied using **random-dot kinematogram (RDK)** stimuli, analogous to the random-dot stereograms described in Chapter 9.

Random-Dot Kinematograms

A random-dot kinematogram stimulus consists of successive frames of spatial patterns projected onto a screen in a rapid stream, typically at 15 to 30 Hz. Each frame is composed of a large number of individual texture elements, referred to as "dots," because in the simplest case the texture elements are simply bright or dark spots of light.

Some of the dots are defined as **motion dots.** These dots have a "lifetime" that lasts a set number of frames. In successive frames during the lifetime of a motion dot, its position is displaced by a fixed

the correspondence question relates to two images that differ in time instead of to images formed in the two eyes. A classic perceptual phenomenon, often called the **Ternus effect** after J. Ternus the German psychologist who made extensive studies of it, in the 1920s is shown in Figure 10.16.

Three dots are presented on a screen briefly (Frame 1), followed (Frame 2) by three dots that

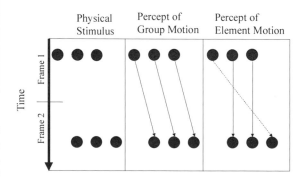

FIGURE 10.16. The fact that two frames of apparent motion, as shown in the left panel, can give rise to two types of motion percepts, as shown in the middle and right panels, is called the Ternus effect.

No Correlation   50% Correlation   100% Correlation

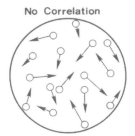 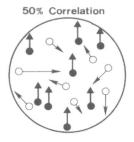 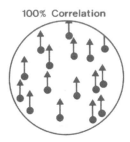

FIGURE 10.17. Motions of individual dots in an RDK display with three different degrees of correlation. Filled circles represent motion dots and open circles represent noise dots. Adapted from W.T. Newsome and E.B. Paré, A selective impairment of motion perception following lesions of the middle temporal visual area (MT). *J. Neurosci.* 8:2201–2211, © 1988, by permission of Society for Neuroscience.

amount in a specified direction. At the end of its lifetime, the motion dot is replaced by a new motion dot at a random location that continues for its lifetime, and so on. The direction and velocity of apparent motion is identical for all of the motion dots in the display.

The rest of the dots are defined as **noise dots.** These can be defined in two ways. Some investigators use noise dots that behave identically to the motion dots except that the direction of each individual noise dot is determined randomly. Other investigators use noise dots that have a shorter lifetime than the motion dots. Each noise dot is replaced by a new noise dot at a new random location on each frame. A stimulus made exclusively of noise dots produced with the second method contains, by chance placement of the dots on successive frames, motion signals of all speeds and directions. Human observers viewing this stimulus report an appearance similar to the visual noise or "snow" seen on a television set that is tuned to a channel that is not being broadcast. A stimulus made exclusively of noise dots produced by the first method produces incoherent motion in all directions but of constant velocity. In general, empirical studies using RDKs based on the two kinds of noise have obtained similar results.

The proportion of the motion and noise dots intermixed in a given RDK is referred to as its **correlation.** The effect of varying correlation is illustrated in Figure 10.17.

The left panel illustrates the conditions for 0% correlation, the middle panel for 50%, and the right for 100%. Under optimal conditions, human and monkey subjects can reliably report the direction of motion of the coherent signal when the correlation is 3% to 5% or more.

An RDK stimulus is a valuable research tool for studying perception of coherent motion. When only a small percentage of the dots are correlated and the lifetime of each dot is brief, the researcher can be certain that detection of coherent motion is not based on local mechanisms. Coherent motion can be detected only based on motion processing that performs a higher-order, **global** analysis of the motions of large numbers of dots.

Short- and Long-Range Motion

The psychologist **Oliver Braddick** made a distinction between **short-range** and **long-range** motion mechanisms based on how people perceive classic apparent motion and coherent motion produced by RDK stimuli. Observers perceive coherent motion in an RDK stimulus only when the spatial separations of the individual elements between frames are less than about 15 minutes of arc. Greater separations produce the appearance of flickering dots but no motion. However, apparent motion produced by classic stimuli consisting of only a small number of elements can be seen with much greater separations. Braddick concluded that classic apparent motion is produced by long-range motion mechanisms and RDK coherent motion by short-range motion mechanisms.

More recent studies suggest that this distinction between short- and long-range motion processing mechanisms may reflect a difference in the response of a single higher-order motion processing mechanism to different kinds of stimuli rather than two separate processing mechanisms. Confronted with some kinds of stimuli, especially simple stimuli consisting of only a few elements, the motion system will work very hard to try to fuse elements over time, even when doing so requires making inferences about large displacements. Under these conditions, the correspondence problem is minimized, because only a few elements are present and these can often be differentiated based on features such as oriented edges. However, for other stimuli, especially complex stimuli such as RDKs with many potential elements that could potentially be fused, the visual system attempts to infer motion only over small regions of space.

Role of Depth and Surfaces in Constraining Motion

One heuristic used by the motion processing system to help solve the correspondence problem is **proximity.** Given two equally good matches, preference is given to the match to the nearest neighbor. This rule is not rigidly adhered to, as demonstrated by perceptual phenomena such as element motion in the Ternus effect illustrated in Figure 10.6, but is generally followed.

One question that has been of interest to perceptual scientists is whether "nearest neighbor" is defined in terms of two-dimensional distance in the retinal image or of three-dimensional distance in the environment. This issue was addressed in a clever study reported by Marc **Green** and **J. Vernon Odom** using the methods illustrated in Figure 10.18.

Subjects viewed a stimulus on a screen with the configuration shown in the left panel. A frame showing four spots of light at the locations shown with solid lines was shown first followed by a frame showing four spots at the locations designated with the dashed lines. Subjects reported seeing apparent motion when viewing this display. The motion was seen as being clockwise about half of the time and

counterclockwise the other half of the time. This is to be expected, since there are equal distances on the retinal image when fusing between pairs of stimuli in the clockwise and counterclockwise directions.

Next, the same stimuli were viewed under conditions of binocular disparity such that each stimulus was seen as being at either a near (N) or a far (F) distance, as illustrated for four of the stimuli in the right panel. Under these conditions, the apparent motion was usually seen as being in the clockwise direction. The lateral distances in the retinal image separating the fused pairs for the clockwise and counterclockwise directions remain equal. However, the distances in perceived three-dimensional space are closer for the clockwise direction. The interpretation of this result is that solutions to the correspondence problem by the motion processing system take into account three-dimensional information that comes from stereovision processing.

Later studies have refined this interpretation. The critical variable is not strictly distance in three-dimensional space but distances along perceived three-dimensional surfaces. This was demonstrated in a study by Zijiang **He** and **Ken Nakayama,** shown in Figure 10.19.

The left panels show results that can be interputed similar to those reported by Green and Odom. Two pairs of stimuli were presented in sequence to produce apparent motion. Results are shown for two observers, SS and ZH. When no depth was introduced to the viewing situation, direction of the motion was ambiguous and was

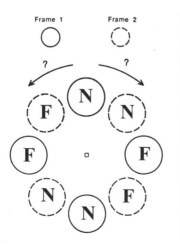
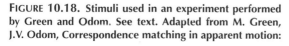
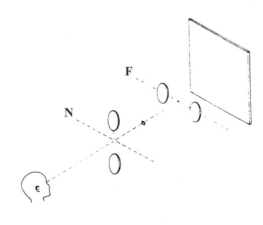

FIGURE 10.18. Stimuli used in an experiment performed by Green and Odom. See text. Adapted from M. Green, J.V. Odom, Correspondence matching in apparent motion:

Evidence for three-dimensional spatial representation. *Science* 233:1427–1429, © 1986, by permission of American Association for the Advancement of Science.

FIGURE 10.19. Stimuli (top panels) and results (bottom panels) for two studies carried out by He and Nakayama. Results from two observers who participated in the first study (SS and ZH) are shown on the left. Results of the same observers in the second study are shown on right. Reproduced from Z.J. He and K. Nakayama, Apparent motion determined by surface layout not by disparity or three-dimensional distance. *Nature* 367:173–175, © 1994, by permission of Macmillan Magazines Ltd.

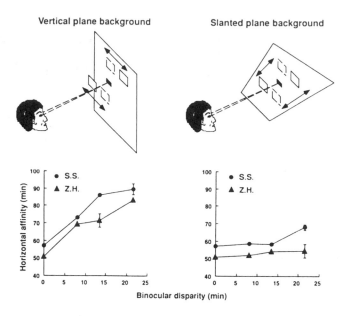

seen about half of the time in the vertical direction and the other half in the horizontal. When binocular disparity was introduced, creating a greater apparent distance in three-dimensional space for the vertical than for the horizontal pairings, there was an affinity for seeing motion in the horizontal direction.

The panels on the right show a follow-up study using the same two observers. The results of this second study demonstrate that horizontal affinity is not exhibited when a slanted plane is introduced into the viewing situation. The elements are now perceived as lying on the slanted surface, and they appear equally likely to move upwards along the surface as horizontally along the same surface.

An interpretation consistent with all of these findings is that basic visual processing mechanisms involved in solving the motion correspondence problem are influenced by information concerning the three-dimensional layout of surfaces. Recall from Chapter 8 that information about surfaces appears to permeate all stages of spatial pattern processing. The same is apparently true for motion perception. Solving the motion correspondence problem is generally considered a relatively low-level, and presumably early, stage of motion processing, yet it is influenced by the properties of perceived surfaces in the scene.

Solutions to the motion correspondence problem most likely do not occur at any single stage of processing. Instead, it appears that solutions arise based on recurrent interactions among several stages of motion processing that occur in parallel and all simultaneously influence one another.

Second-Order (Non-Fourier) Motion

In 1976, George Sperling demonstrated that it is possible to construct stimuli that have motion known to be invisible to neurons with directionally selective receptive fields of the type described in previous sections yet still give rise to perceived motion. The perceived motion produced by these kinds of stimuli is referred to as **second-order motion.** An example of a stimulus that causes second-order motion is shown in Figure 10.20.

The stimulus consists of a set of texture patterns. Texture within some spatial bands moves up, and that in other bands moves down. The separation

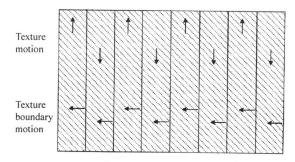

FIGURE 10.20. The configuration of a stimulus that produces second-order motion.

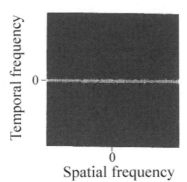

FIGURE 10.21. The spatial and temporal characteristics of a non-fourier motion stimulus in the spatiotemporal domain (left) and in the frequency domain (right). Adapted from T.D. Albright. Form-cue invariant motion processing in primate visual cortex. *Science* 255:1141–1143, © 1992, by permission of the Association for the Advancement of Science.

between the two bands of motion defines a boundary. In the figure, the locations of these boundaries are designated by the vertical lines. However, there is no edge in the actual stimulus – lines are just drawn here to designate where the motion-defined boundaries are located. The motion-defined boundaries are moving to the left, and an observer viewing this stimulus perceives leftward motion.

The motion-defined boundaries are caused by shearing motions of the local texture patterns. However, there is no signal that can be picked up based on luminance differences described in the frequency domain in terms of spatial and temporal frequencies. For this reason, this type of second-order motion stimulus is referred to as producing **non-Fourier motion.**

A characterization of non-Fourier motion in the spatiotemporal and the frequency domains is shown in Figure 10.21. An apparently rightward-moving stimulus is represented on the left in the form of a space-time plot in which the luminance of the stimulus is plotted as a function of time and one spatial dimension (along the direction of motion). The slanted diagonal band on this plot designates motion, as described above in Figure 10.7. However, when this same stimulus is described in the frequency domain, as shown on the right, there is no evidence of cues that could be used to identify stimulus direction or velocity.

A number of other examples that produce second-order motion stimuli have been discovered. These stimuli are based on **motion primitives** other than luminance edges. These include moving color edges, illusory contours, terminators and corners, discontinuities in texture, and discontinuities in depth. It is not clear whether there is a single motion system that can receive inputs about motion from each of these primitives or several different motion subsystems operating in parallel based on these different primitives.

6 Intermediate Levels of Motion Processing in V5

A number of aspects of motion perception cannot be adequately handled by a local motion processing mechanism, such as a neuron in V1 with a directionally-selective visual field. For example these mechanisms cannot detect second-order motion, solve the correspondence problem, combine components to form plaids. These higher order motion perception processing tasks are probably first tackled in the cortical motion-stream. The most extensively studied extrastriate area involved in motion processing is V5.

Directionally Selective Receptive Fields in V5

V5 is the first stage in processing along the geniculostriate pathway at which the majority of the neurons are concerned with motion. The receptive field of a representative V5 neuron is shown the form of a polar plot in Figure 10.22.

The neuron responds vigorously with a burst of spikes when a bar stimulus moves downward and to the left. The receptive field exhibits a tuning curve such that the responses produced for stimuli away from the preferred direction are progressively attenuated. The tuning curve is relatively narrow and no excitatory response is seen to motion in the opposite direction.

V5 neurons show a similar response to RDK stimuli, as illustrated in Figure 10.23.

This neuron responds best to an RDK stimulus that defines coherent motion upward and to the right. This directional selectivity for coherent motion demonstrates that at least some neurons in **V5** are able to integrate across local motion-processing mechanisms to pick up global motion information.

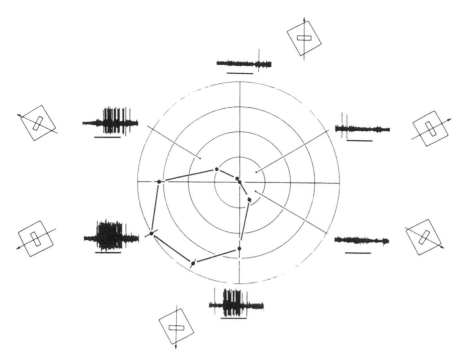

FIGURE 10.22. Receptive field of a motion-sensitive neuron in area V5 of a monkey tested with moving bars of light. Reproduced from J.H.R. Maunsell and D.C. Van Essen, Functional properties of neurons in middle tem- poral visual area of the macaque monkey: I. Selectivity for stimulus direction, speed, and orientation. *J. Neurophysiol.* 49:1127–1147, © 1983, by permission of the American Physiological Society.

V5 Neurons Are Causally Related to Motion Detection

William **Newsome** and colleagues have recently carried out a landmark set of studies in which they

recorded neural activity from single V5 neurons in a monkey while the animal was performing a motion-detection task. The purpose of these studies was to test whether V5 neurons carry signals that support the actual perceptual judg-

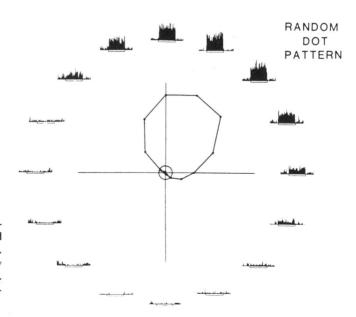

FIGURE 10.23. Receptive field of a motion-sensitive neuron in area V5 of a monkey tested with an RDK stimulus. Reproduced from T.D. Albright, Direction and orientation selectivity of neurons in visual area MT of the macaque. *J. Neurophysiol.* 52:1106–1130, © 1984, by permission of the American Physiological Society.

A

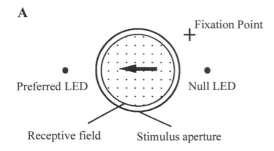

B

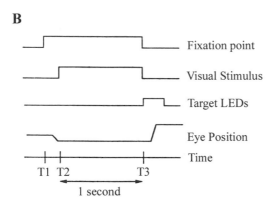

FIGURE 10.24. Experimental paradigm used in studies with monkeys to test the hypothesis that V5 neurons carry signals that support perceptual judgments of motion. See text. Adapted from C.D. Salzman, et al, Microstimulation in visual area MT: Effects on direction discrimination performance. *J. Neurosci.* 12:2331–2355, © 1992, by permission of the Society for Neuroscience.

ments of motion. The procedure is summarized in Figure 10.24.

A fixation point was turned on at time T1 to start a trial. The monkey had been previously trained to fixate on this target when it appeared. Once fixation was established (T2), an RDK stimulus was presented within an aperture positioned over the receptive field of a V5 neuron in the brain of the animal whose output was being recorded. The RDK stimulus was presented for a period of 1 sec. The stimulus parameters were adjusted to match the neuron's preferred velocity and direction, indicated by the arrow. The number of spikes produced by the neuron was recorded during this stimulus period. When the stimulus ended (T3), two light-emitting diodes (LEDs) appeared. One was in the preferred direction from the stimulus and the other in the opposite, or null, direction. The monkey signaled its judgment of perceived direction by looking at one or the other of the LEDs. Correct responses were rewarded with juice. A

number of trials were performed that included a range of correlation values for both directions of stimulus movement.

The number of spikes produced by a V5 neuron in response to any particular stimulus varied from trial to trial, as summarized in the top panel (A) of Figure 10.25.

For low-correlation RDK stimuli, the distribution of responses to motion in the preferred direction (open bars) overlapped with the distribution of responses to motion in the null direction (filled bars). As the degree of correlation increased, the distributions of responses to motion in the two directions became increasingly separated. For the neuron shown in Figure 10.25, the distributions essentially overlap completely when correlation is 0.8% and are almost totally separated when correlation is increased to 12.8%.

Theory of signal detection procedures, as described in Chapter 2, were applied to these distributions to generate **ROC** curves, as illustrated in the lower left panel of the figure (B). Consider first how the curve for the 0.8% condition shown on the ROC curve was generated. A criterion (crit) was systematically varied from a low rate of spikes to a high rate of spikes. For each criterion level, two values were calculated. First, the proportion of the trials in which the spike rate for the preferred direction exceeded the criterion was calculated. This value was plotted on the vertical axis. It corresponds to the hit rate as defined by the theory of signal detection. Second, the proportion of trials in which the spike rate for the null direction exceeded the criterion was calculated and plotted on the horizontal axis. This corresponds to the false alarm rate as defined by the theory of signal detection. These values were plotted for each criterion level to generate the ROC curve for the 0.8% stimulus. Similarly, this entire procedure was repeated for each of the other correlation levels to generate all of the curves shown in B.

Finally, the ROC curves in B were used to generate the neurometric function shown in C. Recall from the discussion in Chapter 2 that the expected percentage correct performance of a diagnostic system for a given stimulus is simply the area under the ROC curve derived from this stimulus. The values for the individual data points in C were each calculated by taking the area under the corresponding ROC curve in B. The smooth curve through the data points in C is the neurometric function. It predicts how well this individual neuron behaves as a diagnostic system for direction of motion.

Neurometric functions are directly analogous to the psychometric functions generated for observers

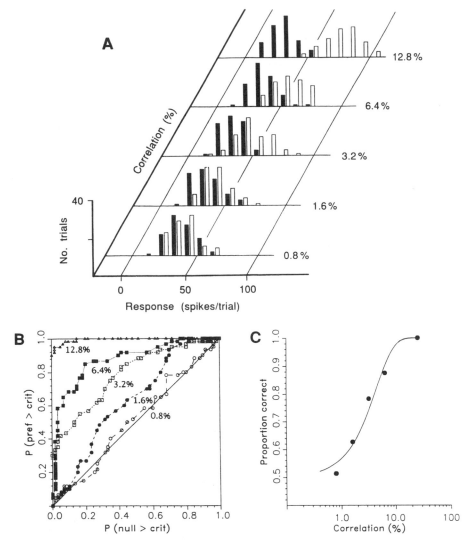

FIGURE 10.25. Illustration of how theory of signal detection methods were used to construct a neurometric function for the spike activity of a single neuron. See text. Reproduced from K.H. Britten, et al, The analysis of visual motion: A comparison of neuronal and psychophysical performance. *J. Neurosci.* 12:4745–4765, © 1992, by permission of the Society for Neuroscience.

during two-alternative forced-choice psychophysical tasks that were described in chapter 2. However, the logic of applying signal detection methods to results from neurons and results from observers proceeds in opposite directions. A psychometric function is obtained behaviorally, and signal detection methods are used to model the underlying neural events that gave rise to the observed performance. In a neurometric function, neural events are measured directly and used to predict behavioral performance of the observer. In Figure 10.25, the prediction about performance of the monkey,

shown in C, is derived by using the output of a neuron in V5.

Note that the prediction was made based on the assumption that the observer's performance is caused exclusively by the firing rate of this single neuron. It is worthwhile to pause here and consider the implication of this prediction. The brain of this monkey contains billions of active neurons that can potentially be used to guide performance on this task. In the analysis just performed, a prediction is made about the performance that would be expected if the monkey's judgments were guided

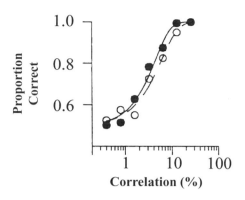

FIGURE 10.26. Comparison from a psychometric function and a neurometric function from a neuron recorded in V5 while a monkey is performing a discrimination of direction of motion. Reproduced from K.H. Britten, et al, The analysis of visual motion: A comparison of neuronal and psychophysical performance. *J. Neurosci.* **12**:4745–4765, © 1992, by permission of the Society for Neuroscience.

similar. The conclusion drawn from these results is that firing rates of individual V5 neurons are sufficient to account for behavioral motion coherence thresholds.

It is not fully understood why behavioral performance should match so closely the performance of the particular neuron that was being recorded from at the time. Prior to this study, many scientists expected that the monkey should have been able to do better than any of its individual neurons. After all, the monkey should be able to pool information from other neurons in its brain to improve performance over that supported by any individual neuron. However, there are two factors that limit the benefits of pooling.

The first is the fact that the stimulus parameters were adjusted to be optimal for the neuron being studied. Within a single hypercolumn that processes information about a small region of space, there are probably only a small number of neurons that process motion information for a given direction and velocity. When the stimulus is made optimal for one of these neurons and lowered to a threshold level of stimulation, the stimulus probably cannot be detected by many other neurons. The second factor is the amount of correlation between the noise present in responses from individual neurons. Statistical analyses have demonstrated that not much benefit can be derived by pooling information from more than one neuron if the noise present in the neurons is highly correlated.

Other scientists, prior to this study, predicted that the monkey should have performed more poorly

simply by the activity of this single neuron and nothing else.

How good is the prediction? Recall that while neural activity was being recorded from single neurons, the monkey was simultaneously making perceptual judgments. Thus, a psychometric function is available, obtained from the same monkey during the same trials, with which to make a direct test. An example of such a direct comparison is illustrated in Figure 10.26.

The filled circles and solid curve show a neurometric function for a particular V5 neuron during motion-detection trials, and the open circles and dashed curve show the psychometric function describing the monkey's performance on the same trials. At low correlations, the monkey correctly chose the direction of stimulus motion on about half of the trials, which is equal to chance performance. As the correlation increased, performance improved towards perfection, the expected result for a psychometric function. The neurometric function is almost identical, quantitatively as well as qualitatively!

A close correspondence between the psychometric and neurometric functions as shown in Figure 10.26 proved true in general, as illustrated in Figure 10.27.

Threshold values were calculated for each pair of psychometric and neurometric functions, and the figure shows the ratio of the neuronal and psychophysical thresholds for 166 neurons. The modal value is near 1 demonstrating the thresholds for the single neuron and for behavior were usually

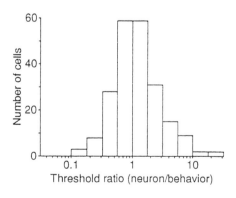

FIGURE 10.27. Comparison of the thresholds for detecting direction of motion based on individual V5 neuron responses and based on the behavioral performance of the monkey. Reproduced from K.H. Britten, et al, The analysis of visual motion: A comparison of neuronal and psychophysical performance. *J. Neurosci.* **12**:4745–4765, © 1992, by permission of the Society for Neuroscience.

than most of the individual neurons in its brain. The stimulus was optimized for the neuron that was being recorded from. However, there are billions of other neurons firing in the brain while the monkey is trying to make the decision, for which the stimulus was not optimized. Thus, the monkey's performance might be expected to have been worse than that of this neuron, since the monkey brain that must make the decision has no way to know that it should pay attention exclusively to this neuron rather than others that are firing at the same time.

The question of exactly why the behavioral performance is so similar to that predicted from analyzing a single neuron in the brain, rather than being substantially better or worse, is currently under extensive psychophysical and computational investigation.

The results described so far imply, but do not conclusively prove, that the monkey's judgments were caused by the activity of the neuron being studied. Newsome and colleagues performed another set of experiments designed to put that idea to a more direct test. An electrode was used to direct electrical microstimulation to the region near the V5 neuron that was being recorded from. Cortical area V5 has a columnar organization such that neurons with similar directional selectivity are clustered together. Thus, when the electrical stimulation was delivered to the area near the neuron being recorded from, it affected that neuron and other nearby neurons with similar directional selectivity. On half the trials, chosen at random, electrical stimulation was delivered for the 1-sec period during which the visual stimulus was presented. Figure 10.28 shows the results.

The open circles and dashed line show performance on trials in which no electrical stimulation was given, and the filled circles and solid line show performance during the stimulation trials. The horizontal axis has been altered to show both positive and negative correlations. Positive correlations indicate motion in the preferred direction of the recorded neurons, and negative correlations indicate motion in the null direction.

For each stimulus correlation, the monkey made more responses in the preferred direction when the electrical stimuli were present than when they were absent. The interpretation of this result is that the effect of electrical stimulation to neurons in V5 is equivalent to adding a particular proportion of real stimulus motion to the display. In Figure 10.28, microstimulation had an effect on perceptual judgment equivalent to an additional 20% of correlation signal being added to the intrinsic signals produced in the brain from the retina.

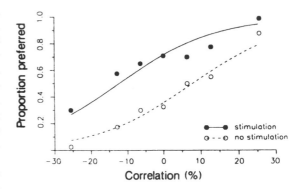

FIGURE 10.28. Comparison of behavioral performance on discriminating direction of motion during normal trials and during trials when microstimulation was delivered to V5 neurons. Reproduced from C.D. Salzman, et al, Cortical microstimulation influences perceptual judgments of motion direction. *Nature* 346:174–178, © 1990, by permission of Macmillan Magazines Ltd.

The Role of V5 in Combining Component Motions

There are three lines of evidence consistent with the idea that processing in V5 is involved in combining component motions detected by local mechanisms into more complex patterns.

The first evidence is that neurons in V5 respond to RDK stimuli in which the motion information is available only based on a global analysis of the individual motions of the elements, as demonstrated in Figure 10.23. The second is that lesions to V5 disrupt the ability to discriminate motion in RDK stimuli. The third line of evidence involves studies using stimuli that form plaids. The psychophysical evidence that human observers, under some conditions, can combine motion from two orthogonal gratings to form a percept of a coherent plaid motion was described earlier. This perceptual ability is not reflected in single neurons in the striate cortex. A directionally selective neuron in the striate cortex responds whenever one of the components moves in the preferred direction but not when the plaid pattern as a whole moves in that direction. Some individual neurons in V5 also respond only to the component motions. However, about 25% of the neurons in V5 respond to the direction of a combined plaid.

Some neurons in V5 also respond to second-order as well as first-order motion. An example of a neuron recorded from V5 of a monkey that shows similar directional tuning for first-order and second-order motion is shown in Figure 10.29.

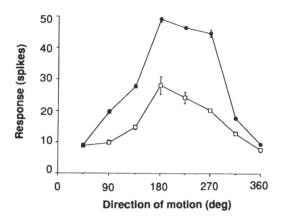

FIGURE 10.29. Responses of a neuron in V5 to first-order (filled symbols) and second-order (open symbols) motion. Reproduced from T.D. Albright. Form-cue invariant motion processing in primate visual cortex. *Science* 255:1141–1143, © 1992, by permission of the Association for the Advancement of Science.

The filled circles show the directional tuning curve for a V5 neuron using traditional motion stimuli defined by luminance. The open circles show a directional tuning curve for the same neuron when tested with non-Fourier motion stimuli. The overall response levels to the second-order motion are smaller than those to first-order motion, but the basic shape of the directional tuning curve is similar for both stimuli. These kinds of higher-order motion properties that are needed for seeing the motions of objects in a complex scene are not seen at earlier stages of neural processing.

7 Higher-Order Processing in the Motion-Stream

The kinds of processing that begin to take place in V5 are able to specify the global motion flow field to at least some degree. Additional processing is needed to extract higher-order motion information from the global flow field. The exact neural stages at which this higher-order motion processing takes place are not well defined but almost certainly include extrastriate cortical areas MST and FST, which receive direct projections from V5. Posterior parietal cortex area VIP, which also gets a direct projection from V5, probably also plays a role. Two kinds of meaningful higher-order motion information thought to be extracted within these stages of

motion processing will be considered: optic flow and structure-from-motion.

Optic Flow

Consider the simple situation in which motion is detected traveling from position X to position Y in the retinal image, as illustrated in the top panel of Figure 10.30.

This retinal motion could have come about in any of three ways. The first is illustrated in cartoon form on the left side of the bottom panel: The head (straight line) and eye (arc) remained steady, and an object in the environment moved from position A to position B, causing the inverted image to move along the retina. The second, illustrated in the middle, is that the object and the eye remained steady, but the head moved from position A to position B, causing identical motion of the image on the retina as in the first example. The third example, on the right, is that the object and the head remained steady, but the eye rotated in the socket from A to B, again causing identical motion of the retinal image. This example illustrates that the motion-processing system is always confronted with an ambiguity when it detects motion of an element on the retina. How does it decide which of these three specific environmental conditions gave rise to the retinal motion that was observed?

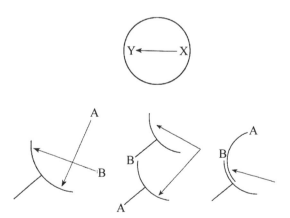

FIGURE 10.30. Retinal image motion from X to Y as illustrated in the top panel could have been produced by any of the three different environmental events shown in the bottom panel. See text. Adapted from R.G. Boothe. Biological perception of self-motion. *Behav. Brain Sci.* 17:314–315, © 1994, by permission of Cambridge University Press.

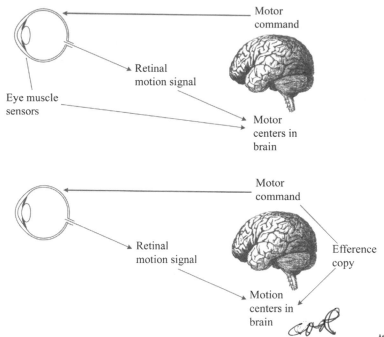

FIGURE 10.31. Illustration of two sources of extraretinal information, inflow (top panel) and outflow (bottom panel) about the fact that the eyes have moved. Adapted from B.A. Wandell, *Foundations of Vision*, © 1995, Sunderland, MA: Sinauer Associates, Inc.

There are two general theories about how this problem is solved by the biological motion processing systems. **Extraretinal theories** are based on nonvisual sources of information that can be used to estimate head and eye movements. These are subtracted from the retinal motion signal, and the remainder is interpreted as movement of objects in the environment. There are two sources of extraretinal information, called **inflow** and **outflow,** as illustrated in Figure 10.31.

The top panel illustrates inflow. Nonvisual sources of information such as stretch receptors in muscles and inputs from the vestibular system provide signals to the motion-processing areas of the brain specifying how the eye and head are moving in space. These signals are subtracted from the retinal motion signal to generate a percept of object motion.

The bottom panel illustrates outflow. Motor centers of the brain generate an extra copy, the **efference copy,** of every signal sent to the muscles to generate movement. The second copy is sent to the motion-processing areas of the brain, where it can be subtracted from the retinal motion signal to generate a percept of object motion. Recall from Chapter 1 that Helmholtz argued for a version of outflow based on the observation that the world appears to move when one presses on the eyeball with a finger, but not when one makes a voluntary eye movement.

In contrast with extraretinal theories are **direct perception theories,** derived from **Gibson,** who asserted that there is no need to postulate extraretinal sources of information. He argued that in the ecological environment, the brain is never confronted with a single element moving from X to Y as in Figure 10.30. In the ecological environment, the fundamental visual percept is movement in a structured environment composed of surfaces. The specific motion flow pattern that is produced when an observer moves with respect to surfaces was illustrated in Figure 9.61, and is called **optic flow.** Gibson argued that optic flow **specifies** self-motion as well as the motion of objects in the surrounding environment.

Current evidence indicates that although Gibson's ideas about optic flow were correct in many respects, he was probably mistaken about the idea that optic flow is sufficient to specify self-motion. This has been demonstrated in experiments in which human subjects try to use optic flow to judge their **direction of heading.**

Consider the example of driving a car. While the driver is looking straight ahead, the visual field appears to expand outward from a focus that indicates the direction the car is heading. Similarly, while the driver is looking out the back window, there is a pattern of contraction indicating the direction from which the car is coming. When the driver looks out of the side window, a complex set of motions is present that varies with viewing distance due to binocular parallax. These are all examples of the optic flow field of an observer moving through an environment.

Gibson claimed that direction of heading is specified by the optic flow field based on the location of the **focus of expansion.** Furthermore, he argued that direction of heading could be picked up by sampling any part of the flow field, as illustrated with an example in Box 10.4.

Box 10.4
How Do We Know Where We Are Headed While Driving?

Gibson argued for an extreme position: that optic flow completely specifies properties such as our direction of heading unambiguously. All we have to do to directly perceive our direction of heading is pick up information from the optic flow. In an anecdote, Gibson claimed to be able to perceive his direction of heading while driving even when his head was turned around looking at the flow field out the back window. (However, he added that he did not actually do this because it made his wife nervous.)

Recent studies have carefully analyzed the optic flow patterns that are produced when observers look in various directions while moving. An example is shown in Figure 10.32.

The left panel (A) shows the flow fields for an observer moving across a ground-level plane. The observer is moving in the direction indicated by the circle while holding the eyes and head fixed. In this

case, the focus of expansion in the flow field does in fact correspond to the direction of heading. However, the right panel (B) shows the same flow field as it appears when the observer moves in the same direction while simultaneously tracking an object moving from left to right. In this case, the flow field is altered in such a way that the focus of expansion no longer corresponds to the direction of heading.

The overall flow field produced when the eye is allowed to rotate while the head is moving through an environment is the combination of two components. One called **expansion**, comes from the forward movement of the head and the other, called **laminar**, from the lateral rotation of the eye, as illustrated in Figure 10.33.

Thus, laminar flow must be subtracted from the combined field to determine direction of heading based on the focus of expansion. Humans can accomplish this based on extraretinal information specifying that the eyes have moved. However, when provided only with the optical flow pattern in the absence of an extraretinal signal that the eyes have moved, observers misjudge their direction of heading.

The receptive field properties of some neurons recorded from in monkey V5 and MST seem ideally suited for extracting optic flow information. These neurons have large receptive fields and generally respond the same regardless of where in the receptive field a moving stimulus is placed. They respond to a variety of types of motion stimuli; stimulus characteristics such as the size of the texture elements or the spacing between them do not seem to matter, and often the receptive fields respond well to complex motions such as expansion and contraction. Neurons discovered in MST may play a role in detecting direction of heading during eye movements these neurons are sensitive to the focus of expansion and make appropriate corrections based on extraretinal information about movements of the eye.

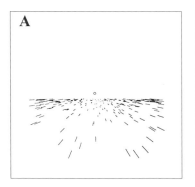

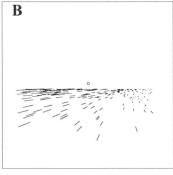

FIGURE 10.32. Illustration of optic flow fields for observer locomotion in the same direction but under two different viewing conditions. Reproduced from C.S. Royden, et al, The perception of heading during eye movements. *Nature* 360:583–585, © 1992, by permission of Macmillan Magazines Ltd.

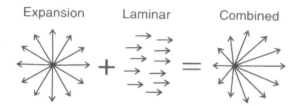

Expansion   Laminar   Combined

FIGURE 10.33. Illustration that optic flow field is composed of two components when the eye rotates during locomotion. Reproduced from D.C. Bradley, et al, Mechanisms of heading perception in primate visual cortex. *Science* 273:1544–1548, © 1996, by permission of the Association for the Advancement of Science.

Structure-from-Motion

Structure-from-motion was studied extensively by **Hans Wallach** and **G. Johansson** and their colleagues in the 1950s. Many of the early studies were done by casting shadows of wire-frame objects. This phenomenon can be demonstrated by taking a coat hanger and bending it into a complex shape, then using a flashlight or light bulb to cast a shadow of the three-dimensional wire frame onto a wall. It is usually impossible to determine the three-dimensional shape by looking at the shadow while the object is held still. However, for many shapes the three-dimensional structure immediately becomes apparent when the object is rotated. This is what is meant by the term "structure-from-motion." A changing two-dimensional form that

corresponds to the shadow cast by a rigid three-dimensional object in motion produces a percept of the three-dimensional shape of the object. The perceptual phenomenon of structure-from-motion is a striking example showing that the percept that is formed is not a function of individual retinal motions but of higher-order combinations of local motions.

More recently, structure-from-motion has been studied using RDK stimuli, as illustrated in Figure 10.34.

When the set of dots shown on the left is presented on a computer display as static elements, no structure is seen. However, if the dots are moved in a complex fashion that causes the motions of the individual dots to mimic the motion that would be seen if they were painted on the surface of a rotating cylinder, human observers perceive a rotating cylinder.

It is instructive to think about what would be likely to happen if the same set of dots were presented to a general-purpose robotic perceptual processing system programmed to draw every possible inference about what might be present in the environment. A huge number of possible environmental forms of stimulation might have formed this complex pattern, for example, shadows on the side of a building cast by a streetlight during a snowstorm.

Human observers usually see only one of the infinite possibilities – the correct one. The visual system must be applying powerful constraints to the problem to force a single solution. An active area

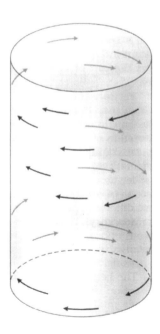

FIGURE 10.34. An RDK stimulus that has no spatial structure when static (left) but produces the percept of a three-dimensional rotating cylinder when the dots move appropriately (right). Reproduced from B.A. Wandell, *Foundations of Vision*, © 1995, Sunderland, MA: Sinauer Associates, Inc., by permission of Sinauer Associates, Inc.

of computational research is driven by trying to discover these constraints. One suggested constraint is **rigidity**, the assumption that the three-dimensional object responsible for the element motions must have a rigid shape. However, it has been demonstrated that the human visual system can derive some sense of structure based on motions produced by nonrigid objects, so it does not employ a rigidity assumption in a strict manner.

A special case of structure-from-motion is referred to as **biological motion.** Johansson made movies of humans or animals engaged in activities while lights were attached to their joints. Then the movies were shown to observers under conditions where only the lights could be seen. Thus, all that is present in this "point light display" is a small number of lights that are in motion. However, observers looking at these displays immediately see a person or animal engaged in the appropriate activity, for example, someone walking, doing push-ups, or riding a bicycle. To perceive these high-level activities based on the sparse information available in the display, the visual system must be applying complicated knowledge about how the joints of animals constrain their motions.

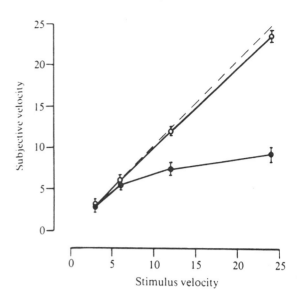

FIGURE 10.35. Velocity perception in human subject LM with damage to the V5 region of the brain. Reproduced from J. Zihl, et al, Selective disturbance of movement vision after bilateral brain damage. *Brain* 106:313–340, © 1983, by permission of Oxford University Press.

The Motion-Stream in Humans

There are a number of case studies of patients with motion-perception deficits that are thought to reflect brain damage that includes V5 and surrounding areas. The most extensively studied patient, LM, reported having difficulty perceiving certain types of motion following a stroke. Brain imaging revealed a bilateral lesion that probably affected V5 as well as surrounding tissue, although some tissue is also spared within this general region. LM has essentially normal function when tested on tasks such as visual acuity. However, she has problems with tasks that require perception of motion. For example, she cannot see coffee flowing from the pot into the cup and often fills the cup to overflowing. She also has difficulty with activities such as crossing a street with oncoming cars, explaining that cars just suddenly appear closer than before. When tested in the laboratory, LM reports abnormally low velocities for moving stimuli compared to what normal subjects perceive, as illustrated in Figure 10.35.

The horizontal axis shows stimulus velocity in degrees per second. The vertical axis shows the observers report of perceived velocity.

Veridical perception of motion would be expected to produce data points that fall along the

dashed line, and this is the basic result obtained from normal observers (open symbols). The results for LM (filled symbols) demonstrate that her perceived velocities are significantly lower than normal.

A second patient, AF, has also been studied extensively. AF is a stroke patient with bilateral damage to extrastriate areas that extend into the posterior parietal and temporal lobes, including V5 and surrounding areas. He has an impaired ability to see coherent motion when viewing RDK stimuli. Patient AF is also somewhat impaired compared to normal controls when tested on structure-from-motion tasks, although the deficits are not as large as would be predicted based on the deficits that are seen in monkeys following lesions to V5. In lesioned monkeys, there is little if any residual ability to recover higher-order global motion from RDK displays. In patient AF, performance is impaired when the lifetimes of the individual dots is short, but approaches normal when lifetimes are lengthened (Figure 10.36). However, in monkeys the lesions are more likely to be complete, while damage to human patients with strokes such as AF may spare some amount of tissue.

When patient AF was tested on biological motion tasks using the point light display stimuli, he was able to quickly identify all of the stimuli presented:

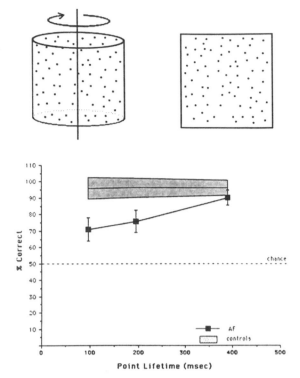

FIGURE 10.36. Performance on a "structure from motion" task by patient AF compared to normal control subjects. Two RDK stimuli were presented side by side: motion of dots in one stimulus specified a rotating cylinder (top panel, left); motion in the other stimulus was random (top panel, right). Lifetimes of individual dots were varied. Patient AF was poorer than normal at detecting the cylinder when lifetimes were short, but approached normal levels of accuracy when lifetimes were long (bottom panel). Reproduced from L.M. Vaina, et al, Intact "biological motion" and "structure from motion" perception in a patient with impaired motion mechanisms: A case study. *Visual Neuroscience* 5:353–369, © 1990, by permission of Cambridge University Press.

someone walking, riding a bicycle, and so on. This has led to the suggestion that biological motion is processed in a pathway separate from the extrastriate motion-stream, perhaps similarly to the way biological perception of faces is processed differently from form perception of other objects.

8 How-Perception and the Posterior Parietal Cortex

Information leaving the motion-stream is used for two purposes. The first serves What-perception by generating motion percepts from which observers obtain their experiences and knowledge about what is happening in the surrounding environment. The second serves How-perception, in which motion information is used to control muscles that produce desired actions in response to what they see.

It is with How-perception that the remainder of this chapter is concerned. The kinds of information previously discussed, in the M-stream and in the motion-stream, have all been expressed in **retinal coordinates,** meaning that the location in space where a receptive field is mapped moves whenever the eye moves. However, How-perception can be accomplished only if the information is transformed from retinal coordinates to muscle coordinates.

Consider the task of reaching for and grasping an object under visual control. The locations of the images on the retina are of limited use for this task unless one knows where the eyes are pointing with respect to the head and what the orientation of the head is with respect to the shoulders. Indeed, the brain has to constantly remap the location of a stimulus that is specified in retinal coordinates whenever the head or eye moves. Conceptually, the coordinate system that is used to specific locations of objects is transformed from one that is **eye-centered,** to one that is **head-centered,** and finally to one that is **limb-centered.** Neurally, this conversion takes place within the How-perception stream in the posterior parietal cortex and frontal lobes. The neural stages do not always map neatly onto the conceptual stages because within many cortical areas neurons are found with receptive fields that reflect various mixtures of intermediate representations.

An example of receptive fields that show a combinations of coordinate representations is illustrated in Figure 10.37.

This figure illustrates the properties of a neuron recorded in the pre-motor cortex (A) of the monkey. This neuron has a bimodal receptive field that responds to both tactile and visual stimulation (B). When the arm of the animal was not in view, the receptive field extended from 45° to 90° and moved with the head. However, when the arm was brought forward into view the receptive field became associated with the position of the arm. For example, when the arm was positioned towards the right side, the receptive field extended from −20° to +50°. However, when the arm was bent towards the left side, the receptive field moved to the left occupying the region from −70° to +50°.

Many neurons in the posterior parietal cortex have receptive fields that still use eye coordinates even though they appear to be involved in reaching and grasping tasks. These neurons may be spe-

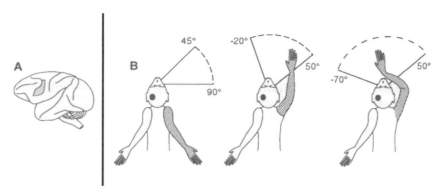

FIGURE 10.37. Neurons in premotor cortex have receptive fields that have been partially transformed from eye-centered to limb-centered coordinates. See text. Reproduced from M.S.A. Graziano, et al, Coding of visual space by premotor neurons. *Science* 266:1054–1057, © 1994, by permission of the Association for the Advancement of Science.

cialized for controlling reaching under conditions in which one looks at the hand. When this strategy is employed, the receptive field remains fixed with respect to the hand even though the hand is moving. The fact that humans have these neurons available perhaps accounts for the annoying fact that it is easier to reach for a coffee cup while reading if one looks toward the cup instead of trying to reach while maintaining fixation where one is reading.

There is evidence that What-perception and How-perception can be dissociated in some patients with neurological damage by using a **posting task,** illustrated in Figure 10.38.

The observer is presented with a slit held in a particular orientation and asked to perform two types of tasks. For the first task, the subject is asked to indicate the orientation of the slit. This might be indicated verbally or by having the subject hold the hand in the same orientation as the slit. For the second task, the observer is asked to place the hand into the slit, as in posting a letter. In some patients with neurological deficits performance is impaired on one of these tasks but not the other. An active line of current research is trying to determine the extent to which deficits on these two tasks map onto localized damage to the neural streams of processing associated with What-perception and How-perception.

Modeling How-Perception with Neural Networks

The classes of neural networks described in Chapter 6 are not very convenient for modeling performance that involves change over time. For the most part, those networks simply subject an input to certain operations, perhaps including some that take place in hidden layers, and generate an output. They do not display any inherent dynamics or ongoing activity.

A class of network models that do exhibit inherent dynamics, called **attractor neural networks (ANNs),** were originally proposed by **J.J. Hopfield** in the early 1980s. Each element in an ANN is a formal neuron, as described in Chapter 6. In the generic case, an ANN is fully interconnected, which means that every formal neuron is connected to every other formal neuron and to itself. An ANN can also have inputs and outputs to and from external sources.

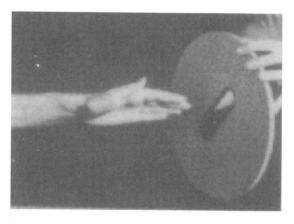

FIGURE 10.38. Posting task used to dissociate How-perception from What-perception. Reproduced from M.-T. Perenin and A. Vighetto, Optic ataxia: A specific disruption in visuomotor mechanisms. Reproduced from *Brain* 111:643–674, © 1988, by permission of Oxford University Press and M.-T. Perenin.

Possible States in a Three-Neuron Attractor Neural Network

State	Neuron #1	Neuron #2	Neuron #3
1	0	0	0
2	0	0	1
3	0	1	0
4	0	1	1
5	1	0	0
6	1	0	1
7	1	1	0
8	1	1	1

FIGURE 10.39. Illustration of fact that an attractor neural network with three neurons can be in any of eight different states at any moment.

A property of ANNs that makes them intuitively appealing to biologically oriented scientists is that they can exhibit spontaneous activity even in the absence of inputs, much like the brain. This feature is modeled in an ANN by random, asynchronous updating of the individual formal neurons. At a particular instant in time, one of the formal neurons in the network is chosen randomly for updating. The inputs to that neuron are collected and used to produce a new output state. Since the ANN is fully interconnected, this new output state is immediately reflected in the inputs of all of the other neurons. At the next instant in time, another formal neuron is selected for update and in the next instant yet another. Consequently, the state of the network changes dynamically from instant to instant, even in the absence of any change in the external inputs.

An ANN has 2^n possible states, where n is the number of neurons in the network. Consider a simple ANN consisting of only three formal neurons. The possible states for this ANN can be illustrated using a notation in which a neuron that is outputting a spike is shown with a "1" symbol and a neuron not outputting a spike with a "0" symbol. The possible network states for an ANN with three neurons would be as shown in Figure 10.39.

In state 1, none of the neurons are firing a spike; in state 2, neuron 3 is firing; in state 7, neurons 1 and 2 are firing; and so on. At any instant, this particular ANN is in one of the eight states shown. However, the ANN also has temporal properties, because from one instant to the next it can change states, as illustrated in Figure 10.40.

Dynamics of an Attractor Neural Network

Time	t_1	t_2	t_3	t_4	t_5	...	t_n
State	6	3	2	7	1	...	5

FIGURE 10.40. An attractor neural network can change states over time, as illustrated here.

The dynamics of an ANN exhibit another interesting property. If the weights of the connections are set appropriately (Chapter 6 described the relationships between the settings of the weights and neural network properties, and Chapter 11 will discuss further how the weights are set during an initial training phase), then certain repeating patterns of output states take on the special property of being **attractors**. Whenever the output states become similar to this attractor, the dynamics are constrained to enter the attractor state. In the simplest case, an attractor is a single state and can be conceptualized by using a landscape metaphor, as described in Chapter 6. Consider the example in Figure 10.41 of an ANN in which state 2 is an attractor.

In this example, the dynamics initially lead the ANN from state 6 to state 3. These initial conditions might have been due to random behavior, or they may have been caused by the inputs to the ANN. Regardless, once the ANN is in state 3 it is similar enough to the attractor (state 2), that it falls into the attractor where it will stay forever unless the input conditions change sufficiently to pull it out.

An ANN can store more than one attractor. For example, an ANN might have its weights set such that it could fall into either attractor state 2 or attractor state 7. In that case, which attractor state is entered into depends on the current state of the ANN (determined by its history) and on its current inputs.

ANN Falls into Simple Attractor

Time	t_1	t_2	t_3	t_4	t_5	...	t_∞
State	6	3	2	2	2	...	2

FIGURE 10.41. An attractor neural network can fall into a simple attractor state.

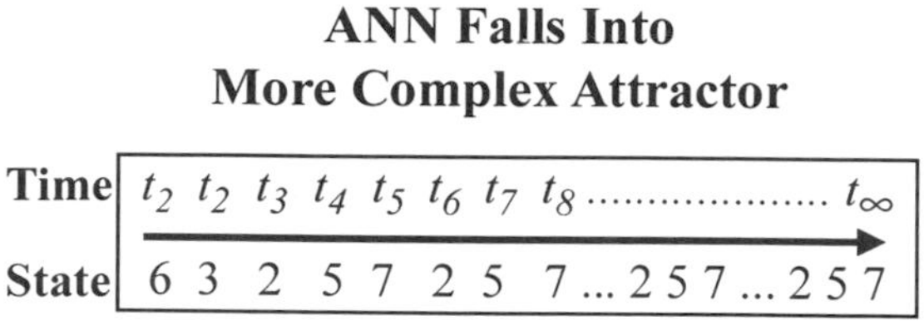

ANN Falls Into
More Complex Attractor

Time	t_2 t_2 t_3 t_4 t_5 t_6 t_7 t_8 .... t_∞
State	6 3 2 5 7 2 5 7 ... 2 5 7 ... 2 5 7

FIGURE 10.42. An attractor neural network can also fall into an attractor that is a repeating pattern.

Attractors do not have to be single states but can also be patterns, as in the ANN in Figure 10.42.

This ANN has fallen into an attractor that is a repeating temporal pattern "2, 5, 7." An attractor can be a finite repeating pattern of any complexity. Attractors in the form of temporal patterns might be particularly useful for evaluating perceptual information that is going to be used for guiding actions. Consider how perception is used to guide an activity such as bouncing a ball. The perceptual system must register a pattern unfolding over time in the environment and generate a range of compensating actions (sequence of motor commands to the arms and other parts of the body). The movements of the arm and hand need to follow an up-down rhythm that is matched to perceptual information about the up-down movement of the ball. This kind of timing can be modeled with an ANN that has oscillations built in as attractor states. The output of such a network will evolve over time to produce a temporal pattern that is in phase with its inputs.

These kinds of networks are robust in the sense that they can maintain the oscillation frequency for a while even if the input signal disappears or misses a beat. Behaviors that are more complex, perhaps even very complex actions such as an outfielder catching a baseball, can, in principle, arise out of interactions among several oscillators.

Summary

Perception is not static but has an inherent dynamic quality. Changes in perceptual input over time are processed to serve higher-order perception functions associated with both What-perception and How-perception.

Neural processing of temporal change begins in the M-stream. Spatiotemporal changes in the retinal image are converted into motion percepts in cortical area V1 by directionally selective neurons. The computational operations that are involved in

extracting motion have been described with two major classes of models, based on autocorrelation and on spatial and temporal derivatives. Both kinds of models depend on introducing a delay for some inputs relative to others. The local motion that is extracted by directionally selective neurons in V1 specifies the motion field but cannot specify higher-order motion because of the aperture problem and the correspondence problem.

Combining local motions to generate motion percepts based on complex patterns is accomplished within the motion-stream. All information passing through the motion-stream passes through extrastriate cortical area V5, which appears to play a special role in motion processing. Humans and monkeys with damage to this area of the brain have impaired motion perception. The electrophysiological activity of single neurons in this brain area is sufficient to account for threshold judgments of motion coherence. Motion information extracted in the motion-stream is used for What-perception in the form of structure-from-motion. It is also used for How-perception in the form of optic flow. Optic flow can specify many aspects of both motion of objects in the environment and self-motion. However, some information about self-motion has to come from extraretinal sources.

Information from the motion-stream that is used to guide action must first pass through the How-perception stream, where the coordinate system is changed from one that is eye-centered to one that is head-centered and then limb-centered. The How-perception stream passes through posterior parietal cortex and portions of premotor cortex in the frontal lobe. Monkeys and human patients with damage to these areas often have problems with visually guided reaching and grasping. The processing that takes place during How-perception can be modeled with attractor neural networks.

Selected Reading List

Adelson, E.H., and Movshon, J.A. 1982. Phenomenal coherence of moving visual patterns. *Nature* 300: 523–525.

Albright, T.D. 1984. Direction and orientation selectivity of neurons in visual area MT of the macaque. *J. Neurophysiol.* 52:1106–1130.

Barlow, H.B., and Levick, R.W. 1965. The mechanism of directional selectivity in the rabbit's retina. *J. Physiol.* 173:477–504.

Braddick, O.J. 1980. Low-level and high-level processes in apparent motion. *Philos. Trans. R. Soc. Lond. B* 290: 137–151.

Britten, K.H., Shadlen, M.N., Newsome, W.T., and Movshon, J.A. 1992. The analysis of visual motion: A comparison of neuronal and psychophysical performance. *J. Neurosci.* 12:4745–4765.

Duffy, C.J., and Wurtz, R.H. 1991. Sensitivity of MST neurons to optic flow stimuli: II. Mechanisms of response selectivity revealed by small-field stimuli. *J. Neurophysiol.* 65:1346–1359.

Goodale, M.A., and Milner, A.D. 1992. Separate visual pathways for perception and action. *Trends Neurosci.* 15:20–25.

Green, M., and Odom, J.V. 1986. Correspondence matching in apparent motion: Evidence for three-dimensional spatial representation. *Science* 233:1427–1429.

He, J., and Nakayama, K. 1994. Apparent motion determined by surface layout not by disparity or three-dimensional distance. *Nature* 367:173–175.

Hildreth, E.C., and Koch, C. 1987. The analysis of visual motion: From computational theory to neuronal mechanisms. *Annu. Rev. Neurosci.* 10:477–533.

Maunsell, J.H.R., and Van Essen, D.C. 1983. Functional properties of neurons in middle temporal visual area of the macaque monkey: I. Selectivity for stimulus direction, speed, and orientation. *J. Neurophysiol.* 49:1127–1147.

Merigan, W.H., Byrne, C.E., and Maunsell, J.H.R. 1991. Does primate motion perception depend on the magnocellular pathway? *J. Neurosci.* 11:3422–3429.

Pasternak, T., and Merigan, W.H. 1994. Motion perception following lesions of the superior temporal sulcus in the monkey. *Cereb. Cortex* 4:247–259.

Reichardt, W. 1969. Movement perception in insects. In *Processing of Optical Data by Organisms and Machines*, ed. W. Reichardt, pp. 465–493. New York/London: Academic Press.

Saito, H.-A., Yukie, M., Tanaka, K., Hikosaka, K., Fukada, Y., and Iwai, E. 1986. Integration of direction signals of image motion in the superior temporal sulcus of the macaque monkey. *J. Neurosci.* 6:145–157.

Sakata, H., Shibutani, H., Ito, Y., and Tsurugai, K. 1986. Parietal cortical neurons responding to rotary movement of visual stimulus in space. *Exp. Brain Res.* 61:658–663.

Salzman, C.D., Murasugi, C.M., Britten, K.H., and Newsome, W.T. 1992. Microstimulation in visual area MT: Effects on direction discrimination performance. *J. Neurosci.* 12:2331–2355.

Schiller, P.H. 1993. The effects of V4 and middle temporal (MT) area lesions on visual performance in the rhesus monkey. *Visual Neurosci.* 10:717–746.

Ship, S., de Jong, B.M., Zihl, J., Frackowiak, R.S.J., and Zeki, S. 1994. The brain activity related to residual motion vision in a patient with bilateral lesions of V5. *Brain* 117:1023–1038.

Smith, A.T. 1992. Coherence of plaids comprising components of disparate spatial frequencies. *Vision Res.* 32:393–397.

Ternus, J. 1926. Experimentelle Untersuchung über phänomenale Identität. *Psychol. Forsch.* 7:81–136. (Translated in *A Source Book of Gestalt Psychology*, ed. W.D. Ellis. New York: Humanities Press, 1967).

Tootell, R.B., Reppas, J.B., Kwong, K.K., Malach, R., Born, R.T., Brady, T.J., Rosen, B.R., and Belliveau, J.W. 1995. Functional analysis of human MT and related visual cortical areas using magnetic resonance imaging. *J. Neurosci.* 15:3215–3230.

Ullman, S. 1979. The interpretation of structure from motion. *Proc. R. Soc. Lond. B* 203:405–426.

Vaina, L.M., Lemay, M., Bienfang, D.C., Choi, A.Y., and Nakayama, K. 1990. Intact "biological motion" and "structure from motion" perception in a patient with impaired motion mechanisms: A case study. *Visual Neurosci.* 5:353–369.

Van den Berg, A.V. 1992. Robustness of perception of heading from optic flow. *Vision Res.* 32:1285–1296.

Wallach, H. 1987. Perceiving a stable environment when one moves. *Annu. Rev. Psychol.* 38:1–27.

Wallach, H., and O'Connell, D.N. 1953. The kinetic depth effect. *J. Exp. Psychol.* 45:205–217.

Wilson, H.R., and Kim, J. 1994. A model for motion coherence and transparency. *Visual Neurosci.* 11:1205–1220.

Wuerger, S., Shapley, R., and Rubin, N. 1996. "On the visually perceived direction of motion" by Hans Wallach: 60 years later. *Perception* 25:1317–1367.

Zihl, J., von Cramen, D., Mai, N., and Schmid, C.H. 1991. Disturbance of movement vision after bilateral posterior brain damage: Further evidence and follow up observations. *Brain* 114:2235–2252.

11
Perceptual Development: Where Does the Information Come from That Allows Perceptual Systems to Become Wired Together in Such a Manner That They Can Perceive?

Questions

After reading Chapter 11, you should be able to answer the following questions.

1. Compare and contrast the kinds of explanations for perception that can be sought from perspectives over developmental and evolutionary time spans.
2. Compare and contrast the empiricist and nativist positions regarding how humans come to know the meaning of environmental stimulation.
3. How do the prenatal and postnatal time courses of visual development compare in monkeys and humans?
4. Summarize the major stages of prenatal brain development.
5. How much does the blur that is present in the retinal images of infants matter?
6. Describe some of the methods that have been designed specifically to assess visual functions in infants.
7. Summarize the perceptual capacities of infants and compare these to the same capacities in adults.
8. What is the fundamental developmental problem that needs to be explained regarding perceptual processing?
9. Describe the effects of rearing animals in altered perceptual environments and relate these effects to clinical disorders in humans.
10. What are the Hebbian rules, and how can they explain some of the effects of environmental influences on perceptual development?
11. Describe how rules based on feedback can be used to allow a neural network to learn how to perform a particular perceptual function.

1 Explanations from a Developmental Perspective

Explanations of biological perception can be derived from perspectives over different time spans. Most of the explanations discussed in

previous chapters have come from one of two extremes. The first has been an **immediate neuroscience perspective,** in which explanations for perception have been sought in terms of concurrent activities of neural tissue. A prototypical example is the concept of a grandmother cell. The explanation for seeing grandmother is sought in the spike train being produced by a grandmother neuron at the same time the experience of grandmother is occurring. The second has been a **perspective over evolutionary time.** An example of this perspective would be an explanation based on arguments about why evolution might have produced species with brains that contain grandmother cells.

The current chapter looks for explanations of perception that involve an intermediate perspective, one that falls between these two extremes – explanations over the life span of the individual observer. This is called a **developmental perspective.** An example of the developmental perspective would be to ask how it was that an individual observer's brain ended up developing in such a manner that a grandmother cell was present.

Empiricism vs Nativism

Issues of how an individual observer comes to know the meaning of sensory stimulation have a long history among philosophers holding one of two positions within a broader context: **empiricism** and **nativism.** The historical nativism position argued that the transformation of sensory information into meaningful percepts is accomplished based on principles built into the observer at birth. The empiricism position argued that the meaning of sensory information must be acquired through experience and learning.

Based on a modern neuroscience appreciation of the mechanisms involved in prenatal and postnatal development of the brain, it is now apparent that it is inappropriate to try to delineate an exact boundary between the two positions. This has led to some amount of confusion in the modern use of these terms. The term "nativism" is still sometimes used to refer to capacities built in at birth. However, other times "nativism" is used to refer to capacities coded for by the genes at conception. Prenatal influences are lumped sometimes into nativism, other times into empiricism.

Keeping these disclaimers in mind, a brief discussion of the historical distinction between nativism and empiricism is nevertheless worthwhile, also commonly referred to in more modern times as the **nature vs nurture debate.** Recall from Chapter 1 that the British empiricist philosophers such as Locke championed an indirect realist position that our knowledge of the world is obtained indirectly, via intervening mechanisms. These advocates of indirect realism were all closely aligned with the empiricist position that ultimately all knowledge is acquired through perceptual experiences.

A powerful nativist position was formulated by the German philosopher **Immanuel Kant.** He developed an elaborate argument pointing out a serious epistemological problem for all indirect theories of perception. His argument stated, essentially, that you can't get there (knowledge about the external world) from here (indirect realism) because perceptual knowledge degenerates into being nothing more than a hypothesis. In order to solve this problem, Kant postulated a preexisting framework, which he named a **synthetic a priori,** within which perception has to take place. This nativist framework exists prior to any experience, and it functions to preserve the causality between the external world and our percepts, thus allowing us to use perception to gain knowledge about our surrounding environment.

An example of a historical nativist approach to scientific study of perception was the Gestalt School of Psychology, whose advocates argued for innate organizing principles such as the Gestalt laws of perception, discussed in Chapter 8. Helmholtz is a good example of a scientist adopting an empiricist position. Working in Germany, Helmholtz had to develop his theory within the context of the dominant influence of Kant's nativism. Helmholtz dealt with this problem by simply accepting, without proof, that an external world causes our perceptions. He asserted that perception takes place via an unconscious inference in which we perceive "as if" the external world exists. For Helmholtz, a nativist innate organization is needed only to produce a few rudimentary reflexes in response to sensory stimulation. These reflexes generate rewards and punishments allowing us to learn the meaning of environmental stimulation.

Helmholtz provides elaborate examples of how complex percepts could arise during development through this simple process of making inferences about properties of the environment based on experience.

William James' Blooming, Buzzing Confusion

Near the turn of the 20th century, William James applied the empiricist position specifically to issues of development, and his ideas have played a prominent role in the theoretical framework of the discipline of developmental psychology. His ideas are often captured with the shorthand phrase that babies come into the world experiencing sensations of **blooming, buzzing confusion.** Experience is supposed to organize these sensations into meaningful percepts.

The mechanism by which experience provides meaning was assumed by James to involve alterations in brain tissue. A newborn brain experiences pure sensation, but as soon as that pure sensation is experienced for the first time it affects the brain. Thus, even if the same exact sensation arises in the next moment, it will not produce the same exact effect, because the brain experiencing the sensation was altered by the first sensation. Consequently, throughout life no two sense impressions will produce identical effects, because each new sensation is interpreted within the context of the history of all the previous sensations experienced.

James argued his position by using the example of a child newly born in Boston who gets a sensation from the candle flame that lights the bedroom. The baby does not locate the flame in longitude 72°W. and latitude 41°N. Nor does he locate it in the third-story bedroom of the house. In fact, he does not even experience the flame to be to the left or right with respect to other sensations received at the same time. The sensation of the flame initially simply fills its own place.

According to James, only after we acquire some experience and begin to notice relationships between sensations do we come to know what seeing a light at a particular location means. For example, when a child later refers to his body, he means only a relationship along the lines of "that place where the pain from the pin is felt." Further, James argued, "It is no more true to say that he locates that pain in his body than to say that he locates his body in that pain. Both are true."

James asserts that as babies get older the original pure sensations start to become organized based on relationships with other sensations. For example, a sensation may take on a quality called its "whereabouts" based on certain relationships and simultaneously a quality of "intensity" based on other relationships. This builds up in complexity as we get older until our perceptual experiences are based mostly on complex relationships called "qualities" that take on names such as "shapes," "colors," "locations," and so on.

James' use of the term "sensations" did not refer to an introspective mental world. Instead, sensations come to newborn babies as objective states of the world, and every sensation refers to something external. James' based his argument for this point on evolution: "A sensation that does not awaken an impulse to produce an outward effect would be useless to a living creature and never have developed." Thus, James argued, our earliest perceptual experiences are of an objective kind but are primitive in the sense of not yet being differentiated into complex perceptual qualities.

2 Development of Visual Neural Pathways

Relating Monkey and Human Prenatal and Postnatal Ages

The remainder of this chapter will intermix results obtained from infant macaque monkeys and from infant humans. The prenatal gestation period is 40 weeks in the human and 24 weeks in the monkey. However, prenatal development is roughly comparable if age is specified in relative units, such as **percentage of gestation**, as illustrated in Figure 11.1.

Infant humans and macaque monkeys follow similar courses of postnatal development for many visual functions, except that monkeys mature four times faster than humans. Thus, postnatal ages when specific developmental milestones occur in monkeys and humans can be made commensurate by using a four-to-one conversion. This has come to be known as the **weeks-to-months rule.** This rule is

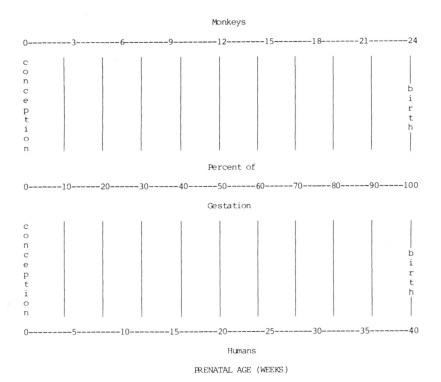

FIGURE 11.1. Prenatal events in humans and monkeys can be roughly equated if age is specified in units of percentage of gestation. Reproduced from R.G. Boothe, Visual development: Central neural aspects. In *Handbook of* *Human Growth and Developmental Biology, vol. I: Part B,* eds. E. Meisami and P.S. Timiras, pp. 179–191. Boca Raton, FL: CRC Press, Inc., © 1988, by permission of CRC Press.

not exact but provides a convenient heuristic. When designating a monkey postnatal age in the remainder of this chapter, we will sometimes use a shorthand phrase that expresses the age in **human equivalent months (HE-months)** or years (HE-years). For example, if a developmental event happens at 3 weeks in the monkey, we can also designate the age at which it occurs in human equivalent terms as 3 HE-months.

Detailed information about development of the brain can be obtained by consulting any standard neuroembryology text. This chapter will simply highlight a few milestone events associated with development of the portions of the brain involved in visual processing. Each area or nucleus of the brain goes through several stages of development, starting with **cell birthdays.**

Cell Birthdays

Every neuron that is present in the adult had a "birthday," defined as the age at which it underwent its last cell division. Most retinal ganglion cells are "born" during the first third of gestation, but a few continue to be born, even beyond midgestation. Neurons in the dLGN are born by one third of the way through gestation. Neurons in the visual cortex continue to be born beyond midgestation but are all present by term.

Migration

Once born, the neuron precursors, called **neuroblasts** at this stage because they have not yet differentiated into neurons, have to migrate from their place of birth to the location that will be their adult home. The end product of this migration forms the major nuclei and areas of the brain.

Neurons always have their birthdays near one of the ventricles of the brain. Neuron precursors that will form the visual cortex are born near the lateral ventricle. They must migrate to the cortical plate at the surface of the developing brain, as illustrated in Figure 11.2.

The initial cells reach the cortical plate by one third of gestation. Migration continues past midgestation but appears to be essentially complete before term.

The neuron precursors that form the thalamic nuclei are born near the third ventricle. Initially, the cells forming the dLGN and the pulvinar are close together, and the borders between these two structures not well defined. Migration to the dLGN is

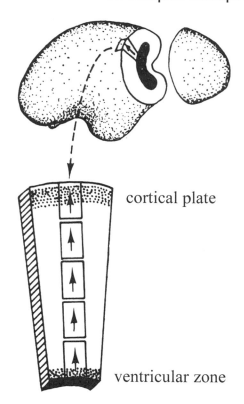

FIGURE 11.2. Neurons that form the visual cortex are born near the lateral ventricle and must then migrate to the cortical plate. Adapted from P. Rakic, Mode of cell migration to the superficial layers of fetal monkey neocortex. *J. Comp. Neurol.* 145:61–84, © 1972, by permission of Wiley-Liss, Inc., a subsidiary of John Wiley & Sons.

completed quickly, within about 3 days. The pulvinar continues to differentiate and expand throughout much of gestation.

Extrinsic Connections

Extrinsic connections from the retina arrive at the dLGN and begin to form synaptic connections by about one third of gestation. Extrinsic connections from the dLGN grow into position just below the developing visual cortex by one third of gestation but do not invade the cortical layers until after midgestation. Feedback extrinsic connections from the cortex to the dLGN are present by mid-gestation. The connections of the pulvinar with extrastriate cortical areas appear to be forming concurrently with the geniculostriate pathway. **Pasko Rakic** and coworkers have speculated that the border between the striate and extrastriate cortical areas may be created by competition between dLGN and pulvinar axons for cells at the border of the two sets of ingrowing fibers.

The basic feedforward connections among extrastriate cortical areas begin to develop during prenatal development and appear to be in place by birth. However, the patterns of feedback connections continue to undergo a period of postnatal refinement.

Cell Death

A common finding in neuroembryology is that many neurons die in a nucleus shortly after extrinsic connections are formed. For example, in a normal monkey or human, about half of the retinal ganglion cells normally die about the same time as their axons form extrinsic connections in the dLGN.

A proposed explanation for this finding is that nature uses the mechanism of cell death to eliminate cells that have not formed good connections. This explanation leads to a testable hypothesis that it should be possible to reduce the amount of cell death by making more potentially good connections available. This has been tested by Rakic and coworkers by enucleating one eye of a monkey fetus, thus making more potential central brain connections available for the ganglion cells of the remaining eye. The results confirm a significant reduction in death of ganglion cells in the remaining eye.

This example highlights the difficulty in the traditional nature versus nurture debate. The fact that cell death takes place is probably programmed genetically. However, the decisions about which neurons die and which survive depend on environmental interactions that influence formation of good connections. Furthermore, this mechanism operates primarily during the prenatal period, so whatever its effect, it is already present when the child is born. One would be hard-pressed to decide to which source to attribute survival of the remaining neurons if the only two alternatives were nature or nurture.

Intrinsic Connections

Most connections between neurons in the brain are intrinsic rather than extrinsic. Recall from Chapter 5 that these are local connections between neurons positioned close to one another, for example, connections between the neurons within a single hypercolumn.

Surprisingly, intrinsic connections do not begin to form in large numbers in the primary visual cortex until near the time of birth, and their formation con-

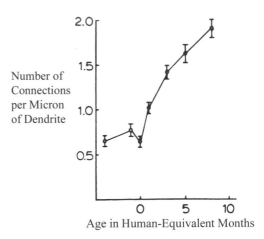

FIGURE 11.3. Most intrinsic connections in the brain form during the first few months after birth. Adapted from R.G. Boothe, et al, A quantitative investigation of spine and dendrite development of neurons in visual cortex (area 17) of *Macaca nemestrina* monkeys. *J. Comp. Neurol.* 186:473–490, © 1979, by permission of Wiley-Liss, Inc., a subsidiary of John Wiley & Sons.

tinues into the postnatal period, as illustrated in Figure 11.3.

This figure shows counts of the number of connections formed with the dendrites of neurons in the striate cortex in a monkey as a function of age.

The environmental influences on brain development prior to this stage have happened while the fetus was in the womb. However, beginning at this stage of development, environmental influences can include what the infant sees while looking around at the environment into which it was born.

Columnar Organization

The basic columnar organization of striate cortex into hypercolumns appears to be already set up at or near the time of birth. Organization into left- and right-eye ocular dominance columns can be demonstrated anatomically on the day of birth, and an ocular dominance organization has been demonstrated physiologically in monkeys within the first few weeks after birth. However, the segregation is not as sharp at birth as it is following visual experience and, as described below, can be dramatically altered by postnatal visual experiences. Similarly, the basic organization into orientation columns has been demonstrated within weeks after birth in normal monkeys, but, as discussed below, orienta-

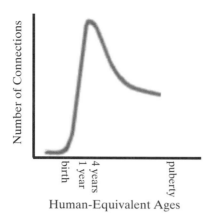

FIGURE 11.4. Formation of connections in the brain proceeds in two stages. The first is an overproduction of connections. This is followed by a pruning stage, in which connections are eliminated down to the adult level.

tion sensitivity can be altered by postnatal visual experiences.

Pruning of Connections

The intrinsic connectivity among neurons in the adult brain is established through two separate developmental processes (Figure 11.4).

The first is the **formation** of intrinsic connections, which begins near the time of birth and continues for about the first year in monkeys (4 HE-years). This is followed by a massive **pruning** of connections, which lasts perhaps until puberty.

An illustration of the effects of forming and pruning connections in monkey striate cortex is shown in Figure 11.5.

The neuron at the left is from the striate cortex of a monkey on the day of birth. The neuron in the middle is from an 8-week-old monkey (8 HE-months) and illustrates the increase in the number of spines on the dendrites that have developed since birth. Each spine is associated with a single excitatory synaptic connection. Finally, the neuron on the right is from the brain of an adult monkey. Many of the connections that were formed during the initial period of brain development have been pruned.

The proposed explanation for pruning is that only "good connections" are maintained into adulthood. This chapter will have more to say about what is meant by "good connections" later, in a discussion of the Hebbian rules.

A massive overproduction of synaptic contacts followed by pruning has been demonstrated in extrastriate cortical areas in addition to in V1.

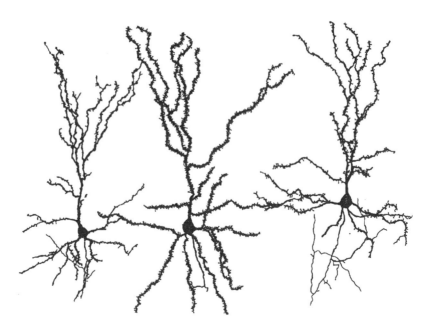

FIGURE 11.5. Illustration of formation of connections in the form of spines on neurons in the visual cortex between birth (left) and 8 HE-months (middle), and of pruning back of connections in the adult (right). Reproduced from J. Lund, et al, Development of neurons in the visual cortex (area 17) of the monkey (*Macaca nemestrina*): A Golgi study from fetal day 127 to postnatal maturity. *J. Comp. Neurol.* 176:149–188, © 1977, by permission of Wiley-Liss, Inc., a subsidiary of John Wiley & Sons.

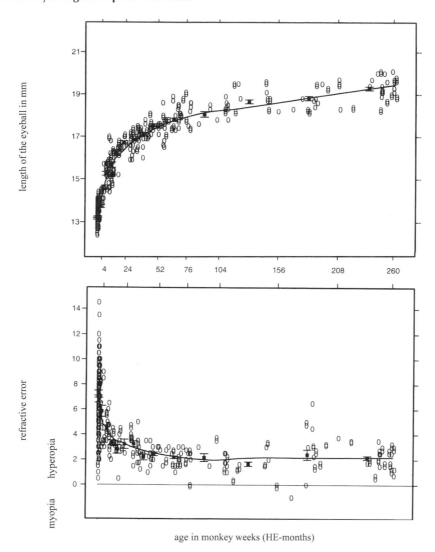

age in monkey weeks (HE-months)

FIGURE 11.6. Time courses for growth of length of the eyeball (top panel) and reduction of neonatal refractive error (bottom panel). Open circles are measurements on individual eyes. Filled circles with error bars are average population values at select ages. The smooth lines are growth curves fit to the data points with a statistical pro- cedure. Reproduced from D.V. Bradley, et al, Emmetropization in the rhesus monkey (*Macaca mulatta*): Birth to young adulthood. *Invest. Ophthalmol. Vis. Sci.* 40:214–229, © 1999, by permission of the Association for Research in Vision and Ophthalmology.

3 Development of Physiological Optics Functions

Emmetropization

The length of the eye is too short relative to the power of its optical components at birth, resulting in hyperopia. **Emmetropization** is the process by which neonatal hyperopia is reduced. It involves the coordination of growth of the eye with matura- tion of its refractive components. Figure 11.6 illus- trates the time course of emmetropization in rela- tion to the concurrent growth of the length of the eye.

At birth, the eyeball is short (top panel), and infants have various amounts of hyperopic refrac- tive error (bottom panel). As the eye begins to grow, the amount of hyperopia is reduced towards emmetropia. In monkeys, the emmetropic state in the adult exhibits about 2 diopters of hyperopia, rather than the expected value of 0 diopters, due to a technical measurement artifact, the details of which go beyond the scope of this discussion. Note

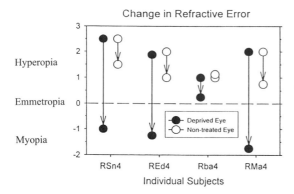

FIGURE 11.7. Eyes that are deprived of form vision during childhood or adolescence grow too long relative to the power of their optics and become myopic. Results are shown for four adolescent monkeys subjected to form deprivation in one eye. Reproduced from E.L. Smith III, et al, form deprivation myopia in adolescent monkeys. *Optom. Vis. Sci.* 76:428–432, © 1999, by permission of the Jippineott, Williams & Wilkins.

that refractive error has stabilized near asymptotic adult levels by two years, but the length of the eye (top panel), as well as other optical parameters (not shown), continues to change for a much longer period.

Maintaining emmetropia in the face of continued growth of the eye and maturation of its optical components requires some regulatory process. It is now well established that the regulatory process includes influences from the visual environment, i.e., what the monkey sees during development. This has been demonstrated by surgical manipulations that prevent one eye of a monkey from receiving normal visual stimulation. When young monkeys are subjected to conditions of monocular visual deprivation, the regulatory process governing emmetropization breaks down and myopia develops in the visually deprived eye, as illustrated in Figure 11.7.

Four individual juvenile monkeys were subjected to visual deprivation in one eye, and all four deprived eyes exhibited a myopic shift. This is called **form-deprivation myopia.**

The mechanisms by which emmetropization is regulated by the visual environment have fascinated scientists for more than a century. These issues have taken on added practical importance in many parts of the world in recent decades because emmetropization is not working properly for large numbers of the world's population.

For example, in some Asian countries myopia among college-age students approaches 90% although there is only a moderate amount of myopia in older adults in these same countries. The reasons for the recent increases in myopia among some populations of humans are not totally understood but appear to be related to the amount of sustained near work, as occurs during reading. The visual deprivation studies performed with monkeys are aimed at elucidating the physiological mechanisms responsible for regulating growth of the eye during development. Recent studies with monkeys have demonstrated that rate of eye growth is sensitive to retinal image blur, becoming faster than normal if defocus is unusually hyperopic and slower if myopic.

Accommodation

As discussed in Chapter 4, hyperopic refractive error does not necessarily cause the retinal image to be blurred because it can be eliminated through accommodation. However, direct measurements of accommodation in humans and monkeys demonstrate that the accommodative response is immature at birth and does not reach asymptotic adult levels until about 4 HE-months. This is illustrated with results from infant monkeys in Figure 11.8.

The horizontal axis shows the fixation target distance in diopters. The vertical axis is the fixation distance of the monkey. The dashed line indicates the performance that would be expected from an observer with perfect accommodation. Adults typically do not respond perfectly but do show responses that are parallel to, although slightly below, the dashed line. The left panel shows results for five individual infant monkeys ranging in age from 2 to 25 days. The right panel shows results from five individual monkeys that were 5 to 10 weeks of age. The results demonstrate that monkeys less than 4 HE-months of age have poor accommodation, but beyond 4 HE-months, infants have essentially adultlike accommodation.

This suggests that, at most, hyperopic defocus will cause blur for only about the first 4 HE-months. After that age the hyperopia will be eliminated by an appropriate accommodative response. Furthermore, as covered in the discussion of development of acuity and contrast sensitivity below, it is unlikely that infant brains are even aware of a small amount of hyperopic blur during the neonatal period.

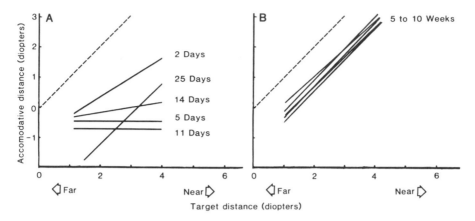

FIGURE 11.8. The accommodative response is shown for monkeys during the first four weeks after birth (A) and for monkeys older than four weeks (B). Reproduced from H. Howland, et al, Accommodative defocus does not limit development of acuity in infant *Macaca nemestrina* monkeys. *Science* 215:1409–1411, © 1982, by permission of American Association for the Advancement of Science.

4 How Does One Know What an Infant Can See?

General methods for asking an observer what it can see were discussed in Chapter 2. This section describes briefly a few additional specialized methods that have been designed or adapted for the express purpose of testing visual function of infants.

Habituation Methods

In experiments using habituation methods, a baby is placed in front of a visual display. A stimulus is presented, and a researcher measures how long the infant looks at it. Then the same stimulus is presented repeatedly and the time the baby spends looking is measured on each trial. Following several presentations of the same stimulus, looking time declines, a phenomenon referred to as **habituation.**

Once habituation has reached some criterion level, such as half of the duration seen on the initial trials, a new stimulus can be substituted. If the stimulus is sufficiently different from the original, looking time jumps back to near the initial levels, a phenomenon called **dishabituation.**

The following rationale allows habituation methods to be used to test whether a baby can discriminate between two stimuli. Following habituation to one stimulus, changing to a stimulus that cannot be discriminated from the original is not expected to result in dishabituation. By similar logic, if a new stimulus leads to dishabituation, it can be inferred that the infant can discriminate between the original and the new stimulus.

Preferential Looking

In the 1960s, the developmental psychologist Robert **Fantz** developed a procedure called **preferential looking** (PL) to assess vision in babies. Two visual stimuli are presented side by side in front of an infant, and the experimenter notes whether the infant looks preferentially towards one side of the display. These procedures have been adapted for testing infant monkeys as well as infant humans, as illustrated in Figure 11.9.

One person, the **holder,** holds the infant in the proper position. A second person, the **judge,** looks at the infant's face while the stimuli are being presented. The testing shown in Figure 11.9 is being carried out informally, but in an actual experiment is typically carried out in an apparatus constructed of homogeneous gray walls so there is nothing to distract the infant. Also, the judge views the face through a small peephole in the center of the stimulus card so as not to distract the infant. The judge's view of the infant through the peephole is illustrated in the Figure 11.10.

The top panel illustrates the view of the infant's face that will be seen when a strongly preferred stimulus is presented to the infant's right. The bottom panel shows the view when the same stimulus is presented on the infant's left.

The procedure can be repeated several times with the left-right positions of the stimuli varied randomly. If it can be demonstrated that an infant

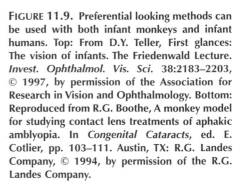

FIGURE 11.9. Preferential looking methods can be used with both infant monkeys and infant humans. Top: From D.Y. Teller, First glances: The vision of infants. The Friedenwald Lecture. *Invest. Ophthalmol. Vis. Sci.* 38:2183–2203, © 1997, by permission of the Association for Research in Vision and Ophthalmology. Bottom: Reproduced from R.G. Boothe, A monkey model for studying contact lens treatments of aphakic amblyopia. In *Congenital Cataracts*, ed. E. Cotlier, pp. 103–111. Austin, TX: R.G. Landes Company, © 1994, by permission of the R.G. Landes Company.

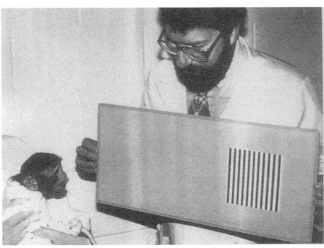

prefers one stimulus to the other, it can be concluded that the infant can discriminate between the two stimuli. A number of specific measures have been devised to quantify visual preferences, including total duration spent looking at each side during some fixed period, duration of the first look to each side, and number of looks to each side.

In the 1980s, the psychologist Davida **Teller** described a modified version of this procedure, called **forced-choice preferential looking (FPL)**, that brings its underlying rationale within the domain of Class A observations carried out with forced-choice psychophysical procedures as described in Chapter 2. The judge is kept uninformed about which side contains the stimulus and is forced to make a judgment about its physical location based on viewing the face and eyes of the infant. If the judge's performance is better than predicted by chance, it has to be because the infant responded differentially to the stimulus, and this can be taken as evidence that the infant could also discriminate between the stimuli.

Preferential looking methods depend on a built-in tendency of newborn infants to look towards some stimuli more than others, so these methods are of no value if there is no inherent preference. However, it has been discovered empirically that human and monkey infants have a natural tendency to look at any stimulus that can be seen and stands out from the background. For example, if one side of a gray screen has a grating stimulus and the other side of the screen has a stimulus composed of a homogeneous gray of matching mean luminance, the screen infants prefer the grating if it is visible. This can be used to measure acuity or contrast sensitivity by varying spatial frequency or contrast, respectively, to find the threshold for eliciting a preference. Similarly, color vision can be tested by presenting a pattern on one side of the screen defined only by the wavelength of the stimulus, and so on.

Neonates and young infants continue to show the same preference over many trials of stimulus presentations. However, at ages beyond three to six

FIGURE 11.10. View as seen through the peephole in the stimulus card by an experimenter carrying out the preferential looking procedure when the stimulus is on the infant's right (top panel) and left (bottom panel). Reproduced from R.G. Boothe, Experimentally induced and naturally occurring monkey models of human amblyopia. Reproduced from *Comparative Perception, vol. 1, Basic Mechanisms*, eds. M. Berkley and W. Stebbins, pp. 461–486, © 1990, by permission of John Wiley & Sons, Inc.

months in monkeys and beyond one to two years in humans, infants often become bored after a few trials, and other methods must be used, such as habituation or **operant reinforcement.**

Operant Reinforcement Methodologies

Neonatal monkeys can be taught to perform an operant task to indicate what they see within days after birth if raised in a specially designed **face mask cage** as illustrated in Figure 11.11.

The monkey is allowed to roam freely inside the cage. It can initiate a trial at any time by approaching the wall and placing its face in the mask. The presence of the face in the mask breaks a photocell beam that turns on a visual display. The monkey is trained to pull the bar corresponding to the side of the display that contains the correct visual stimulus. Correct bar pulls result in delivery of milk for very young babies or juice for older infants. Incorrect responses result in a time-out period, signaled by an audible tone, that lasts several seconds, during which further responses are ignored. Infant monkeys can be trained to perform a two-alternative forced-choice psychophysical procedure with these methods within the first few weeks after birth.

Human infants do not have the motor skills to perform operant tasks that require a motor response, such as pulling a bar. However, a modification of PL tasks allows operant testing methods to be used with children starting at about 6 months of age, as illustrated in Figure 11.12.

During an initial training period, the observer waits until the child looks towards the side of the screen containing the stimulus and then reinforces the response by activating a toy animal for a few seconds. Children learn to look towards the stimulus in order to receive the toy reinforcement. Following training, standard FPL methods can be used, except that whenever the judge's judgment is correct the toy is activated. By interspersing a sufficient number of easy trials that the infant can see, operant reinforcement maintains behavior even if there is no inherent preference for looking at the stimulus.

5 Development of Perceptual Functions

The descriptions of brain development and physiological optics functions in previous sections of this chapter lead to the expectation that some visual functions in infants should be immature compared to those of adults.

Absolute Thresholds

The improvement in absolute sensitivity to light with age is illustrated in Figure 11.13.

This figure summarizes the results of a number of studies carried out with human infants by Anne

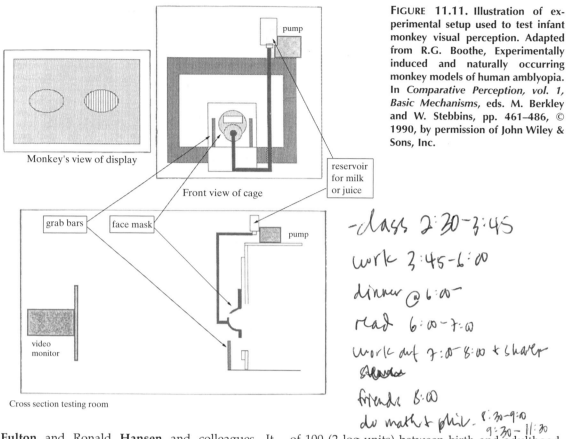

FIGURE 11.11. Illustration of experimental setup used to test infant monkey visual perception. Adapted from R.G. Boothe, Experimentally induced and naturally occurring monkey models of human amblyopia. In *Comparative Perception, vol. 1, Basic Mechanisms*, eds. M. Berkley and W. Stebbins, pp. 461–486, © 1990, by permission of John Wiley & Sons, Inc.

Fulton and Ronald **Hansen** and colleagues. It demonstrates that absolute sensitivity to light improves by about a factor of 10 (1 log unit) during the first 4 months after birth and by about a factor of 100 (2 log units) between birth and adulthood. These improvements are likely to be mostly due to increases in the lengths of photoreceptor outer segments, described further in the next section.

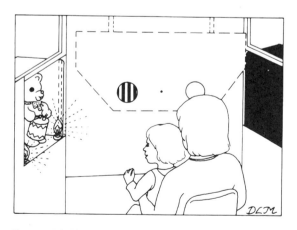

FIGURE 11.12. Illustration of experimental setup used to test infant human visual perception. Reproduced from D.L. Mayer and V. Dobson, Assessment of vision in young children: A new operant approach yields estimates of acuity. *Invest. Ophthalmol. Vis. Sci.* 19:566–570, © 1980, by permission of Association for Research in Vision and Ophthalmology.

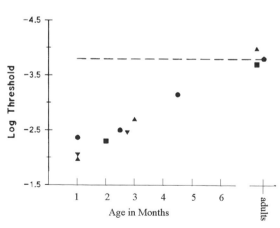

FIGURE 11.13. Absolute thresholds for detecting light improve by about 2 log units from birth to adulthood. Adapted from R.M. Hansen and A.B. Fulton, Development of Scotopic retinal sensitivity. In *Early Visual Development, Normal and Abnormal*, ed. K. Simons, pp. 130–142, © 1993, by permission of Oxford University Press.

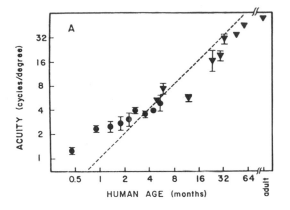

FIGURE 11.14. Acuity improves from near 1 cy/deg at birth to about 50 cy/deg in adulthood in both humans (A) and monkeys (B). Reproduced from D.Y. Teller, The devel-opment of visual acuity in human and monkey infants. *Trends Neurosci.* 4:21–24, © 1981, by permission of Elsevier Science.

Acuity and Contrast Sensitivity

Acuity improves from birth by about a factor of 50, as illustrated in Figure 11.14.

The left panel (A) illustrates results obtained from humans and the right panel (B) results from monkeys. The age axes have been scaled by months for humans and by weeks for monkeys. The similarity of the results on these axes is a demonstration of the weeks-to-months rule. Acuity starts out at about 1 cy/deg and does not reach adult levels of 50 cy/deg until about 4 HE-years of age. The dashed lines in each figure demonstrate a useful mnemonic: that acuity of infants and toddlers specified in cy/deg is approximately equal to age.

Contrast sensitivity improves by about a factor of 20 over this same age range. Representative contrast sensitivity functions (CSFs) obtained longitudinally from a single infant monkey are illustrated in Figure 11.15.

The numbers in the boxes superimposed on each CSF curve designate the age in HE-months when the results were obtained. The CSF curve shifts upward and to the right with age. This is because sensitivity to low spatial frequencies reaches adult levels and stops improving at young ages, but sensitivity to high spatial frequencies continues to improve for longer periods before it reaches adult levels.

Four kinds of factors have been considered that might account for the poor acuity and contrast sensitivity in infants compared to those in adults. The first is optical factors.

Optical Limitations on Infant Acuity

One hypothesis about why infants cannot see patterns of high spatial frequency or low contrast is that the optics of the infant eye do not form good enough images on the retina.

One potential optical factor might be lack of an accommodative response. However, that explanation cannot account, even in principle, for poor acuity beyond 4 HE-months of age, because accommodation is adultlike by then, as described in the previous section (Figure 11.8).

Another possible role for optical factors is the optical modulation transfer function (MTF) of the eye that was described in Chapter 8. If the quality of the optics of the infant eye were sufficiently poor, that fact could account for the poor behavioral

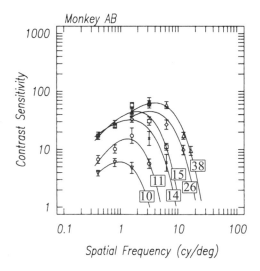

FIGURE 11.15. Development of contrast sensitivity in infant monkeys from 10 to 38 weeks of age. Adapted from R.G. Boothe, et al, Development of contrast sensitivity in infant *Macaca nemestrina* monkeys. *Science* 208:1290–1292, © 1980, by permission of American Association for the Advancement of Science.

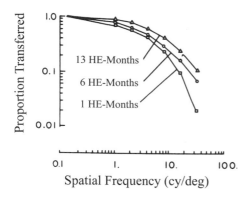

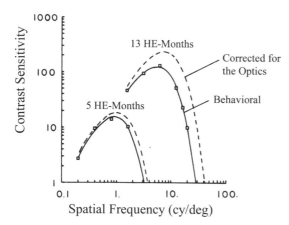

FIGURE 11.16. Development of the optical MTF in infant monkey eyes (top panel) and the implications of this development for interpreting behavioral improvements in contrast sensitivity (bottom panel). See text. Adapted from R.A. Williams and R.G. Boothe, Development of optical quality in the infant monkey (*Macaca nemestrina*) eye. *Invest. Ophthalmol. Vis. Sci.* 21:728–736, © 1981, by permission of Association for Research in Vision and Ophthalmology.

acuity and contrast sensitivity results. However, direct measurements of the optical MTF of infant monkeys reveals that optical quality does not pose any major limit to behavioral sensitivity at young ages. This analysis is demonstrated in Figure 11.16.

The top panel illustrates the optical MTF measured in monkeys at 1, 6, and 13 HE-months. These results demonstrate that there are moderate developmental improvements in the transfer by the optics of medium to high spatial frequency information from the environment to the retinal image, but little improvement in the transfer of low spatial frequency information.

The bottom panel demonstrates that these developmental changes in the optical MTF have little influence on developmental changes in the behavioral CSF. The symbols and solid lines show the

behaviorally measured CSF of monkeys at 5 and 13 HE-months of age. The dashed curves show how much behavioral performance would have improved at each age if there had been no attenuation by the optics of the eye. The effects of poor optics at the younger age are particularly small. At this early age, the neural visual system is sensitive only to very low spatial frequencies, at which the effects of poor optics on contrast in the retinal image are small. Thus, even an inadequate accommodative response is not likely to have much effect at these early ages. At the older age, the neural visual system is sensitive to higher spatial frequencies, and the optics play a larger role in attenuating contrast in the retinal image of these spatial frequencies. However, the overall magnitude of the behavioral improvement in the CSF during development is about the same whether or not the effects of the optics are taken into account.

Effects of Receptor Sampling on Infant Acuity

The second set of factors that might account for poor acuity and contrast sensitivity are related to receptor sampling. The foveal cones are immature at birth, as illustrated in Figure 11.17.

These drawings illustrate how the shapes of cones in the foveal region of the human retina change between mid-gestation and adulthood. In the adult there is a long, thin outer segment (OS) containing photopigment, and the inner segment (IS) is also thin, allowing the cones to be packed tightly together. The horizontal cells and bipolar cells with which the receptors make synaptic contact are not present in the foveal region. This accounts for the long strand called the **fiber of Henle** (FH) that extends from each receptor body to the **cone pedicle** (CP) located at the edge of the fovea, where synaptic contacts are formed. In the newborn, the outer segment is very short and consequently contains little photopigment; this accounts for the lack of sensitivity to light shown in Figure 11.13. In addition, the inner segments are thick and do not allow the cones to be packed closely together. Note also that there is no long fiber of Henle present; these develop only later, as the cone's inner segments migrate into the central fovea.

As a result of this immaturity of foveal cones, Martin **Banks** and colleagues have estimated that the neonatal fovea absorbs only one out of every 350 photons that would be absorbed in the adult fovea. They have also made quantitative predictions, based on an ideal observer analysis, about what effects the factors of changes in receptor size

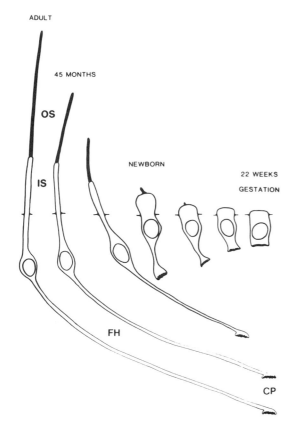

FIGURE 11.17. Development of cone photoreceptors. Adapted from A.E. Hendrickson, Development of the primate retina. Reproduced from *Handbook of Human Growth and Developmental Biology, vol. 1: Part B*, eds. E. Meisami and P.S. Timiras, pp. 165–177, 1988, by permission of Boca Raton, FL: CRC Press, Inc.

and spacing, in combination with overall changes in size of the eye, should have on the CSF.

The analysis demonstrates that these immaturities are not sufficient to account for the poor contrast sensitivity of neonates. For example, there is a twenty-two-fold discrepancy between their prediction about how much these factors would be expected to affect contrast sensitivity at 5 cy/deg and actual performance.

Role of Neural Processing and "Psychological" Factors in Infant Acuity

Beyond the optics and receptors, two remaining classes of factors could potentially play a role in the poor contrast sensitivity and acuity of infants. The first is immaturities in neural processing in early visual pathways. The second is more general factors, such as lack of motivation on the part of infants to look at every spatial pattern they can see

when tested behaviorally. These kinds of factors can be loosely described as being "psychological."

Recordings from single neurons in the dLGN and in V1 by Colin **Blakemore** and Français **Vital-Durand** and colleagues demonstrate that the neurons with the best spatial resolution are only about a factor of 2 better than behavioral acuity. This suggests that immaturities in neural processing at early stages of the geniculostriate pathway are the primary factor accounting for poor spatial resolution, and that more general "psychological" factors play only a small role.

Color Perception

Spectral Sensitivity

The scotopic and photopic spectral sensitivity curves of infants look essentially like those of adults. For example, visual evoked potentials from the scalp have been used to assess photopic spectral sensitivity in infant and adult subjects using the same methods. Results plotted on the same axes are illustrated in Figure 11.18.

The filled circles are from adult subjects and the open circles from infants. The smooth curve is the theoretical photopic spectral sensitivity function that would be expected if the neural pathways are carrying a signal related to the weighted sum of the L-, M-, and S-cone pigments described in Chapter 7.

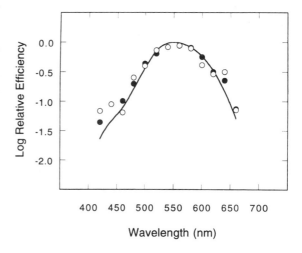

FIGURE 11.18. Photopic spectral sensitivity functions of infants (open symbols) and adults (filled symbols). Modified from M. Bieber, et al, 1995. Spectral efficiency measured by heterochromatic flicker photometry is similar in human infants and adults. *Vision Res.* 35:1385–1392, © 1995, by permission of Elsevier Science.

acuity – depth globhoular vbtn (handwritten)

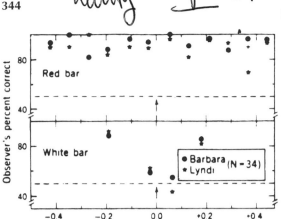

FIGURE 11.19. Results of wavelength discriminations of red from white light for two infant subjects. Reproduced from D.R. Peeples and D.Y. Teller, Color vision and brightness discrimination in two-month-old human infants. *Science* 189:1102–1103, © 1975, by permission of American Association for the Advancement of Science.

Wavelength Discriminations

As illustrated in Figure 11.19, FPL methods have been used to demonstrate that infants can discriminate between at least some wavelengths and white light. In this example, two 2-month-old human infants, Barbara and Lyndi, are presented with a stimulus in the form of a white or a red bar on the left or the right side of a white surround. The bottom panel reveals that both infants prefer the stimulus when it differs in brightness from the surround, but performance falls to chance near the adult brightness match indicated by the arrow. The infants can discriminate every luminance of red from white, including, by inference, the brightness match. This demonstrates that 2-month-old infants are at least dichromats based on the rationale discussed in Chapter 7. Most infants can discriminate red, blue, and green from white and from each other. However, there are deficiencies for yellows and yellow-greens in human infants until about 3 months.

Infant monkeys have been tested for discriminations of wavelengths from white light using the face mask cage operant method. When first tested at 5 months, monkeys could discriminate wavelengths from throughout the visible spectrum from white light over a wide range of relative luminances and thus monkeys appear to be trichromatic by that age.

Motion

The overall shape of the temporal CSF function in the infant is similar to that of the adult but shifted

downwards in sensitivity by about a factor of 20, as illustrated in Figure 11.20.

The extrapolated high-frequency cutoff of the temporal CSF provides an estimate of the limit of temporal resolution, called the Critical Flicker Fusion (CFF) threshold as described in Chapter 10. The CFF in adults is about 55 Hz. The CFF is not very much affected by vertical shifts of the temporal CSF, due to the steepness of the high-frequency portion of the curve. The CFF of a 1-month-old is about 40 Hz, and this increases to 50 Hz at 2 months.

Sensitivity to direction of motion does not appear until about 2 months, and the initial response to direction of motion is asymmetrical when viewed with one eye, with a stronger response to motion in the nasal than to that in the temporal direction.

Stereovision

Infant monkeys respond to depth, most likely using monocular cues, within days after birth. This has been demonstrated by using a **visual cliff,** in which an infant is placed on a ramp in the middle of a piece of glass. One side (shallow side) of the ramp has a patterned stimulus close to the glass surface, and the other (deep side) has the same pattern well below the glass surface. Measurements are made of how frequently the monkey leaves the ramp to walk on the glass on the shallow side and the deep

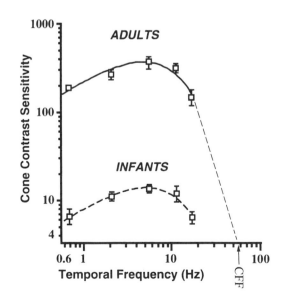

FIGURE 11.20. Temporal contrast sensitivity for adult and infant subjects. Adapted from K.R. Dobkins, et al, Infant color vision: Temporal contrast sensitivity functions for chromatically-defined stimuli in 3-month-olds. *Vision Res.* 37:1–18, © 1997, by permission of Elsevier Science.

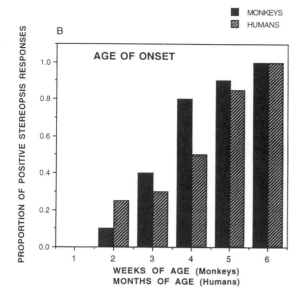

FIGURE 11.21. Age at onset of stereopsis for infant monkeys and humans. Reproduced from C. O'Dell and R.G. Boothe, The development of stereoacuity in infant rhesus monkeys. *Vision Res.* 37:2675–2684, © 1997, by permission of Elsevier Science.

side. The shallow side is chosen most often even at 3 days of age, and the deep side is almost never chosen by animals over 10 days old.

Binocular responses to depth based on stereopsis do not begin to emerge until 2 to 3 HE-months, as illustrated in Figure 11.21.

Once stereopsis has emerged, **stereoacuity**, the smallest amount of disparity that can be seen, improves rapidly and by a large amount as illustrated in Figure 11.22.

This figure shows stereoacuity measured longitudinally for a single infant monkey from birth until 75 days of age. During the first three weeks no stereopsis could be detected even when using stereograms with a disparity of 1,800 seconds of arc. However, once stereopsis emerged, the stereoacuity values improved rapidly to disparity values of less than 100 seconds within six weeks. Somewhat surprisingly, the rate of maturation of stereoacuity once stereopsis emerges is about the same in monkeys and humans on an absolute time scale rather than on a scale adjusted for the weeks-to-months rule that applies to development of most visual functions. This suggests that two biological clocks are responsible for development of stereopsis. One, responsible for its emergence, runs about four times faster in monkeys than in humans and thus follows the weeks-to-months rule. The other, responsible for fine tuning of stereoacuity once stereopsis has emerged, appears to run at the same rate in monkeys and humans.

Two factors could be responsible for the emergence of stereopsis. One is a lack of disparity-sensitive neurons in the striate or extrastriate cortex. Yuyo **Chino** and colleagues have reported the presence of disparity sensitive neurons in V1 in neonatal monkeys, so it appears the neural machinery necessary for disparity processing is already present at the earliest stages of binocular process-

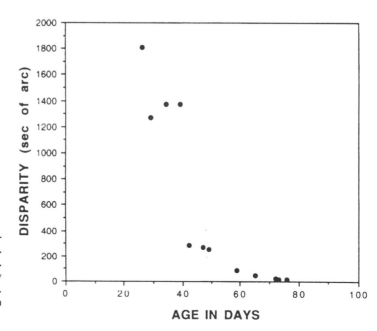

FIGURE 11.22. Improvement of stereoacuity with age in a single infant monkey. Reproduced from C. O'Dell and R.G. Boothe. The development of stereoacuity in infant rhesus monkeys. *Vision Res.* 37:2675–2684, © 1997, by permission from Elsevier Science.

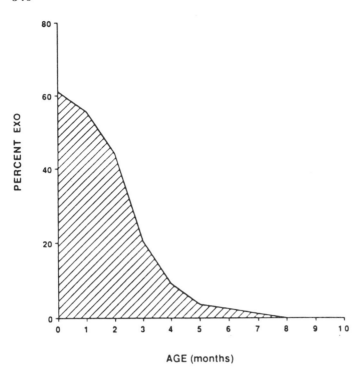

FIGURE 11.23. Most newborn infants' eyes exhibit exotropia over the first few months after birth. Adapted from S.M. Archer, et al, Strabismus in infancy. *Ophthalmol.* 96:133–137, © 1989, by permission of Elsevier Science.

ing. The maturity of disparity-sensitive neurons in higher-order extrastriate areas of infant monkeys has not been reported.

Another factor that could account for lack of stereopsis in infants is lack of motor fusion. Most infant humans and monkeys are exotropic (walleyed) at birth, as illustrated in Figure 11.23.

The prevalence of exotropia decreases during the first few months prior to the time when stereopsis emerges. At these early ages, infants also do not exhibit vergence eye movements in response to disparity stimuli. Overall, these findings suggest that emergence of stereopsis depends on maturation of appropriate binocular eye movements rather than on neural machinery for sensory processing of disparity. However, this issue is not yet resolved, because it has been reported that human infants do not exhibit stereopsis at younger ages even when stimuli are used that do not require precise eye alignment.

Davida Teller's Hazy, Blurry Blandness

As described above, near the beginning of the 20th century the psychologist **William James** described the perceptual world of infants as resembling a blooming, buzzing confusion. Near the end of the 20th century, the psychologist **Davida Teller** summarized the state of our current knowledge of infant perception with a somewhat different conception. In her view, infants are more likely to experience the **haziness** of low contrast, the **blurriness** of low-pass filtering, and the **blandness** of monochrome. In addition, the dynamic and three-dimensional qualities of percepts are weak or immature compared to those experienced by adults.

6 The Fundamental Developmental Problem

At the instant of conception, each human being was simply a single cell (Figure 11.24). Somehow, the process of development had to turn this single cell into a complex organism with a brain that can perceive. Where did the information come from that allowed the brain to be wired properly to accomplish perception? This is the **fundamental developmental problem**.

Solutions Based on Design

One potential solution to the developmental problem can be derived from an analogy to artificial complex information-processing devices. Consider the example of a computer that can carry out

FIGURE 11.24. The conceptual developmental problem.

A person starts out at conception as a single cell

Development happens

The person has an adult brain

Developmental Problem:
What controls what happens during develement so that the adult brain can carry out perception?

complex algorithms by virtue of the connections made between its individual processing elements. A similar question could be asked about where the information came from that allowed the connections in the computer to be made properly. The answer in the case of a computer is that the information came from a wiring diagram. In the factory where the computer was built, there was a wiring diagram that specified that element X should be connected to element Y and so forth.

Solutions Based on Nature

Extending the computer analogy to biological brains, one possible solution to the developmental problem is to propose that there is something **analogous to a wiring diagram** specifying the connections formed in our brain. The most likely candidate as a mechanism for storing this wiring diagram is the **genetic code.** Genes are units made up of DNA molecules, which are constructed out of long strands of four basic building blocks, the bases adenine (A), thymine (T), guanine (G), and cytosine (C). In order to use genes to specify connections in the brain, a code would have to be constructed, perhaps something along the lines of ATGC = connect 1 to 2, AGCT = connect 1 to 3, and so on.

At a detailed level of description, the wiring diagram analogy is demonstrably inadequate, because there is simply not enough information available in the genetic code to specify each individual connection that is formed in the human brain. Current estimates are that humans have only about 30,000 genes. An adult brain has in excess of 5×10^{14} connections that need to be specified, a number too immense even to imagine and orders of magnitude too large to be specified in the genetic code.

However, at a more general level of description, the idea that the genes encode the information that is used to construct a brain is correct. Over evolutionary time, individual organisms that could not differentiate food from poison or prey from predator did not survive to produce offspring. Consequently, species that have survived to the present day are composed of individuals whose genes encode information about how to build brains that can pick up meaningful aspects of stimulation from the environment in which the species evolved.

However, instead of specifying the connections point by point, the genes set up certain programs that are carried out during development via interactions with the environment. These programs allow the specific connections that form to be fine tuned by sensory stimulation from the environment in which the brain is developing.

Solutions Based on Nurture

Primate species have been able to adapt to a variety of environments. Some of this ability to adapt has come about because the process of constructing brain circuits includes a capacity for **developmental neural plasticity,** in which the connections of the individual neural elements are formed to some extent based on the prevailing conditions in the environment in which development takes place.

One experimental method that has been used to study developmental neural plasticity has been to raise animals in altered perceptual environments and then examine the brain changes produced by such rearing.

7 Effects of Rearing in Altered Perceptual Environments

Effects of Form Deprivation

The visual processing parts of the brain can be altered by visual deprivation that is initiated during

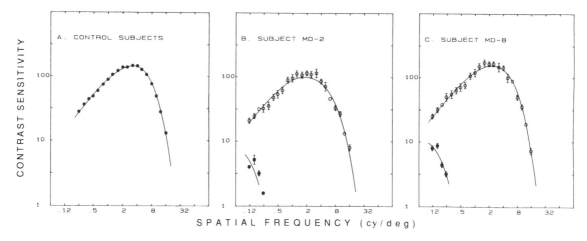

FIGURE 11.25. Behavioral deficits in contrast sensitivity in two monkeys raised under conditions of monocular form deprivation compared to contrast sensitivity in normal monkeys. Adapted from R.S. Harwerth, et al, Functional effects of bilateral form deprivation in monkeys. *Invest. Ophthalmol. Vis. Sci.* 32:2311–2327, © 1991, by permission of the Association for Research in Vision and Ophthalmology.

a sensitive period of postnatal brain development. A **classic set of deficits** can be produced in monkeys by any rearing conditions in which one eye does not receive normal visual stimulation. The rearing conditions that have been studied include use of an opaque contact lens to occlude all light from an eye; surgical removal of the lens from the eye such that the retinal image is blurred; or surgical closing of the eyelids so that only diffuse light reaches the retina.

Two major **physiologic deficits** have been described in such animals. First, very few binocular neurons are present in the visual cortex, in contrast to the case in normal animals, in which the majority of neurons can be driven by either eye. Second, most of the remaining, monocular neurons respond only to the untreated eye, not to the deprived eye. This also differs from the case in normal monkeys, in which about half of the neurons respond to the left eye and the other half to the right eye.

There are also characteristic **anatomic changes**. In the normal animal, the ocular dominance columns in V1 associated with the left and with the right eyes are the same size. However, following visual deprivation, the columns for the untreated eye expand at the expense of the deprived eye.

Other anatomic effects can be seen in other areas of the brain, such as the dLGN, where the size of cell somas is smaller in the layers receiving input from the normal than those for the deprived eye. However, in general, these effects are secondary to the primary changes that take place in V1. The changes in size of the somas in the dLGN simply reflect the sizes of their terminal fields in the cortex.

Two characteristic **behavioral defects** have been reported. First, there are deficits in binocular functions, such as stereopsis. This is thought to be related to the loss of binocular neurons. Second, there is impaired visual function in the deprived eye when it is tested on tasks such as visual acuity or contrast sensitivity. This is thought to be related to the physiological finding that very few neurons in the cortex process information from the deprived eye.

Figure 11.25 shows results of a study carried out by **Ronald Harwerth** and colleagues. Contrast sensitivity functions are shown for a group of normal monkeys and for two monkeys that had been form deprived in one eye during infancy. Results for the deprived eyes (filled symbols in panels B and C) show severe deficits in contrast sensitivity compared to the nondeprived eyes in the same animals (open symbols in panels B and C), and compared to eyes of normally reared monkeys (panel A).

There is a postnatal period lasting about 6 HE-months during which the anatomical sizes of the columns in layer 4 of V1 can be altered. Alterations in physiological function of monocular and binocular neurons in V1 can be produced for a somewhat longer period, probably extending to at least 4 HE-years.

The deleterious effects of visual deprivation of one eye can be reversed if the formerly deprived eye is given normal stimulation and the opposite eye is deprived. However, this reversal has to take place within the sensitive period of brain development. To completely reverse the anatomical effects within V1, this has to be done within 3 HE-months. Partial reversal can be obtained up to about 8 HE-months. Reversal of the functional effects of deprivation as assessed physiologically by recording from neurons in V1, or as measured behaviorally, can be obtained at somewhat later ages. Surprisingly, simply providing the formerly deprived eye with good stimulation but not depriving the other eye yields little or no recovery.

Another finding that is somewhat surprising is that form deprivation of both eyes has less dramatic effects than deprivation of only one eye. Binocular neurons disappear following deprivation of both eyes, and consequently binocular functions such as stereopsis are lost. However, the effects on receptive field properties and on behavioral functions such as acuity and contrast sensitivity are less severe than those of monocular deprivation.

Two general principles have been developed based on these findings from studies of visual deprivation. The first is the concept of a **sensitive period of postnatal brain development** during which visual stimulation can influence connectivity and function. This sensitive period lasts for an extended postnatal period – at least the first couple of years in monkeys and, by extrapolation, at least eight years in humans. The sensitive period does not end suddenly but slowly tails off, making it hard to establish the end point, but the tail end probably extends out at least to puberty. Any time visual stimulation is abnormal during this period, the brain is at risk of developing altered, perhaps inappropriate, connections.

The second general principle has to do with the neural mechanism that is responsible for allowing the separate visual inputs from the left and right eyes to influence brain connections. The basic mechanism, first described by Hubel and Wiesel, is called **binocular competition.** The essence of this mechanism is that since only a finite number of potential connections can be made in the brain, inputs from the left and right eyes must compete for these connections. In the normal animal, the two eyes are able to compete equally, and both end up with about half of the connections. A deprived eye is unable to compete against a normal eye, and this explains why it ends up with fewer connections. A mechanism that might be able to account for these findings is described later in the section "Models Based on Hebbian Rules."

Deprivation of Oriented Contours

Monkeys have been reared under conditions that deprived them only of contours of a particular orientation. The methods used to achieve this rearing condition are illustrated in Figure 11.26.

The optical conditions during rearing are illustrated in the top panel. A cylindrical lens was placed in front of one eye to cause horizontal and vertical contours to be focused at different planes. Contours oriented parallel to the axis of the cylindrical lens were focused in front of the retina, where they were always blurred. Contours oriented orthogonal to the axis were focused slightly behind the retina, where the animal could bring them into focus by accommodating.

Two groups of monkeys were reared, some with vertical and others with horizontal contours out of

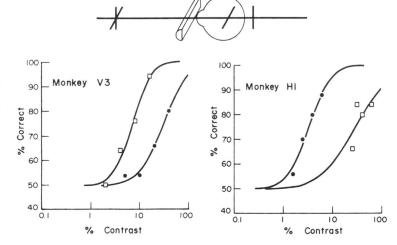

FIGURE 11.26. The top panel shows the rearing methods used to selectively deprive monkeys of contours of a particular orientation. The bottom panels show the effects of these rearing conditions for two monkeys. See text. Adapted from R.G. Boothe and D.Y. Teller, Meridional variations in acuity and CSF's in monkeys (*Macaca nemestrina*) reared with externally applied astigmatism. *Vision Res.* 22: 801–810, © 1982, by permission of Elsevier Science.

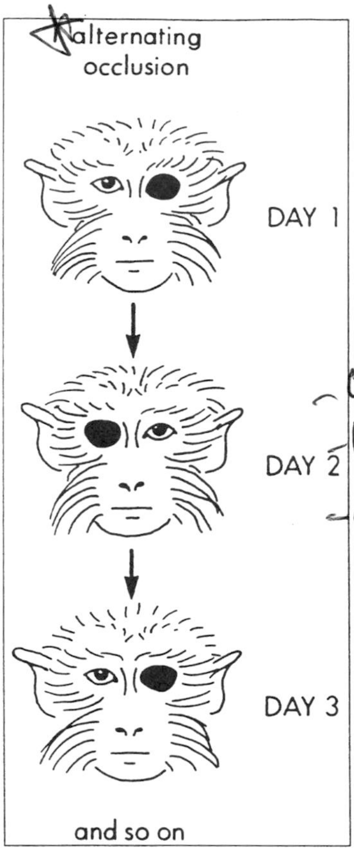

[handwritten top margin: How are those binocular ~~neurons~~ neurons used?]

FIGURE 11.27. Illustration of methods used to rear monkeys such that each eye receives visual stimulation but the two eyes never receive simultaneous binocular stimulation. Reproduced from L. Tychsen, et al, Neural mechanisms in infantile esotropia: What goes wrong? *Am. Orthop. J.* 46:18–28, © 1996, by permission of The University of Wisconsin Press.

[handwritten left margin: -tight- areas of the brain larger.]

[handwritten margin: -cornea lens contacts]

[handwritten left margin: ale eye super acuity]

circles). The opposite pattern of results is shown for monkey H1, reared with only horizontal contours in good focus.

Pure Binocular Deprivation

A slightly different effect on brain connectivity is produced by conditions in which both eyes receive visual stimulation but the stimulation is discordant in the two eyes. One way to produce this rearing condition experimentally in an infant monkey is with alternating monocular occlusion, as illustrated in Figure 11.27.

An extended-wear contact lens that has been dyed opaque to serve as an occluder is placed on one eye on the day of birth. The following day the occluder is moved to the opposite eye, and so on, alternating between eyes on alternate days.

Following the rearing period, binocular neurons are eliminated as occurs following unilateral deprivation and binocular functions, such as stereopsis, are disrupted. However, unlike the case with unilateral deprivation, about half of the monocular neurons respond to the left eye and the other half to the right eye. Consequently, basic functions such as acuity are the same in each eye.

Monkey Deprivation Effects Have a Human Analog

The deficits in visual function seen in monkeys following visual deprivation rearing are similar to the visual losses seen in human patients who have a clinical condition called **amblyopia,** often referred to as **lazy eye.** For many decades, two sets of independent facts accumulated in the scientific literature. One set was in the discipline of clinical ophthalmology and concerned characteristics of humans with amblyopia. The other was in the discipline of developmental neuroscience and concerned the anatomical, physiological, and behavioral effects of visual deprivation rearing in animals.

These two literatures have now been synthesized, as it has become apparent that amblyopia is the human analog of the visual deprivation effects that can be produced in monkeys. There are a number of subtypes of amblyopia, each named according to the type of visual deprivation that was present (or assumed to be present) during infancy and is now understood to be the causative agent.

One implication of this modern synthesis is that the causal factors and neural mechanisms that

focus. During rearing, it was established that visual acuity for seeing the contour in best focus was better than acuity for the contour that was out of focus. This is not very interesting, as it simply reflects optical blur. The more interesting finding was obtained when the lenses were removed. Then both orientations were in equally good focus on the retina. However, the animals continued to show deficits in their ability to see the orientation that had been defocused during rearing, as illustrated in the bottom panels of Figure 11.26. Monkey V3 was reared with only vertical lines in good focus. When tested with a vertical grating, this animal could correctly identify the stimulus on 75% of the trials when only 7% contrast was present (open squares). When it was tested with a horizontal grating, similar performance required 31% contrast (filled

[handwritten bottom margin: Space the argument - what if you lose one eye when you are older - you would want to because to see some.]

underlie various types of human amblyopia can be studied in the monkey model. An example of using a monkey model to study amblyopia associated with congenital cataracts is shown in Box 11.1.

An example of a naturally occurring condition in children that might cause deprivation of contours of certain orientations is astigmatism, an optical condition described in Chapter 3. An example of a

Box 11.1
Using Infant Monkeys to Study Treatments of Congenital Cataracts in Human Babies

intraocular

Worldwide, tens of thousands of human babies are born with cataracts each year. These children will go blind unless the cataracts are removed and rehabilitation treatment instituted at an early age. However, there is uncertainty about exactly what forms of treatment are most effective. Infant monkeys have been used as a model to address some of these questions. A contact lens made out of a diffusing material that simulates a cataract is placed on one eye at birth. At designated ages, surgery is performed to remove the lens from the eye, as would be done

for a child to remove a cataract enmeshed within the lens (Figure 1).

The surgery to remove the lens, rendering the eye aphakic, is carried out in sterile conditions under anesthesia, using the same procedures as would be used with a human child. The optical power of the natural lens is replaced by some combination of an intraocular lens implanted surgically and/or an extended-wear contact lens (Figure 2).

The infant monkey shown in this photograph is wearing an extended-wear contact lens in its

cause of cataracts patiens?

• ultraviolet light. •

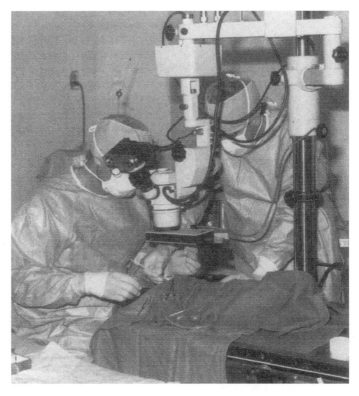

FIGURE 1. Surgery is performed on an infant monkey under anesthesia using methods that are identical to those that would be applied to a human infant with a cataract. Reproduced from RG Boothe, A monkey model for studying contact lens treatments of aphakic amblyopia. In *Congenital Cataracts*, ed. E. Cotlier, pp. 103–111. Austin, TX: R.G. Landes Company, © 1994, by permission of the R.G. Landes Company.

Continued

Box 11.1 *Continued*

FIGURE 2. Monkeys provide models for studying the treatments that can potentially be applied to human babies with cataracts. Reproduced from J.A. Gammon, et al, Extended-wear soft contact lenses for vision studies in monkeys. *Invest. Ophthalmol. Vis. Sci.* 26:1636–1639, ©, 1985, by permission of Association for Research in Vision and Ophthalmology.

right eye following surgery to remove its natural lens. This monkey is also wearing an opaque contact lens in its left eye to simulate patching therapy commonly used to treat this condition in human children.

There are a number of advantages to studying these treatments in an animal model, because there are limitations to what can be learned from studies of human babies treated for this disorder. It is usually difficult to determine the characteristics of the population from which the reported human cases are being drawn, and published studies of individual human cases are more likely to include successes than mediocre or poor outcomes. Even in the best-designed human studies, there are problems such as associated ocular abnormalities in addition to the infantile cataract. Other problems are incomplete medical histories during the neonatal period, difficulty ensuring patient follow-up, and inability to directly monitor compliance with prescribed therapy. Ethical considerations often make it difficult to assign human subjects randomly to all treatment groups, especially when untreated controls are included. Finally, information about underlying neuropathology cannot be obtained from human children. These limitations can all be overcome or minimized by using a monkey model.

condition leading to pure binocular deprivation would be strabismus. Each eye is receiving visual stimulation, but since the two eyes are not pointing in the same direction, the stimulation is different in the two eyes.

Other conditions that can cause form deprivation amblyopia in one or both eyes of humans include swollen eyelids, due to conditions such as trauma or infection, and high refractive errors in one or both eyes. These conditions, even when present for only short periods, have the potential to lead to abnormal brain connections and consequent poor visual function.

The earlier conditions that cause visual deprivation are detected and corrected, the better the chances for a good outcome (Figure 11.28). This figure summarizes visual acuity results from a number of studies of children who received surgery to remove a congenital cataract. Children in whom

the cataract was detected shortly after birth and treatment including surgery to remove the cataract was initiated early ended up with better acuity than children whose treatment was delayed. The children who received the earliest treatment achieved acuity near normal values of 20/20 Snellen. The children who had the longest delays before treatment had poor acuity or rudimentary vision, such as an ability only to count fingers (CF) or to detect hand motion (HM) or light perception (LP).

As in the animal studies, simply removing the original deprivation condition is usually not sufficient to restore function. The clinical treatment usually includes part-time patching of the fellow eye. This forces usage of the formerly deprived eye and stimulates the brain to form connections with that eye. Of course, this patching therapy must be carefully monitored so as not to cause damage to the visual pathways processing information from

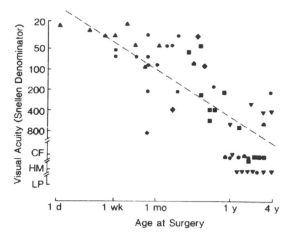

FIGURE 11.28. The outcome in terms of visual acuity of children with congenital cataracts is correlated with the age at which surgery is performed to eliminate the cataract. Each symbol is the result for one patient. This figure summarizes the results of a number of previous studies, and the different symbols reflect results reported in the different studies. Reproduced from E.E. Birch and D.R. Stager, Prevalence of good visual acuity following surgery for congenital unilateral cataract. *Arch Ophthalmol.* 106:40–43, © 1988, by permission of the American Medical Association.

the patched eye. An evaluation of the potential tradeoffs between the benefits and risks of various amounts of patching has been carried out in the monkey model, as illustrated in Figure 11.29.

Results from monkeys in which an aphakic eye was optically corrected but no patching was done on the fellow eye are shown on the left of the graph. These animals ended up with normal acuity in the fellow eye but a reduction in acuity of the aphakic eye by about a log unit, which would correspond to a Snellen acuity of 20/200. Animals that received small amounts of daily patching for up to 50% of the daylight hours showed essentially the same results. Patching of 75% had a beneficial effect on the acuity of the aphakic eye and no detrimental effect on the patched eye. However, when patching was carried out full-time (100%), a reversal occurred, with the aphakic eye now having normal acuity and the fellow eye impaired by as much as 2 log units.

These findings were obtained from monkeys by performing studies in which amount of patching was varied systematically while all other factors were held constant. The question of how much patching should be done in human children could not be resolved based on the human clinical literature for the reasons discussed in Box 11.1. This

raises an ethical issue about whether or not it is appropriate to carry out visual deprivation studies with monkeys or other animals. This issue is discussed in Box 11.2.

Box 11.2
A Comment About Issues of "Animal Rights"

Many of the findings and discoveries reported in this chapter, and indeed in this entire book, have been obtained from studies using animals. Some argue that collection of these data is unethical, because it violates the rights of the animals used in the studies. It is true that if animal research were totally halted, some animals would be spared having to undergo invasive and sometimes terminal experiments. However, it is also true that if animal research were halted, another consequence would be that some human babies would go blind who would otherwise be able to be treated with new methods derived from the animal research. There is disagreement among reasonable individuals about the relative importance of the benefits and costs of animal research.

A closely related issue is "animal welfare." Whether or not there is agreement about animals' having "rights," all should agree that it is the responsibility of humans who use animals for any purpose, including research, to assume responsibility for their welfare.

---

8 Models Based on Hebbian Rules

Many of the effects of rearing infants in altered visual environments on brain connectivity can be explained, in principle, by neural models based on **Hebbian rules,** first formulated by **Donald Hebb,** a psychobiologist, in the 1940s. His general idea was that changes in strength of synaptic connections are a function of simultaneous activity between connected neurons. He formalized this idea by asserting that whenever a pair of neurons, one presynaptic and one postsynaptic, are active simultaneously, the strength of the connection between them is increased.

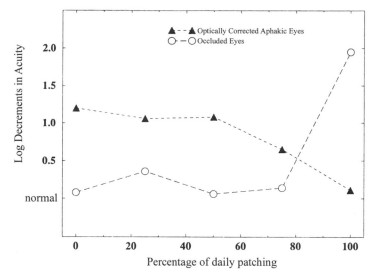

FIGURE 11.29. There is a tradeoff between the benefits and risks associated with patching therapy. Reproduced from R.G. Boothe, Visual development following treatment of a unilateral infantile cataract. In *Infant Vision*, eds. F. Vital-Durand, J. Atkinson, and O.J. Braddick, pp. 401–412. Oxford: Oxford University Press, © 1996, by permission of Oxford University Press.

The rules formulated by Hebb have been elaborated on and further formalized by subsequent investigators. A modern formulation of the Hebbian rules is:

- If the presynaptic neuron fires a spike and the postsynaptic neuron fires a spike at the same time, then the strength of the connection between the two neurons is increased.
- if the presynaptic neuron fires a spike and the postsynaptic neuron remains silent, then the strength of the connection between the two neurons is decreased.
- if the postsynaptic neuron fires a spike and the presynaptic neuron is silent, then the strength of the connection between the two neurons is decreased.

Application of Hebbian Rules to Visual Development

To illustrate how these rules might operate, suppose the brain of a hypothetical observer has just reached the developmental period when pruning of connections has to take place. Consider what will happen to a single neuron in the brain of this observer that is illustrated in Figure 11.30.

This neuron reached the developmental stage of pruning with four connections, two carrying signals from the left eye and two from the right. Obviously, this example is in the form of a simplified model rather than being biologically realistic. An actual neuron receiving inputs from both eyes would receive perhaps 10,000 input connections,

and these would transmit information from the two eyes that had passed through neurons located at intermediate stages of processing rather than coming directly from the eyes as shown here. Nevertheless, this simplified model can be used to elucidate the general principles by which the Hebb rules operate, without becoming bogged down with the kinds of details that would make the model more biologically realistic. Also, it is not relevant here to spell out how these four particular connections happened to form on this neuron. Perhaps the axons carrying inputs from the two eyes were just programmed to grow randomly into this general region of the brain and connect with the first neuron they encountered, and it happened to be this one. This example is concerned with trying to understand what happens from this point forward as the pruning stage begins.

Assume that this neuron has built-in information, perhaps in the genetic code, that specifies it has to prune back 50% of its connections, perhaps as conceptualized in Figure 11.31.

Since it is starting out with four connections, two will have to be removed and the other two can be retained into adulthood. Thus, there are three possible outcomes for how this neuron will end up connected. It might end up connected only with the left eye, only with the right eye, or with both. Which of these outcomes occurs is going to be determined entirely by the Hebbian rules.

In this example, the Hebbian rules will be implemented in the following manner: A tally sheet will be kept by the postsynaptic neuron for each of its input connections. Whenever a presynaptic neuron

FIGURE 11.30. Simplified model used to demonstrate the operation of the Hebbian rules. See text.

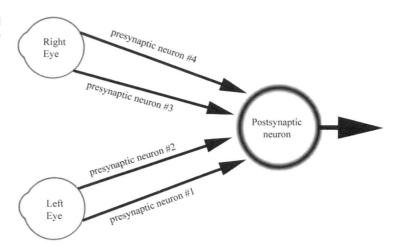

providing the input to a connection and the postsynaptic neuron fire a spike at the same time, one brownie point will be tallied for that connection. Whenever either a presynaptic neuron providing the input to a connection or the postsynaptic neuron fires alone, a demerit will be tallied for the connection. Each connection will accumulate a running score obtained by adding up its brownie points and subtracting its demerits. When pruning has to occur, the connections that have the lowest scores are the ones to be pruned.

There is one additional fact that needs to be made explicit. In general, activity in a single presynaptic neuron is insufficient to strongly activate a postsynaptic cell. Simultaneous spikes from several presynaptic neurons are much more likely to cause a postsynaptic neuron to produce a spike of its own. In order to capture this biological fact, this example assumes that the postsynaptic neuron requires simultaneous inputs from two or more presynaptic neurons to generate a spike.

Predictions for Normal Development

The simple model described in the previous section is sufficient to explain the outcomes of each of the types of postnatal visual deprivation described earlier in this chapter. Consider first what this model predicts will happen in a normal visual environment. Suppose the observer looks at a light. Both eyes will be aimed towards the light, and all four presynaptic neurons will be activated to produce spikes. This simultaneous activity from all four presynaptic neurons will cause the postsynaptic neuron to also produce spikes. Thus, many brownie points will be accumulated by all four con-

nections. When two of the connections have to be pruned, which connections have the lowest scores will be based simply on chance. Thus, the expectation is that the hypothetical brain will end up with some neurons with connections from only the left eye, some with connections from only the right eye, and some with binocular connections from both eyes, somewhat along the lines shown in Figure 11.32.

This prediction is consistent, at least to a first-order approximation, with the distributions of neurons in V1 in normally reared monkeys.

Predictions for Monocular Visual Deprivation

Next, consider what connections are likely to remain present if this hypothetical observer is reared under conditions in which the visual input is blocked from the left eye, as by a dense cataract. Suppose the observer looks at a light. This will cause both neurons carrying inputs from the right eye to fire simultane-

Rules Stored in Genes of Hypothetical Neuron

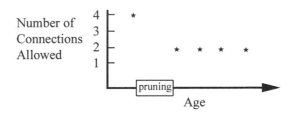

FIGURE 11.31. The kinds of rules that might be built into a neuron by a genetic code. See text.

Possible Outcomes Following Pruning	Receptive Field Properties	Expected Occurrence Following Normal Rearing	
L_1L_2	Left Eye Monocular	~17%	~34%
R_1R_2	Right Eye Monocular	~17%	
L_1R_1	Binocular		
L_1R_2	Binocular	~67%	
L_2R_1	Binocular		
L_2R_2	Binocular		

FIGURE 11.32. Predictions of the Hebbian model about the brain connections that should be formed during rearing in a normal visual environment. See text.

ously. This simultaneous spike activity in the two right eye neurons will be sufficient to generate spike activity in the postsynaptic neuron. Thus, the two connections with right eye neurons will generate many brownie points. The neurons from the left eye will be active only because of spontaneous activity. Occasionally one of these left eye neurons will be active due to spontaneous activity, but this activity will not be sufficient to drive the postsynaptic neuron by itself. Consequently, this activity will generate mostly demerits. This spontaneous activity is unlikely to occur at the same moment as the visually modulated activity coming from the right eye, so every time the right eye neurons cause the postsynaptic neuron to fire, the connections with left eye neurons will generate more demerits.

When the times comes for this neuron to prune 50% of its connections, it will most likely be the two left eye connections that are pruned and the two right eye connections that are maintained. The same conditions will hold for all other neurons in the brain. Thus, as an adult this individual will end up with a brain whose neurons are connected along the lines illustrated in Figure 11.33.

The binocular neurons are missing, and the remaining monocular neurons are all connected to the right eye. These predictions correspond as a first order approximation to the distributions of neurons in V1 following monocular deprivation rearing in monkeys.

The same brain, programmed with the exact same rules, might grow up in a different visual

environment. This time, suppose the visual input that is blocked is from the right eye. It should be obvious that under these conditions the pattern of connections in the individual's adult brain will be the mirror image of what it would be following deprivation of the left eye.

Predictions for Pure Binocular Deprivation

Next, consider what will happen if the same individual is raised in an environment with discordant binocular input, such as from alternating monocular deprivation or from strabismus. Sometimes the right eye will look at a light, generating simultaneous activity in its neurons that will be sufficient to activate the postsynaptic neuron. At this moment, the other eye will not be activated by the same light. In this instance, the two right eye connections will receive brownie points and the two left eye connections will receive demerits. Moments later the left eye may look at the same light, generating simultaneous inputs sufficient to generate activity in the postsynaptic neuron. This results in brownie points for the left eye connections and demerits for the right eye connections. The expected results when it is time to prune two connections are illustrated in Figure 11.34.

Statistically, the two left eye connections are likely to have very similar scores. Similarly, the two right eye connections are likely to have very similar

Possible Outcomes Following Pruning	Receptive Field Properties	Expected Occurrence Following Normal Rearing	
L_1L_2	Left Eye Monocular	~ 0%	~100%
R_1R_2	Right Eye Monocular	~100%	
L_1R_1	Binocular		
L_1R_2	Binocular	~ 0%	
L_2R_1	Binocular		
L_2R_2	Binocular		

FIGURE 11.33. Predictions of the Hebbian model about the brain connections that should be formed during rearing in an environment in which the left eye is deprived of normal visual stimulation.

FIGURE 11.34. Predictions of the Hebbian model about the brain connections that should be formed during rearing in an environment in which each eye receives stimulation but the two eyes never receive input at the same time.

Possible Outcomes Following Pruning	Receptive Field Properties	Expected Occurrence Following Normal Rearing	
L_1L_2	Left Eye Monocular	~ 50%	~100%
R_1R_2	Right Eye Monocular	~ 50%	
L_1R_1	Binocular		
L_1R_2	Binocular	~ 0%	
L_2R_1	Binocular		
L_2R_2	Binocular		

scores. Depending on chance factors, the right eye scores might be higher or lower than those from the left eye for an individual neuron. In cases in which the right eye scores happen to be higher, the neuron will end up with two right eye input connections, and vice versa for higher left eye scores. Across the population of neurons in the brain, there are likely to be about equal numbers of neurons that receive two left inputs and two right inputs. However, it is very unlikely that any given neuron will end up with one input from the left eye and one from the right. Thus, in this case the brain will not have a normal complement of binocular neurons. These predictions correspond roughly to the distribution of neurons in the brains of monkeys reared with alternating monocular occlusion or with experimentally induced strabismus.

Predictions for Oriented Contour Deprivation

Finally, consider what would happen if contours of a particular orientation were eliminated, as would occur with astigmatism. Figure 11.35 shows the locations on the retina of five ganglion cells, each of

which projects to the same central neuron in the hypothetical observer's brain.

For this example, assume that pruning will leave only three of the original five connections. It should be obvious that if vertical contours are optically eliminated from the retinal image, the remaining horizontal contours will frequently cause retinal ganglion cells #2, #5, and #4 to fire simultaneously. Other groups of three neurons will seldom be activated simultaneously. Thus, the postsynaptic neuron is likely to end up connected to ganglion cells #2, #5, and #4 and as a result will be more efficient in the adult observer at detecting horizontal lines than vertical ones. Recall that this prediction was born out in monkeys reared with experimentally induced astigmatism described earlier in this chapter.

Broader Implications of Hebbian Rules

These examples have been using the Hebbian rules to explain the connectivity that results in the brain following different kinds of visual deprivation rearing. However, the experimental paradigm of visual deprivation rearing simply reveals the oper-

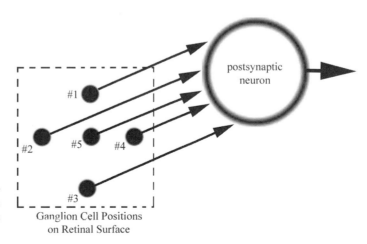

FIGURE 11.35. A Hebbian explanation for the effects of rearing in an environment in which an eye is deprived of contours of a particular orientation.

Ganglion Cell Positions on Retinal Surface

ation of these rules. Thus, it is interesting to ask what the Hebbian rules accomplish in a context more general than studies of visual deprivation.

The Hebbian rules have the effect of wiring up a brain that can efficiently detect the forms of sensory stimulation that were highly correlated with one another during development. As a result, the adult brain will be connected in a manner that is finely tuned to the spatiotemporal patterns that were correlated with one another in the environment in which development took place.

This has profound implications! Human beings' brains end up being finely tuned to what they experienced while growing up, without the genes' ever having had to specify how to connect the brains in order to accomplish this. The genes just specify a few very simple rules that operate during interactions with the environment during development. The operation of these rules allows the environment to fine tune the connectivity of each brain. The end result is brains that are very efficient at perceiving the correlated sense impressions caused by objects and events in the various environments in which humans are raised.

9 Using Neural Networks to Study Developmental Neural Plasticity

Chapters 6 and 10 described the design of neural networks that can serve as models of perceptual functions. Recall that these models must undergo an initial training stage during which the weights are set appropriately before the models can be used to evaluate specific hypotheses about perception. The details about how this training is done have implications within two domains. First, this topic is important as a practical matter for scientists who need to know how to design neural networks that can model particular perceptual activities. However, this topic also has implications for issues of developmental neural plasticity. One way to understand how brains develop their connections under the influence of their environments is to use the training stage of neural networks as a model of how information from the environment can be used to mold functional connections. There may be important analogies between the ways scientists use training to create a network that can perform specific functions and the way nature uses environmental influences during development to wire up functional brains.

We will begin by describing the task that must be accomplished when using training to create a network that can perform some specific function. Every network will have some range over which it can perform a particular function, ranging from very poor to very good. The exact level of performance that will be attained within this range is determined by how the weights of the network are set, as described in Chapter 6. This can be stated formally as a cost function, which is simply a function that specifies how performance will change as the weights are changed:

$$P = F(w)$$

Where P is the accuracy of performance of the network on the given task, w is a particular setting of the weights, and F is some unknown function that specifies the relationships between the setting of the weights and the accuracy of the performance.

The values of P generated by function F can be conceptualized as in Figure 11.36. This graph is oversimplified, because it shows the height of P as though it were changing along only a single dimension. In actuality, the height is defined in a multidimensional space with as many dimensions as there are weights. Nevertheless, this graph serves the heuristic function of giving an intuitive feeling for the fact that different values of weights produce differing levels of accuracy of performance. Points along the function where P is greater than for nearby surrounding points, like a, c, d, e, f, and g, are called **local maxima.** Point b is the **global maximum** across all settings, and this is the setting of the weights that is needed to be sure the network is exhibiting optimal performance on the task.

Thus, the task during training is simply to set the weights so that the network is performing at the global maximum. If one knew the shape of the cost function, it would be a simple matter to just look at the graph and find the global maximum (b) and then set the weights to values corresponding to this maximum. Unfortunately, the shape of the cost function is generally not known in network problems,

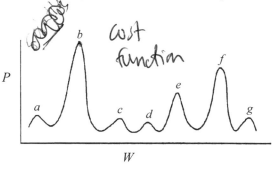

FIGURE 11.36. A neural network can be conceptualized as carrying out a cost function. See text.

and the purpose of training is to find optimum settings of the weights based on interactions of the network with its environment.

Two basic strategies can be used. The first is based on Hebbian rules and the second on feedback.

Using Hebbian Rules to Train a Neural Network

Hebbian rules can be used to adjust the weights of a neural network in a manner that will allow it to perform certain useful perceptual tasks. While the network is responding to inputs from its environment, each weight is simply increased slightly whenever the input and output element that are connected are both active and decreased slightly whenever either is active alone, just as described above for the example of Hebbian rules operating during visual deprivation.

Hebbian rules result in a form of **associative learning** in which the network becomes **self-organized** in such a manner that it responds to correlated activities in the environment. This form of associative learning is strictly local. Each connection, in effect, adjusts its own weight based on local conditions without being influenced by whether the output being produced by the network as a whole is correct or wrong. Other types of training of network connections depend on feedback, as described in the next section.

Training Based on Learning Rules and Feedback

Training a neural network based on environmental feedback requires some mechanism that acts as a **teacher.** The teacher has knowledge about the goals the network is trying to achieve and monitors the outputs of the network. The teacher then uses this global information to adjust the weights to better accomplish the goals.

When implementing learning based on feedback, the neural network is initially allowed to generate an output signal based on its current weights, which may have been simply set randomly or indiscriminately. This initial output is evaluated in terms of the magnitude of its error compared to some desired output. Then the weights are adjusted based on this evaluation to try to reduce the error. This procedure is continued in an iterative manner until the weights are adjusted appropriately to minimize the error to below some acceptable level.

When conceptualized as optimizing a cost function, the training has the goal of trying to find a global maximum of the function. Training can be thought of as traveling in the horizontal direction along a function, as shown in Figure 11.36. Recall that this figure shows only a one-dimensional function but that in actuality, the function is multidimensional. A slightly more general way of conceptualizing what is happening is to consider the cost function as being represented on a two-dimensional landscape and the rules are applied to try to find the highest mountain peak on the plain. Conceptualized in this manner, the general rules that are used to train a network are sometimes referred to as rules for **hill-climbing.** Local and global maxima (peaks) are discovered by a search of possible settings directed by the hill climbing rules. These procedures are analogous to the process of trying to find the peak of a mountain by walking around in complete darkness. Since one cannot see the peak, search strategies must be based on moving in some direction based on one's experience of the piece of ground one is standing on, perhaps combined with memory of recent locations where one has stood, and moving in directions that result in higher elevation. A number of specific hill-climbing rules have been developed to try to achieve the goal of finding the highest peak.

Random Searches *? actual hill climbing*

In **random hill-climbing searches**, one starts at some random value for the weight settings, called W1, and evaluates its performance, called P1. Another point, W2, is then picked randomly in the near vicinity of W1 and its performance, P2, evaluated. One moves to W2 if P2 is greater than P1 and then repeats this process indefinitely. This technique guarantees an uphill climb and will therefore reach at least a local minimum.

Gradient Descent and the Delta Rule

A more efficient procedure, called **gradient descent,** examines what happens when a single step is taken in every direction from the current location. One chooses not just an uphill step but the particular uphill step with the steepest slope. A formula called the **delta rule** can be used to calculate the amount by which each synaptic weight in the network should be changed to achieve a gradient descent.

Note that neither the random procedure nor gradient descent will ever be able to escape from a local maximum if there are no other local maxima having a higher value within the size of the individual steps that are taken. A number of modifications have been proposed to increase the chances of finding a global maximum rather than remaining trapped in a local maximum.

Probabilistic Random Searches

In **probabilistic random searches**, also called **stochastic hill-climbing**, the rule for taking a step is probabilistic rather than determined. A step that will result in a steep ascent has a high probability of being chosen, but a step that results in a smaller ascent, or even a descent, still has at least a small probability of being taken. This provides a probabilistic method of escaping from a local maximum by traveling downhill long enough to reach the next peak.

The exact probabilities to be used at each stage of the search can be set with a variable called **temperature** (T). At high temperatures, the search is about as likely to move downhill as uphill. Random hill-climbing, described above, is simply a special case of stochastic hill-climbing when the temperature is very high. At low temperatures, the search is less likely to move downhill.

A variation on stochastic hill-climbing is **simulated annealing,** in which the temperature starts out high and is gradually reduced over a number of trials to a low level. This allows the entire landscape to be sampled early on, but then gradually reduces the range of the search to promising areas. Much active computational research is concerned with trying to determine optimal cooling rates that give the highest likelihood of finding the global maximum.

Genetic Learning

Genetic algorithms are based on some general principles of evolution. Given a population of organisms and variety within that population, some will do better at survival and reproduction than others. Mechanisms of heredity allow descendants to inherit some of the structure of their parents along with some amount of variation and diversification. Genetic algorithms simulate this evolutionary process.

Genetic hill-climbing treats the weight settings as a genetic string, or "chromosome." Initially, two particular weight settings (each specified as a vector and often called a "string") are chosen, perhaps randomly, and the cost function evaluated for each. These are called the "parents." Then a genetic "crossover," or "mutation," is performed. A crossover allows some part of the specific weight

settings of one parent to be swapped for the corresponding values from the other parent, with no new values being introduced. A "point mutation" changes one value in the string to a new value. Point mutations can be combined with crossovers.

The new generation can be evaluated to find the most successful variants. This process, when iterated over perhaps hundreds of thousands of generations, finds the global maximum for some classes of problems.

what does this do?

Summary

The developmental approach searches for explanations of perceptual phenomena in terms of events that take place during the life span of a single observer. Some of the properties of perception in an individual organism have been built in, and others are learned. Traditionally, the nativist position emphasized the information that is built in, while the empiricist position emphasized the role of learning. Modern neuroscience has revealed that treating this issue as either/or is generally not helpful, since most properties of perception can be understood in terms of built-in rules that produce their effects via interactions with the environment.

Most of the neural machinery involved in visual processing develops during the prenatal period. However, formation and pruning of intrinsic connections between neurons in central visual processing pathways takes place mostly during the postnatal period. The implication of this finding is that the connectivity of these portions of the brain can be influenced by postnatal perceptual experiences.

The optics of the eye and basic reflexes such as accommodation mature relatively quickly after birth. However, newborns exhibit major immaturities in most basic visual functions, including absolute sensitivity to light, acuity and contrast sensitivity, color vision, temporal vision, and stereovision, some of which do not reach adult levels for years.

A fundamental problem in studying perceptual development is to understand where the information comes from that allows the neural processing portions of the brain to become wired together properly. Some of the information used to accomplish this is built in genetically, based on perceptual features of the environment that have been stable over evolutionary time. Other information comes from interactions with the environment during the lifespan of the organism.

Influences of the environment on brain connectivity and function have been studied by carrying out various forms of visual deprivation rearing in monkeys. In addition to providing us with knowledge about the mechanisms by which the environment influences the connectivity of our brains, these studies have clinical relevance to human disorders that lead to blindness in babies.

Two kinds of abstract mechanisms can be used to explain how a biological brain or, equivalently, a neural network, learns how to connect itself together to accomplish perceptual functions. Hebbian rules can explain how neural circuits become organized to respond efficiently to correlated patterns of activity that is from objects and events in the environment. Rules based on feedback from a teacher demonstrate how connections can be adjusted to optimize performance on particular perceptual functions.

Selected Reading List

Banks, M.S., and Shannon, E. 1992. Spatial and chromatic visual efficiency in human neonates. In *Carnegie-Mellon Symposium on Cognitive Psychology*, ed. C. Granrud, pp. 1–46. Hillsdale, NJ: Erlbaum.

Boothe, R.G. 1988. Visual development: Central neural aspects. In *Handbook of Human Growth and Developmental Biology*, Vol. I: Part B, eds. E. Meisami and P.S. Timiras, pp. 179–191. Boca Raton, FL: CRC Press.

Boothe, R.G. 1996. Visual development following treatment of a unilateral infantile cataract. In *Infant Vision*, eds. F. Vital-Durand, J. Atkinson, and O.J. Braddick, pp. 401–412. Oxford: Oxford University Press.

Boothe, R.G., Greenough, W.T., Lund, J.S., and Wrege, K. 1979. A quantitative investigation of spine and dendrite development of neurons in visual cortex (area 17) of *Macaca nemestrina* monkeys. *J. Comp. Neurol.* 186:473–490.

Bradley, D.V., Fernandes, A., Lynn, M., Tigges, M., and Boothe, R.G. 1999. Emmetropization in the rhesus monkey (*Macaca mulatta*): Birth to young adulthood. *Invest. Ophthalmol. Vis. Sci.* 40:214–229.

Dobson, V. 1990. Behavioral assessment of visual acuity in human infants. In *Comparative Perception*, Vol. 1, Basic Mechanisms, eds. M. Berkley and W.C. Stebbins, pp. 487–521. New York: Wiley.

Gunderson, V.M., and Sackett, G.P. 1984. Development of pattern recognition in infant pigtailed macaques (*Macaca nemestrina*). *Dev. Psychol.* 20:418–426.

Harwerth, R.S., Smith E.L. III, Paul, A.D., Crawford, M.L.J., and von Noorden, G.K. 1991. Functional effects of bilateral form deprivation in monkeys. *Invest. Ophthalmol. Vis. Sci.* 32:2311–2327.

Hendrickson, A.E. 1988. Development of the primate retina. In *Handbook of Human Growth and Developmental Biology*, Vol. 1: Part B, eds. E. Meisami and P.S. Timiras, pp. 165–177. Boca Raton, FL: CRC Press.

Linsker, R. 1986. From basic network principles to neural architecture: Emergence of orientation-selective cells. *Proc. Natl. Acad. Sci. U. S. A.* 83:8390–8394.

Mates, S.L., and Lund, J.S. 1983. Spine formation and maturation of type 1 synapses on spiny stellate neurons in primate visual cortex. *J. Comp. Neurol.* 221:91–97.

Maurer, D., and Lewis, T.L. 1993. Visual outcomes after infantile cataract. In *Early Visual Development, Normal and Abnormal*, ed. K. Simons, pp. 454–484. New York: Oxford University Press.

Mayer, D.L., and Dobson, V. 1980. Assessment of vision in young children: A new operant approach yields estimates of acuity. *Invest. Ophthalmol. Vis. Sci.* 19:566–570.

O'Dell, C., and Boothe, R.G. 1997. The development of stereoacuity in infant rhesus monkeys. *Vision Res.* 37:2675–2684.

Rakic, P., and Goldman-Rakic, P.S. 1982. Development and modifiability of the cerebral cortex. *Neurosci. Res. Prog. Bull.* 20:429–611.

Rodman, H.R., and Moore, T. 1997. Development and plasticity of extrastriate visual cortex in monkeys. In Cerebral Cortex, Vol. 12: Extrastriate Cortex, eds. J.H. Kaas, K. Rockland, and A. Peters, pp. 639–671. New York: Plenum Press.

Teller, D.Y. 1979. The forced-choice preferential looking procedure: A psychophysical technique for use with human infants. *Inf. Behav. Dev.* 2:135–153.

Teller, D.Y. 1997. First glances: The vision of infants. The Friedenwald Lecture. *Invest. Ophthalmol. Vis. Sci.* 38:2183–2203.

free will:
neural network
↑ look-up.

algorithms well defined +
guaranteed to
solve a certain
class of
problems. ?–

12
Higher-Order and Subjective Aspects of Perception: How Are Low-Level Stimulus Properties Transformed into High-Level Percept Qualities?

Questions

After reading Chapter 12, you should be able to answer the following questions:

1. Give some examples of perceptual qualities that are likely to be derived primarily from the observer instead of the stimulus.
2. Describe a relatively low-level representation of sensory information that can be measured in the dLGN. Compare this with representations seen at later stages of cortical processing.
3. What role do illusions play in allowing scientists to study neural correlates of percepts?
4. Summarize some of the empirical evidence regarding where along the neural pathways correlates can be found for higher-order processes such as attention and memory.
5. Describe some neural correlates of bistable percepts.
6. What brain areas are activated during visual imagery?
7. Summarize the empirical results of studies that utilized electrical stimulation of the brains of humans and monkeys, and the implications of those results for theories trying to explain the neural correlates of conscious perceptual experiences.

8. What is blindsight, and what relevance does it have to issues of neural correlates of consciousness?

9. What is the dynamic core hypothesis regarding a neural correlate of conscious perceptual experiences, and how does this hypothesis differ from bridge locus hypotheses?

10. Summarize the arguments based on general principles about whether or not monkeys or machines can have perceptual experiences.

11. Describe some methods based on empirical validation that might be used to decide whether or not a monkey or a machine has perceptual experiences.

12. Summarize the overall scheme of perception that has been presented in this book.

As signals travel from the receptors through the various stages of neural processing, a transformation takes place in the information being coded. Initially, neural signals reflect primarily **low-level properties of the proximal stimulus.** However, at later stages of processing, signals begin to take on more and more **high-level qualities of percepts.** This idea was introduced in the section on Information Processing Theories in Chapter 1 (Figure 1.12). Subsequent chapters have given repeated empirical examples of the emergence of perceptual qualities during processing:

• Absorbed photons of various wavelength are transformed into color.
• Discontinuities in luminance across space are transformed into perceived shape.
• Discontinuities in luminance across space and time are transformed into perceived motion.
• Disparities in the images in the two eyes are transformed into a single binocular percept of a three-dimensional world.

These examples suggest that perceptual processing has the effect of adding something to the percept over and above what was provided by the sensory information available at the sense organ.

Another set of examples leading to the same suggestion involves illusions, cases in which the percept differs in some substantial way from the stimulus. These illusory qualities of percepts reflect characteristics of the observer rather than of the stimulus. A third example involves the subjective conscious experiences that humans frequently describe as being an integral part of their percepts. This final chapter focuses on these "value-added," higher-order qualities that are sometimes conferred onto percepts during perceptual processing.

1 Neural Correlates of Percept Qualities

Early stages of processing in the visual system primarily reflect low-level properties of the retinal image. However, as neural signals travel along the visual streams of processing, an edited representation of the visual world begins to emerge. Information deemed irrelevant to the observer is filtered out; Information deemed important or associated with the observer's current focus of attention are accentuated; Information about objects whose presence is remembered or inferred is added in.

Early Processing Primarily Reflects Low-Level Image Properties

Examples of neural representation related closely to low-level image properties can be found in the retina and dLGN. Figure 12.1 illustrates an example in which the mean firing rates of individual neurons in the dLGN appear to code a simple representation of the relative amounts of light present at each location of the retinal image.

The visual scene shown in the left panel was imaged onto the retina of an anesthetized monkey while physiological activity was recorded from neurons in the dLGN. The right panel shows a grayscale map in which portions of the image that are represented by a high rate of spikes are shown

Stimulus projected onto retina

Physiological response

FIGURE 12.1. Comparison of intensity values of the retinal image (left panel) and the rate of neural activity in the dLGN of a monkey (right panel). Adapted from R. Srinivasan, Reading a neural code: How images are encoded by neurons of the macaque monkey's visual system. Doctoral dissertation Graduate Program in Neuroscience at Emory University, Atlanta, Georgia, USA, 2000, by permission of R. Srinivasan and J. Wilson.

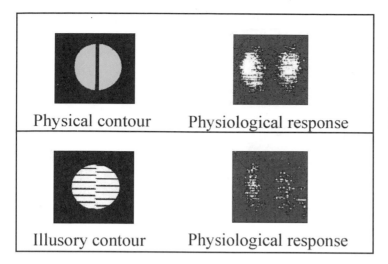

FIGURE 12.2. A neuron in V2 responds to an illusory percept of a vertical line as well as to a physical vertical line. Adapted from R. von der Heydt and E. Peterhans, Mechanisms of contour perception in monkey visual cortex: I. Lines of pattern discontinuity. *J. Neurosci.* 9:1731–1748, © 1989, by permission of Society for Neuroscience.

in white and those with a low rate in black. Comparisons between intensities in the image and amounts of neural activity in the topographic map reveal a close correspondence.

When this experiment was repeated in cortical area V1, the relationship was much less obvious, revealing that as early as V1, information is being reencoded in forms less closely related to image intensities. The following sections discuss several examples in which the neural representations seen in striate and extrastriate cortical areas are related more closely to the percept than to the stimulus.

Neural Correlates of Illusory Percepts

In cases in which perception is veridical, qualities of the percept resemble properties of the stimulus, making it difficult for scientists to disentangle whether properties of the neural representation derive from the stimulus or from the observer. For this reason, scientists interested in studying sensory neural coding often study conditions that are known to cause illusory perception. Although these conditions may not be particularly relevant to understanding how perception operates in ecologically valid conditions, they have the advantage that neural representations of the stimulus and the percept are easier to distinguish from one another.

The term **illusion** refers to dramatic and easily demonstrable examples in which an observer's percept of a stimulus is distorted or properties are perceived that are not present in the stimulus. An example of an illusory perceptual phenomenon is subjective visual contours, discussed in Chapter 8.

A stimulus that gives rise to an illusory subjective contour is shown in the lower left corner of Figure 12.2.

While viewing this stimulus, human subjects report seeing a vertical contour running down the middle of the figure. However, close examination of the stimulus reveals that the physical contours present in the stimulus are all oriented horizontally; there is no vertical contour present. Thus, the vertical contour that is perceived is illusory.

Figure 12.2 also shows the responses of a neuron in extrastriate cortical area V2 of a monkey to this subjective contour. The top panel reveals that this neuron responds vigorously to a vertically oriented line swept across its receptive field. Additional testing of this neuron, not represented in this figure, demonstrated that this neuron has strong orientation tuning and does not respond to a horizontally oriented line. The bottom panel shows that the response to the subjective vertical contour is qualitatively similar to that elicited by a physical vertical contour. In other words, the physiological response of this neuron is more closely related to the subjective quality of the percept (vertical contour) than to the physical property of the retinal image (horizontal contours).

Influences of Attention and Memory

Visual attention is used to direct limited processing resources toward information about the aspects of the environment that seem most relevant to the individual observer and away from information that seems less relevant. There are several kinds of visual attention. One major type, called **directed**

attention, is involved in shifting attention from one object to another. This can be reflexive and under the control of bottom-up processes, as when a new object appears in the scene, as discussed in Chapter 4. However, attention can also be under voluntary control of top-down processes, as when searching for an object having certain characteristics in a cluttered environment consisting of many different objects.

A neural correlate of directed attention is found in neurons that respond more strongly to a stimulus when it is being attended than to the identical stimulus when it is not being attended to. Studies in humans and monkeys have discovered neurons with these properties. Directed attention can result in small modulations of neural activity as early as V1 and in stronger modulations of activity in other areas of the extrastriate cortex, including V2 and V4.

Another kind of directed attention is involved in continuous tracking of more than one event. An example is the visual processing involved in playing a team sport like basketball. A player must continuously monitor the locations of opponents and teammates. This aspect of attention has been studied in the laboratory by having humans perform on a multiple-tracking task in which a number of randomly moving "bouncing balls" are displayed on a screen. The individual balls are identical in color and shape and cannot be distinguished from one another except by their position. While the observer is viewing the display, three or four individual balls are tagged by changing their color for a short period. Then the balls revert to being all the same color, and the subject must continue tracking the positions of the tagged balls.

Brain imaging during this task revealed no modulations in activity in V1 during this form of directed attention. However, moderate, but significant, modulations of neural activity were found in the human brain areas associated with the extrastriate motion-processing stream. Greater effects were seen in a number of parietal areas and in the frontal cortex.

Attention can also be maintained on visual stimuli that disappear from sight temporarily, as, for example, when a stimulus being attended to passes behind another object or when the lights in a room are extinguished for a short period. These higher-level aspects of perceptual attention are often referred to as reflecting object permanence. An observer usually does not perceive that an object goes out of existence when it disappears briefly and then magically come back into existence when it reappears. Instead, a percept is formed of a permanent object that remains present when it goes out of view temporarily. Neural correlates of these activities have been reported in extrastriate visual areas and in higher-order portions of the How-perception stream.

Maunsell and colleagues studied the responses of neurons in the brains of monkeys performing on tasks in which they had to attend to a target that disappeared behind an occluding object. They discovered many neurons in extrastriate cortical areas such as MST that not only responded to the moving stimulus while it was visible but also remained active while the unseen stimulus was moving behind an occluding object. However, this occurred only if the monkey was performing on a task that required paying attention to the stimulus. Otherwise the neuron stopped responding to the stimulus whenever it disappeared from view.

A similar finding has been obtained by **Michael Graziano** and colleagues from neurons in later stages of the How-perception stream in premotor cortex. Some neurons in these areas respond to seen objects that are within grasping distance. If the monkey is attending to an object that is causing one of these neurons to fire, the neuron will continue to respond in the dark if the room lights are turned off as if the object were still visible.

Neural Activity Correlated with Bistable Percepts

Binocular rivalry can be induced in humans by having them view stimuli under dichoptic viewing conditions arranged such that they are seen to move in one direction by one eye and in the opposite direction by the other eye. Observers viewing these stimuli report that during some periods motion is perceived in the direction of the stimulus seen with the left eye and during other periods in the direction of the stimulus seen with the right eye. Because conscious perception changes over time while the stimuli remain constant, this paradigm offers a way to distinguish between neural activity related to the physical stimulus and neural activity related to conscious experience.

Nikos Logothetis and colleagues took advantage of this perceptual phenomenon to examine neural correlates of percepts. Monkeys were trained to report the direction of perceived motion under these viewing conditions, and simultaneously electrical activity was recorded from directionally selective units in a number of visual processing areas of the brain. Most neurons in V1 respond to stimuli based on properties of retinal stimulation, irrespective of the direction of perceived motion reported by the monkey. However, in later stages of processing in the visual motion-stream and in other visual areas dealing with coding of higher-order properties, the activity of many neurons reflected the monkey's reported perception of motion direction. An example of the response of a neuron in the extrastriate motion-stream is shown in Figure 12.3.

The top panel shows the response to a nonrivalrous stimulus in which the motion is in the same direction for both eyes. The bottom panel is for a rivalrous stimulus in which the motion seen by the left eye is upward and that seen by the right eye downward. The left panels show responses

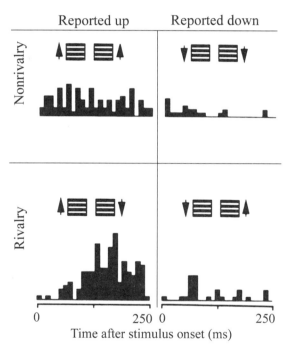

FIGURE 12.3. A neuron in an extrastriate cortical area of the motion processing stream was recorded from while a monkey discriminated the direction of motion of stimuli during normal viewing conditions (top panel), and during binocular rivalry (bottom panel). The neuron's response is shown in the form of histogram showing the average number of spikes during the first 250 msec following stimulus onset. This neuron gives a stronger response to upward than downward motion during normal viewing (top panel). During rivalry the response reflects the monkey's reported direction. Adapted from N. Logothetis and J.D. Schall, Neuronal correlates of subjective visual perception. *Science* 245:761–763, © 1989, by permission of the American Association for the Advancement of Service.

recorded during periods when the animal reported a percept of upward motion and the right panels those recorded during reports of downward motion. The histogram below each stimulus condition indicates the average number of spikes generated during presentation of the stimulus.

Examination of the histograms in the top panels demonstrates that this neuron gives a stronger response for upward than for downward motion when measured under nonrivalrous conditions. The bottom panel illustrates that under rivalrous conditions in which the retinal stimuli are moving in both directions, there is a stronger response during periods when the reported percept is of upward motion.

Measurements of brain waves of human subjects during binocular rivalry have also reported evidence of a neural correlate of the percept rather than of the stimulus. For example, **Giulio Tononi** and colleagues had human subjects view a binocularly rivalrous stimulus in the form of a vertical grating presented to one eye and a horizontal grating presented to the other. The gratings flickered, but at different rates. Brain activity was monitored, and it was determined that responses at the same temporal frequencies as the gratings could be recorded over widely distributed areas of the occipital, temporal, and frontal lobes of the brain. The subjects reported which grating, vertical or horizontal, was being consciously perceived, and the brain waves associated with each percept were compared. The modulation of the brain waves at the temporal frequency of each grating was 30% to 60% higher during periods when the grating was being consciously perceived than when it was not being perceived.

Activation of Visual Processing Areas During Imagery

Brain imaging studies performed in human subjects have found that striate, extrastriate, and higher-order processing areas in the parietal-occipital and temporal-occipital lobes sometimes become activated during visual imagery even when no retinal stimulation is present.

For example, **Stephen Kosslyn** and colleagues performed a study in which human subjects were instructed to close their eyes and visualize an object while brain activity was being measured with positron emission tomography (PET). They found increased brain activity in V1 that was localized somewhat differently depending on whether the

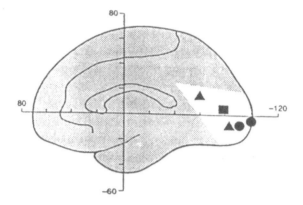

FIGURE 12.4. Summary of PET results of human subjects while visualizing small (circle), medium (square), and large (triangle) stimuli. The activation while small images were visualized was restricted to the most posterior region. Activation for medium and large images spread anteriorly. Reproduced from S.M. Kosslyn, et al, Topographical representations of mental images in primary visual cortex. *Nature* 378:496–498, © 1995, by permission of Macmillan Magazines, Ltd.

subject was visualizing a small, medium, or large object, as illustrated in Figure 12.4.

There was activation towards the posterior pole while subjects visualized small objects, and this activation spread anteriorly when they visualized larger objects. The interpretation of this finding is that the foveal representation of V1 is near the posterior pole, and visualizing a small object as straight ahead would produce retinal stimulation near the fovea. Visualizing a larger object would produce stimulation over a larger region of retina around the fovea and would be expected to activate more anterior portions of V1 due to its topographic mapping.

2 Neural Correlates of Conscious Perceptual Experiences

Bridge locus theories make the assertion that there are specific anatomical loci where neural activities give rise to conscious perceptual experiences. As discussed in Chapter 1, these ideas have a long history, going back at least to Descartes, who thought the bridge locus was located in the pineal gland. Similar ideas have continued to influence our thinking to the present day. For example, the concept of **cortical streams of processing** is often treated, at least implicitly, by modern neuroscience texts as though these form bridge loci. For example, the concept of a motion-stream is often not only

used to refer to brain regions that process motion information but also assumed to give rise to motion experiences, and similarly for streams associated with color, form, and depth and for the higher-order streams associated with What- and How-perception. This underlying belief in bridge loci provides the rationale for studies of patients with selective damage to these areas of the brain. It is assumed that these patients should have predictable impairments related to specific kinds of conscious perceptual experiences. Previous chapters have provided several examples of this approach. This chapter gives attention to more general questions about whether certain anatomical structures are privileged, in the sense of being necessary and sufficient for producing conscious perceptual experiences.

Classic Studies of Penfield

One attempt to directly manipulate and study causal relationships between brain activity and conscious experiences was carried out by the neurosurgeon **Wilder Penfield** and colleagues during and prior to the 1960s. They studied a series of patients who were about to undergo removal of brain tissue to eliminate neurological problems such as epileptic seizures. The patients were awake and sat in the operating room with the head numbed by a local anesthetic while the brain was stimulated electrically with a small electrode.

When primary sensory cortical areas were stimulated, patients reported simple sensations. For example, electrical stimulation of the primary visual cortex sometimes evoked a sensation of a point of light in a specific region of space, and stimulation of the primary auditory cortex evoked a sensation of a tone. Stimulation of other cortical sites elicited more complex experiences, as illustrated in Box 12.1.

A summary of the stimulated sites that gave rise to experiential responses in this group of patients is shown by the dots in Figure 12.5.

Reports of simple visual sensations were elicited by stimulation of sites in the occipital lobe. The majority of the sites giving rise to complex experiences were in the temporal lobe.

These findings are controversial and difficult to interpret, because it is not exactly clear whether the patients were reporting "perceptual experiences," "experiences of memories," "hallucinatory or dreamlike experiences," or some combination of these. However, what is clear is that during electrical stimulation some patients reported an

Box 12.1
Electrical Stimulation of the Human Brain
Causes Perceptual Experiences

Patient RB received stimulation repeatedly at the locations marked "5" and "7" on the temporal lobe (Figure 1). A transcript of his responses during this stimulation is given below.

5. Patient did not reply.
5. Repeated. "Something."
5. Patient did not reply.
5. Repeated. "Something."
5. Repeated again. "People's voices talking." When asked, he said he could not tell what they were saying. They seemed to be far away.
5. Stimulation without warning. He said, "Now I hear them." Then he added, "A little like in a dream."

7. "Like footsteps walking – on the radio."
7. Repeated. "Like company in the room."
7. Repeated. He explained "it was like being in a dance hall, like standing in the doorway – in a gymnasium – like at the Kenwood High School." He added, "If I wanted to go there it would be similar to what I heard just now."
7. Repeated. Patient said. "Yes. Yes, yes." After withdrawal of the stimulus, he said it was "like a lady was talking to a child. It seemed like it was in a room, but it seemed as though it was by the ocean – at the seashore."
7. Repeated. "I tried to think." When asked whether he saw something or heard something, he said, "I saw and heard. It seemed familiar, as though I had been there."
5. Repeated (20 minutes after last stimulation at 5). "People's voices." When asked, he said, "Relatives, my mother." When asked if it was over, he said, "I do not know." When asked if he also realized he was in the operating room, he said "Yes." He explained it seemed like a dream.
5. Repeated. Patient said, "I am trying." After withdrawal of the electrode he said. "It seemed as if my niece and nephew were visiting at my home. It happened like that many times. They were getting ready to go home, putting their things on – their coats and hats." When asked where, he said, "In the dining room – the front room – they were moving about. There were three of them and my mother was talking to them. She was rushed – in a hurry. I could not see them clearly or hear them clearly."

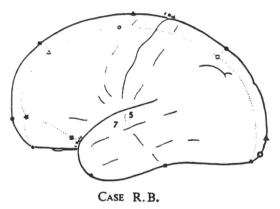

CASE R.B.

FIGURE 1. The brain regions marked "5" and "7" were stimulated repeatedly in patient R.B. Reproduced from W. Penfield and P. Perot, The brain's record of auditory and visual experience. *Brain* 86:595–696, © 1963, by permission of Oxford University Press.

(Reproduced from page 614 of W. Penfield and P. Perot, the Brains Record of auditory and visual experience. *Brain* 86:595–696, 1963, by permission of Oxford University Press.

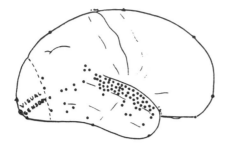 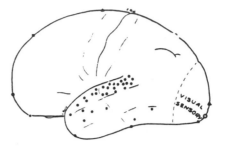

FIGURE 12.5. Summary of all the points on the lateral surfaces of the two hemispheres of the brain that, upon stimulation, resulted in experiential responses. Reproduced from W. Penfield and P. Perot, The brain's record of auditory and visual experience. *Brain* 86:39–696, © 1963, by permission of Oxford University Press.

experiential component that was described in perceptual terms such as "I see such and such" or "I hear so and so." Any theory of the neural correlates of conscious perceptual experiences will have to account for these facts.

Motion Detector Single Units as Bridge Loci

Bridge locus theories, when stated in their strongest form, assert that a conscious perceptual experience of a complex object such as one's grandmother can be caused by the activity of a single neuron, a grandmother cell. The strong version concept of a bridge locus neuron was formulated as a theoretical idea, but recent neuroscience studies support the possibility that hypotheses regarding bridge loci are amenable to empirical tests.

The essential characteristics that would make a neuron a bridge locus are as follows: 1) The neuron should fire in response to retinal stimulation from whatever object in the environment is related to the experience. 2) If the object is present in the environment but does not lead to firing of the neuron, the experience should not occur. 3) If the neuron is caused to fire although the object is not present in the environment, the experience should nevertheless occur.

No studies performed to date have been able to meet all of these criteria to establish a neural correlate for a higher-order percept as complicated as "grandmother." However, the studies reported by Newsome and colleagues, discussed in Chapter 10, appear to meet these rudimentary criteria for establishing the existence of a bridge locus for a mid-level percept: an experience of coherent motion associated with a particular part of the visual field. These studies have established that outputs recorded from neurons in the motion-processing

extrastriate area V5, are both necessary and sufficient for producing a report, albeit by a monkey, that a percept of coherent motion in a particular direction is present. Any theory regarding neural correlates of conscious perceptual experiences, particularly any theory that denies the existence of bridge locus neurons, is going to have to account for these empirical results.

Blindsight as Evidence of a Special Role for V1

Blindsight is a neurological disorder in which individuals with damage to V1 have implicit knowledge about certain explicit facts regarding "what is out there" even though they are not aware they possess that knowledge, and in fact deny it when asked. Some have used the clinical phenomenon of blindsight as evidence to argue that V1 is a necessary anatomical structure for producing conscious awareness of visual stimuli in humans.

Blindsight was initially documented by Ernst **Poppel** and colleagues in 1973 and labeled with the term blindsight by Lawrence **Weiskrantz** and colleagues in a published case study of patient DB in 1974. This patient began to experience headaches at the age of 14. The headaches were so severe that DB was often confined to bed and unable to carry out any activities except sleeping for periods up to 48 hours. When DB was in his twenties, the frequency of these episodes increased to about once every 3 weeks, severely hampering his ability to live a normal life. Brain imaging revealed a malformation in the right occipital lobe. Surgery was performed to remove the malformation, and the headaches stopped. The malformation was localized in the striate cortex, and the surgery had the effect of removing the right striate cortex while leaving the remainder of the occipital lobe intact. This is a

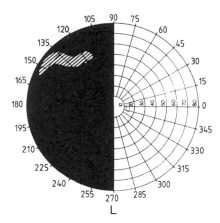
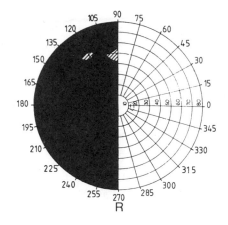

FIGURE 12.6. The visual fields for each eye of DB, a patient with blindsight, measured with standard perimetry methods in the clinic. The dark portions of the field demarcate the "blind" portions of the fields. The left panel shows the visual field while DB was viewing through the left eye and the right panel the field viewed with the right eye. Adapted from L. Weiskrantz, et al, Visual capacity in the hemianopic field following a restricted occipital ablation. *Brain* 97:709–728, © 1974, by permission of Oxford University Press.

very unusual occurrence. Typically, patients who undergo this type of surgeries have damage to multiple cortical areas.

Following surgery, patient DB was "blind" in the left hemifield, as illustrated in Figure 12.6.

This diagram shows a standard visual field plot obtained in a clinical setting. The patient fixates at the center of a screen. Small spots of light are presented, one at a time, at various locations on the screen, and the patient reports whether each spot is seen. The plot on the left shows the results while DB was viewing with the left eye, that on the right the results for the right eye. For spots of light presented to the right of midline, DB had essentially normal fields in both eyes. However, DB was not able to see spots of light presented in most of the left hemifield. The only exception was within some small islands in the periphery of the upper hemifield. This is accounted for by the fact that a small portion of the striate cortex having a topographic projection from this part of the field was spared by the surgery.

The visual fields shown for patient DB are as expected for a patient with extensive damage to the right occipital cortex. These results have been demonstrated countless times in the neurology literature, and if Weiskrantz and colleagues had not decided to do additional testing, nothing remarkable would have been noted when testing this patient.

However, Weiskrantz and colleagues went on to do some nontraditional assessments of patient DB using psychophysical methods based on a combination of Class A and Class B observations, as defined in Chapter 2. In one example, DB sat in front of a screen on which "X" or "O" could be flashed in the left visual field. When asked to report whether he could see the flashed stimuli, DB reported "no," as expected. However, when forced to choose between the two alternatives in a forced-choice task, DB was correct almost all of the time. In other words, when asked to describe his perceptions with Class B observations, DB appeared to be blind, but Class A observations collected with a forced-choice procedure demonstrated that he was able to use his eyes to extract considerable information from the environment. This paradoxical observation is the essence of blindsight.

Subject DB was tested for acuity by presenting a stimulus that was either a grating or a homogeneous field of the same mean luminance. DB was simply asked to guess on each trial whether the stimulus was a grating. His acuity was about 15 cy/deg (~20/40 Snellen), which is a factor of two worse than normal (see Chapter 8), but nowhere near being blind. In fact, patient DB could qualify for a driver's license in most places based on this level of acuity.

A similar set of methods was used to demonstrate residual How-perception in addition to an unconscious awareness of some aspects of What-perception. Subject DB was asked to point towards the location of a stimulus projected onto the "blind" field. Once again, DB claimed that he could not perform this task, but when he was forced to guess and reach out with his arm, the results were as shown in Figure 12.7.

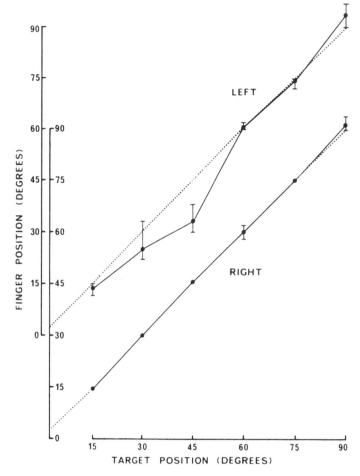

FIGURE 12.7. Accuracy of reaching to small spots of light in the "blind" (left) and "normal" (right) hemifields. Adapted from L. Weiskrantz, et al, Visual capacity in the hemianopic field following a restricted occipital ablation. *Brain* 97:709–728, © 1974, by permission of Oxford University Press.

The reaches into the "blind" hemifield were not quite as accurate as those into the "normal" hemifield but nevertheless remarkably accurate based on the fact that DB claimed to be simply guessing.

An initial concern of the investigators of this study was that DB was malingering. They conducted several control experiments and concluded, "... D.B. throughout convinced us of his reliability." One control study took advantage of the fact that the surgery had spared a small island of vision in the periphery of the upper left visual field. Weiskrantz et al. comment: "If the projected stimulus happened to fall on [a portion of his visual field that fell within this island] he reported this promptly." Another way in which DB convinced the experimenters that he was not malingering was his reaction when he was told about or shown his own results: "[DB] expressed surprise and insisted several times that he thought he was just 'guessing'. When he was shown a video film of his [performance] he was openly astonished...." There are

now published reports of several additional patients with striate cortex damage who exhibit blindsight.

Attempts have recently been made to establish that the syndrome of blindsight also occurs in monkeys in which the striate cortex has been removed surgically on one side of the brain. One paradigm that has been used to search for a phenomenon analogous to human blindsight in monkeys is illustrated in Figure 12.8.

This figure depicts the test situations for a monkey with the left striate cortex removed, resulting in a "blind" right hemifield and a "normal" left hemifield. Two paradigms have been used to test vision in this monkey. In paradigm 1, shown in the left panel, the monkey is trained to fixate at location F. Then a single target is presented at one of four locations on the test screen and the monkey trained to reach out and touch the target. This is a forced-choice task. Monkeys in which the striate cortex has been removed on one side of the brain perform the

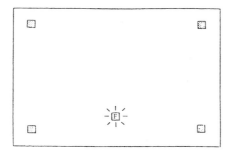 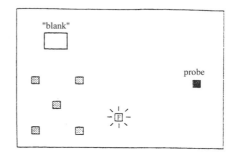

FIGURE 12.8. Experimental paradigms for testing blindsight in monkeys. See text. Adapted from A. Cowey and P. Stoerig, Blindsight in monkeys. *Nature* 373:247–249, © 1995, by permission of Macmillan Magazines, Ltd.

same on this task as normal monkeys and respond appropriately to stimuli in both the "blind" and the "normal" hemifield.

In paradigm 2, shown in the right panel, the monkey is again trained to fixate at F, and while it is fixating, a target is presented at one of five fixed locations within the "normal" field on some trials but not on other trials. The monkey's task is to reach for the target on trials in which one is presented but to reach towards the location of a box in the upper part of the "normal" hemifield on those in which no target is presented. In other words, reaches to the box are used to respond to a blank trial on which no stimulus was presented. When the monkey has learned this task, a single "probe" stimulus is presented in the "blind" hemifield during fixation. When these probe trials are included for a normal monkey that has been trained on this task, the monkey immediately reaches towards the probe. However, monkeys with lesions to the striate cortex reach for the box, indicating a blank trial, whenever the target is presented to the "blind" hemifield.

These results are interpreted as support for the hypothesis that monkeys lacking the striate cortex, like humans with blindsight, have no conscious awareness of a stimulus presented in the "blind" field. When forced to choose, the monkeys respond correctly, but when allowed to respond with an option designating whether or not a stimulus is present, the monkeys report that nothing is seen.

The phenomenon of blindsight has led to considerable speculation about the role of the striate cortex in shaping the properties of percepts. Some have interpreted blindsight as evidence that only neural signals passing through the striate cortex can produce percepts that convey conscious awareness. The deficits in blindsight patients reflect either the absence of processing in the striate cortex or lack of processing in higher-order visual processing areas

that depend on input from the striate cortex. The residual visual capacities of patients with blindsight must reflect neural processing in portions of the brain that still function following removal of the striate cortex.

Two neural pathways have been proposed as the ones providing the residual capacities of patients with blindsight. The first involves projections from the eyes that pass through midbrain structures such as the superior colliculus. In species with more primitive brains, such as frogs, these other structures are the main visual processing areas of the brain. It is only in mammals, and particularly in primates, that the massive geniculostriate projection has evolved. Observers in which no neural processing takes place in the striate cortex, be they primates with damage to the striate cortex or species in which a striate cortex never evolved, may have certain similarities in the properties of their percepts. Consider the percept formed when a fly buzzes in front of a frog's eye. The visual system of the frog provides it with information about the environment that allows its tongue to shoot from its mouth and obtain lunch. However, lacking a geniculostriate system, the frog may not have any conscious perceptual experience that a fly was present.

A second hypothesis about the neural processing areas of the brain that are responsible for residual capacities in blindsight involves extrastriate cortical areas that receive input from the retina that is not derived from V1. Recall from Chapter 5 that most visual processing in primates passes through V1, which acts as a bottleneck, before fanning out to extrastriate cortical areas. However, there are exceptions to this general principle. Some neural signals that pass through brainstem structures are relayed through the pulvinar to reach several extrastriate areas, including V5. There is also a small anatomical projection from the dLGN that bypasses the striate cortex and projects directly to the extrastriate cortex in primates, including humans.

The two hypotheses about what anatomical areas are responsible for the residual capacities seen in blindsight have somewhat different implications regarding the role of the striate cortex in conscious awareness. If only midbrain areas are responsible for the residual capacities seen in blindsight, then the lack of conscious awareness could be the result of either missing striate cortex or missing inputs to extrastriate areas. However, if extrastriate areas are responsible for the residual capabilities of blindsight, that implies that neural processing in the striate cortex in particular is needed for conscious awareness of visual percepts. The implication is that the extrastriate areas responsible for processing specific kinds of information about the environment, such as motion, color, shape, and depth, that still function during blindsight to provide residual visual capabilities are not sufficient to provide conscious awareness. These issues are currently unresolved. Any theories of neural correlates of conscious perceptual awareness will have to take the phenomenon of blindsight into account.

The Dynamic Core Hypothesis

Tononi and colleagues have argued for a neural correlate of conscious perceptual experience called a **dynamic core,** which they characterize in terms of brain activity rather than as an anatomical location. They point out that conscious perceptual experiences have the property of being *highly integrated and unified* at any moment but also *highly differentiated* over time.

Stating that a momentary conscious state is **highly integrated and unified** means that it cannot be broken down into components. Consider, for example, the perceptual experience that results from viewing ambiguous figures, such as the Necker cube in Figure 8.9. While viewing the Necker cube stimulus, normal human observers can experience two incongruous percepts sequentially over time but are unable to experience the two percepts simultaneously. Another example illustrating the fact that a momentary conscious state is highly integrated and unified is the fact that humans cannot make more than one conscious decision within a short interval of a few hundred milliseconds.

Stating that conscious states are **highly differentiated** over time emphasizes that when one partic-

ular state rather than another occurs at a given point in time, this fact is informative to an observer in the sense described by Shannon, as discussed in Chapter 6. This is one criterion that can be used to differentiate conscious perceptual experiences occurring in a human brain from states existing in simpler kinds of hardware. Consider why the differentiation between light and dark made by a human is associated with conscious experience, while that made by a photodiode is not. One critical difference may be the amount of information generated in the two cases. The discrimination by the photodiode of dark from light conveys a minimal amount of information, in the sense of the degree of reduction of uncertainty. In a human observer, the experiences of lightness or darkness are highly informative since they reduce uncertainty regarding an enormous repertoire of possible per-ceptual experiences.

Tononi and colleagues have argued that a neural process responsible for conscious experience should exhibit these same two properties: The neural process should be both highly integrated and capable of exceptionally informative differentiation. They have argued that this could be achieved in the form of a unified neural process whose dynamic operation is distributed over various portions of the brain rather than in a single place. This neural process is what they refer to as the "dynamic core."

A dynamic core comes into existence rapidly, within no more than a few hundred milliseconds, by integrating, or binding, synchronous correlated responses from distributed neuronal groups.

For example, suppose someone is experiencing a conscious percept of a freshly baked chocolate chip cookie as he picks it up and brings it towards his mouth. Somatosensory portions of his brain are responding to the touch of the cookie. Olfactory parts of his brain are responding to its smell. Within the visual system, signals regarding the cookie's shape and color are reverberating through the What-perception stream. Signals guiding the arm muscles that are bringing the cookie towards the mouth are reverberating through the How-perception stream. Simultaneously, higher-order, top-down, attentional processes are riveted on the cookie! These neural signals will be distributed over large portions of the brain, but their dynamic activities are all correlated with one another. Over a period of no more than a few hundred milliseconds, a dynamic core of neural activity evolves, based on reentrant long-range extrinsic connections among brain areas. This dynamic core serves to bind all of the correlated signals from all over the brain into a unified activity that can be described in words as a percept of a cookie.

3 What Kinds of Observers Can Produce Conscious Perceptual Experiences?

In addition to information that can be used for What-perception and How-perception, percepts convey to humans perceptual experiences of which they are consciously aware. It is worth considering whether the perceptual systems of other classes of observers, such as robotic machines, might also output **conscious perceptual experiences.** Some argue that this issue can be decided based on general principles. Others argue that it can be decided empirically.

Arguments Based on General Principles

The arguments about what kinds of physical systems can support perceptual experiences based on general principles cover an immense range. At one extreme is the **liberal position** that, in principle, any physical system can support perceptual experiences. This position was argued by the philosopher **David Hartley,** writing in the 18th century:

"... Matter, if it could be embued with the most simple Kinds of Sensations, might also arrive at all that Intelligence of which the human Mind is possessed. ..."

A modern version of the liberal position, called **strong artificial intelligence,** was described in Chapter 6. Strong AI makes the radical claim that any physical system can generate conscious experiences just by virtue of implementing the proper algorithm. It is not supposed to make any difference what physical substance is used to implement the algorithm.

The opposite **extreme conservative position** is that the causal power to generate full-fledged conscious experiences belongs only to living human brains. A historical example of this position would be that of Descartes as described in Chapter 1. The general idea that humans are special and that the capacity to have conscious perceptual experiences does not extend beyond humans remains pervasive to the current day among many philosophers and scientists who study perception.

An **intermediate conservative position** is that conscious perception extends beyond humans, but only to biological organisms with brains similar to our own. This intermediate position is grounded in the ancient philosophical position of **vitalism,** which asserts that the forces involved in living bodies are different from those in the inorganic world.

In a recent debate about these issues, the philosopher **John Searle** adopted a modern variation on the intermediate position, arguing that any physical system capable of causing consciousness will have to have **causal powers equivalent to those of biological brains.** Philosophers **Patricia Churchland** and **Paul Churchland** responded with a rejoinder, arguing that Searle's argument begs the question until we know what those equivalent causal powers are. Furthermore, they argued, it seems highly unlikely that all causal powers present in biological brains would be necessary in order to achieve consciousness, for example, the causal power to produce a bad smell when rotting.

Arguments Based on Empirical Validation

Since there is currently no consensus about whether or not, in principle, physical systems other than human brains can generate conscious perceptual experiences, others have argued for an empirical approach. Such an approach tries to validate claims for conscious perceptual experiences in an observer based on some observable aspect of behavior. A general premise of all of these strategies is that phenomena in the form of **verbal reports of adult human observers** serve as the **gold standard** for what is meant by the term "conscious perceptual experiences." The performance of other observers is evaluated by measuring it in some way against this gold standard reference.

Validation Based on Correlation with Actions

One strategy is to find a secondary behavioral response in the form of an **action** that appears to be **highly correlated** with reports of a particular perceptual experience in verbal humans. An example would be the stereotypical reaction to a **looming object** that was described in the section on measuring How-perception in Chapter 2. In adult humans, whenever this stereotypical action is elicited, there is a corresponding experience of a looming object. Thus, it has been argued that if a nonverbal observer exhibits the same action, it can be inferred that the observer is *experiencing* a looming object.

Willingness to accept results obtained by this strategy as evidence for perceptual experiences is heavily influenced by the characteristics of the observer. For example, many who accept such evidence as demonstrating a perceptual experience of "looming" when applied to human infants show strong resistance to accepting the same argument for a robot programmed to make similar stereotypical movements to a looming object.

Validation Based on a Turing Test

A potential strategy that might be helpful for devising an operational definition of perceptual experiences in machines comes from a solution proposed to solve a similar problem addressed in the artificial intelligence literature:

How would one decide whether a machine is intelligent in the same sense in which the term "intelligence" is applied to humans?

Alan Turing, a pioneering scientist working in the field of artificial intelligence, devised an operational definition of intelligence, now commonly called the **Turing test,** to address this question. The issues involved in evaluations of intelligence and of perceptual experience are similar enough to warrant a look at Turing's test.

Turing had an important insight. Any judge trying to decide whether an agent is intelligent has access to two kinds of information: information about the agent being evaluated and information about functions the agent performs. It is difficult to disentangle these two sources of information. For example, many human judges will be biased against the notion that a machine can be intelligent. Consequently, they will be willing to accept evidence obtained from a human agent as demonstrating intelligence but will reject the exact same evidence when informed it came from a machine. There is a similar bias against attributing conscious perceptual experiences to a machine.

Turing tried to solve this problem with respect to questions about intelligence by keeping the judge masked as to the identity of the agent. The judge holds a conversation with two agents, one human and the other machine. Then the judge must decide which agent is the human. Turing argued that if the judge attributes intelligence to the human and could not distinguish between these two agents, then the judge should also be compelled to attribute intelligence to the machine.

In fact, Turing even biased his test in such a way that if the two agents were really of equal intelligence, the judge would more likely pick the human.

He did this by having both agents claim to be humans. Thus, the human agent being interrogated just has to be truthful. However, the machine has to pull off an act of deception that makes the judge believe it is a human.

A similar rationale can be used to propose the following **Turing perception test** as an operational definition for perceptual experience. A judge is allowed to interrogate and/or watch the behavior of two observers, only one of which is a human, while each claims to be having a particular type of perceptual experience. The judge's task is to identify the human observer. The test must be arranged in such a way that the judge cannot tell by outward appearances which observer is the human. The exact details of how this is accomplished are not important as long as the judge has access to the relevant performance of the two observers. If the judge cannot reliably pick out which observer is the human, then the judge will be forced to attribute to the nonhuman observer the same level of perceptual experience as to the human. In other words, if the judge accepts that the human has conscious perceptual experiences based on an evaluation of the performance, then the judge will also have to attribute conscious perceptual experiences to the other observer.

Only nonlanguage forms of this proposed Turing perception test are applicable to human infants or to animals. However, there is nothing, in principle, that precludes a machine with sufficient language (symbolic communication) capabilities from passing a language version of this test as well at some date in the future.

Validation Based on Correlation with Neural Processing

Another approach that can be used when evaluating animals with brains similar to human brains is to define perceptual experiences in a reductionist manner in terms of **patterns of neuronal activity** that are correlated with conscious perceptual experience in adult humans. This book includes several examples in which noninvasive brain imaging methods allow one to monitor simultaneously the action of relevant neuronal populations and behavior. For example, brain activity can be measured while a human views a particular visual stimulus and simultaneously gives a verbal report describing the associated perceptual experience. Suppose a particular part of the brain reacts in a characteristic manner if and only if a human reports a certain perceptual experience. Brain activity can then be measured in animals with similar brains, such as

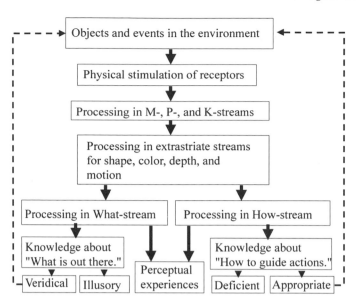

monkeys, under the same viewing conditions to
look for this same pattern of neuronal activity.

4 An Overall Conceptual Scheme of Perception

As this text comes to a close, it is worthwhile to consider one last time the overall conceptual scheme of
human perception that has been adopted, as characterized in Figure 12.9.

Objects and events in the environment (distal
stimuli) cause activation of receptors in the retina
(proximal stimuli). Signals generated by these receptors carry information that is analyzed as it passes
through a number of parallel streams of neural processing. Within each stream, processing is organized
into stages, although extensive feedback circuits (not
shown) have the effect that information being
processed at most stages is heavily influenced by
what is happening simultaneously at other stages.
Processing passes through the M-, P-, and K-streams
and then through various extrastriate processing
streams, and finally fans out to the What- and How-
perception streams. At the end of perceptual processing, percepts are formed that convey to the
observer **knowledge about what** is present in the
immediate surrounding environment, **knowledge
about how** to interact with this environment, and
perceptual experiences related to the environmental stimulation. In most cases the percepts associated
with What-perception are veridical, by which is
meant that the observer's perceptual knowledge
about "what is out there" corresponds to empirical

information derived from instruments. In some
cases, What-perception can be shown to be in error,
or illusory. Similarly, motor responses guided by
How-perception are usually appropriate and help
humans accomplish the goals of finding food, avoiding danger, and so on. In some cases performance is
deficient; for example, an outfielder fails to catch a
fly ball. Perceptual errors seldom occur in the ecological visual environment in which our species
evolved, presumably because the individuals with
perceptual systems that led to frequent errors did not
survive. However, when perceptual errors are
studied in artificial laboratory environments, they
provide rich information about the forms of
processing that operate in transforming low-level
properties of sensory stimulation into high-level
qualities of percepts.

Summary

Physical stimulation of the visual sense organ produces signals that are processed along the geniculostriate pathway, then the extrastriate cortical
streams, and finally the What-perception and How-
perception streams. As information travels along
these stages of processing, it is transformed so that
instead of representing primarily low-level physical properties of sensory stimulation, it represents
higher-order qualities of percepts. This transformation has been demonstrated in numerous studies
that compare properties of the neural activity at
various stages of processing with properties of the
sensory stimulation and qualities of the percepts.

Studies at relatively early stages of processing, such as those that take place in the dLGN, have found evidence for representations of relatively low-level stimulus properties, such as the intensity at each point in the image. Studies at later stages, in cortical areas, have found evidence for neural representations that relate more closely to perceptual qualities, including illusions and bistable percepts, and reflect the influences of higher-order processes, such as attention and memory, on percepts.

Bridge locus theories about the neural correlates of conscious perceptual experiences assert that a conscious experience is caused at a specific anatomical locus. Bridge locus theories can be weak, asserting that large regions of the brain are involved, or strong, asserting in the limiting case that a single neuron is responsible for a single perceptual experience. Accumulating evidence is consistent with both weak and strong forms of bridge locus theories. Theories that argue against the idea of a bridge locus will have to find other ways of accounting for these empirical findings.

An approach to neural correlates of conscious perceptual experiences that is similar in some ways to bridge locus theories but does not propose a specific anatomical location is the dynamic core hypothesis, which asserts that the neural correlate takes the form of distributed dynamic neural activity. This neural dynamic core binds synchronous neural responses from widely distributed regions of the brain to form a conscious percept.

It is an unresolved question whether observers other than humans can have conscious perceptual experiences. Many scientists are willing to accept that animals with humanlike brains, such as monkeys, have conscious perceptual experiences similar to those of humans. However, much skepticism remains about attributing conscious perceptual experiences to other types of observers, such as robotic machines. Some argue that issues about what kinds of observers can have conscious perceptual experiences should be decided based on general principles. Others argue that these issues can be validated empirically.

Selected Reading List

Boring, E.G. 1933. *Dimensions of Consciousness*. New York: Appleton-Century Crofts, Inc.

Bourassa, C.M. 1986. Models for sensation and perception: A selective history. *Human Neurobiol.* 5:23–36.

Churchland, P.M., and Churchland, P.S. 1990. Could a machine think? *Sci. Am.* 262:32–37.

Cowey, A., and Stoerig, P. 1991. The neurobiology of blindsight. *Trends Neurosci.* 4:140–145.

Culham, J.C., Brandt, S.A., Cavanagh, P., Kanwisher, N.G., Dale, A.M., and Tootell, R.B.H. 1998. Cortical fMRI activation produced by attentive tracking of moving targets. *J. Neurophysiol.* 80:2657–2670.

Desimone, R., and Duncan, J. 1995. Neural mechanisms of selective visual attention. *Annu. Rev. Neurosci.* 18:193–222.

Graziano, M.S.A., Hu, X.T., and Gross, C.G. 1997. Coding the locations of objects in the dark. *Science* 277: 239–241.

Heinze, H.J., Mangun, G.R., Burchert, W., Hinrichs, H., Sholz, M., Munte, T.F., Gos, A., Scherg, M., Johannes, S., Hundeshagen, H., Gazzaniga, M.S., and Hillyard, S.A. 1994. Combined spatial and temporal imaging of brain activity during selective attention in humans. *Nature* 372:543–546.

Kosslyn, S.M., Thompson, W.L., Kim, I.J., and Alpert, N.M. 1995. Topographical representations of mental images in primary visual cortex. *Nature* 378:496–498.

Logothetis, N.K., and Schall, J.D. 1989. Neuronal correlates of subjective visual perception. *Science* 245:761–763.

Maunsell, J.H.R. 1995. The brain's visual world: Representation of visual targets in cerebral cortex. *Science* 270:764–769.

Moore, T., Rodman, H.R., and Gross, C.G. 1998. Man, monkey, and blindsight. *The Neuroscientist* 4:227–230.

Moran, J., and Desimone, R. 1985. Selective attention gates visual processing in the extrastriate cortex. *Science* 229:782–784.

Motter, B.C. 1993. Focal attention produces spatially selective processing in visual cortical areas V1, V2, and V4 in the presence of competing stimuli. *J. Neurophysiol.* 70:909–919.

Penfield, W., and Perot, P. 1963. The brain's record of auditory and visual experience. *Brain* 86:595–696.

Poppel, E., Held, R., and Frost, D. 1973. Residual visual function after brain wounds involving the central visual pathways in man. *Nature* 234:295–296.

Tononi, G., and Edelman, G.M. 1998. Consciousness and complexity. *Science* 282:1846–1851.

van Voorhis, S.T., and Hillyard, S.A. 1977. Visual evoked potentials and selective attention to points in space. *Percept. Psychophys.* 22:54–62.

von der Heydt, R., and Peterhans, E. 1989. Mechanisms of contour perception in monkey visual cortex: I. Lines of pattern discontinuity. *J. Neurosci.* 9:1731–1748.

Weiskrantz, L., Warrington, E., Sanders, M., and Marshall, J. 1974. Visual capacity in the hemianopic field following a restricted occipital ablation. *Brain* 97:709–728.

Index

Visual acuity
  20/20 vision, as determined
    by a Snellen acuity
    chart, 232
  upper limit of spatial
    resolution, for the
    human eyeball, 232
Visual angle, relating to the
  size of a retinal image,
  72
Visual attention, 364–365
  possible role of pulvinar in,
    357–358
Visual cliff, for evaluating
  infant depth
  perception, 344–345
Visual cortex. *See* Striate
  cortex; Cortex;
  Extrastriate cortex
Visual direction, as a local
  sign carried by a
  percept, 266–267
  change in, during
    stereopsis, 276
Visual fields, of primates,
  259–261
Visual information. *See*
  Information;
  Information processing
Visual processing, brain areas
  involved in, 128–151
  See also Neural processing;
    Information processing
Vital-Durand, François, 343
Vitalism, defined, 374
Vitamin A, role in sight,
  115–116
Vitreous fluid, of the posterior
  chamber of the eyeball,
  103
Voltages, sensory coding of
  signals in terms of, 123
  See also Electrical activity
Volumetric descriptions,
  provided by geons,
  252–253

W
Wallach, Hans, 305, 321
Wavelengths, of light, 57–59
  complementary, 199
  complex distribution of,
    reducing to color, 185,
    217
  discrimination of, in infants,
    344
  perceived as color, 193
  primary, 208
Waves, light in the form of, 56
Weak forms, of bridge locus
  theories, 9
Weak signals, detecting, aided
  by noise, 188
  See also Signal-to-noise ratio;
    Signal detection theory
    (SDT)
Weber-Fechner law, 81–83
Weber fraction, 82–83
Weber's law, 81–83
Weeks-to-months rule, for
  comparing human and
  monkey ages, 331–332
Weights, in neural networks,
  adjusting, 177–178
  See also Training stage
Weiskrantz, Lawrence, 369
Well-behaved mathematic
  function, 176
Wertheimer, Max, 300–301
What-perception
  contributions to, from
    information that
    changes over time, 297
  defined, 26
  involvement of stereovision
    processing in, 263
  measuring, 26–32
  percepts conveying
    information about, 3,
    376
What-perception stream,
  158–159, 217, 248–249,
  263

Wheatstone, C., 279, 281, 289
White light, wavelength
  mixtures perceived as,
  193
Wiesel, Torsten, 126, 146–147,
  167, 287, 349
Wiring of brain connections,
  effect of Hebbian rules
  on, 358
Wiring diagram, biological,
  solutions to the
  fundamental
  developmental problem
  based on, 347
Words, describing explicit
  facts, 3
Wundt, Wilhelm, 16

X
X chromosome, genes for M-
  cone and L-cone
  pigments on, 209

Y
Yellow, wavelengths perceived
  as, 193
Yes-no psychophysical
  procedures, 29
  defined, 30
Young, Thomas, 199

Z
Zero crossings, as primitive
  operations involved in
  form vision processing,
  243–245
  as inputs for motion
    processing, 304
  utilizing in the coarse-to-
    fine cooperative model
    of stereopsis, 286
Zero neurons, tuned for depth
  at the plane of the
  horopter, 287
Zonule ligaments, 103